Molecular Basis of Apomixis in Plants

Molecular Basis of Apomixis in Plants

Editor

Diego Hojsgaard

MDPI • Basel • Beijing • Wuhan • Barcelona • Belgrade • Manchester • Tokyo • Cluj • Tianjin

Editor
Diego Hojsgaard
Department of Biodiversity,
Systematics, and Evolution of Plants
Georg-August-University of Göttingen
Göttingen
Germany

Editorial Office
MDPI
St. Alban-Anlage 66
4052 Basel, Switzerland

This is a reprint of articles from the Special Issue published online in the open access journal *Genes* (ISSN 2073-4425) (available at: www.mdpi.com/journal/genes/special_issues/plants_apomixis).

For citation purposes, cite each article independently as indicated on the article page online and as indicated below:

LastName, A.A.; LastName, B.B.; LastName, C.C. Article Title. *Journal Name* **Year**, *Volume Number*, Page Range.

ISBN 978-3-0365-1508-3 (Hbk)
ISBN 978-3-0365-1507-6 (PDF)

Contents

About the Editor

Diego Hojsgaard

Diego Hojsgaard (achieved a Ph.D. degree in 2010 from the National University of the Northeast, Argentina) is a population geneticist with special interest in plant reproduction, polyploidy and hybridization. For his Ph.D., he studied the genetic system and genomic region responsible for apomixis in species of Paspalum. Since then, he has been convinced that thorough knowledge of the multiple facets of apomixis is crucial to unravel its genetic and molecular basis, as well as to understand its evolution. Diego is a senior scientist and guest researcher at the Department of Biodiversity, Systematics, and Evolution of Plants, Georg-August-University of Göttingen.

genes

Editorial

Molecular Basis of Apomixis in Plants

Diego Hojsgaard

Department of Systematics, Biodiversity and Evolution of Plants, Albrecht-von-Haller Institute for Plant Sciences, University of Göttingen, Untere Karspüle 2, 37073 Göttingen, Germany; diego.hojsgaard@biologie.uni-goettingen.de

Sexual reproduction in plants is a complex, stringently regulated process that leads to the creation of diaspores for a new generation: sexual seeds. Traditionally, sexuality is exploited to segregate or selectively assemble desired genes and traits during the creation of new crop varieties. However, the exploitation of sexuality also imposes constraints on plant breeding, which include high seed costs and time-consuming methods. Most of these limitations can be mitigated by sequentially exploiting sexuality and apomixis during plant breeding.

Apomixis is the consequence of a concerted mechanism that harnesses the sexual machinery and acts in a way that coordinates developmental steps in the ovule to produce an asexual (clonal) seed. Altered sexual developments involve widely characterized functional and anatomical changes in the meiosis, gametogenesis and embryo and endosperm formation. The ovules of apomictic plants skip meiosis and form unreduced female gametophytes whose egg cells develop into a parthenogenetic embryo, and the central cells may or may not fuse to a sperm to develop the seed endosperm. Thus, functional apomixis involves at least three components, apomeiosis + parthenogenesis + endosperm development, modified from sexual reproduction that have to be coordinated at the molecular level to progress through the developmental steps and form a clonal seed. Despite recent progress uncovering specific genes related to apomixis-like phenotypes and clonal seed formation, the molecular basis and regulatory network of apomixis is still unknown. This is a central problem underlying the current limitations of apomixis breeding.

This Special Issue collects twelve publications addressing different topics around the molecular basis of apomixis, illustrating recent discoveries and advances towards understanding the genetic regulation of the trait and discussing the possible origin of apomixis and the remaining challenges for its commercial deployment in plants.

Since theories that apomixis is a phenomenon based on gain- or loss-of-function mutations over sexuality remain open, Barcaccia et al. [1] reassess the evolutionary origin of apomixis in angiosperms and their alternative developmental pathways and present phylogenetic and genetic evidence supporting that apomixis evolved from sex and is a consequence of the molecular disruption of key players in sexual development. Furthermore, Schmidt [2] gives an overview of the molecular aspects of apomixis in higher plants and provides a clear interpretation of the regulatory complexity involved in apomictic development, emphasizing the active role of DNA- and RNA-binding proteins, as well as non-coding RNAs, in the activation and repression of developmental programs through epigenetic regulatory mechanisms. Similarly, Ortiz et al. [3] summarize the vast information on apomixis in a case study on *Paspalum* spp. and provide details on the key aspects of apomictic development in the genus and the various genetic analyses used, including the molecular characterization of genomic loci, functional characterization of three reproductive candidate genes (*ORC3*, *QGJ* and *TGS1*) and a roadmap for further genome-based studies.

Further molecular details on apomixis have been derived from different plant species. Mateo de Arias et al. [4] use genetic and cytoembryological analyses combined with stress treatments in five species to present substantial evidence in support of a polyphenic

Citation: Hojsgaard, D. Molecular Basis of Apomixis in Plants. *Genes* **2021**, *12*, 576. https://doi.org/10.3390/genes12040576

Received: 1 April 2021
Accepted: 13 April 2021
Published: 16 April 2021

condition for meiosis and apomeiosis determined by metabolic states. The study lays the groundwork for a new avenue of research and future studies focused on integrating metabolism into the current understanding of apomixis and on unraveling the role that metabolic homeostasis plays in reproductive switches. Likewise, Selva et al. [5] use gene expression and cytoembryological analyses combined with water stress experiments to pinpoint genes that modulate the expression of sexuality in the apomictic grass species *Eragrostis curvula*, and to identify genes with more specific roles in the stress response and control of the reproductive mode. Pellino et al. [6] take advantage of gene expression analyses on microdissected ovules to disclose genes that are differentially expressed between sexual and apomictic individuals of the *Ranunculus auricomus* complex, owing to effects of transgressive gene expression, parent of origin and ploidy. Three genes probably involved in the reproductive steps are highlighted. In *Poa pratensis*, Marconi et al. [7] use detailed genetic and in situ hybridization analyses, as well as protein prediction and molecular modelling tools, to characterize a collection of apomixis-related alleles identified in relation to *APOSTART*, the putative gene with an active role in the development of apomixis in this grass species.

Parthenogenesis is a central component of apomixis and is often associated with the development of the endosperm, the least studied feature in apomictic plants. van Dijk et al. [8] collect cytoembryology, flow cytometry and molecular marker data from experimental crosses in *Taraxacum officinale* and advance the genetic characterization of autonomous endosperm formation in the Asteraceae by providing evidence supporting a three-locus model for apomixis, with parthenogenesis and diplospory controlled by independent dominant loci and autonomous endosperm development under complex regulatory control. Focusing on parthenogenesis, Zhang et al. [9] take a further step toward inducing apomixis in sexual plants. These colleagues transferred the *PsASGR-BABYBOOM*-like gene (*PsASGR-BBML*) that confers parthenogenesis in the apomict grass *Pennisetum squamulatum* (and other monocot species) into tobacco plants and confirmed, for the first time, that the *PsASGR-BBML* gene regulated by egg cell-specific promoters enables parthenogenesis in a dicotyledonous species. Another substantial advance is derived from Henderson et al. [10], who introduced a *Cas9* construct in an apomictic *Hieracium* species via *Agrobacterium*-mediated leaf disk transformation and evaluated the efficiency of CRISPR/Cas9 editing to target the endogenous *PHYTOENE DESATURASE* (*PDS*) gene. The study—featured in this collection—demonstrates, for the first time, that gene editing tools can be effectively used in apomixis research and inaugurates a new stage for the identification of apomixis genes.

Finally, an opinion piece summarizes the relevant molecular data and the three currently accepted but divergent hypotheses explaining the nature of apomixis to point to a probable single molecular event with a multigenic effect on reproductive development implicated in the origin of apomixis, remarking on the need to find a unifying molecular model to fully exploit apomixis technology [11]. In addition, Scheben and Hojsgaard [12] discuss alternative methods from the gene-editing toolbox and their feasible use to induce apomixis in sexual plants depending on the type of molecular model of the genetic control of apomixis that is chosen, and stress the importance of understanding the molecular basis of apomixis and its natural variation in apomictic plants for trait breeding and optimization in sexual ones.

Each of the plant species used as model systems in apomixis research has its own characteristics and challenges. The present collection covers plant species that exhibit diverse changes in the sexual machinery and a range of methodologies, from embryology and genomics to genetic modification and bioinformatics methods, that exemplify such particularities and expose the efforts and ingenuity that researchers devote to uncovering the molecular details of apomixis in plants. Each of the contributions, many of which point to new research directions, paves the way toward a better understanding of apomixis and guides the advancement of this scientific topic by bringing us closer to resolving the molecular basis of apomixis and its biotechnological application.

Funding: This research received no external funding.

Conflicts of Interest: The authors declare no conflict of interest.

References

1. Barcaccia, G.; Palumbo, F.; Sgorbati, S.; Albertini, E.; Pupilli, F. A Reappraisal of the Evolutionary and Developmental Pathway of Apomixis and Its Genetic Control in Angiosperms. *Genes* **2020**, *11*, 859. [CrossRef] [PubMed]
2. Schmidt, A. Controlling Apomixis: Shared Features and Distinct Characteristics of Gene Regulation. *Genes* **2020**, *11*, 329. [CrossRef] [PubMed]
3. Ortiz, J.P.A.; Pupilli, F.; Acuña, C.A.; Leblanc, O.; Pessino, S.C. How to Become an Apomixis Model: The Multifaceted Case of *Paspalum*. *Genes* **2020**, *11*, 974. [CrossRef] [PubMed]
4. Mateo de Arias, M.; Gao, L.; Sherwood, D.A.; Dwivedi, K.K.; Price, B.J.; Jamison, M.; Kowallis, B.M.; Carman, J.G. Whether Gametophytes are Reduced or Unreduced in Angiosperms Might Be Determined Metabolically. *Genes* **2020**, *11*, 1449. [CrossRef] [PubMed]
5. Selva, J.P.; Zappacosta, D.; Carballo, J.; Rodrigo, J.M.; Bellido, A.; Gallo, C.A.; Gallardo, J.; Echenique, V. Genes Modulating the Increase in Sexuality in the Facultative Diplosporous Grass *Eragrostis curvula* under Water Stress Conditions. *Genes* **2020**, *11*, 969. [CrossRef] [PubMed]
6. Pellino, M.; Hojsgaard, D.; Hörandl, E.; Sharbel, T.F. Chasing the Apomictic Factors in the Ranunculus auricomus Complex: Exploring Gene Expression Patterns in Microdissected Sexual and Apomictic Ovules. *Genes* **2020**, *11*, 728. [CrossRef] [PubMed]
7. Marconi, G.; Aiello, D.; Kindiger, B.; Storchi, L.; Marrone, A.; Reale, L.; Terzaroli, N.; Albertini, E. The Role of APOSTART in Switching between Sexuality and Apomixis in *Poa pratensis*. *Genes* **2020**, *11*, 941. [CrossRef] [PubMed]
8. van Dijk, P.J.; den Camp, R.O.; Schauer, S.E. Genetic Dissection of Apomixis in Dandelions Identifies a Dominant Parthenogenesis Locus and Highlights the Complexity of Autonomous Endosperm Formation. *Genes* **2020**, *11*, 961. [CrossRef] [PubMed]
9. Zhang, Z.; Conner, J.; Guo, Y.; Ozias-Akins, P. Haploidy in Tobacco Induced by *PsASGR-BBML* Transgenes via Parthenogenesis. *Genes* **2020**, *11*, 1072. [CrossRef] [PubMed]
10. Henderson, S.W.; Henderson, S.T.; Goetz, M.; Koltunow, A.M.G. Efficient CRISPR/Cas9-Mediated Knockout of an Endogenous *Phytoene Desaturase* Gene in T1 Progeny of Apomictic *Hieracium* Enables New Strategies for Apomixis Gene Identification. *Genes* **2020**, *11*, 1064. [CrossRef] [PubMed]
11. Hojsgaard, D. Apomixis Technology: Separating the Wheat from the Chaff. *Genes* **2020**, *11*, 411. [CrossRef] [PubMed]
12. Scheben, A.; Hojsgaard, D. Can We Use Gene-Editing to Induce Apomixis in Sexual Plants? *Genes* **2020**, *11*, 781. [CrossRef] [PubMed]

genes

MDPI

Review

A Reappraisal of the Evolutionary and Developmental Pathway of Apomixis and Its Genetic Control in Angiosperms

Gianni Barcaccia [1,*], Fabio Palumbo [1], Sergio Sgorbati [2], Emidio Albertini [3] and Fulvio Pupilli [4]

1 Department of Agronomy Food Natural Resources Animals Environment, University of Padova, Campus of Agripolis, Viale dell'Università 16, Legnaro, 35020 Padova, Italy; fabio.palumbo@unipd.it
2 Department of Environmental and Territory Sciences, University of Milano-Bicocca, Piazza della Scienza 1, 20126 Milano, Italy; sergio.sgorbati@unimib.it
3 Department of Agricultural, Food and Environmental Sciences, University of Perugia, 06121 Perugia, Italy; emidio.albertini@unipg.it
4 Research Division of Perugia, Institute of Biosciences and Bioresources, National Research Council (CNR), Via Madonna Alta 130, 06128 Perugia, Italy; fulvio.pupilli@ibbr.cnr.it
* Correspondence: gianni.barcaccia@unipd.it

Received: 3 July 2020; Accepted: 27 July 2020; Published: 28 July 2020

check for updates

Abstract: Apomixis *sensu stricto* (agamospermy) is asexual reproduction by seed. In angiosperms it represents an easy byway of life cycle renewal through gamete-like cells that give rise to maternal embryos without ploidy reduction (meiosis) and ploidy restitution (syngamy). The origin of apomixis still represents an unsolved problem, as it may be either evolved from sex or the other way around. This review deals with a reappraisal of the origin of apomixis in order to deepen knowledge on such asexual mode of reproduction which seems mainly lacking in the most basal angiosperm orders (i.e., Amborellales, Nymphaeales and Austrobaileyales, also known as ANA-grade), while it clearly occurs in different forms and variants in many unrelated families of monocots and eudicots. Overall findings strengthen the hypothesis that apomixis as a whole may have evolved multiple times in angiosperm evolution following different developmental pathways deviating to different extents from sexuality. Recent developments on the genetic control of apomixis in model species are also presented and adequately discussed in order to shed additional light on the antagonist theories of gain- and loss-of-function over sexuality.

Keywords: agamospermy; basal angiosperms (ANA-grade); sporocyteless; polycomb-group proteins; reproductive systems; apomixis evolution

---

1. Introduction and Background Information

Apomixis *sensu stricto* (agamospermy) is asexual reproduction by seed [1]. For eukaryotes in general, apomixis is life cycle renewal through gamete-like cells that give rise to maternal embryos but without sexuality and sex, that is, without ploidy reduction (i.e., meiosis) and ploidy restitution (i.e., syngamy). Both meiosis and syngamy are canalisations of complex molecular processes that have remained conserved among single-celled and, later, multi-celled species of eukaryotes since eukaryogenesis [2–8]. Apomixis, like sex, is kingdom ubiquitous occurring in thousands of species across eukaryotes [9–11].

Since apomictic reproduction involves the development of an embryo from an apomeiotic or somatic cell with a maternal genome, there are several ways to produce seeds of apomictic origin.

This process can have a nucellar or integumental initiation, depending on the tissue of the ovule from which the maternal cell with embryonic competence differentiates.

The simplest pathway, known as adventitious embryony, avoids the production of a gametophyte and one or more vegetative embryos form within the nucellus or the integument. This phenomenon seems to have evolved more frequently in tropical than in temperate flora [12,13] and to be more represented in diploid species, while other forms of apomixis are more frequent in polyploids. Adventitious embryony is found in several non-agriculturally important species, with the exception of several *Citrus* species and mango (*Mangifera indica*) [14].

Another pathway, known as gametophytic apomixis, occurs when the maternal embryo originates from an apomeiotic egg cell differentiated into an unreduced embryo sac [15] arisen from a somatic nucellar cell that acquires the developmental program of a functional megaspore (apospory) or from a megaspore mother cell with suppressed or modified meiosis (diplospory). It is worth nothing that the gametic cell fate in apomictic plants can be activated either in somatic cells (apospory) or in unreduced megaspores (diplospory) as surrogate for meiotic products [6,16].

Sexual reproduction is based on the alternation of a diploid (sporophytic) and haploid (gametophytic) generation, both of which are fringed by events that lead to a change in ploidy, i.e., meiosis and fertilization. In gametophytic apomixis, both edge events are short-circuited as the egg cells originate through mitosis (apospory) or restitutional meiosis (diplospory), preserve a maternal genomic composition, and the embryos develop autonomously without any contribution of a spermatic nucleus (parthenogenesis). This combination was referred to as recurrent apomixis [15] as the original genotype may be indefinitely reiterated over generations.

Scaling up and down the ploidy level by means of genome accumulation or limitation can eventually take place by unreduced egg cell fertilization or reduced egg cell parthenogenesis respectively. These reproductive strategies have been referred to as non-recurrent apomixis [15]. Although not offering a stable means for genotype propagation, apomictic variants have likely been an important player in the evolution of polyploid species.

2. The Evolution of Apomixis

From the evolutionary point of view, the nature and persistence of asexual reproduction remains one of the most challenging phenomena [17]. In plants, asexuality (apomixis) either resurfaces in sexual lineages or it is derived by mutations from sexuality (amphimixis), and not only has it originated independently multiple times between different species but it has also resurfaced or evolved recurrently within certain species. Hence, genomic shocks, such as hybridization, polyploidization or both, have repeatedly switched from sexual to asexual reproduction. Despite the hypothesized disadvantages associated with apomixis, including limited genetic variation and mutation accumulation, asexually reproducing plants are highly adaptable, surprisingly stress tolerant and stable from an evolutionary perspective, and therefore questions regarding the origin and evolution of asexuality are still a matter of debate among population geneticists.

Species characterized by gametophytic apomixis are mainly polyploid, whereas their sexual relatives are usually diploid [18]. It is well known that many wild apomictic species are characterized not only by polyploidy, but also by hybridity [19]. In fact, apomicts usually belong to complex taxa whose members regularly undergo intra- and interspecific hybridization patterns. However, it is not yet clear what the relative contributions of hybridization and polyploidization models are, due to origin and evolution of the asexual lineage, since both phenomena can have different regulatory consequences [20–22], which could presumably lead to coordinated deregulation of stress perception and adaptation genes as well as the sexual path in a sexual ancestor. Moreover, an asexual pathway is obligate infrequently: most apomictic species are facultative, as individuals can set seeds through both sexual and apomictic reproductive modes.

Apomixis may now be regarded as a consequence of sexual failure (i.e., loss-of-function) rather than as a recipe for clonal success (i.e., gain-of-function). There is increasing evidence that apomixis is a modification of the normal sexual developmental pathway (reviewed by [1,23]). Most of the events that characterize sexual reproduction may be retained both structurally and functionally in apomictic reproduction, with the exceptions that the reduced egg cell is replaced by an unreduced egg cell, with absent or modified meiosis (i.e., apomeiosis), and the seed development does occur without egg cell fertilization (i.e., parthenogenesis). In addition, it is clear that residual sexual function is retained in pseudogamous apomixis; in fact, if it is true that seed development may occur without fertilization of either the egg cell or the central cell (i.e., autonomous apomixis), it is also true that fertilization may be required to form the endosperm in many apomictic plants [15].

Comprehension of parallel or convergent evolution of apomixis may be crucial for seeking the causal gene/s, as they represent distinctive labels for its independent resurfacing or origin among populations or species. The rationale usually suggests that, given a plant phenotype, the underlying genetic factors and mechanisms could be considered similar in closely related species or different in distantly related species. With particular reference to the mode of reproduction, i.e., apomixis vs. sexuality, the same reproductive strategy might have theoretically resurfaced or evolved among populations even within species by changes in different gene/s or among distantly related species by changes in the same gene/s through convergent or parallel evolution. As a general perspective, nowadays in the post-genomic era, reconsidering and understanding what we have learned about the genetic control of apomixis should help us to critically determine the evolution of this trait, including both apospory and diplospory, at least in the most studied model species.

If parallel evolution has occurred for apomixis, this reproductive strategy is expected to have evolved independently in different populations even in closely related species. In fact, different populations or species showing similar patterns of development are likely modified in similar ways if subjected to similar selection pressures. Hence, "parallelism" refers to independent developmental modifications of the same kind that gives rise to similar phenotypes. As a matter of fact, because closely related species have similar developmental programs, parallelism is frequent among phylogenetically related organisms [24]. With convergent evolution, different developmental pathways may generate similar phenotypes. This means that "convergence" includes independently evolved features that are similar as manifested trait, but have arisen by different developmental pathways [24]. If so, apomixis is expected to have arisen from different genetic modifications and developmental programs that have then evolved similarly even in distantly related organisms.

Apomixis has been detected in ferns (where this is better referred as to apogamy, i.e., development of a sporophyte from a gametophyte without fertilization), is rare in gymnosperms and very common in angiosperms [25]. Apomictic fern lineages have been documented only in recent years [26], distinguishing between two pathways to diploid spore production (premeiotic endomitosis, more common, and meiotic first division restitution, less frequently observed). Each of these two alternative spore-generating pathways yields chromosomally unreduced spores (i.e., diplospores) which then germinate. Within leptosporangiate ferns, nearly half of the families contain one or more apomictic taxa [26]. Interestingly, the frequency of apomixis seems significantly correlated with species diversity, but any significant relationship was found between apomixis and diversification rates [27]. These findings are in agreement with studies showing that apomictic lineages are generally youthful, with initial estimates placing the ages of extant apomictic ferns in relatively recent evolutionary time, most having appeared within 8 Ma [27] or 15 Ma [28].

In flowering plants, apomixis has been detected and documented in at least 79 families and 292 distinct genera [10] (Figure 1).

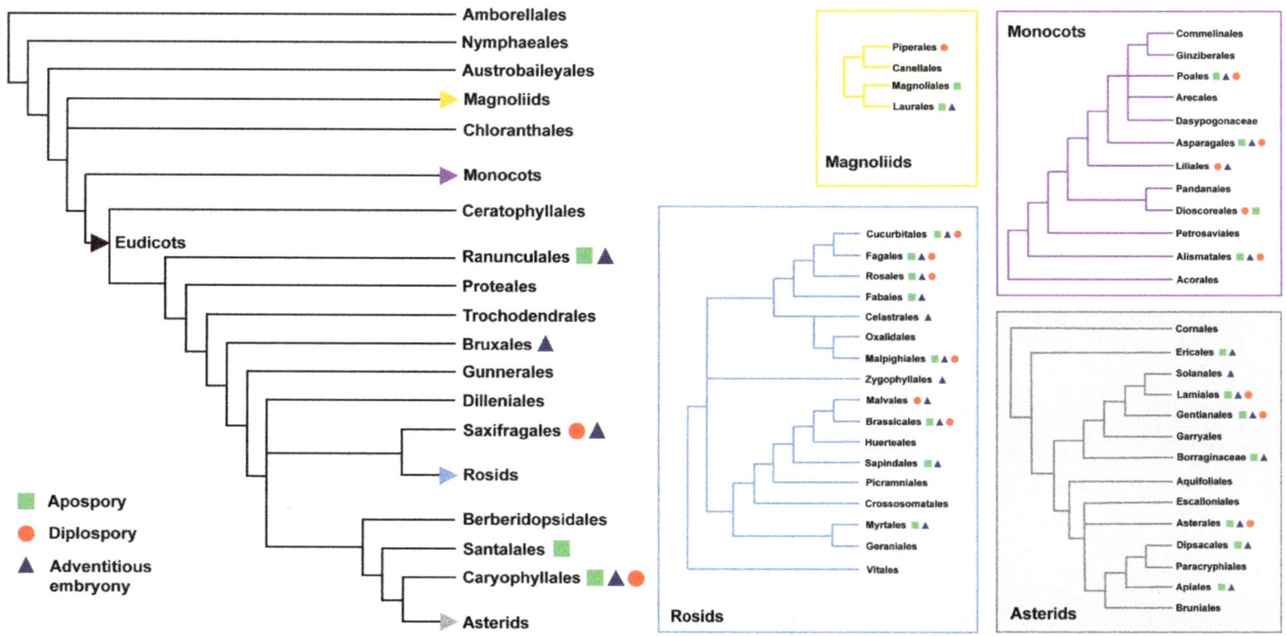

Figure 1. Distribution of the main types of apomixis (i.e., apospory, diplospory and adventitious embryony) among the main clades, orders and families of flowering plants. Detailed information for subclusters belonging to Magnoliids, Monocots, Rosids and Asterids are reported in the coloured boxes. Experimental evidences on apomictic pathways for taxonomic units are derived from [10,29] while phylogenetic trees are modified from [30].

Moreover, there are certain angiosperm families that show a great deal of apomixis affecting several genera: the outstanding examples are in the Rosaceae, Poaceae (Graminae) and Asteraceae (Compositae). Because, in these families, gametophytic apomixis, by means of either apospory or diplospory, is phylogenetically clustered above the genus level, it is argued that some clades, including closely related species with common ancestry within eudicots and monocots, may be preadapted and inclined to let apomixis evolve more easily [25]. With particular reference to Rosaceae, the phylogenetic origin and taxonomic distribution of apomixis have recently being investigated in great details in tribe Potentilleae [31]. Regular sexuality was distinguished from apomixis, as well as the zygotic versus parthenogenic origin of embryos and the pseudogamous versus autonomous origin of the endosperm. Records on the reproductive mode were provided for the genus *Farinopsis*, including 29 species belonging to five genera and seven series of *Potentilla*. Regular sexuality was observed in *Aphanes*, *Argentina*, *Comarum*, *Dasiphora*, *Drymocallis*, *Farinopsis*, *Fragaria*, *Horkeliella*, *Potentilla* and *Sibbaldia*, whereas apomixis was restricted to two evolutionary lineages: the *Potentilla* core group and *Alchemilla*/*Aphanes*. Early evolutionary divergence of these lineages (approximately 50 Ma), characterized by pseudogamous and autonomous apomictic seed formation, respectively, suggested parallel origins of apomixis [31]. Such a parallel origin would be the consequence of a repeated evolution of the same phenotype or genotype in different populations. Moreover, apomixis is shown to be taxonomically widespread in the whole Northern Hemisphere distribution range of *Potentilla*, a pattern that is explained by interspecific hybridization/introgression events and repeated intercontinental dispersals.

Furthermore, apomicts and their sexual relatives are often sympatric and thus frequently have overlapping ranges of adaptation. More specifically, compared to their sexual relatives, apomictic plants typically have wider geographical ranges, which extend into higher latitudes and altitudes, and better abilities to re-colonize regions after glaciation [32]. Despite a number of factors, i.e., the Pleistocene origins of apomixis—together with hybridity and polyploidy, unidirectional gene flow, niche targeting by asexual clones and limited biotic interactions in regions of glaciations—probably may be accounted for the geographical differences between sexual and apomictic plants; the relative influence of each is probably species-specific [33].

In the case the two main different variants of apomixis, i.e., apospory and diplospory, these do not have a common ancestor for the two major angiosperm lineages, i.e., monocots and eudicots; as it seems most likely because of the very differentiated developmental routes and taxonomic entities, then apomixis has arisen independently many times in the evolution of angiosperms. In addition, considering the two most widespread routes to originate the unreduced gametophyte, it is also apparent that the diplosporous condition involves a less radical departure from the normal sexual pattern that does the aposporous one (reviewed by [23]).

In the last 20 years, fossil records and morphological and molecular analyses involving extant taxa have clarified the position of many clades inside the angiosperms phylogenetic tree [34]. Three clades, Amborellaceae, Nymphaeales and Austrobaileyales (ANA-grade), are now considered "basal angiosperms" or "basalmost angiosperms", sisters to all other angiosperms or "mesangiosperms", representing an important information to infer ancestral morphological, chemical and genomic attributes to the earliest angiosperms [35–39]. The extant basal angiosperms may not be representative of the earliest, now extinct lineages. However, the large diversification (woody and terrestrial, herbaceous and aquatic, annual and perennial, lacking vessels or not, with different types of leaves, phyllotaxis, flowers and reproductive biology) of the relative few extant and basalmost angiosperms (about 200 species), in addition to fossil records, could be representative of a much larger diversity of early and now extinct lineages [34]. Particularly interesting is the presence, inside the ANA grade, of three different patterns of female gametophyte development: *Amborella*–type, *Nuphar-Schizandra* type and *Polygonum* type [40]. A large variety of breeding systems is also present: *Amborella trichopoda* is dioecious, *Trithuria* (Hydatellaceae, 12 species) possesses a mixture of dioecious, monoecious and bisexual breeding systems, whereas the remaining Nymphaeales (91 species) show bisexual flowers.

In addition, within Austrobaileyales, *Austrobaileya scandens* is bisexual, *Trimenia* (8 species) contains both andromonoecious and bisexual species, whereas in the Schizandraceae family *Illicium* (44 species) is fully bisexual, *Schisandra* (25 species) contains a mixture of monoecious and dioecious species and *Kadsura* (16 species) is predominantly monoecious, with a tendency toward dioecious behaviours [41]. In general, outcrossing systems are prevalent, some species have common adaptation to promote selfing, whereas only two species (*Trithuria filamentosa* and *T. incospicua*, Hydatellaceae) may putatively hide apomictic features [42,43], even if several other modes of asexual propagation are exhibited, including vegetative proliferation by tubers, stolons and other foliar parts [44,45].

Polyploidy by whole genome duplication (WGD), followed by gene loss and diploidization is spread along almost all fundamental lineages of the angiosperm phylogenetic tree and is generally considered to be a common mode of speciation that has far-reaching consequences for plant macroevolution and ecology [46,47]. Intragenomic syntenic analysis within Spermatophyta provides clear structural evidence of an ancient genomic duplication event shared by all flowering plants in a (difficult to estimate) time in 300–200 Mya, leading to the most recent common ancestor of extant angiosperms [48,49]. Establishment of a newly arisen apomictic lineage is often fostered by side-effects of polyploidy [50]. As for basal angiosperms (ANA grade), no evidence of WGD and apomictic mechanisms were found in the *Amborella* species and Austrobayleales lineages, whereas a WGD event was reported in water lilies (Nymphaeales) lineage [51] that would reveal the putative presence of apomictic species in *Trithuria* genus [43].

In conclusion, based on available experimental evidences, the three main types of agamospermy occur in all major clades of flowering plants, adventitious embryony being the most frequent form (148 genera), followed by aposporous apomixis (110 genera) and diplosporous apomixis (68 genera) [29]. Since the vast majority of historical records only reported the most prevalent type, we cannot rule out the coexistence of at least two if not all three main types of apomictic forms, because this is known as a common feature in apomictic plants [52]. In addition, with specific reference to the most basal angiosperm, *Amborella trichopoda*, apomixis is not occurring in this species and it was never documented in other non-flowering seed plant lineages, including gymnosperm species [29]. The only exception is represented by *C. dupreziana* (Pinophyta, Cupressaceae), where a particular case of paternal apomixis is thought to have evolved from sexuality in response to the reduction of population size, apparently limited to a few hundreds of individuals [53]. However, this type apomixis, where the embryo seems to result from the development of diploid pollen, is hardly comparable to the forms of apomixis observed within the Angiosperms clade.

3. The Genetic Control of Apomixis: A General Overview

One of the major challenges of population biology is to understand which are the genetic determinants that control the maintenance of sexual reproduction under natural selection. In this view, the fundamental components of amphimixis, such as genetic recombination and gamete fusion, allow the single individuals of a population to experience new allelic combinations and interactions leading to diversification and adaptation. Nevertheless, alternative routes of reproduction that circumvent sex, such as apomixis, gained significant evolutionary success [54]. In the offspring of apomictic plants, genetic diversity is avoided or minimized, as the embryos retain the maternal genotype and their development is independent from both meiotic reduction and egg cell fertilization in ovules. Fertilization of the central cell is often required for endosperm formation (pseudogamy).

During the last two decades, many scientists have worked on the isolation of the genetic determinants of the apomictic pathway with the perspective to induce apomictic reproduction in crop plants by genetic engineering (for review, see [16,18]). More recently, although artificially induced apomictic rice has been obtained [55] using a synthetic approach (e.g., by engineering key regulator genes of sexual development), additional research is required to determine stability of induced apomictic reproduction in field conditions. Consequently, even now using the modern tools of genomics, understanding the molecular pathway leading to apomixis in natural apomicts is necessary, but more complicated than expected.

Nowadays, new insights have contributed to shed light into the structural and functional feature of apomixis; these include the structural parallelism between the apomixis controlling region (ACR) in several natural apomicts and the Y-chromosome of dioecious plants [18], the silencing mechanism of a specific apomixis-linked genetic factor [56] and the functional validation of a genetic determinant of parthenogenesis in *Pennisetum squamulatum* [57].

Recent advances, based on sequencing, in silico mapping and in vitro expression analysis of selected apomixis-linked genes allowed the identification of a chromosome area common to *Sorghum bicolor*, *Setaria italica*, *Brachypodium distachyum*, rice and maize syntenic to the apomixis locus of *Paspalum simplex* [58]. This synteny group revealed different extents of gene collinearity with the apomixis locus, including genes with well-defined annotations for biological processes and molecular functions. Most importantly, apomixis-linked genes were expressed as both sense and antisense mRNAs and both transcripts proved to be more abundant in sexual compared to apomictic ovules, indicating a putative silencing effect of the apomixis-linked alleles on their sexual-specific counterparts in these cells [58]. This finding could act in favour of apomixis surfacing or evolved by silencing sex genes.

More specifically, it would appear that there are at least two distinct elements in the control of gametophytic apomixis: the production of unreduced embryo sacs, originating through apospory or diplospory, and the subsequent development of the embryos through parthenogenesis. However, other elements may need to be incorporated into a model explaining/miming the genetic control of apomixis, including pseudogamy and, in some apomicts, the autonomous development of endosperm.

Apomeiosis is rather a rare phenomenon when uncoupled with parthenogenesis while it is frequently observed as formation of unreduced gametes to overcome sterility of inter-specific hybrids leading to the origin and evolution of polyploid forms [59–61]. Conversely, when apomeiosis is coupled with parthenogenesis it attains regular elevated expression in natural apomicts [62]. Furthermore available data suggest that in some cases apomeiosis, by either apospory or diplospory, may be functionally and genetically independent from parthenogenesis and autonomous endosperm formation [16,18]. Finally, in most apomicts, both apospory and diplospory proved to be simply inherited in populations segregating for apomixis and a complex genetic model based on delicate interactions between initiators and repressors of both apomeiosis and parthenogenesis has been proposed for several species [63,64].

Analysis of genetic and molecular studies of apomicts is provided by several reviews, including those of Ozias-Akins and Conner [65], Hojsgaard [66], Whitton et al. [12], Barcaccia and Pupilli [18] Hand and Koltunow [1] and Schmidt [67]. Both naturally occurring and induced mutants holding individual components of apomixis have been identified (e.g., [68,69]), meaning that many taxa can potentially express apomixis-related traits, and that each component is under independent control and regulation. In several natural apomicts, the existence of genotypes which express only one component of apomixis or suppress the other (reviewed by [16]), further supports the hypothesis that distinct genetic factors control apospory, diplospory and parthenogenesis. It has long been recognized that apomixis is under control of single-dose dominant alleles (if hypothetically, apomixis were to be controlled by recessive alleles, then multiple-copies would be necessary in polyploids). Indeed, apomixis is inherited as a dominant trait in several apomictic species (see [70], for a review). However, even in species with simple inheritance patterns it is doubtful whether a single gene controls apomixis. As a matter of fact, the occurrence of genotypes that form embryos either from fertilized apomeiotic egg cells or by parthenogenic development of meiotically reduced egg cells have been documented by cyto-histological and flow-cytometric analyses, suggesting that apomeiosis and parthenogenesis may be uncoupled [16]. Recombinants for apomixis components that lack either apomeiosis or parthenogenesis have been reported in *Taraxacum officinale* [71], *Erigeron annuus* [72], *Poa pratensis* [73,74], *Hypericum perforatum* [75], Ranunculus [76] and *Cenchrus* species [77]. Finally, autonomous endosperm formation segregates independently from the other components of apomixis in *Hieracium*, [78,79]. On the whole, these findings suggest that apomixis may be controlled by a complex of closely linked genes.

A theoretical scenario for the origin and evolution of a two-gene apomixis system was proposed by Van Dijk and Vijverberg [45], including two dominant mutations that occur in a population of outcrossing hermaphrodites. A plant is changed by a first mutation from meiotic into apomeiotic, and by a second mutation from zygotic to parthenogenic embryo development. The chance that these two mutations would occur soon after each other within a nascent apomictic population seems unlikely [12]. There is, however, another possibility. Since most apomicts are of hybrid polyploid origin, perhaps the two mutations needed for the functioning of apomixis may be separately present in polyploid parent stocks and brought together by hybridization, leading to functional apomicts [23]. For the rise of apomixis in natural populations, an additional hypothesis calls the stepwise evolution model into question, as proposed by Hojsgaard and Hörandl [50]. It could be shown that sexual diploids have latent alleles for parthenogenesis with little or no penetrance, which become important benefiting from high expressivity if apomictic behaviour was introduced by specific mutations, giving rise to the development of apomeiotic egg cells. Moreover, for successful hybridization, a strict parthenogenic plant cannot function as a seed parent, since these mutations can only be combined in crosses between an apomeiotic seed parent and a parthenogenic male parent. This automatically results in polyploid apomictic hybrids, suggesting a direct relationship between gametophytic apomixis and polyploidy. New apomictic plants can function as a pollen donor in crosses with sexual plants, thereby generating new, secondary apomictic clones. This way would help explaining high clonal and genetic diversities, commonly found in populations of apomicts [80]. However, the main problem remains with this evolutionary scenario that the mutations for apomeiosis and parthenogenesis are individually deleterious and so they are expected to be selected against [81].

As reported by Briggs and Walters [23], devising a universal model of apomixis may be unrealistic: clearly this asexual mode of reproduction occurs in different forms and variants, and in many unrelated families of monocots and eudicots, suggesting that apomixis as a whole has evolved multiple times in angiosperm plant evolution following different developmental pathways, which perhaps are controlled by distinct genetic factors.

4. The Comparative Genomics of Master Genes

Nowadays, much is known about the formation of germlines in sexual plants, including genes specifying the cyto-genetic competence of sporocytes and driving the post-embryonic development. For instance, the earliest gene controlling this process in *Arabidopsis* is *SPOROCYTELESS* (*SPL*)/*NOZZLE* (*NZZ*). It encodes a transcriptional regulator of sporocyte development: loss of *SPL*/*NZZ* function changes the cell fate and abolishes the commitment to and initiation of sporogenesis in both male and female organs [82,83]. Since the gametophytic phase of the life cycle of all vascular plants (from ferns to angiosperms) starts with the development of haploid spores, this gene—known as essential for both male and female meiosis during sex organ development—is a master regulatory element of sexual plant reproduction.

Unfortunately, a similar gene that controls the primary step of apomixis is unknown and little is known about the mechanism that switches the reproductive process from sexual (i.e., meiotic) to asexual (i.e., apomeiotic). Many genes responsible for the formation of egg cells and unreduced embryo sacs, or involved in the development of the endosperm or embryo have been described, but a common apomictic pathway applicable to all crop plants has not been identified yet [16,18]. Hence, after two decades of substantial studies conducted in several laboratories and model plants, apomixis still seems to be an unsolved puzzle.

We know, for example, that *POLYCOMB GROUP* (*PcG*) genes, whose proteins exhibit some structural and functional conservation among higher plants, mammals and insects [84,85], are crucial for the development of multicellular organisms. Several variants well studied belong to the MEDEA-FERTILIZATION-INDEPENDENT ENDOSPERM (MEA–FIE) complex [86] that regulates cell proliferation during reproductive development, as meiotic products in flowering plants do not directly differentiate into gametes, but rather, they form the gametophytes, multicellular structures producing the gametes.

The endosperm and embryo developmental processes are both controlled by the female gametophyte at two different stages: repression of embryo/endosperm development in the absence of fertilization through imprinting, and expression of factors that are necessary after fertilization. In the absence of fertilization, the Arabidopsis FIE/FIS2/MEA complex—constituted by the FIS class PcG proteins MEDEA (MEA), FERTILIZATION-INDEPENDENT ENDOSPERM (FIE) and FERTILIZATION-INDEPENDENT SEED2 (FIS2)—suppresses endosperm development by regulating negatively the transcription of the genes directly involved in this process. Moreover, it has been shown that all mutations of the *fis* class of genes caused aberrant embryo and endosperm development if fertilized and exhibit autonomous endosperm development if unfertilized (reviewed by [18]). In any case, autonomously developed embryos and endosperms abort irrespectively from the paternal contribution. The proteins encoded by these genes mediate chromatin remodelling during seed formation. Imprinting, or parent-specific expression of genes, is a mechanism by which early stages of seed development are controlled by the female gametophytes.

Here we investigated the *SPL/NZZ* gene homologs in plant species representing the two major groups of angiosperms, monocotyledons and dicotyledons, including different basal angiosperm lineages belonging to the families *Amborellaceae* and *Nymphaeaceae*. Based on their putative orthology (BLASTp; [87]) with the well-characterized SPL protein belonging to *Arabidopsis thaliana* (AT4G27330), we selected 11 amino acid sequences from *Zea mays* (KY110964.1), *Oryza sativa* (LOC_Os01g11430.1), *Hypericum perforatum* (apomictic species, OBUPD-D1 Hpctg51499), *Brassica oleracea* (Bol013057), *Malus domestica* (MD11G1234600), *Prunus persica* (Prupe.4G192500.1), *Vitis vinifera* (VIT_219s0014g03940.1), *Lactuca sativa* (Lsat_1_v5_gn_0_5400.1), *Amborella trichopoda* (XP_006833114.1), *Nymphaea colorata* (XP031473161.1) and *Nymphaea thermarum* (KAF3782288.1). Protein sequences were aligned (Geneious software v7.1.5, Biomatters, Ltd., Auckland, New Zealand) using MUSCLE [88] to highlight any conserved structures (Figure 2A). A similarity-based neighbour-joining analysis was also performed (Figure 2B).

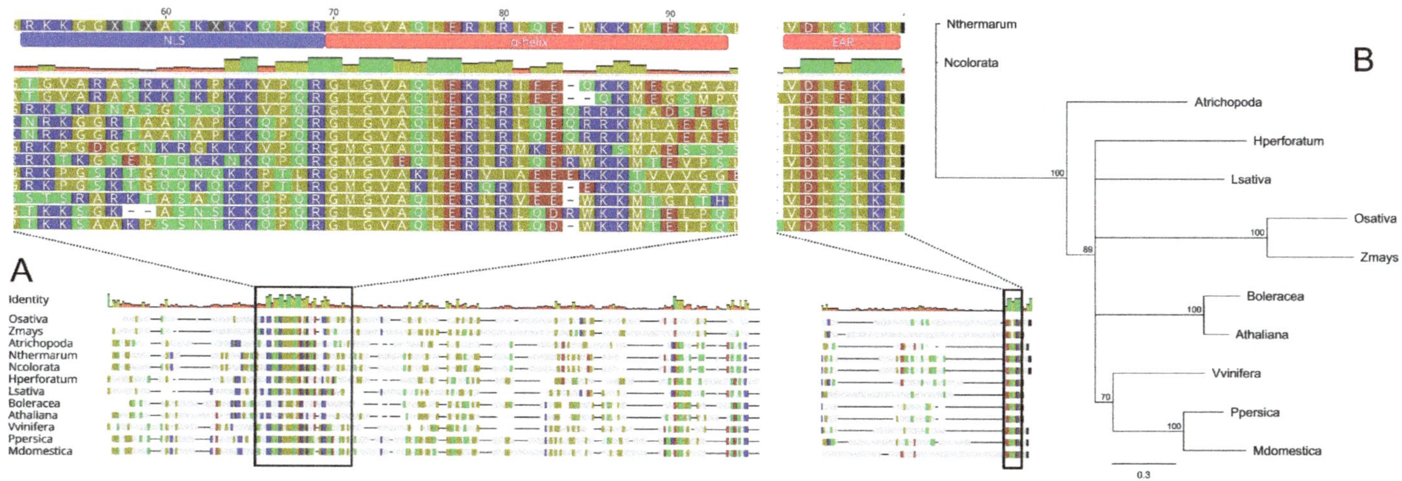

Figure 2. Bioinformatic analysis of the *SPL/NZZ* gene homologs in plant species representing the two major groups of angiosperms, monocotyledons and dicotyledons, including different basal angiosperm lineages belonging to the families Amborellaceae and Nymphaeaceae. (**A**) MUSCLE-based amino acid alignment of 11 putative SPL/NOZZLE sequences retrieved from *Zea mays* (KY110964.1), *Oryza sativa* (LOC_Os01g11430.1), *Hypericum perforatum* (apomictic species, OBUPD-D1 Hpctg51499), *Brassica oleracea* (Bol013057), *Malus domestica* (MD11G1234600), *Prunus persica* (Prupe.4G192500.1), *Vitis vinifera* (VIT_219s0014g03940.1), *Lactuca sativa* (Lsat_1_v5_gn_0_5400), *Amborella trichopoda* (XP_006833114.1), *Nymphaea colorata* (XP031473161.1), *Nymphaea thermarum* (KAF3782288.1), putative orthologues of SPL protein of *Arabidopsis thaliana* (AT4G27330). On the top of the panel the three main functional domains that resulted conserved among the 12 proteins are highlighted. NLS is a basic region rich in Arginine (R) and Lysine (K) that is thought to represent a putative nuclear localization signal; the α-helix sequence (also known as SPL-motif) is crucial for the constitution of homodimers and heterodimers, binding and inhibiting CINCINNATA (CIN)-like TEOSINTE BRANCHED1/CYCLOIDEA/PCF (TCP) transcription factors (TF); the EAR motif in the C-terminal region recruits TOPLESS/TOPLESS-RELATED (TPL/TPR) proteins to co-suppress the activity of the CIN-like TCP family. (**B**) Similarity-based neighbor-joining analysis performed using the 12 amino acid sequences with bootstrap values supporting all major nodes.

Overall, sequences length ranged from 288 (*L. sativa*) to 386 amino acids (*Z. mays*) while the identity percentages resulting from all possible pairwise comparisons among proteins, varied from 10.7% (between classes, e.g., *Z. mays* vs. *B. oleracea*) to 97.8% (within genera, e.g., *N. thermarum* vs *N. colorata*, see Figure S1). Despite that most of the SPL-like sequences displayed high levels of inter-variability, the three main functional regions resulted in being conserved between dicots and monocots species. In the N-terminal region, as originally observed in *Arabidopsis* by Yang et al. [82] and Schiefthaler et al. [89], we detected a conserved basic region rich in Arginine (R) and Lysine (K) that is thought to represent a putative nuclear localization signal (NLS, [89]). A short α-helix sequence (also known as SPL-motif) immediately downstream of NLS and crucial for the constitution of homodimers and heterodimers both in vitro and in vitro [83] resulted also conserved among the sequences analysed. This region seems to be involved in binding and inhibiting CINCINNATA (CIN)-like TEOSINTE BRANCHED1/CYCLOIDEA/PCF (TCP) transcription factors (TF) whose activities are pivotal for both leaf development [90,91] and normal ovule development [92].

Finally, in the C-terminal region, a DLxLKL consensus sequence, previously described as ethylene-responsive element binding factor-associated amphiphilic repression (EAR) motif [83], characterized all the protein sequences here taken into consideration. Wei et al. [92] and Chen at el. [83] hypothesized that EAR motif is crucial to recruit TOPLESS/TOPLESS-RELATED (TPL/TPR) proteins and together they co-suppress the activity of the CIN-like TCP family, control the expression of other *TCP* genes and stimulate megasporocytes differentiation during ovule development.

Overall, the high conservation degree found in the three functional domains of the 12 SPL-like proteins analysed suggests a common role of this master gene in promoting megasporocytes differentiation both in monocotyledons and dicotyledons. Moreover, the identification of *SPL-like* sequences in *A. trichopoda*, *N. colorata* and *N. thermarum* and the lack of evidences of apomictic reproduction in these basalmost angiosperms species [29,93], support the hypothesis that apomixis may have evolved from sex. In the event that apomixis may not have stemmed from sexuality, the total absence of the SPL/NZZ gene was expected in apomictic species, as found for the *Hypericum perforatum* genome used as model, since the role of this gene in the initiation of the sexual pathway is now well consolidated.

Although originally thought to represent an ancient evolutionary predisposition manifested by specific taxa [19], a detailed phylogenetic analysis coupled with apomixis distribution recently performed by León-Martínez and Vielle-Calzada [29] suggests a multiple independent emergence and rapid spreading of apomixis among large families.

5. The Genetic Control of Apomixis

Apomictic reproduction has been detected in 78 out of 460 families of the flowering plants mainly in Rosaceae, Asteraceae and Poaceae and analysis of apomixis in current phylogenetic trees lead to the hypothesis that this trait evolved independently multiple times among plant families. In particular, in grasses for which genome collinearity has been largely demonstrated and apomixis is quite common, it has been hypothesized that this reproductive trait might be under the control of the same genes. It has been reported that apomixis controlling regions in grasses share intrinsic characteristics such as block of recombination and presence of high rate of mutations and TEs [18]. Due to the recombination repression the apomixis locus is highly conserved within species, slightly divergent between species within genera and highly divergent between genera [65]. On the basis of comparative mapping using multiple grass genomes as reference, no chromosome areas were identified as syntenic to the apomixis locus that is common to the several apomictic species studied to date. In fact, the apomixis locus of *Paspalum simplex* (ACL) was syntenic, at the map level, to a conserved chromosome area that is located in telomeric position on chromosome 12, 8, 3 and 4 of rice, *Sorghum*, *Setaria* and *Brachypodium*, respectively, and near to the centromere on chromosome 1 of maize [58]. Conversely the apomixis locus in *Pennisetum squamulatum* (ASGR) showed frequently small- but not large-scale synteny with chromosomal regions of rice, *Setaria* and *Sorghum* genomes [94]; in *Tripsacum dactyloides*, RFLP markers linked to apomixis mapped to chromosome 6 of maize and in *Cenchrus ciliaris* to chromosome 6 of *Sorghum* [95] and in *Brachiaria brizantha,* to chromosome 5 of maize [96]. More recent comparative mapping studies in *Brachiaria* showed that, although high levels of synteny and collinearity were found between the ASGR of *P. squamulatum* and the apomixis locus of this genus, the apomixis carrier chromosome was identified as homologous to chromosome 1 and 5 of *Setaria* in *Brachiaria humidicola* and *B. decumbens*, respectively [97]. Furthermore, fine mapping of ASGR-carrier chromosome outside the ASGR of *P. squamulatum* showed a clear synteny with chromosome 2 of *Setaria* and *Sorghum* [98]. Taken together, these information strongly indicate that the apomixis locus has been evolved as a single event within an evolutionary lineage and was spread among plant families through hybridisation or phylogenetic diversification [99]. Nevertheless, at least two independent events have been identified within Panicoideae subfamily one in *Paspalum* and the other in *Pennisetum/Cenchrus/Brachiaria*.

Do different routes of evolution reflect functional differences in apomixis? Although apomixis in *Paspalum* and *Pennisetum* are both of the aposporous type, followed by the parthenogenic development of the egg cell, substantial differences do exist for the development of endosperm. In both cases, fertilization of the central cell is necessary for the formation of viable endosperm and seed; however, whereas in *Paspalum,* both unreduced polar nuclei are fused so as to get a C ratio of embryo:endosperm of 2:5, considering a reduced sperm to fertilize central cell, in *Pennisetum/Cenchrus* only one polar nucleus is fertilized yielding a 2:3 C ratio, as well as in sexual races of the same complex. This means that in apomictic *Paspalum* spp., the endosperm balance number of 2:1 of maternal:paternal genome ratio in the endosperm that has been proven to be essential for proper seed development, especially in grasses [100,101], might be relaxed in these species. As a matter of fact, apomictic tetraploid strains of *P. notatum* tolerated several assortments of genomes in the endosperm whereas sexual strains yielded viable seeds only when endosperm held the canonical 2m:1p ratio of parental genomes in the endosperm [102]. On the basis of findings that rare triploid individuals do occur in regions in which apomictic tetraploids are grown in sympatry with sexual diploids [103] and of the above reported studies of comparative mapping, it has been hypothesized that the ACL of *Paspalum* originated in an unstable chromosomal region of the ancestral grass genome where: (i) sex related genes were grouped by gene migration in the same genome context during speciation, (ii) a polyploidization event (through an intermediate triploid bridge) induced locally further small scale rearrangements that, in turn, (iii) caused lack of chromosome pairing and local sequence divergence and a block of recombination. In this view gene migration and polyploidization are critical steps toward emerging apomixis in this genus. In one recently proposed model to explain the successful plant gene mobility in angiosperms, Bennetzen and Wang [104] highlighted the criteria for a protein-encoding gene to conserve its function after transfer to a new location of the genome: these are its (i) small size, (ii) conservation of cis regulatory elements and (iii) landing in a no repressive genomic context (Small Insulated Genes Move Around (SIGMAR)). Furthermore, the most likely mechanism for insertional gene mobility is retroduplication by LTR retrotransposons, involving a reverse transcription of a transcript that causes elimination of introns. Among the apomixis-linked genes analysed by Galla et al. [58], *PsORC3* fits the SIGMAR condition for gene mobility well, as it is a small gene (2 Kb), it expresses a functional protein and completely lacks introns [56]. The ortholog of this gene was neither in the syntenic area of rice nor even in the same chromosome. Similarly, in *Brachypodium* it is far from the homology area of the apomixis locus though in the same chromosome. Conversely, in the *Panicoideae* subfamily (*Zea mays, Setaria, Sorghum*), the *PsORC3* ortholog was in the same syntenic group as the apomixis-linked BACs of *Paspalum*. Furthermore, since a variant allele of this gene was found in the sexual counterpart of the ACL of *P. notatum* (Ortiz et al. in preparation) this indicates that migration of this gene preceded the development of apomictic reproduction in *Paspalum*. Another intrinsic characteristic of the ACL of *Paspalum* probably peculiar of this genus is that the genes contained in this locus are transcribed in a coordinated manner as an operon-like gene cluster, so as these genes are constitutively transcribed [105] and expressed in reproductively committed cells and tissues as sense and antisense transcripts [56,58]. Although the conventional view of gene action in biology has centred on the dogma DNA→mRNA→protein, there are mounting evidences that many processes related to development are regulated by portions of the genome not necessarily linked to their coding capacity [106]. In particular, mRNAs that are transcribed from the reverse strand of a gene (natural antisense transcripts, NATs) are traditionally considered nonfunctional noise [107]. More recently, the development of the next sequence generation technology and the gaining of knowledge on the gene silencing, contribute to change this opinion revealing that an extensive number of NATs have a functional role. NATs have been surveyed in Arabidopsis [108], rice [109], wheat [110] and several legumes [111] and involved in response of abiotic stresses [112,113] and developmental transitions [114–116]. The mechanism of action of NATs are still not fully understood [117], although in many cases their role is in the repression of their sense cognate transcripts [118]. Our findings show that apomixis-linked genes are expressed as sense and antisense transcripts in cell lineages of sexual ovules, whereas their expression of both transcripts was strongly down regulated in the same cells of apomictic ovules. Coordinated down or up regulation of both sense and antisense transcripts derived from NAT pairs has been reported under water stress conditions in maize [112], suggesting that "sexual" and "apomictic" alleles may interact each other as NAT gene pairs (i.e., gene pairs that express transcripts that are overlapping and complementary). The nature of this interaction is generally poorly understood, although a possible link between antisense transcripts and chromatin modification has been proposed [108,119]. Again, we take *PsORC3* as a case study. This gene exists in three copies In *P. simplex* of which one, (*PsORC3a*) is specific of the ACL and is expressed constitutively as sense and antisense transcripts in nucellus and polar nuclei. *PsORC3a* might act a dominant negative regulator of the gene machinery responsible of the correct genome set up in the endosperm through antisense silencing of downstream acting genes that are common between sexual and apomictic genotypes of the *P. simplex* agamic complex. If this hypothesis will be confirmed we then have a further evidence of the Koltunow theory according to which apomixis acts dominantly over the sexual state [120] as well as the male phenotype is superimposed over female condition in some dioecious systems [121]. Finally, as Polegri et al. [105] pointed out, the genetic determinants of apomixis (i.e., *PsORC3a*) will likely be able to trigger apomixis only if the sexual recipient genome is preadapted to regulate expression of genes acting downstream (other copies of *PsORC3*) of the apomixis-linked factors. In a more general view, evolutionary, structural and functional findings serve modulate strategies aimed at introgressing apomictic reproduction in sexual crops.

6. Concluding Remarks

Apomixis, understood as asexual reproduction by seed (agamospermy), is thought to be repeatedly emerged in sexual lineages and eventually derived from sexual genotypes by means of mutations or modifications of genes or genomes. Despite being faced by several researchers over many years, nowadays the origin of apomixis still represents an unsolved problem, as it may be either evolved from sex [29] or the other way around [122]. Despite a number of disadvantages associated with apomixis, including narrow genetic variation and high mutation accumulation, asexually reproducing plants are highly adaptable and stable from an evolutionary perspective, mainly in terms of stabilization of polyploid genomes. Nowadays, an increasing body of evidence suggests that apomixis may be regarded as a consequence of sexual breakdown, rather than as a recipe for clonal success. As a matter of fact, ancestral sexual traits, such as meiosis and syngamy, show strong phylogenetic continuities among either closely related or distant taxa. The latter does not seem to be the case for apomixis since there are great discontinuities within orders, families within orders, genera within families, and even species within genera. Considering the two most widespread routes to originate unreduced gametophytes, i.e., apospory and diplospory, they do not appear to have a common ancestor in monocots and eudicots. In addition, it is also evident that the diplosporous condition involves a less radical departure from the normal sexual pattern that does the aposporous one. Hence devising a universal evolution model of apomixis may be unrealistic suggesting as a consequence that its different routes are likely controlled by distinct genetic factors showing distinct molecular functions.

Here we have considered the basal angiosperm lineages, the ANA grade, as through the analysis of their reproduction systems and barriers is possible to re-evaluate the significance of apomixis in the evolution of angiosperm plants. As a matter of fact, within the basal ancestors and early-branching angiosperms, such as *Amborellaceae*, *Nymphaeales* and *Austrobaileyales* outcrossing systems are prevalent, some species have common adaptations to promote selfing, whereas only two species, *Trithuria incospicua* and *T. filamentosa* have been documented as putative agamospermous, even if several other modes of asexual propagation are exhibited, including vegetative proliferation by tubers, stolons and other propagules. We know that gametophytic apomixis, by means of either apospory or diplospory, is especially common in genera of the families of *Rosaceae*, *Asteraceae* and *Poaceae*. Because in these families apomixis is phylogenetically clustered above the genus level, it has been already argued that some clades, including closely related species that originated from a common ancestor within eudicots and monocts, may be preadapted and inclined to let apomixis evolve more easily. Increasing access to plant genome sequences have offered us the opportunity to compare genes related to germline initiation in sexual and apomictic species of modern genomes with reconstructed founder ancestors of flowering plants (emerged around 214 million years ago during the late Triassic era). In particular, we investigated master regulatory elements of sexual plant reproduction, including genes essential for both male and female meiosis during sex organ development (i.e., genes that control functional changes of the cell fate of sporocyte initials and the commitment to and initiation/progression of sporogenesis in both male and female organs). Since the formation of haploid spores marks the initiation of the gametophytic phase of the life cycle of all vascular plants, ANA grade genomic data assisted us to understand the evolutionary forces that have shaped this master gene in sexual plant genomes and allowed us to gain insight into how it is organized and structured in apomictic plant genomes. As we know that functional apomixis requires not only apomeiosis but also parthenogenesis and, in plants, endosperm formation, the unreduced embryo sacs should possess egg cells that are believed to be epigenetically programmed, as apomicts, for the other two features. Therefore, we also investigated the studies meant to identify the epigenetic mechanisms that may control the reproductive switch.

In conclusion, we carried out a re-evaluation of the origin of apomixis in order to deepen knowledge on such asexual mode of reproduction, which seems only supposed in one basal angiosperm family (Hydatellaceae), while it clearly occurs in different forms and variants as well as in many unrelated families of monocots and eudicots. Overall findings strengthen the hypothesis that apomixis as a whole may have evolved multiple times in angiosperm plant evolution following different developmental pathways deviating to different extents from sexuality. Recent developments on the genetic control of apomixis in model species allowed us to shed additional light on the antagonist theories of gain- and loss-of-function over sexuality.

Supplementary Materials: The following are available online at http://www.mdpi.com/2073-4425/11/8/859/s1, Figure S1: Pairwise identity percentages between SPL protein from *Arabidopsis thaliana* (AT4G27330) and other 11 SPL-like amino acid sequences retrieved, from *Zea mays* (KY110964.1), *Oryza sativa* (LOC_Os01g11430.1), *Hypericum perforatum* (OBUPD-D1 Hpctg51499), *Brassica oleracea* (Bol013057), *Malus domestica* (MD11G1234600), *Prunus persica* (Prupe.4G192500.1), *Vitis vinifera* (VIT_219s0014g03940.1), *Lactuca sativa* (Lsat_1_v5_gn_0_5400), *Amborella trichopoda* (XP_006833114.1), *Nymphaea colorata* (XP031473161.1), *Nymphaea thermarum* (KAF3782288.1).

Author Contributions: Conceptualization, G.B.; methodology, G.B. and F.P. (Fulvio Pupilli); software, F.P. (Fabio Palumbo); investigation, G.B., F.P. (Fabio Palumbo), S.S., E.A. and F.P. (Fulvio Pupilli); writing—original draft preparation, G.B., F.P. (Fabio Palumbo), S.S. and F.P. (Fulvio Pupilli); writing—review and editing, G.B., F.P. (Fabio Palumbo), S.S., E.A. and F.P. (Fulvio Pupilli); supervision, G.B. All authors have read and agreed to the published version of the manuscript.

Funding: This research received no external funding.

Conflicts of Interest: The authors declare no conflict of interest.

References

1. Hand, M.L.; Koltunow, A.M.G. The genetic control of apomixis: Asexual seed formation. *Genetics* **2014**, *197*, 441–450. [CrossRef]
2. Bernstein, H.; Bernstein, C. Evolutionary Origin and Adaptive Function of Meiosis. In *Meiosis*; Bernstein, C., Ed.; InTechOpen: Rijeka, Croatia, 2013; pp. 41–76.
3. Bernstein, H.; Bernstein, C.; Michod, R.E. Sex in microbial pathogens. *Infect. Genet. Evol.* **2018**, *57*, 8–25. [CrossRef] [PubMed]
4. Cavalier-Smith, T. Origin of the cell nucleus, mitosis and sex: Roles of intracellular coevolution. *Biol. Direct* **2010**, *5*, 1–78. [CrossRef] [PubMed]
5. Hanson, S.J.; Schurko, A.M.; Hecox-Lea, B.; Mark Welch, D.B.; Stelzer, C.P.; Logsdon, J.M. Inventory and phylogenetic analysis of meiotic genes in monogonont rotifers. *J. Hered.* **2013**, *104*, 357–370. [CrossRef] [PubMed]
6. Hörandl, E.; Hadacek, F. The oxidative damage initiation hypothesis for meiosis. *Plant Reprod.* **2013**, *26*, 351–367. [CrossRef] [PubMed]
7. Schurko, A.M.; Logsdon, J.M. Using a meiosis detection toolkit to investigate ancient asexual "scandals" and the evolution of sex. *Bioessays* **2008**, *30*, 579–589. [CrossRef]
8. Speijer, D.; Lukeš, J.; Eliáš, M. Sex is a ubiquitous, ancient, and inherent attribute of eukaryotic life. *Proc. Natl. Acad. Sci. USA* **2015**, *112*, 8827–8834. [CrossRef]
9. Bilinski, C.A.; Marmiroli, N.; Miller, J.J. Apomixis in Saccharomyces cerevisiae and Other Eukaryotic Micro-organisms. *Adv. Microb. Physiol.* **1990**, *30*, 23–52.
10. Hojsgaard, D.; Klatt, S.; Baier, R.; Carman, J.G.; Hörandl, E. Taxonomy and Biogeography of Apomixis in Angiosperms and Associated Biodiversity Characteristics. *CRC Crit. Rev. Plant Sci.* **2014**, *33*, 414–427. [CrossRef]
11. Suomalainen, E.; Saura, A.; Lokki, J. *Cytology and Evolution in Parthenogenesis*; CRC Press, Inc.: Boca Raton, FL, USA, 1987.
12. Whitton, J.; Sears, C.J.; Baack, E.J.; Otto, S.P. The dynamic nature of apomixis in the angiosperms. *Int. J. Plant Sci.* **2008**, *169*, 169–182. [CrossRef]
13. Firetti, F. Apomixis in Neotropical Vegetation. In *Vegetation*; Sebata, A., Ed.; InTechOpen: Rijeka, Croatia, 2018; pp. 129–148.
14. Naumova, T. *Apomixis in Angiosperms: Nucellar and Integumentary Embryony*; CRC Press, Inc.: Boca Raton, FL, USA, 1992.
15. Nogler, G.A. Gametophytic Apomixis. In *Embryology of Angiosperms*; Johri, B.M., Ed.; Springer: Berlin, Germany, 1984; pp. 475–518.
16. Barcaccia, G.; Albertini, E. Apomixis in plant reproduction: A novel perspective on an old dilemma. *Plant Reprod.* **2013**, *26*, 159–179. [CrossRef] [PubMed]
17. Bell, G. *The Masterpiece of Nature: The Evolution and Genetics of Sexuality*; Routledge Revivals: Abingdon-on-Thames, UK, 1982.
18. Pupilli, F.; Barcaccia, G. Cloning plants by seeds: Inheritance models and candidate genes to increase fundamental knowledge for engineering apomixis in sexual crops. *J. Biotechnol.* **2012**, *159*, 291–311. [CrossRef]
19. Richards, A.J. Apomixis in flowering plants: An overview. *Philos. Trans. R. Soc. Lond. Ser. B Biol. Sci.* **2003**, *358*, 1085–1093. [CrossRef] [PubMed]
20. Comai, L.; Madlung, A.; Josefsson, C.; Tyagi, A.; Chen, Z.J. Do the different parental "heteromes" cause genomic shock in newly formed allopolyploids? *Philos. Trans. R. Soc. Lond. Ser. B Biol. Sci.* **2003**, *358*, 1149–1155. [CrossRef] [PubMed]
21. Osborn, T.C.; Pires, J.C.; Birchler, J.A.; Auger, D.L.; Chen, Z.J.; Lee, H.S.; Comai, L.; Madlung, A.; Doerge, R.W.; Colot, V.; et al. Understanding mechanisms of novel gene expression in polyploids. *Trends Genet.* **2003**, *19*, 141–147. [CrossRef]
22. Swanson-Wagner, R.A.; Jia, Y.; DeCook, R.; Borsuk, L.A.; Nettleton, D.; Schnable, P.S. All possible modes of gene action are observed in a global comparison of gene expression in a maize F1 hybrid and its inbred parents. *Proc. Natl. Acad. Sci. USA* **2006**, *103*, 6805–6810. [CrossRef] [PubMed]

23. Briggs, D.; Walters, S.M. *Plant Variation and Evolution*; Cambridge University Press: Cambridge, UK, 2016; ISBN 9781139060196.
24. Futuyma, D.J. *Evolutionary Biology*; Sinauer Associates Inc.: Sunderland, MA, USA, 1997.
25. van Dijk, P. Apomixis: Basics for Non-Botanists. In *Lost Sex: The Evolutionary Biology of Parthenogenesis*; Schon, I., Martens, K., Eds.; Springer: Dordrecht, The Netherlands, 2009; pp. 47–62.
26. Grusz, A.L. A current perspective on apomixis in ferns. *J. Syst. Evol.* **2016**, *54*, 656–665. [CrossRef]
27. Liu, H.-M.; Dyer, R.J.; Guo, Z.-Y.; Meng, Z.; Li, J.-H.; Schneider, H. The Evolutionary Dynamics of Apomixis in Ferns: A Case Study from Polystichoid Ferns. *J. Bot.* **2012**, *2012*, 1–11. [CrossRef]
28. Tanaka, T.; Isaka, Y.; Hattori, M.; Sato, T. Ecological and phylogenetic approaches for diversification of apogamous ferns in Japan. *Plant Syst. Evol.* **2014**, *300*, 2041–2050. [CrossRef]
29. León-Martínez, G.; Vielle-Calzada, J.P. Apomixis in Flowering Plants: Developmental and Evolutionary Considerations. In *Current Topics in Developmental Biology*; Elsevier Inc.: Amsterdam, The Netherlands, 2019; Volume 131, pp. 565–604. ISBN 9780128098042.
30. Byng, J.W.; Smets, E.F.; van Vugt, R.; Bidault, E.; Davidson, C.; Kenicer, G.; Chase, M.W.; Christenhusz, M.J.M. The phylogeny of angiosperms poster: A visual summary of APG IV family relationships and floral diversity James. *Glob. Flora* **2018**, *1*, 4–35.
31. Dobeš, C.; Lückl, A.; Kausche, L.; Scheffknecht, S.; Prohaska, D.; Sykora, C.; Paule, J. Parallel origins of apomixis in two diverged evolutionary lineages in tribe Potentilleae (Rosaceae). *Bot. J. Linn. Soc.* **2015**, *177*, 214–229. [CrossRef]
32. Bierzychudek, P. Patterns in Plant Parthenogenesis. In *The Evolution of Sex and Its Consequences*; Stearns, S.C., Ed.; Springer: Basel, Switzerland, 1987; pp. 197–217.
33. Hörandl, E. The complex causality of geographical parthenogenesis. *New Phytol.* **2006**, *171*, 525–538. [CrossRef] [PubMed]
34. Soltis, D.; Soltis, P.; Endress, P.; Chase, M.; Manchester, S.; Judd, W.; Majure, L.; Mavrodiev, E. *Phylogeny and Evolution of the Angiosperms*; University of Chicago Press: Chicago, IL, USA, 2018.
35. Doyle, J.A.; Endress, P.K. Integrating early cretaceous fossils into the phylogeny of living angiosperms: Magnoliidae and eudicots. *J. Syst. Evol.* **2010**, *48*, 1–35. [CrossRef]
36. Soltis, D.E.; Smith, S.A.; Cellinese, N.; Wurdack, K.J.; Tank, D.C.; Brockington, S.F.; Refulio-Rodriguez, N.F.; Walker, J.B.; Moore, M.J.; Carlsward, B.S.; et al. Angiosperm phylogeny: 17 genes, 640 taxa. *Am. J. Bot.* **2011**, *98*, 704–730. [CrossRef] [PubMed]
37. Drew, B.T.; Ruhfel, B.R.; Smith, S.A.; Moore, M.J.; Briggs, B.G.; Gitzendanner, M.A.; Soltis, P.S.; Soltis, D.E. Another look at the root of the angiosperms reveals a familiar tale. *Syst. Biol.* **2014**, *63*, 368–382. [CrossRef]
38. Wickett, N.J.; Mirarab, S.; Nguyen, N.; Warnow, T.; Carpenter, E.; Matasci, N.; Ayyampalayam, S.; Barker, M.S.; Burleigh, J.G.; Gitzendanner, M.A.; et al. Phylotranscriptomic analysis of the origin and early diversification of land plants. *Proc. Natl. Acad. Sci. USA* **2014**, *111*, E4859–E4868. [CrossRef]
39. Zeng, L.; Zhang, Q.; Sun, R.; Kong, H.; Zhang, N.; Ma, H. Resolution of deep angiosperm phylogeny using conserved nuclear genes and estimates of early divergence times. *Nat. Commun.* **2014**, *5*, 4956. [CrossRef]
40. Friedman, W.E.; Ryerson, K.C. Reconstructing the ancestral female gametophyte of angiosperms: Insights from Amborella and other ancient lineages of flowering plants. *Am. J. Bot.* **2009**, *96*, 129–143. [CrossRef]
41. Anger, N.; Fogliani, B.; Scutt, C.P.; Gâteblé, G. Dioecy in Amborella trichopoda: Evidence for genetically based sex determination and its consequences for inferences of the breeding system in early angiosperms. *Ann. Bot.* **2017**, *119*, 591–597.
42. Rudall, P.J.; Remizowa, M.V.; Beer, A.S.; Bradshaw, E.; Stevenson, D.W.; MacFarlane, T.D.; Tuckett, R.E.; Yadav, S.R.; Sokoloff, D.D. Comparative ovule and megagametophyte development in hydatellaceae and water lilies reveal a mosaic of features among the earliest angiosperms. *Ann. Bot.* **2008**, *101*, 941–956. [CrossRef]
43. Smissen, R.D.; Ford, K.A.; Champion, P.D.; Heenan, P.B. Genetic variation in Trithuria inconspicua and T. filamentosa (Hydatellaceae): A new subspecies and a hypothesis of apomixis arising within a predominantly selfing lineage. *Aust. Syst. Bot.* **2019**, *32*, 1–11. [CrossRef]
44. Wiersema, J.H.S. Reproductive Biology of Nymphaea (Nymphaeaceae). *Ann. Mo. Bot. Gard.* **1988**, *75*, 795–804. [CrossRef]

45. van Dijk, P.J.; Vijverberg, K. The Significance of Apomixis in the Evolution of the Angiosperms: A Reappraisal. In *Plant Species-Level Systematics: New Perspectives on Pattern & Process*; Bakker, F.T., Chatrou, L.W., Gravendeel, B., Pelser, P.B., Eds.; Gantner Verlag: Vaduz, Germany, 2005; pp. 101–116.
46. Soltis, P.S.; Marchant, D.B.; Van de Peer, Y.; Soltis, D.E. Polyploidy and genome evolution in plants. *Curr. Opin. Genet. Dev.* **2015**, *35*, 119–125. [CrossRef] [PubMed]
47. Clark, J.W.; Donoghue, P.C.J. Whole-Genome Duplication and Plant Macroevolution. *Trends Plant Sci.* **2018**, *23*, 933–945. [CrossRef] [PubMed]
48. Jiao, Y.; Wickett, N.J.; Ayyampalayam, S.; Chanderbali, A.S.; Landherr, L.; Ralph, P.E.; Tomsho, L.P.; Hu, Y.; Liang, H.; Soltis, P.S.; et al. Ancestral polyploidy in seed plants and angiosperms. *Nature* **2011**, *473*, 97–100. [CrossRef]
49. Van De Peer, Y.; Mizrachi, E.; Marchal, K. The evolutionary significance of polyploidy. *Nat. Rev. Genet.* **2017**, *18*, 411–424. [CrossRef]
50. Hojsgaard, D.; Hörandl, E. The rise of apomixis in natural plant populations. *Front. Plant Sci.* **2019**, *10*, 358. [CrossRef]
51. DePamphilis, C.W.; Palmer, J.D.; Rounsley, S.; Sankoff, D.; Schuster, S.C.; Ammiraju, J.S.S.; Barbazuk, W.B.; Chamala, S.; Chanderbali, A.S.; Determann, R.; et al. The Amborella genome and the evolution of flowering plants. *Science* **2013**, *342*, 1241089.
52. Asker, S.; Jerling, L. *Apomixis in Plants*; CRC Press, Inc.: Boca Raton, FL, USA, 1992; ISBN 9780849345456.
53. Pichot, C.; El Maâtaoui, M.; Raddi, S.; Raddi, P. Surrogate mother for endangered Cupressus. *Nature* **2001**, *412*, 39. [CrossRef]
54. Spillane, C.; Steimer, A.; Grossniklaus, U. Apomixis in agriculture: The quest for clonal seeds. *Sex. Plant Reprod.* **2001**, *14*, 179–187. [CrossRef]
55. Wang, C.; Liu, Q.; Shen, Y.; Hua, Y.; Wang, J.; Lin, J.; Wu, M.; Sun, T.; Cheng, Z.; Mercier, R.; et al. Clonal seeds from hybrid rice by simultaneous genome engineering of meiosis and fertilization genes. *Nat. Biotechnol.* **2019**, *37*, 283–286. [CrossRef]
56. Siena, L.A.; Ortiz, J.P.A.; Calderini, O.; Paolocci, F.; Cáceres, M.E.; Kaushal, P.; Grisan, S.; Pessino, S.C.; Pupilli, F. An apomixis-linked ORC3-like pseudogene is associated with silencing of its functional homolog in apomictic Paspalum simplex. *J. Exp. Bot.* **2016**, *67*, 1965–1978. [CrossRef]
57. Conner, J.A.; Mookkan, M.; Huo, H.; Chae, K.; Ozias-Akins, P. A parthenogenesis gene of apomict origin elicits embryo formation from unfertilized eggs in a sexual plant. *Proc. Natl. Acad. Sci. USA* **2015**, *112*, 11205–11210. [CrossRef] [PubMed]
58. Galla, G.; Siena, L.A.; Ortiz, J.P.A.; Baumlein, H.; Barcaccia, G.; Pessino, S.C.; Bellucci, M.; Pupilli, F. A Portion of the Apomixis Locus of Paspalum Simplex is Microsyntenic with an Unstable Chromosome Segment Highly Conserved Among Poaceae. *Sci. Rep.* **2019**, *9*, 1–12. [CrossRef] [PubMed]
59. Peloquin, S.J.; Yerk, G.L.; Werner, J.E.; Darmo, E. Potato breeding with haploids and 2n gametes. *New Dev. Plant Breed.* **1989**, *31*, 1000–1004. [CrossRef]
60. Ramsey, J.; Schemske, D.W. Pathways, mechanisms, and rates of polyploid formation in flowering plants. *Annu. Rev. Ecol. Syst.* **1998**, *29*, 467–501. [CrossRef]
61. Barcaccia, G.; Tavoletti, S.; Mariani, A.; Veronesi, F. Occurrence, inheritance and use of reproductive mutants in alfalfa improvement. *Euphytica* **2003**, *133*, 37–56. [CrossRef]
62. Noyes, R.D. Inheritance of apomeiosis (diplospory) in fleabanes (Erigeron, Asteraceae). *Heredity* **2005**, *94*, 193–198. [CrossRef]
63. Holsinger, K.E. Reproductive systems and evolution in vascular plants. *Proc. Natl. Acad. Sci. USA* **2000**, *97*, 7037–7042. [CrossRef]
64. Matzk, F.; Prodanovic, S.; Bäumlein, H.; Schubert, I. The inheritance of apomixis in Poa pratensis confirms a five locus model with differences in gene expressivity and penetrance. *Plant Cell* **2005**, *17*, 13–24. [CrossRef]
65. Ozias-Akins, P. Apomixis: Developmental characteristics and genetics. *CRC Crit. Rev. Plant Sci.* **2006**, *25*, 199–214. [CrossRef]
66. Hojsgaard, D. Apomixis technology: Separating the wheat from the chaff. *Genes* **2020**, *11*, 411. [CrossRef] [PubMed]
67. Schmidt, A. Controlling apomixis: Shared features and distinct characteristics of gene regulation. *Genes* **2020**, *11*, 329. [CrossRef] [PubMed]

68. Curtis, M.D.; Grossniklaus, U. Amphimixis and apomixis: Two sides of the same coin! In *Apomixis. Evolution, Mechanisms and Perspectives*; Horandl, E., Grossniklaus, U., Van Dijk, P., Sharbel, T., Eds.; Koeltz Scientific Books: Wien, Austria, 2007; pp. 37–62.
69. Ravi, M.; Marimuthu, M.P.A.; Siddiqi, I. Gamete formation without meiosis in Arabidopsis. *Nature* **2008**, *451*, 1121–1124. [CrossRef]
70. Grossniklaus, U.; Nogler, G.A.; van Dijk, P.J. How to avoid sex: The genetic control of gametophytic apomixis. *Plant Cell* **2001**, *13*, 1491–1497. [CrossRef]
71. van Dijk, P.J.; Bakx-Schotman, J.M.T. Formation of Unreduced Megaspores (Diplospory) in Apomictic Dandelions (Taraxacum officinale, s.l.) Is Controlled by a Sex-Specific Dominant Locus. *Genetics* **2004**, *166*, 483–492. [CrossRef]
72. Noyes, R.D.; Rieseberg, L.H. Two independent loci control agamospermy (apomixis) in the triploid flowering plant Erigeron annuus. *Genetics* **2000**, *155*, 379–390.
73. Barcaccia, G.; Mazzucato, A.; Falcinelli, M. Inheritance of apomictic seed production in kentucky bluegrass (*poa pratensis* L.). *J. New Seeds* **2001**, *2*, 43–58. [CrossRef]
74. Albertini, E.; Porceddu, A.; Ferranti, F.; Reale, L.; Barcaccia, G.; Romano, B.; Falcinelli, M. Apospory and parthenogenesis may be uncoupled in Poa pratensis: A cytological investigation. *Sex. Plant Reprod.* **2001**, *14*, 213–217. [CrossRef]
75. Schallau, A.; Arzenton, F.; Johnston, A.J.; Hähnel, U.; Koszegi, D.; Blattner, F.R.; Altschmied, L.; Haberer, G.; Barcaccia, G.; Bäumlein, H. Identification and genetic analysis of the *APOSPORY* locus in *Hypericum perforatum* L. *Plant J.* **2010**, *62*, 773–784. [CrossRef]
76. Barke, B.H.; Daubert, M.; Hörandl, E. Establishment of apomixis in diploid F2 hybrids and inheritance of apospory from F1 to F2 hybrids of the Ranunculus auricomus complex. *Front. Plant Sci.* **2018**, *9*, 1–12. [CrossRef] [PubMed]
77. Conner, J.A.; Gunawan, G.; Ozias-Akins, P. Recombination within the apospory specific genomic region leads to the uncoupling of apomixis components in Cenchrus ciliaris. *Planta* **2013**, *238*, 51–63. [CrossRef] [PubMed]
78. Catanach, A.S.; Erasmuson, S.K.; Podivinsky, E.; Jordan, B.R.; Bicknell, R. Deletion mapping of genetic regions associated with apomixis in Hieracium. *Proc. Natl. Acad. Sci. USA* **2006**, *103*, 18650–18655. [CrossRef]
79. Ogawa, D.; Johnson, S.D.; Henderson, S.T.; Koltunow, A.M.G. Genetic separation of autonomous endosperm formation (AutE) from the two other components of apomixis in Hieracium. *Plant Reprod.* **2013**, *26*, 113–123. [CrossRef]
80. van Dijk, P.J. Ecological and evolutionary opportunities of apomixis: Insights from Taraxacum and Chondrilla. *Philos. Trans. R. Soc. Lond. Ser. B Biol. Sci.* **2003**, *358*, 1113–1121. [CrossRef]
81. Mogie, M. *Evolution of Asexual Reproduction in Plants*; Springer Netherlands: Houten, The Netherlands, 1992; ISBN 978-0-412-44220-9.
82. Yang, W.C.; Ye, D.; Xu, J.; Sundaresan, V. The SPOROCYTELESS gene of Arabidopsis is required for initiation of sporogenesis and encodes a novel nuclear protein. *Genes Dev.* **1999**, *13*, 2108–2117. [CrossRef] [PubMed]
83. Chen, G.H.; Sun, J.Y.; Liu, M.; Liu, J.; Yang, W.C. SPOROCYTELESS is a novel embryophyte-specific transcription repressor that interacts with TPL and TCP proteins in Arabidopsis. *J. Genet. Genom.* **2014**, *41*, 617–625. [CrossRef] [PubMed]
84. Reyes, J.C.; Grossniklaus, U. Diverse functions of Polycomb group proteins during plant development. *Semin. Cell Dev. Biol.* **2003**, *14*, 77–84. [CrossRef]
85. Ringrose, L.; Paro, R. Epigenetic Regulation of Cellular Memory by the Polycomb and Trithorax Group Proteins. *Annu. Rev. Genet.* **2004**, *38*, 413–443. [CrossRef]
86. Köhler, C.; Hennig, L.; Bouveret, R.; Gheyselinck, J.; Grossniklaus, U.; Gruissem, W. Arabidopsis MSI1 is a component of the MEA/FIE Polycomb group complex and required for seed development. *EMBO J.* **2003**, *22*, 4804–4814. [CrossRef]
87. National Center for Biotechnology Information BLASTp—Standard Protein BLAST 2020. Available online: https://blast.ncbi.nlm.nih.gov/Blast.cgi/ (accessed on 15 July 2020).
88. Edgar, R.C. MUSCLE: Multiple sequence alignment with high accuracy and high throughput. *Nucleic Acids Res.* **2004**, *32*, 1792–1797. [CrossRef] [PubMed]

89. Schiefthaler, U.; Balasubramanian, S.; Sieber, P.; Chevalier, D.; Wisman, E.; Schneitz, K. Molecular analysis of NOZZLE, a gene involved in pattern formation and early sporogenesis during sex organ development in Arabidopsis thaliana. *Proc. Natl. Acad. Sci. USA* **1999**, *96*, 11664–11669. [CrossRef] [PubMed]
90. Palatnik, J.F.; Allen, E.; Wu, X.; Schommer, C.; Schwab, R.; Carrington, J.C.; Weigel, D. Control of leaf morphogenesis by microRNAs. *Nature* **2003**, *425*, 257–263. [CrossRef] [PubMed]
91. Nath, U.; Crawford, B.C.W.; Carpenter, R.; Coen, E. Genetic Control of Surface Curvature. *Science* **2003**, *299*, 1404–1407. [CrossRef]
92. Wei, B.; Zhang, J.; Pang, C.; Yu, H.; Guo, D.; Jiang, H.; Ding, M.; Chen, Z.; Tao, Q.; Gu, H.; et al. The molecular mechanism of SPOROCYTELESS/NOZZLE in controlling Arabidopsis ovule development. *Cell Res.* **2015**, *25*, 121–134. [CrossRef]
93. Hörandl, E.; Hojsgaard, D. The evolution of apomixis in angiosperms: A reappraisal. *Plant Biosyst.* **2012**, *146*, 681–693.
94. Conner, J.A.; Goel, S.; Gunawan, G.; Cordonnier-Pratt, M.M.; Johnson, V.E.; Liang, C.; Wang, H.; Pratt, L.H.; Mullet, J.E.; DeBarry, J.; et al. Sequence analysis of bacterial artificial chromosome clones from the apospory-specific genomic region of Pennisetum and Cenchrus. *Plant Physiol.* **2008**, *147*, 1396–1411. [CrossRef]
95. Bowers, J.E.; Abbey, C.; Anderson, S.; Chang, C.; Draye, X.; Hoppe, A.H.; Jessup, R.; Lemke, C.; Lennington, J.; Li, Z.; et al. A high-density genetic recombination map of sequence-tagged sites for Sorghum, as a framework for comparative structural and evolutionary genomics of tropical grains and grasses. *Genetics* **2003**, *165*, 367–386.
96. Pessino, S.C.; Ortiz, J.P.A.; Leblanc, O.; Do Valle, C.B.; Evans, C.; Hayward, M.D. Identification of a maize linkage group related to apomixis in Brachiaria. *Theor. Appl. Genet.* **1997**, *94*, 439–444. [CrossRef]
97. Worthington, M.; Heffelfinger, C.; Bernal, D.; Quintero, C.; Zapata, Y.P.; Perez, J.G.; de Vega, J.; Miles, J.; Dellaporta, S.; Tohme, J. A Parthenogenesis Gene Candidate and Evidence for Segmental Allopolyploidy in Apomictic Brachiaria Decumbens. *Genetics* **2016**, *203*, 1117–1132. [CrossRef]
98. Sapkota, S.; Conner, J.A.; Hanna, W.W.; Simon, B.; Fengler, K.; Deschamps, S.; Cigan, M.; Ozias-Akins, P. In silico and fluorescence in situ hybridization mapping reveals collinearity between the Pennisetum squamulatum apomixis carrier-chromosome and chromosome 2 of sorghum and foxtail millet. *PLoS ONE* **2016**, *11*, 1–13. [CrossRef] [PubMed]
99. Worthington, M.; Ebina, M.; Yamanaka, N.; Heffelfinger, C.; Quintero, C.; Zapata, Y.P.; Perez, J.G.; Selvaraj, M.; Ishitani, M.; Duitama, J.; et al. Translocation of a parthenogenesis gene candidate to an alternate carrier chromosome in apomictic Brachiaria humidicola. *BMC Genom.* **2019**, *20*, 1–18. [CrossRef] [PubMed]
100. Lin, B.Y. Ploidy barrier to endosperm development in maize. *Genetics* **1984**, *107*, 103–115. [PubMed]
101. Haig, D.; Westoby, M. Genomic imprinting in endosperm: Its effect on seed development in crosses between species, and its implications for the evolution of apomixis. *Philos. Trans. R. Soc. Lond. Ser. B Biol. Sci.* **1991**, *333*, 1–13.
102. Quarin, C.L. Effect of pollen souce and pollen ploidy on endosperm formation and seed in pseudogamous apomitic Paspalum notatum. *Sex. Plant Reprod.* **1999**, *11*, 331–335. [CrossRef]
103. Urbani, M.H.; Quarin, C.L.; Espinoza, F.; Penteado, M.I.O.; Rodrigues, I.F. Cytogeography and reproduction of the Paspalum simplex polyploid complex. *Plant Syst. Evol.* **2002**, *236*, 99–105. [CrossRef]
104. Bennetzen, J.L.; Wang, X. Relationships between Gene Structure and Genome Instability in Flowering Plants. *Mol. Plant* **2018**, *11*, 407–413. [CrossRef]
105. Polegri, L.; Calderini, O.; Arcioni, S.; Pupilli, F. Specific expression of apomixis-linked alleles revealed by comparative transcriptomic analysis of sexual and apomictic Paspalum simplex Morong flowers. *J. Exp. Bot.* **2010**, *61*, 1869–1883. [CrossRef]
106. Mattick, J.S. RNA regulation: A new genetics? *Nat. Rev. Genet.* **2004**, *5*, 316–323. [CrossRef]
107. Struhl, K. Transcriptional noise and the fidelity of initiation by RNA polymerase II. *Nat. Struct. Mol. Biol.* **2007**, *14*, 103–105. [CrossRef]
108. Wang, H.; Chung, P.J.; Liu, J.; Jang, I.C.; Kean, M.J.; Xu, J.; Chua, N.H. Genome-wide identification of long noncoding natural antisense transcripts and their responses to light in Arabidopsis. *Genome Res.* **2014**, *24*, 444–453. [CrossRef] [PubMed]
109. Lu, T.; Cui, L.; Zhou, Y.; Zhu, C.; Fan, D.; Gong, H.; Zhao, Q.; Zhou, C.; Zhao, Y.; Lu, D.; et al. Transcriptome-wide investigation of circular RNAs in rice. *RNA* **2015**, *21*, 2076–2087. [CrossRef] [PubMed]

110. Coram, T.E.; Settles, M.L.; Chen, X. Large-scale analysis of antisense transcription in wheat using the Affymetrix GeneChip wheat genome array. *BMC Genom.* **2009**, *10*, 1–11. [CrossRef] [PubMed]
111. Bardou, F.; Merchan, F.; Ariel, F.; Crespi, M. Dual RNAs in plants. *Biochimie* **2011**, *93*, 1950–1954. [CrossRef] [PubMed]
112. Xu, J.; Qi, W.; Freeling, M.; Xuecai, Z.; Xu, Y.; Mao, Y.; Xin, T.; Wu, F.; Lan, H.; Cao, M.; et al. Natural antisense transcripts are significantly involved in regulation of drought stress in maize. *Nucleic Acids Res.* **2017**, *45*, 5126–5141. [CrossRef] [PubMed]
113. Postnikova, O.A.; Nemchinov, L.G. Natural Antisense Transcripts Associated with Salinity Response in Alfalfa. *Plant Genome* **2015**, *8*, 1–5. [CrossRef]
114. Fedak, H.; Palusinska, M.; Krzyczmonik, K.; Brzezniak, L.; Yatusevich, R.; Pietras, Z.; Kaczanowski, S.; Swiezewski, S. Control of seed dormancy in Arabidopsis by a cis-acting noncoding antisense transcript. *Proc. Natl. Acad. Sci. USA* **2016**, *113*, E7846–E7855. [CrossRef]
115. Swiezewski, S.; Liu, F.; Magusin, A.; Dean, C. Cold-induced silencing by long antisense transcripts of an Arabidopsis Polycomb target. *Nat. Lett.* **2009**, *462*, 799–802. [CrossRef]
116. Liu, F.; Marquardt, S.; Lister, C.; Swiezewski, S.; Dean, C. Targeted 3′ Processing of Antisense Transcripts Triggers Arabidopsis FLC Chromatin Silencing. *Science* **2010**, *327*, 94–97. [CrossRef]
117. Bouchard, J.; Oliver, C.; Harrison, P.M. The distribution and evolution of Arabidopsis thaliana cis natural antisense transcripts. *BMC Genom.* **2015**, *16*, 1–9. [CrossRef]
118. Pelechano, V.; Steinmetz, L.M. Gene regulation by antisense transcription. *Nat. Rev. Genet.* **2013**, *14*, 880–893. [CrossRef] [PubMed]
119. Li, C.H.; Chen, Y. Targeting long non-coding RNAs in cancers: Progress and prospects. *Int. J. Biochem. Cell Biol.* **2013**, *45*, 1895–1910. [CrossRef] [PubMed]
120. Koltunow, A.M.G.; Johnson, S.D.; Rodrigues, J.C.M.; Okada, T.; Hu, Y.; Tsuchiya, T.; Wilson, S.; Fletcher, P.; Ito, K.; Suzuki, G.; et al. Sexual reproduction is the default mode in apomictic Hieracium subgenus Pilosella, in which two dominant loci function to enable apomixis. *Plant J.* **2011**, *66*, 890–902. [CrossRef] [PubMed]
121. Bergero, R.; Charlesworth, D. The evolution of restricted recombination in sex chromosomes. *Trends Ecol. Evol.* **2008**, *24*, 94–102. [CrossRef]
122. Albertini, E.; Barcaccia, G.; Carman, J.G.; Pupilli, F. Did apomixis evolve from sex or was it the other way around? *J. Exp. Bot.* **2019**, *70*, 2951–2964. [CrossRef]

genes

MDPI

Review

Controlling Apomixis: Shared Features and Distinct Characteristics of Gene Regulation

Anja Schmidt

Department of Biodiversity and Plant Systematics, Centre for Organismal Studies, Heidelberg University, D-69120 Heidelberg, Germany; anja.schmidt@cos.uni-heidelberg.de; Tel.: +49-6221-54-4683

Received: 3 February 2020; Accepted: 18 March 2020; Published: 20 March 2020

check for updates

Abstract: In higher plants, sexual and asexual reproduction through seeds (apomixis) have evolved as alternative strategies. As apomixis leads to the formation of clonal offspring, its great potential for agricultural applications has long been recognized. However, the genetic basis and the molecular control underlying apomixis and its evolutionary origin are to date not fully understood. Both in sexual and apomictic plants, reproduction is tightly controlled by versatile mechanisms regulating gene expression, translation, and protein abundance and activity. Increasing evidence suggests that interrelated pathways including epigenetic regulation, cell-cycle control, hormonal pathways, and signal transduction processes are relevant for apomixis. Additional molecular mechanisms are being identified that involve the activity of DNA- and RNA-binding proteins, such as RNA helicases which are increasingly recognized as important regulators of reproduction. Together with other factors including non-coding RNAs, their association with ribosomes is likely to be relevant for the formation and specification of the apomictic reproductive lineage. Subsequent seed formation appears to involve an interplay of transcriptional activation and repression of developmental programs by epigenetic regulatory mechanisms. In this review, insights into the genetic basis and molecular control of apomixis are presented, also taking into account potential relations to environmental stress, and considering aspects of evolution.

Keywords: apomixis; evolution; germline; gene regulation; sporogenesis; plant reproduction; ribosome; RNA helicase; sexual development; stress response

1. Plant Reproduction Is Characterized by Developmental Flexibility Including Sexual and Asexual Formation of Seeds (Apomixis)

Reproduction is an elementary process in the life cycles of all living species. In order to accomplish successful reproduction, propagation and adaptation, land plants adapted versatile strategies marked by commendable developmental flexibility. Apart from vegetative reproduction giving rise to offspring directly from tissues of the dominant sporophytic generation, sexual reproduction through seeds and apomixis are also common strategies. In contrast to sexual reproduction, apomixis leads to the formation of clonal offspring fully maintaining the genetic constitution of the mother plant. As this would allow the fixation of advantageous and complex genotypes, it has an outstanding potential for crop seed production. However, although apomixis is phylogenetically distributed in all major groups of angiosperms and occurs in more than 400 species [1–7], it is largely absent in major crop species. Therefore, engineering of apomixis for harnessing in agriculture is a longstanding aim [8–10]. To fully accomplish this, a detailed understanding of the genetic basis and the molecular mechanisms governing apomixis will be a prerequisite. So far, despite longstanding interest and research on apomixis, the underlying gene regulatory programs and their evolutionary origins are not well understood.

In this review, the developmental processes of sexual plant reproduction and apomixis are briefly outlined. The focus lies on a comprehensive description of current knowledge about the genetic basis and gene regulatory processes governing apomixis in different species and distinct types of apomixis. Thereby, it is intended to also propose and discuss new aspects and views to stimulate the scientific discussion on the topic. In addition, aspects of evolution and potential involvement of environmental conditions and stress regulations in apomixis control are presented. Detailed descriptions on gene regulatory programs underlying sexual reproduction can be found in other recent reviews [11–13].

From a developmental perspective, sexual reproduction and apomixis are related processes. In both cases, the female and male reproductive lineages (germlines) form in the reproductive tissues of the flower. These are the female ovules developing enclosed in the pistil and the male anthers, respectively. Germline specification and development proceeds in two consecutive steps, with mega- and microsporogenesis being the formation of mega- and microspores from selected female megaspore mother cells (MMCs) or male pollen mother cells (PMCs), respectively (Figure 1). Subsequent gametogenesis denotes the development of the female and male gametophytes (Figure 1). In higher plants they are reduced to a few cells only. During sexual reproduction, typically single sporophytic cells in the ovule and anther tissues are selected as MMCs or PMCs and determined for meiotic fate. The MMC is specified in a specialized domain of the developing ovule referred to as nucellus (Figure 1A). After meiosis, three of the four megaspores that have been formed undergo apoptosis and only one functional megaspore (FMS) survives as the founder cell of the gametophytic lineage. In the majority of angiosperms, a Polygonum-type mature female gametophyte (embryo sac) is formed by three rounds of mitosis in a syncytium and subsequent cellularization [14]. The mature gametophyte comprises seven cells and four distinct and specialized cell types (Figure 1A): the two female gametes, which are the egg cell and central cell that give rise to the embryo and the nourishing endosperm upon double fertilization, two synergid cells important for pollen tube guidance and reception, and three antipodal cells potentially playing a role in nourishing the gametophyte. Unlike in the female reproductive lineage during pollen development, all four meiotically formed microspores survive (Figure 1B). They undergo a first asymmetric mitotic division (pollen mitosis I) to form a two celled pollen with a generative cell engulfed in the vegetative cell. During a second mitotic division (pollen mitosis II), the two sperm cells derive from the generative cell.

Compared to sexual reproduction, apomixis represents alterations of the developmental program. This concerns mainly a few steps during the formation and development of the female germline (Figure 2). It is commonly accepted that apomixis derived several times independently and that distinct types of apomixis are represented in higher plants [6,15]: First, sporophytic apomixis is distinguished from gametophytic apomixis, as the apomictic embryos either originate directly from sporophytic cells or from the egg cell formed in the gametophyte, respectively. Sporophytic apomixis, also known as adventitious embryony, is widespread throughout the plant kingdom and is particularly frequent e.g., in *Citrus* and *Orchidaceae* [6,16]. In sporophytic apomixis, one or more adventitious embryos derive from sporophytic cells of the nucellus, which is surrounding the sexually formed gametophyte (Figure 2B). Unlike sexual reproduction, which usually leads to the formation of a single embryo per seed, adventitious embryony is frequently marked by polyembryony [6]. Both the sexually derived embryo and its asexual siblings compete for resources of the endosperm (Figure 2B). While the formation of more than one embryo in a single seed is a feature of sporophytic apomixis, polyembryony can also occasionally result from gametophytic apomixis. However, polyembryony alone is not a clear indication for apomixis, as it also rarely occurs through sexual reproduction. This is the case in particular in gymnosperms, where apomictic reproduction appears to be largely absent [1].

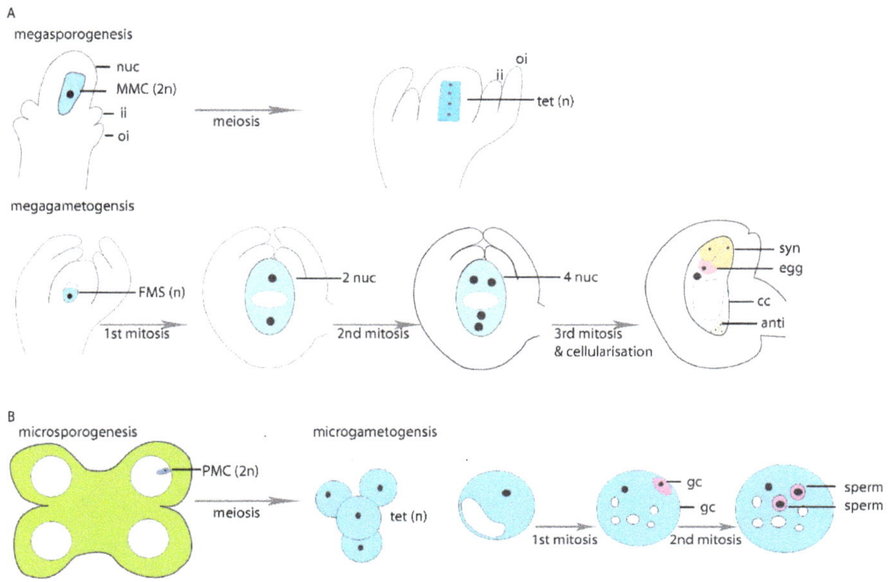

Figure 1. Development of the female (**A**) and male (**B**) reproductive lineages in sexual higher plants. (**A**) Formation of the female reproductive lineage initiates with the selection of a single diploid sporophytic cell in the nucellus (nuc) tissue of the ovule. This cell specifies as megaspore mother cell (MMC). Before meiosis of the MMC, the inner- and outer integuments of the ovule (ii and oi, respectively) are starting to grow. The MMC undergoes meiosis to give rise to a tetrad (tet) of haploid megaspores. Dependent on their position in the nucellus, three of the megaspores undergo apoptosis. Only the surviving functional megaspore (FMS) initiates gametogenesis. It undergoes three rounds of mitoses and cellularization to form the mature gametophyte harboring the two synergids (syn), the egg cell (egg), central cell (cc) and the antipodals (anti). (**B**) Formation of the male reproductive lineage initiates with selection of a single sporophytic cell, which is the pollen mother cell (PMC) that is committed to meiosis. Each of the four microspores of the tetrad (tet) survives and develops into a mature pollen by two mitotic divisions. During pollen mitosis I a generative cell (gc) engulfed in the vegetative cell (vc) is formed. During pollen mitosis II the generative cell divides to give rise to two haploid sperm cells.

Unlike through sporophytic apomixis, during gametophytic apomixis embryo and endosperm derive from the female gametes. To maintain the full genetic composition of the mother plant in the offspring, meiotic reduction and recombination need to be circumvented, as well as fertilization and thus the paternal contribution to the embryo. Developmental pathways of gametophytic apomixis are classified as diplospory or apospory [9,15,17] (Figure 2A). In diplosporous plants, the first cell of the female germline is an apomictic initial cell (AIC) developing in place of the MMC, but having a different fate. It undergoes a modified meiosis or it omits meiosis to give rise to an unreduced FMS (Figure 2A). Thereby, omission of meiosis and direct acquisition of gametophytic fate by the AIC holds true in Antennaria-type apomixis, while unreduced FMS are formed by restitution nucleus in Taraxacum-type and Ixeris-type of apomixis [10,18–20]. The Antennaria-type also referred to as mitotic diplospory has a wide systematic distribution [19]. In the Taraxacum-type of diplospory the MMC/AIC enters meiotic prophase I. As the chromosomes persist as univalent, this results in restitution nuclei remaining genetically identical to the sporophytic cells of the mother plant [21]. Also, the Ixeris-type of diplospory leads to formation of an unreduced FMS as consequence of restitution nucleus. In addition, 2n megaspores are formed in Allium-type of apomixis by premeiotic chromosome doubling [22]. Unlike in diplosporous apomicts, in aposporous apomicts, sexual and apomictic germlines initiate development in the same ovule (Figure 2A). One or several additional sporophytic cells adjacent to the sexual MMC directly give rise to the gametophytic lineage without intervening meiotic or mitotic divisions. Thereby, a competition of sexually and apomictically formed gametophytes can arise, or the development of the sexual gametophyte gets repressed by the apomictic germline lineage [22]. Independent of the developmental origin of the apomictic FMS, an embryo sac harboring egg cell and central cell is subsequently formed in all cases. To initiate seed development, the unfertilized egg cell then develops into an embryo parthenogenetically. The endosperm can form either autonomously without paternal contribution or by pseudogamy dependent on fertilization. Pseudogamy is prevalent in most apomicts. In contrast to the female germline, the male germline in apomicts may form either reduced or unreduced pollen [17].

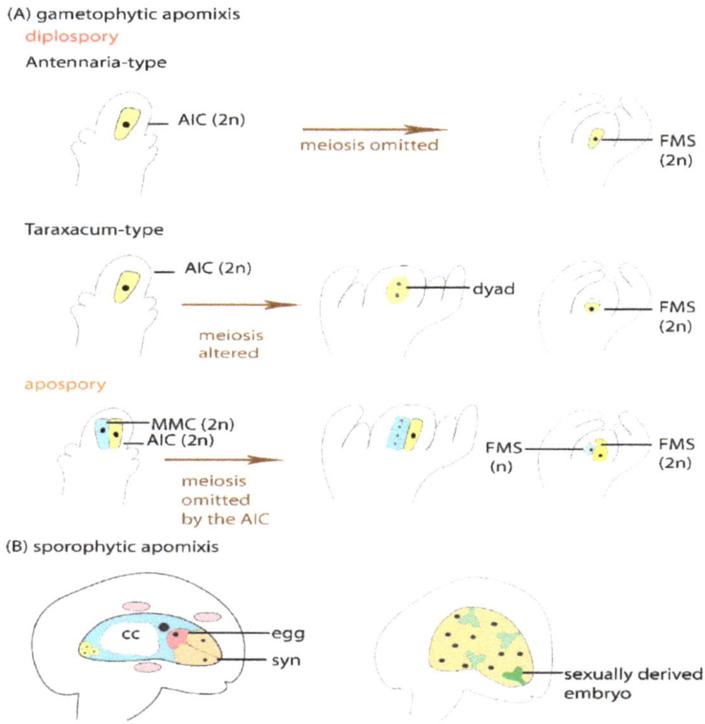

Figure 2. Major types of apomixis. (**A**) Different types of gametophytic apomixis are classified based on the origin and fate of the first cell of the germline lineage. In the Antennaria-type of diplospory the apomictic initial cell (AIC) directly specifies into an unreduced functional megaspore (FMS). In the Taraxacum-type of diplospory meiosis of the AIC is altered to give rise to a dyad of unreduced megaspores of which only one survives as the FMS. During apospory, an additional sporophytic cell in the ovule specifies adjacent to the sexual MMC. This cell omits meiosis to give rise to the FMS. While the sexual germline lineage typically gets repressed by the apomictic germline lineage, also the MMC can undergo meiosis resulting in the formation of two gametophytic lineages, one sexual, one apomictic, in the same ovule. (**B**) During sporophytic apomixis, the sexual gametophyte forms and additional sporophytic cells in the surrounding ovule tissues acquire the competency for embryogenesis (depicted in light red). After fertilization this typically leads to polyembryony, with the sexually derived embryo (dark green) and the somatic embryos (light green) competing for resources.

Taken together, distinct developmental adaptations lead to formation of clonal offspring in the different apomictic systems. To date it is largely unclear if similar regulatory mechanisms are shared in all apomicts, if related mechanisms lead to apomixis, or if different pathways mediate apomictic reproduction in different taxa. Gaining insights into these questions is not only of interest from a scientific point of view, but also relevant for potential applications of apomixis.

2. The Developmental Flexibility of Plant Reproduction might Hold Evolutionary Advantages

As outlined above, apomixis and sexual reproduction represent alternative strategies of reproduction that are developmentally related. From an evolutionary point of view, concepts on the origin and advantages of the different reproductive modes and their co-existence are still under debate and partially contradictory. Sexual reproduction is most commonly regarded as original mode of reproduction and asexuality as derived. However, this view has recently been questioned by new hypotheses proposing that both might represent evolutionary ancient concepts [18].

So far this puzzle has not unequivocally been resolved by phylogenetic analyses. While asexual reproduction is common in ~10% of species in fern, apomixis only occurs in less than 1% of angiosperms [1]. The representation of apomixis is broadly scattered in the angiosperms and has rapidly spread in the large families of *Poaceae*, *Asteraceae*, and *Rosaceae* [3]. Apart from the occurrence of apomixis in the genera *Draba*, *Erysimum*, and *Parrya*, in *Brassicaceae* apomixis is only represented in *Boechera* and the related genus *Phounicaulis* [23]. The absence of apomixis in *Amborella* as basal sister clade of angiosperms might rather indicate that it is derived, however, this alone cannot be taken as sufficient indication [6]. Moreover, the distinctiveness of the different types of apomixis like sporophytic and gametophytic apomixis, but also of apospory and diplospory, suggests that they represent non-homologous mechanisms that arose independently [24]. Consistently, apomictic lineages are commonly regarded as evolutionarily young. This is in line with the perception of apomixis being an evolutionary dead end due to an accumulation of deleterious mutations resulting from the absence of meiotic recombination [25]. Already in 1939 Darlington proposed that apomixis purely represents an escape from sterility that is often caused by polyploidization or hybridization typically associated with apomixis [26]. While polyploidy has long been regarded as a precondition for apomixis, the identification of diploid apomicts, e.g., in the genus *Boechera* has changed this view [27]. In *Boechera*, apomixis arose recurrently by hybridization and intra-specific crosses, and also outside *Boechera* evidence for hybrid origin is given for an increasing number of apomictic taxa [6,28].

Importantly, hybridization and polyploidization are likely to cause genome-wide effects resulting in alterations of gene regulation. It has long been hypothesized that apomixis derived by temporal and spatial deregulation of the gene regulatory pathway governing sexual reproduction [4,17]. Alternatively, or as a consequence of mutation accumulation in asexual species, apomixis might be caused by mutations in genes regulating sexual reproduction. However, this view is challenged by new findings and hypotheses [6,18]. In some systems like *Hieracium* and likely also in *Paspalum*, apomixis is superimposed on sexual reproduction and dominantly silencing sexual reproduction [22,29]. Interestingly, thereby the sexual pathway can be re-established, suggesting that the gene regulatory program underlying sexual reproduction remains intact. Thus, apomixis and sexual reproduction might indeed represent distinct and evolutionary ancient concepts [18]. In contrast to the classical view of apomixis as evolutionary dead end [25], recent studies suggest that apomixis might rather represent an evolutionary opportunity. Furthermore, sexual reproduction and apomixis might be regarded as competing strategies, as, for e.g., discussed for the facultative tetraploid aposporous apomict *Paspalum malacophyllum* [18,30]. Apomixis can be advantageous as it leads to the fixation of beneficial genotypes and it can be beneficial to overcome sterility and incompatibility effects [24]. Furthermore, in many systems studied to date, apomixis is largely facultative, setting the basis for a broad developmental flexibility. Together with the potential occurrence of reversions from apomictic reproduction to sexuality this allows to purge deleterious mutations from the genome [1,25,31–33].

The understanding of the evolutionary origin of apomixis will largely affect the routes of research taken to identify the gene regulatory basis of apomixis. An important aspect to elucidate is, if similar molecular mechanisms are underlying shared elementary features like parthenogenesis or the acquisition of gametophytic fate without preceding meiosis in different apomictic systems. From an evolutionary point of view, it remains astonishing that repeatedly all major components of apomixis could establish simultaneously in the different apomicts. Activation of any element of apomixis alone would have deleterious effects for the plants, e.g., apomeiosis without parthenogenesis would lead to polyploidization, parthenogenesis without preceding apomeiosis would lead to haploidization, and even the uncoupling of autonomous embryo development and endosperm development would prevent successful reproduction. Nevertheless, taking the apparent differences in the reproductive systems into account, it appears likely that also control mechanisms are diverse or similar mechanisms established by convergent evolution. Although the possibility cannot fully be ruled out that apomixis represents an ancient mechanism as an alternative to sexual reproduction, the non-homologues developmental pathways resulting in apomictic reproduction rather suggest independent origins. Different associated gene regulatory programs are likely to be required for certain developmental processes, in particular with respect to megasporogensis, which differs considerably in the different types of apomixis. While formation of an unreduced embryo sac by diplospory represents an alteration of the fate of the MMC/AIC, during apospory both the sexual and the apomictic germline lineage initiate their development so that a tight coordination and cross-talk between the two germline lineages is required. In aposporous *Hieracium* subgenus *Pilosella*, for example, the specification of the sexual MMC is preceding the formation of the aposporous initial cell, while the apomictic reproductive lineage subsequently suppresses further development of the sexual lineage [22]. Nevertheless, certain developmental flexibility and the formation of two gametophytes in the same ovule occurs in *Hieracium praeltum* and also in apomictic *Boechera*, providing the opportunity to reinforce the facultative nature of apomixis [33,34]. Also the occurrence of both, diplospory and apospory, demonstrates a striking flexibility of developmental concepts in *Boechera* [33]. Increasing attention is recently also given to the question, whether and how environmental factors are modulating the regulation of the reproductive programs or if they might even be sufficient to determine the reproductive mode.

3. Genetic Loci Linked to Apomixis Typically Represent Hemizygous Heterochromatic Regions

Genetically in all taxa studied so far apomixis is heritable. This has been revealed by genetic analysis using an apomict as the male and a sexual plant as female parent [35]. Direct identification of the genes and genomic elements comprised on these apomixis linked loci however has proven difficult. This is because they are commonly recombination-suppressed and flanked by repetitive regions, interfering with sequencing approaches and map-based cloning [10,35–41]. In several species studied, apomixis linked loci represent chromosomal regions that largely diverged from corresponding sexual loci. They presumably originate from chromosomal rearrangements and transposable element activity consequently resulting in the frequently observed reduction or loss of recombination [42,43]. One locus is typically underlying each of the major components of apomixis, namely apomeiosis, parthenogenesis, and the developmental adaptations needed for endosperm formation [10,35,39,44].

Inheritance of diplospory as single dominant locus has been described for *Erigeron anuus* and for *Taraxacum officinale* [21,45]. In *Taraxacum*, two unlinked dominant loci control diplospory (DIPLOSPOROUS, DIP) and fertilization independent development of an embryo from the egg cell (PARTHENOGENESIS, PAR) [46]. The DIP locus thereby maps to the distal arm of one nucleolar organizer region (NOR) chromosome [46]. Unlike for most apomixis loci identified, recombination occurs in the hemizigous *Taraxacum* DIP locus that has been fine mapped to about 0.6 cM estimated to cover about 200–300 Kb [46,47]. Also, the apospory-specific genomic regions (ASGR) of *Pennisetum spamulatum* and *Cenchus ciliaris* are located on hemizygous heterochromatic regions on single chromosomes [48–50]. The apomixis-controlling locus (ACL) in *Paspalum simplex* is a single non-recombining hemizygous region [51]. In *Hieracium*, three loci have been identified to control apospory (LOSS OF APOMEIOSIS, LOA), parthenogenesis (LOSS OF PARTHENOGENSIS, LOP), and autonomous endosperm development (AutE) [52]. Also LOA was mapped to a recombination suppressed distal arm of a single chromosome and is surrounded by complex repeats and transposable elements that however are not essential for the function of the locus [53,54].

Further evidence for hemizygous and heterochromatic segments of chromosomes is given from investigations on *Boechera*, where the presence of the largely heterochromatic B-like chromosomes (Het and Del) has been observed in apomictic accessions by karyotype analyses [55–58]. Also in the closely related *Boecheraea* genus *Phoenicaulis* a largely heterochromatic Het chromosome is present in triploid and tetraploid cytotypes, unlike in diploid [23]. While it has been hypothesized that these chromosomes might be relevant for apomixis expression and in particular possible implications for diplospory have been discussed, transmission of a Het chromosome alone is not sufficient for apomixis to arise [23,59]. Interestingly, the Het chromosomes and heterochromatic chromosomal regions linked to apomixis resemble features of Y-chromosomes in animals and dioecious plants for sex determination with respect to typical accumulation of transposable elements and gene loss [29,46,60]. Therefore, it is tempting to speculate that similarly to well known mechanisms in Y-chromosomes, epigenetic regulatory mechanisms might be major driving forces in regulation of apomixis. Epigenetic regulatory mechanisms in general are involved in controlling gene activity by DNA methylation, introducing repressing or activating histone modifications, as well as modulation of overall chromosome structure.

Genes Located on Apomixis Loci Suggest That Different Regulatory Pathways Are Involved in Controlling Apomixis

Despite the similar features presented by the apomixis linked loci so far investigated, knowledge about the genes encoded and their roles in controlling apomictic development is scarce to date. In order to confer successful reproduction, it is essential that all elements of apomixis and associated developmental processes take place in a coordinated manner. To allow this, these regions might include master or key regulatory genes activating a downstream cascade controlling all major aspects of apomixis, or many linked genes encoded on the apomixis linked loci might be required [27]. So far, different genes have been identified to be linked to the apomixis loci in different species (Table 1). Based on these findings, several distinct regulatory mechanisms appear to be involved in controlling apomixis. These include the activity of transcription factors, but also degradation of nucleobases and control of protein turnover, and the modulation of gene activity by mechanisms involving non-coding RNAs. In particular long non-coding RNAs including antisense RNAs are increasingly recognized as important players involved in the regulation of reproduction and a range of developmental decisions [61].

For apomeiosis, a small number of candidate genes from a few loci have been proposed: In *Hypericum perforatum* the Hypericum Apospory- (HAPPY-)locus is co-segregating with apospory but not with parthenogenesis [44] (Table 1). This locus contains a truncated allele of the homologue of *Arabidopsis thaliana ARIADNE7* (*ARI7*), encoding for an E3 ligase Ring-finger protein involved in regulatory processes and protein degradation [44]. Consistent with the dominant nature of the HAPPY-locus, its simplex constitution has been confirmed in tetraploid plants [44]. Recently, sequencing approaches allowed the annotation of 33 predicted genes located on the HAPPY-locus, 24 of which were expressed in pre-meiotic nucellus tissues of the ovule [62]. In *Boechera*, two different candidates have been identified for regulation of female and male apomeiosis. As a candidate for female apomeisis, the APOmixis Linked LOcus (*APOLLO*) gene has been identified that is higher expressed in apomictic ovules at apomeiosis as compared to sexual ovules [63,64] (Table 1). Likewise, comparative transcriptome analyses of anthers containing pollen mother cells in sexual and apomictic *Boechera* identified that the activity of *UPGRADE (UPG)* is correlated with apomixis [65]. While it is inviting to hypothesize that *APOLLO* and *UPG* are likely to be localized on the heterochromatic HET and DEL chromosomes, a direct proof has so far not been presented [66]. *APOLLO* encodes for an Aspartate Glutamate Aspartate Aspartate Histidine exonuclease and is heterozygous for apomixis specific alleles in apomicts [63]. These alleles contain 20-nucleotide polymorphisms in the 5′ untranslated region (5′ UTR) [63]. Interestingly, *UPG2* represents a long non-coding RNA that has been proposed as candidate for the formation of unreduced pollen [64,65] (Table 1). Further evidence for roles of long non-coding RNAs in the regulation of (apo)meiosis comes from *Paspalum notatum* [67]. In *P. notatum* a long non-coding RNA related to a gene encoding mitogen-activated protein kinase kinase kinase (N46) is linked to the ACR [67] (Table 1). N46 is named *QUI-GON JINN* (*QGJ*) as a member of the YODA family. It is not only differentially expressed in flowers from sexual as compared to apomictic plants, but also its downregulation mediates a reduction of the rate of aposporous embryo sac formation [67]. Taken together the investigations on different apomicts provide increasing evidence for different regulatory mechanisms to control aspects of apomeiosis. Importantly however, as molecular mechanisms involved in the control of apomeiosis, mainly changes in gene regulation appear to be important that are enforced e.g., by changes in regulatory elements and the activity of non-coding RNAs. These findings do not provide evidence for the idea that elements of apomixis might derive from genetic mutations in coding regions that lead to alterations in protein function. Nevertheless, future investigations will be needed to more comprehensively understand the regulatory processes controlling apomeiosis.

Interestingly, pathways including the activity of non-coding RNAs might also be involved in the regulation of parthenogenesis and endosperm formation in certain apomicts. From the ACL of *Paspalum simplex*, expression of antisense transcripts for three genes has been identified [51,68]. This has led to the hypothesis that the ACL region modulates epigenetic processes regulating parthenogenesis and endosperm formation. This is consistent with the finding that parthenogenesis is superimposed on sexual reproduction in this system and that DNA demethylation affects parthenogenesis but not apomeiosis [29,51]. In particular the homologue of subunit 3 of the ORIGIN RECOGNITION COMPLEX (ORC3), which is functional in sexual plants, is regulated by an apomixis specific antisense pseudogene [68] (Table 1). The precise regulation of *ORC3* activity in apomicts appears to be relevant for formation of functional endosperm with a ratio of maternal to paternal contributions alternating from 2n: 1n [68].

Unlike the regulatory mechanisms involving non-coding RNAs, from the ASGR of *Pennisetum* and *Cenchrus BABY BOOM* (*BBM*)-like genes have been identified as promising candidates for parthenogensis based on the similarities to *BBM* of *Brassica napus* [49,69] (Table 1). *BBM* and *BBM*-like genes belong to a family of transcription factors characterized by two conserved APETALA2 (AP2) binding domains and a bbm-1 domain with functional implications for somatic embryogenesis [70]. From studies in *A. thaliana* BBM acts upstream of major regulators of totipotency and embryonic identity [71]. Originally identified as a gene involved in controlling somatic embryogenesis in microspore cultures, embryo development from somatic cells of *A. thaliana* leaves can be triggered by expression of *Brassica napus BBM* [72]. This supports its strong potential for inducing the gene regulatory program relevant to acquire the competence for embryogenesis. Evidence for the functional importance of *ASGR-BBML* for parthenogenesis was further substantiated by the identification of a *C. ciliaris* recombinant that retained apospory but lost parthenogenesis along with the *BBML* containing fragment of the ASGR [69]. Also, when expressed under its native promoter and terminator, expression of *ASGR-BBML* in egg cells of *Pennisetum squamulatum* is sufficient to trigger parthenogenesis in sexual plants [73]. Furthermore, studies from apomictic *Brachiara decumbens* suggest the importance of *ASGR-BBML* genes for parthenogenesis in *Poaceae* [74]. A recent study supports the broader validity of *BBM* genes to trigger embryogenesis and parthenogenesis by demonstrating that even in *Oryza sativa* expression of *BBM1* in egg cells is sufficient to allow autonomous embryo development in the absence of fertilization [75]. Strikingly, *BBM1* as the gene triggering embryogenesis behaves as an imprinted gene in young embryos, as only the paternal but not the maternal allele is expressed [75]. Imprinting in general describes a control mechanism that allows activity of one parental allele, while the allele from the other parent is silenced due to epigenetic regulation. The mechanism outlined for controlling *BBM1* activity elegantly explains the requirement for fertilization for seed development during sexual reproduction. As so far implications of BBM and BBML for parthenogenesis have only been described in monocotyledons, to date evidence is lacking for a broader importance of this mechanisms to repress embryogenesis of the egg cell in the absence of fertilization also in dicotyledons.

Table 1. Candidate genes for apomixis encoded from apomixis linked loci.

Gene	Type of Apomixis	Element of Apomixis	Locus	Plant Family	Species	Publication
ARI7	gametophytic	apospory	HAPPY	*Hypericaceae*	*Hypericum perforatum*	[44,62]
APOLLO	gametophytic	female apomeiosis	-	*Brassicaceae*	*Boechera ssp.*	[63,64]
UPG2	gametophytic	male apomeiosis	-	*Brassicaceae*	*Boechera ssp.*	[65]
QGJ	gametophytic	apospory	-	*Poaceae*	*Paspalum notatum*	[67]
ORC3	gametophytic	endosperm formation	ACL	*Poaceae*	*Paspalum simplex*	[51,68]
BBM(L)	gametophytic	parthenogensis	ASGR	*Poaceae*	*Pennisetum squamulatum* *Cenchrus ciliaris* *Brachiara decumbens*	[49,69,70,73,74]
RKD	sporophytic	somatic embryogenesis	-	*Rutaceae*	*Citrus*	[5,76]

From studies of somatic embryogenesis in *Citrus*, another transcription factor has been proposed to be relevant for the regulatory control. The genomic locus linked to somatic embryogenesis has first been described to comprise ~380 kb and it could further be fine-mapped to a genomic region of 80 kb harboring the sequences of 11 genes [5,16]. From this region, *CiRKD1* is recognized as candidate gene for polyembryony and somatic embryogenesis [5,76] (Table 1). *RKD* genes encode RWP-RK domain-containing transcription factors. In *A. thaliana* the five members of the family are predominantly expressed in the egg apparatus (egg cell and synergid cells) and are important regulators of gametogenesis and acquisition of egg cell fate [77–79]. Interestingly, in the egg apparatus of the triploid apomict *Boechera gunnisoniana*, *RKD* genes are present only at low levels [80]. This suggests that the gene family might play a role in maintaining egg cell identity in the absence of fertilization during sexual reproduction. Studies on *Marchantia polymorpha* with only a single *RKD* homologue represented in the genome indeed support the evidence of RKD to be an evolutionary conserved factor in plants important to acquire egg cell identity and to keep the egg cell in a developmentally repressed state in the absence of fertilization [81,82]. The regulation of acquiring the competence for embryogenesis and to activate this program might be more complex in *Citrus*. In *Citrus* studies from satsuma mandarin have recently revealed the presence of two *CiRKD1* alleles with one of them containing a miniature inverted-repeat transposable element (MITE)-like insertion in the upstream region [76]. Increased expression of this allele in the tissues where somatic embryogenesis occurs was observed and antisense silencing of *CiRKD1* in transgenic sweet orange leads to loss of somatic embryogenesis [76]. Interestingly, like in the case of *BBM* and *BBML*, differences in activity and regulation of one (type of) transcription factor(s) appear to be sufficient to acquire the competence for embryogenesis and to allow embryogenesis from a sporophytic cell or the egg cell in the absence of fertilization.

4. Transcriptional Analysis Identifies Genes Differentially Regulated during Sexual and Apomictic Reproduction

Genetic studies identified only few apomixis linked loci suggesting that a limited number of genes might be required for apomixis. However, transcriptional studies often suggest a more global deregulation of the gene regulatory program underlying sexual reproduction in apomicts. This discrepancy might potentially be explained by master regulators which control complex programs of gene activity. Transcriptional analyses to identify genes differentially expressed in sexual as compared to apomictic plants have been presented for a variety of species including *Pennisetum ciliare* and *Pennisetum glaucum* [83,84], *Panicum maximum* [85,86], *Poa pratensis* [87,88], *Brachiaria brizantha* [89,90], *Paspalum notatum* and *Paspalum simplex* [91–94], *Eragostris curvula* [95], *Medicago falcata* [96], *Boehmeria tricuspis* [97], *Hypericum perforatum* [62,98], *Hieracium* [99,100], *Boechera* [101–103], and also for *Citrus* [76] (Table 2). These studies provide evidence for temporal deregulation of the gene regulatory processes governing sexual reproduction in apomicts and identify large numbers of up to hundreds of genes to be differentially expressed.

Given the large numbers of genes identified as differentially regulated, it remains difficult to identify the genes that are relevant for the determination of the reproductive mode or developmental processes governing apomictic reproduction. As most of the studies are based on ovule or floral tissues it is likely that a large fraction of the genes identified is differentially expressed in sporophytic tissues rather than in the developing reproductive lineages. The overabundance of sporophytic tissues in the samples can mask the regulatory profiles controlling germline formation and development. To overcome this difficulty, cell and tissue type-specific transcriptome analysis, i.e., by combining laser assisted microdissection (LAM) with microarray analysis or RNA-Seq have proven to be powerful approaches [104–106]. Novel insights have already been gained into the gene regulatory pathways governing the development of the sexual MMC and the cells of the mature female gametophyte in *A. thaliana* [79,105–107], and the corresponding cells in the related triploid apomict *Boechera gunnisoniana* [80] (Table 2). Tissue type-specific transcriptome analysis targeting AIC/MMC and surrounding nucellus tissues furthermore allowed comparative analyses of gene expression and pathways relevant for megasporogenesis in different sexual as compared to apomictic *Boechera* accessions and in sexual versus aposporous *Hypericum perforatum* [62,103] (Table 2). LAM in combination with RNA-Seq recently has also shed light onto the cell type specification of the aposporous initial cell (AIC) as compared to early developing embryo sacs and somatic ovule tissues in *Hieracium praealtum*. These studies suggest advanced acquisition of gametophytic fate by the AIC [108,109] (Table 2).

Consistently, comparative transcriptome analysis in sexual and apomictic *Boechera* [103], *Bohemeria tricuspis* [97], and *Hypericum perforatum* [62] suggest that differential activity of genes involved in cell-cycle regulation, hormonal pathways, signal transduction, ubiquitinylation and protein degradation, and epigenetic regulatory pathway are involved in determining and sustaining megasporogenesis in either reproductive mode (Figure 3). To narrow down the number of candidate genes and to disentangle part of the effects of ploidy and species differences, differential expression analysis was recently applied to compare four apomictic versus two sexual *Boechera* accessions [103]. Thereby LAM and RNA-Seq have been combined to analyze gene expression in reproductive nucellus tissues harboring the AIC or MMC [103]. This has identified 45 genes to be consistently differentially expressed in all samples from sexual as compared to apomictic accessions [103]. This study supports the importance of genes involved in cell-cycle regulation, protein degradation and hormonal pathways for distinguishing sexual from apomictic reproduction, and also suggests functions related to stress and redox regulation to be relevant [103] (Figure 3). Taken together, evidence for the involvement of these pathways in regulation of megasporogenesis is consistently given from different apomicts. Apart from these investigations, additional transcriptional studies focused on mature gametophytes in sexual and apomictic plants and the transition to early stages of seed development would be beneficial to allow a more comprehensive understanding of the gene regulatory processes distinguishing apomixis from sexual reproduction.

Table 2. Transcriptional analyses on reproductive tissues to identify genes involved in apomixis regulation.

Plant Family	Species	Type of Apomixis	Tissues Profiled	Methods of Analysis	References
Poaceae	*Pennisetum ciliare*	gametophytic	unpollinated ovaries	modified differential display	[83]
Poaceae	*Pennisetum glaucum*	gametophytic	spikelets at 4 developmental stages (pre-meiosis, meiocyte, gametogenesis, mature gametophyte)	suppression subtractive hybridization	[84]
Poaceae	*Panicum maximum*	gametophytic	flower buds	cDNA library	[85]
Poaceae	*Panicum maximum*	gametophytic	spikelets (pre-meiosis)	RNA-Seq (Illumina HiSeq2500)	[86]
Poaceae	*Panicum maximum*	gametophytic	immature pistils	custom microarray	[110]
Poaceae	*Paspalum notatum*	gametophytic	inflorescences at 4 developmental stages (early premeiosis; late premeiosis/ meiosis; postmeiosis; anthesis)	RNA-Seq (Roche 454)	[94]
Poaceae	*Paspalum notatum*	gametophytic	florets at different developmental stages	cDNA-AFLP	[92]
Poaceae	*Paspalum notatum*	gametophytic	inflorescences	differential display analysis	[91,111]
Poaceae	*Eragostris curvula*	gametophytic	panicles	differential display analysis	[112]
Poaceae	*Eragostris curvula*	gametophytic	spikelets with embryo sacs at all developmental stages	RNA-Seq (Roche 454)	[95]
Poaceae	*Poa pratensis*	gametophytic	florets at 4 developmental stages (pre-meiosis; meiosis; post-meiosis; anthesis)	cDNA-AFLP	[87,88]
Hypericaceae	*Hypericum perforatum*	gametophytic	nucellus tissues harboring MMC or AIC before (apo)meiosis	RNA-Seq (Illumina NextSeq500)	[62]
Hypericaceae	*Hypericum perforatum*	gametophytic	pistils	custom microarray	[98]
Hypericaceae	*Hypericum perforatum*	gametophytic	whole flowers at range of developmental stages	cDNA libraries	[113]
Asteraceae	*Hieracium praealtum Hieracium aurantiacum; parthenogenesis incapable accession lop138*	gametophytic	ovules and ovaries at different developmental stages isolated by manual microdissection	RNA-Seq (Illumina HiSeq2000)	[100]
Asteraceae	*Hieracium praealtum*	gametophytic	AIC, developing female gametophytes (2–4 nucleate), and somatic ovule cells isolated by LAM	RNA-Seq (Roche 454; Illumina HiSeq2000)	[108,109]
Asteraceae	*Hieracium praealtum*	gametophytic	ovaries	RNA-Seq (Illumina HiSeq2000)	[114]
Urticaceae	*Boehmeria tricuspis*	gametophytic apomixis	Flowers at 4 developmental stages (MMC; FMS, embryo sac, mature embryo)	RNA-Seq (Illumina HiSeq4000)	[88]
Brassicaceae	*Boechera*	gametophytic	nucellus tissues harboring MMC or AIC isolated by LAM	RNA-Seq (Illumina NextSeq500)	[103]

Table 2. *Cont.*

Plant Family	Species	Type of Apomixis	Tissues Profiled	Methods of Analysis	References
Brassicaceae	*Boechera gunnisoniana*	gametophytic	AIC, egg cell, central cell, synergids isolated by LAM	ATH1 microarray, RNA-Seq (SOLiD V4)	[81]
Brassicaceae	*Boechera*	gametophytic	ovules isolated by manual microdissection	RNA-Seq (Roche 454); custom microarray	[63]
Brassicaceae	*Boechera*	gametophytic	antherheads at pollen mother cell stage	custom microarray	[65]
Brassicaceae	*Boechera*	gametophytic	ovules isolated by manual microdissection at 4 developmental stages (early premeiosis; late premeiosis; FMS, gametophyte)	SuperSAGE	[101]
Brassicaceae	*Boechera*	gametophytic	ovules isolated by manual microdissection pooled flower stages	SuperSAGE; RNA-Seq (Roche 454)	[102]
Fabaceae	*Medicago sativa*	apomeiotic mutant	flower buds at 4 developmental stages (pre-meiosis, initial meiosis, final meiosis, and post-meiosis)	cDNA-AFLP	[96]
Rutaceae	*Citrus*	somatic embryogensis	fruits 15, 30, 45, and 60 d after flowering	custom microarray	[76]
Rutaceae	*Citrus*	somatic embryogensis	leaves, ovules, seeds, fruits	RNA-Seq (Illumina Genome Analyzer)	[5]
Rutaceae	*Citrus*	somatic embryogensis	ovaries at anthesis and at 3, 7, 17, 21, and 28 d after flowering	RNA-Seq (Illumina Genome Analyzer)	[115]

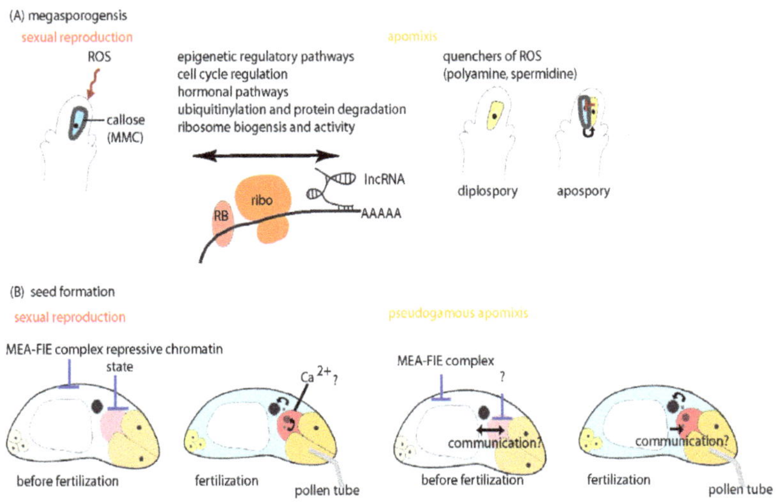

Figure 3. Molecular mechanisms differentially regulated during sexual reproduction and apomixis. (**A**) During megasporogenesis players in several pathways are differentially regulated implementing control of gene and protein activity. This likely involves also the activity of specialized ribosomes (ribo) in conjunction with RNA binding proteins (RB) like RNA helicases and products of non-canonical open reading frames including long non-coding (lnc) RNAs. Also stress and stress response appear to differentially affect megasporogenesis in the different reproductive modes. During meiosis in sexual reproduction the MMC is enclosed by callose potentially as a response to reactive oxygen species (ROS). In contrast in apomicts high activity of polyamine biosynthesis and spermidine metabolism allows quenching of ROS. Furthermore, diplospory involves alterations in the meiotic program, while during apospory communication between the sexual and apomictic germline is required. (**B**) During sexual reproduction proliferation of the female gametes is repressed in the absence of fertilization by the activity of the MEA-FIE PRC2 complex in the central cell and a repressive chromatin state in the egg cell. Double fertilization initiates seed formation involving fusion of the two sperm cells with each of the female gametes. Likely rise in Ca^{2+}-levels is involved in activation of the egg cell. In addition, in pseudogamous apomicts both female gametes need to remain repressed in the absence of fertilization of the central cell. Only the central cell nucleus fuses with sperm nucleus. Communication between the egg- and central cell is required to coordinate development.

5. Different Layers of Regulation Are in Place to Control Development during Sexual and Apomictic Megasporogenesis

From the different apomicts a number of candidate genes for apomeiosis which are linked to apomixis loci have been determined (Table 1). As these genes are involved in diverse regulatory pathways they might be relevant for several aspects of development during megasporogenesis in the different types of apomixis. Thereby, potentially similar genes and molecular mechanisms might be required for the determination of the reproductive mode. However, complex regulatory programs should be in place controlling the distinct developmental programs associated with sexual and apomictic reproduction. To regulate megasporogenesis in diplosporous and aposporous apomicts, this should involve the determination of meiotic versus mitotic fate, the acquisition of germline identity, and the control and cell-cell communication to decide how many cells per ovule can make this fate transitions. In contrast for parthenogenesis, certain transcription factors appear to be crucial to acquire embryonic potency and to activate the regulatory program underlying embryogenesis.

5.1. Specialized Ribosomes and Associated Factors Emerge as Novel Players in Gene Regulation

Both in sexual plants and apomicts, germline formation and development requires complex and tight regulatory systems [12,105,116]. Different molecular machineries are in place to control gene expression and translation, protein activity and turnover, and cell-cell communication. Especially during megasporogenesis and preparation of the MMC/AIC for (apo)meiosis, processes related to translation and ribosome biogenesis are enriched both in sexual *A. thaliana* and apomictic *Boechera* [103,107]. Further evidence for the importance of ribosome biogenesis and function, and nucleosome assembly for megasporogenesis in apomicts comes from transcriptome analysis of *Hieracium* subgenus *Pilosellum*, as related functions are enriched in the AIC as compared to developing embryo sacs [109]. While it has long been noticed that a cycle of ribosome degradation and reassembly is associated with plant meiosis [117], the relevance of this processes for sexual and apomictic reproduction has not been described in detail to date. Nevertheless, functions related to ribosome assembly are over-represented in transcripts from the ASGR of *Pennisetum squamulatum*, suggesting that the control of ribosome activity is playing a role for apomixis [118]. A further hint in this direction is presented from the localization of DIP locus on a NOR chromosome in *Taraxacum* [46].

While ribosomes are more historically thought to have constitutive functions in mRNA translation, recently the notion has emerged of specialized ribosomes as an additional layer of gene regulation that has so far been largely overlooked. It might represent a mechanism to regulate a switch of developmental fate such as the determination of meiosis or apomeiosis by co-regulation of a larger number of relevant target genes. The ribosome based regulatory machinery involves specific mRNA regulatory elements such as internal ribosome entry sites which are mostly located in the 5′ UTR regions of mRNAs [119]. Recent findings also point towards the importance of non-canonical open reading frames (ORFs) including non-coding RNAs in the regulation of ribosome biogenesis and function [120,121] (Figure 3). While estimates suggest that e.g., in *A. thaliana* up to 80–90% of the genome is transcribed at least at a certain developmental time point, less than half of these transcripts appear to be coding proteins or peptides [122]. The non-protein coding RNAs comprise housekeeping RNAs like ribosomal RNAs, tRNAs, small nuclear and nucleolar RNAs, and small RNAs involved in epigenetic regulatory processes. Also long non-coding RNAs are increasingly perceived as important players in the regulation of gene activity [122] (Figure 3). It is tempting to speculate that *UPG2* and the long non-coding RNA related to *QGJ* might act in target gene regulation by association to ribosomes. The identification of long non-coding RNAs with potential relevance for apomeiosis uncovers a new layer of complexity of regulatory processes. While the molecular mechanisms of their activities have so far not comprehensively been elucidated, involvement of long non-coding RNAs in the control of meiosis appears to be a conserved feature of eukaryotes. Evidence for the importance of non-canonical ORFs in general is given in yeast, where they have high occupancy in meiotic cells [119]. In the fission yeast *Schizosaccharomyces pombe*, the polyadenylated long-non coding RNA meiRNA forms a nuclear body in meiotic cells and is involved in the regulation of the entry into meiosis, homologous pairing and chromosome retention [123]. In plants, evidence for high abundance of long non-coding RNAs in meiocytes comes from a study from sunflowers suggesting their importance during meiosis [124]. Taken together, this provides accumulating evidence for the importance of such regulatory processes controlling RNA abundance and activity for germline specification and likely for discrimination of plant meiosis and apomeiosis. However, experimental prove for this is largely lacking to date.

Importantly, ribosome assembly and function is typically correlated with the activity of RNA helicases and other types of RNA binding proteins often associated to ribonucleoprotein complexes [125,126] (Figure 3). Members of the large gene family of RNA helicases are involved in basically any aspect of RNA metabolism, storage and degradation. They are of crucial importance for gene regulation to control developmental processes, and they are involved in epigenetic processes. In addition, RNA helicases are central players in transforming stress induced signals into regulatory responses [127]. From studies in sexual *A. thaliana*, the abundant activity of RNA helicases in the MMC has previously been determined, reminding of the crucial and conserved roles of RNA helicases for germline development in animals [107]. Thereby, the RNA helicase *MNEME* (*MEM*) has been uncovered that plays a role to restrict germline fate to allow the specification of only one MMC per ovule [107]. In *A. thaliana* plants carrying a mutant allele of *MEM* frequently AIC-like cells form adjacent to the sexual MMC which give rise to formation of presumably unreduced gametophytes, closely resembling apospory [107]. Furthermore, comparative cell type-specific transcriptome analyses of MMC versus AIC and the cells of the mature embryo sac in sexual *A. thaliana* versus the triploid diplosporous apomicts *Boechera gunnisoniana* point towards a deregulation of *MEM*. However, a consistent differential gene expression in reproductive nucellus tissues prior to apo(meiosis) could not be identified in different sexual and apomictic *Boechera* accessions [80,103]. Given the broad developmental flexibility during apomictic germline formation in *Boechera* [33], a role in regulation of developmental processes relevant for apomixis cannot be ruled out by this finding. Apart from the identification of *MEM*, further studies provide evidence of RNA helicases to be likely involved in regulation of apomictic development, as in *Brachiaria brizantha* and in *Hypericum perforatum*, *BrizHELIC* and a homologue of *MATERNAL EFFECT EMBRYO ARREST29*, respectively, are differentially expressed in tissues of sexual and apomictic plants [10,128].

5.2. *Epigenetic Regulatory Pathways Are Involved in Regulation of Germline Development*

RNA helicases and non-coding RNAs are also players in epigenetic regulatory pathways. Epigenetic regulatory processes which modify gene activity based on DNA methylation, histone modifications, and modulation of chromatin structure are involved in controlling diverse developmental and cell fate decisions. Epigenetic regulation is increasingly recognized to be important during plant germline development [129]. While long non-coding RNAs can act in modulating DNA methylation and histone modifications to regulate gene activity [130], future investigations are required to elucidate their specific importance during reproductive development in detail.

First evidence for the involvement of epigenetic regulatory pathways in controlling components of apomixis comes from the heterochromatic nature of apomixis loci. Furthermore, alteration in epigenetic regulations might be a consequence of polyploidization and hybridization. It has been hypothesized that apomixis might be superimposed on sexual reproduction by epigenetic control mechanisms. This is suggested from a study in *Paspalum ssp.*, where treatment of apomictic plants with the demethylation agent 5′-azacytidine leads to reduction in the frequencies of parthenogenesis [29]. This is in line with the indications for roles of epigenetic regulation for apomixis coming from comparative transcriptional analyses. Also functional evidence supports this notion, as mutations in certain epigenetic regulators lead to induction of elements of apomixis in sexual plants. Studies in *A. thaliana* and maize revealed phenotypes reminiscent of apospory or diplospory for mutants in different players in small RNA and DNA-methylation pathways [131–134]. This included *ARGONAUTE9* (*AGO9*) and additional genes involved in the RNA directed DNA-methylation pathway [132,134]. AGO proteins act by binding different types of small RNAs, such as microRNAs (miRNA), small interfering RNAs (siRNA), and PIWI-associated RNAs (piRNAs) [135]. Knowledge about their roles in natural apomicts is so far limited.

To gain insights into the possible involvement of small RNAs in apomixis control, their activity has been studied in different natural apomicts including *Eragostris curvula* [136], *Paspalum notatum* [29,137], *Hieracium* subgenus *Pilosella* [119], and *Boechera* [138,139]. Differential representation of small RNA reads in *Paspalum notatum* points towards their involvement in regulation of meiosis, cell cycle control, transcriptional regulation and hormonal signaling [117]. In contrast, in *Hieracium* only small numbers small RNA targets were identified as differentially expressed [114]. From comparisons of sexual and apomictic *Boechera* ovules differential activity of the small RNAs miR156/157 has been identified which is relevant for the regulation of the transcription factor *SQUAMOSA PROTEIN BINDING PROTIEN LIKE 11* (*SPL11*) [139]. As epigenetic regulations are highly dynamic, future studies focusing on elucidating DNA modifications and small RNAs at cell and tissue type-specific resolution will be relevant to gain additional meaningful insights into the contribution of epigenetic regulation to apomixis control. Also, while the investigations on mutant lines of sexual *A. thaliana* and maize indicate that the two modes of reproduction are related, it is currently not well understood how. In the mutants, elements of apomixis establish in sexual species by disabling major players in epigenetic regulatory pathways. This suggests that certain components of apomixis can at least derive from mutations leading to a loss of gene function. However, if apomixis loci epigenetically control the sexual pathway, it remains to be elucidated if suppression of sexual reproduction alone is sufficient for apomixis to establish.

5.3. Cell-Cycle Control and Regulation of Meiosis is Differentially Regulated during Meiosis and Apomeiosis

Apart from mutations in genes involved in epigenetic regulations, also certain mutations or combinations of mutations in core meiotic genes can lead to apomeiosis instead of meiosis. In *A. thaliana* apomeiosis by a diplospory-like mechanism has been observed in plants carrying mutations in *DYAD/SWITCH*. While leading to sterility at high penetrance, also formation of triploid offspring retaining parental heterozygocity occurs at very low frequencies below 1% [140]. Also in maize *ameiotic1* mutants, which is an orthologue of *SWITCH*, designated MMCs undergo a mitosis-like division instead of meiosis [141]. In addition, triple mutants of *sporulation 11-1* (*spo11-1*), *omission of second division 1* (*osd1*), and *recombination 8* (*rec8*) or the A-type cyclin *cyc1;2/tardy asynchronous meiosis* (*tam*) lead to mitotic division instead of meiosis in MiMe1 and MiMe2, respectively [142,143]. This has first been shown for *A. thaliana* and subsequently been used to generate clonal offspring in *A. thaliana* and rice. Clonal offspring has been obtained by combining the meiotic mutants with a manipulation of the centromere-specific histone variant CENH3 that leads to an elimination of the paternal genome [144,145]. Recently, in rice, generation of clonal seeds using MiMe in combination with editing of *MATRILINEAL* has also been demonstrated [146]. *MATRILINEAL* encodes a sperm specific phospholipase and has previously been identified as haploid inducer in maize [147].

While these studies provide a prove of concept that engineering of clonal crop plants is feasible, it remains unclear if and how similar mechanisms are involved in natural apomicts. From transcriptional studies strong evidence is provided for differences in regulation of meiosis and cell cycle to play a role during megasporogenesis in sexual and apomictic plants [80,103]. However, it is important to note that a consistent deregulation of core meiotic genes in all studied sexual versus apomictic accessions has not been observed in *Boechera* [103]. From studies of the *Hieracium praeltum* AIC no expression of 14 selected meiotic genes has previously been observed, consistent with the cellular fate destined to mitosis [108]. The lack of consistency likely relates to the developmental flexibility during germline formation: As apospory and diplospory both occur in *Boechera* at different frequencies [33], this also results in different frequencies of the determination and meiosis of the sexual MMC in addition to the AIC. Therefore, future studies focusing on proteins at cellular level will be required to disentangle their involvement in regulation of apomeiosis.

5.4. Signal Transduction, Cell-Cell Communication and Hormonal Pathways Appear to Be Involved in Regulation of Apomixis

The developmental flexibility during germline specification and transition to gametogenesis further suggests that involvement of cell-cell communication is required in the regulation of these processes. During sexual reproduction, acquisition of reproductive fate of additional sporophytic cells in the ovule is typically suppressed. As previously determined in maize, rice and *A. thaliana*, this involves the activity of signaling pathways, including the *A. thaliana* Leucine rich repeat receptor kinases *SOMATIC EMBRYOGENESIS RECEPTOR-LIKE KINASE1/2* (*SERK1/2*) [116]. From studies in *Poa pratensis*, *PpSERK* has been proposed as candidate that is activated in apomictic nucelli to enable development of the apomictically derived gametophyte [10]. The SERK signaling pathway might interact with auxin hormonal pathways involving APOSTART [10]. Strong evidence for the importance of signaling and hormonal pathways for specification of the apomictic germline lineages also comes from transcriptional analyses of sexual and apomictic *Boechera*. From the analysis of genes with evidence of expression in the *B. gunnisoniana* AIC and not in the *A. thaliana* MMC, an enrichment of "MAP kinase kinase" activity was observed in addition to gene and protein families related to auxin transport and signaling [80]. When comparing transcriptional profiles of *Boechera* nucelli harboring MMCs or AICs furthermore the homologue of *A. thaliana GRETCHEN HAGEN3.6* (*GH3.6*) is consistently higher expressed in apomicts as compared to sexual plants [103]. It encodes a GH3 family protein involved in modulation of auxin response. *GH3* was further identified as higher expressed in AICs than in embryo sacs of *Hieracium praealtum*, further suggesting its importance for megasporogenesis in apomicts [109]. Also the homologue of the nonethylene receptor *HISTIDINE KINASE1* (*HK1*) is higher expressed in apomictic as compared to sexual nucelli in *Boechera*. In *A. thaliana* HK1 is involved in abscisic acid signal transduction in response to salt and drought stress [103]. A homologue of HK1 is furthermore located on the *Hypericum* HAPPY-locus. The potential functional relevance of HK1 activity for apomixis has not been described to date. Nevertheless, the reports provide strong evidence for the importance of cell communication and hormonal pathways for apomixis regulation.

5.5. Regulation of Seed Development in Apomicts Might Require both Repression and Activation of Gene Activity

To allow for successful seed development, a precise coordination of development of the embryo, endosperm, and also the seed coat is required. In most species also the maintenance of precise ratios of parental contributions and controlled activation of maternal or paternal alleles for certain regulators of seed development by imprinting is critical [148]. During sexual plant reproduction, double fertilization of the two female gametes (egg cell and central cell) with the two sperm cells initiates formation of the embryo and its nourishing tissue, the endosperm (Figure 3B). It is well understood that double fertilization occurs almost simultaneously [149]. Prior to fertilization, the female gametes are arrested in the cell cycle with the egg cell presumably at G1 of mitosis and the central cell at G2 [150]. At the time of fusion, cell cycle synchronicity between male and female gametes appears critical for initiation of embryo and endosperm development [150]. In sexual species, an increase in Ca^{2+}-concentration has been proposed to be the signal for activation of the zygotic program in vertebrates and potentially also in plants [10,151] (Figure 3B). Furthermore, in the sexual mature embryo sac prior to fertilization, the egg cell chromatin is highly condensed and thus in a repressive and transcriptionally silent state [151] (Figure 3B). This might be relevant for the acquisition of potency to allow embryogenesis [151]. However, it might also represent a mechanism to prevent premature or autonomous egg cell activation in the absence of fertilization.

To allow seed development in apomicts, several aspects of the regulation in sexual species need to be altered. For parthenogenesis, the repressive state of the egg cell needs to be either omitted by precocious activation of embryogenesis or relieved, potentially by chromatin remodeling as the underlying mechanism [151]. Similarly, for autonomous endosperm development, the repression of central cell proliferation needs to be overcome. In sexual species, this repression in the absence of fertilization requires the activity of Polycomb group proteins (Figure 3B). These are interacting in Polycomb Repressive Complexes 2 (PRC2) to control target gene activity by introduction of histone modifications and repressive H3K27me3 marks [152]. In *A. thaliana* mutant alleles of genes encoding components of the MEA-FIE PRC2 complex, in particular *MEDEA* (*MEA*), *FERTILIZATION INDEPENDENT SEED2* (*FIS2*), *FERTILIZATION INDEPENDENT ENDOSPERM* (*FIE*), and *MULTICOPY SUPRESSOR OF IRA1* (*MSI1*), lead to fertilization independent initiation of endosperm development [152]. Implications of FIE for endosperm development have been discussed also for apomictic *Hieracium*; however, the composition of the PRC2 complex including FIE appears to be different from the one identified in *A. thaliana* [153].

It is feasible that the coordination of all components of seed development is under dual or more complex control. This likely involves signals keeping the quiet state of the gametes and repressing their development, while other signals are needed to activate the developmental programs. Evidence for such twofold control mechanism is given from the "Salmon system" used for haploid production in wheat by activating autonomous embryogenesis [151]. Thereby two nuclear genes are involved which are the inducer *Parthenogenesis gain* (*Ptg*) under sporophytic control and the repressor *Supressor of Parthenogenesis* (*Spg*) under gametophytic control. This control mechanism appears to be in contrast to findings that activation of BBM and BBM-like genes in the egg cell of rice alone is sufficient to trigger parthenogenesis [75]. However, future investigations and closer understanding of the regulation of *BBM* and *BBML* genes both in sexual and apomictic plants might resolve this puzzle. It is interesting to note that BBM belongs to the group of *APETELA2/ETHYLENE RESPONSIVE FACTOR* (*AP2/ERF*) transcription factors. Indications have been found that members of this group are regulated by histone modifications [154]. For the *BBM1* homologue of *Coffea canephora* evidence of epigenetic regulation based on DNA-methylation and histone modifications is given [155]. Strikingly, strong evidence indicates an involvement of H3K27me3 marks in the regulation of *BBM1* [155]. This raises the question, if epigenetic repression based on PRC2 activity and DNA-methylation are responsible for the repression of the maternal allele as observed in rice. This could be a mechanism that safeguards to keep the egg quiet in the absence of fertilization and thus prohibits parthenogenesis to occur in sexual plants.

It is likely that the molecular machineries controlling development of all components of the seed in different natural apomicts are more complex. In pseudogamous apomicts, which depend on fertilization of the central cell for endosperm development, also parthenogenesis appears to remain repressed in the absence of fertilization as recently shown for *Boechera gunnisoniana* [80] (Figure 3B). The underlying molecular control is likely not involving a fusion of the egg and sperm cells nuclei. In contrast to animals, where sperm dependent parthenogenesis is a common mechanism that still requires fertilization for embryogenesis without paternal contribution [151], similar mechanisms have so far not been observed for plants. From a recent study in apomictic *Boechera*, the second sperm cell nucleus typically does not fuse with the egg cell nucleus [156]. Still the occasional formation of BIII hybrids by fertilization of an unreduced egg cell demonstrates that this fusion is not strictly prevented in the apomicts [43]. Interestingly, differential regulation of CENH3 in sexual and apomictic gametophytes as observed in comparative transcriptional analyses including sexual *A. thaliana* and apomictic *B. gunnisoniana* suggests that this might serve to safeguard embryogenesis without paternal contribution [80]. This would imply that only the maternal genome is retained during early embryogenesis. Similarly, mechanisms of depletion of the paternal genome after occasional fertilization might be reinforced by mechanisms related to the heterochromatic B-like chromosomes in *Boechera*, as previously demonstrated for jewel wasps [157]. In the jewel wasp *Nasonia vitripennis* paternally inherited B-chromosomes promote their own transmission at the expense of other paternal chromosomes which are eliminated to form a haploid embryo [157]. Future studies will be needed to uncover, if similar pathways are active in plants.

Taken together, current knowledge suggests that complex regulatory networks act upon germline specification and development both in sexual plants and apomicts. The observed developmental flexibility of apomictic reproduction might thereby represent a trait off from the deregulation of developmental programs resulting in incompletely established or leaky mechanisms, however it might also be a mechanism enforcing new evolutionary options. While in different plant systems different candidate genes have been proposed to be relevant for apomixis, common features of regulatory machineries emerge that can serve as a starting point for future functional and evolutionary investigations. Importantly, genetic analyses of apomictic linked loci also consistently indicate a considerable divergence as compared to sexual loci. Thus it cannot be excluded that important regulators of apomixis might so far have been overlooked due to largely restricting the search to described and annotated genes and genomic elements. While due to the heterozygotic or polyploid nature of apomicts genome assembly and annotation remains challenging so far, especially state of the art sequencing technologies increasingly allow to obtain longer sequence reads. In conjunction with further studies focusing on genome evolution in apomicts and related sexual species, this will set the basis for a deeper understanding of the origin of apomixis and its genetic basis.

6. Are Stress Signal and Nutritional State Triggering the Determination for Sexual Reproduction or Apomixis?

Sexual reproduction and apomixis are classically viewed as two alternative types of reproduction with their own evolutionary histories. Recent ideas and insights challenge this perception and consider that both reproductive strategies might be polyphenic [18]. In this view, both modes of reproduction can be temporarily activated based on environmental conditions, stress, and nutritional state [18]. Evidence for the potential of a stress induced switch from apomixis to sexuality is given for a number of apomictic systems including *Boechera*, *Paspalum*, *Ranunculus*, and *Eragrostis* [18]. This might resemble an ancient mechanism. It has been hypothesized that the evolution of sexuality is a result of reactive oxygen species (ROS) that were generated starting with the development of primitive mitochondria at the basis of the evolution of eukaryotes [158,159]. From this perspective, the necessity for sex would be the consequence of ROS induced DNA damages as it allows purging of deleterious mutations from the genome [158]. Oxygen based DNA damage might have been also a requirement for meiosis to evolve and thereby lay the foundation for sexual reproduction [160]. It has been hypothesized that interactions of oxidized DNA and the core meiotic gene *SPO11* has enabled double strand breaks and meiotic recombination to occur [160]. Curiously, however, in maize anthers, low levels of ROS promotes acquisition of meiotic fate [161]. From a recent study in *A. thaliana*, repression of the homeobox gene *WUSCHEL* (*WUS*) is important for the acquisition of meiotic fate by the MMC [162]. Relieving the repression of *WUS* activity in the MMC causes mitotic divisions before the cells eventually enter meiosis [162]. Interestingly, in the shoot apical meristem WUS activity is activated by ROS [163]. It might be speculated that also in the maize anthers *WUS* regulation might be involved in determination of meiotic fate in response to ROS levels, if similar to regulatory processes are active in reproductive tissues. Furthermore, ROS has an impact on epigenetic regulatory systems and global DNA-methylation, suggesting the integration of stress signals and epigenetic regulation to control reproduction [18].

Consistent with a role of ROS to trigger meiosis, recent evidence suggests that redox regulation differs in sexual MMCs and the AICs (Figure 3A). In *Boechera gunnisoniana*, enrichment of polyamine and spermidine synthesis is a characteristic feature of the AIC [80]. This is in line with the identification of spermine/spermidine synthase from the ASGR in *Pennisetum squamulatum* [49]. The importance of the polyamine spermidine to protect the DNA from oxidative damage by scavenging of free radicals arising mostly from ROS has long been described [164]. Potentially, the importance of detoxification of ROS in the AIC is a consequence from the absence of meiosis. Such mechanisms to prevent deleterious mutations to arise by oxidative stress appear to be relevant particularly in the founder cells of the apomictic germline lineages. Interestingly, in *Boechera* nucelli tissues an UDP-glycosyltransferase superfamily protein is significantly higher expressed in all sexual as compared to all apomictic accessions analyzed [103]. While the functional role of this gene has not been investigated, it might be involved in synthesis of callose, as shown for certain members of this gene family [165]. Callose deposition is promoted by ROS [166], further supporting the idea that redox stress plays different roles for meiosis and apomeiosis, particularly as callose is typically not enclosing the AIC in contrast to the MMC [167,168] (Figure 3A). Thereby, callose deposition around the MMC might either shield the surrounding cells from ROS and its effects, or might be effective in protection of the meiocyte from disturbances. Future molecular studies are required to shed light onto this question. Nevertheless, the connection of ROS and callose further supports the idea of the importance of redox state for mode of reproduction.

Apart from ROS other types of stress like nutritional starvation and abiotic stress conditions including cold and heat have a great and versatile impact on meiosis and reproduction. Thereby, not only a shift from apomixis to sexual reproduction occurs, but also alterations of meiosis concerning recombination frequencies or the formation of unreduced or aneuploid gametophytes [169]. A central integrator of nutrient, energy, and stress related signals to regulate cell growth and development in eukaryotes is the target of rapamycin (TOR) kinase. In the yeast *Schizosaccharomyces pombe* nutritional starvation triggers the onset of meiosis and sexual reproduction dependent on the activity of TOR pathways [170]. Recent evidence suggests an evolutionary conservation of these pathways, as application of glucose at a certain developmental time point leads to features of apomeiosis in sexual *A. thaliana* [171]. The underlying molecular mechanism remains to be investigated in detail. An interesting question will be if *WUS* activity in the MMC is elevated by application of glucose, a mechanism described for the shoot apical meristem [172]. If so, this might be the molecular mechanism of obtaining mitotic divisions of the MMC similar to previous reports on de-repression of WUS activity in the MMC [162].

Interestingly the TOR pathway also coordinates ribosome activity [173], implying a connection between stress, nutritional state, reproduction and cell cycle. Further ribosome biogenesis factors like RNA helicases are not only involved in the regulation of gene activity and developmental decisions, but also in mediating stress response and growth regulation [126]. Functional implications in stress response have in particular been described for a number of RNA helicases, including *AtRH36* involved in regulation of gametogenesis and *ENHANCED SILENCING PHENOTYPE3* that has previously been described to be active in the AIC in *Boechera gunnisoniana* unlike in the *A. thaliana* MMC [80,126]. Also heat shock proteins are stress responsive proteins tightly associated to ribosome function, as they typically assist folding of newly derived polypeptide sequences to proteins as chaperones. Evidence for roles of heat shock proteins in apomixis regulation comes from different types of apomixis, including apospory in *Hieracium prealtum* [34], apomixis in *Paspalum notatum* [92], somatic embryogenesis in *Citrus* [76], and apogamy in the fern *Drypteris affinis* [174]. In addition, it is interesting to note that AP2/ERF transcription factors are important players in the integration of hormonal pathways and stress responses to control developmental decisions [154].

The regulation of reproductive development related to environmental factors and nutrition represents a conserved mechanism in eukaryotes. It can easily be envisioned that particularly in largely facultative apomictic systems such factors allow us to modify the frequencies of apomixis. Nevertheless, the heritability of apomixis and the identification of the genetically linked loci suggests the requirement of certain genetic elements for apomixis.

7. Brief Summary and Conclusions

Despite longstanding interest in apomixis, the gene regulatory processes and molecular mechanisms underlying apomixis are currently not fully understood. It remains an unresolved puzzle, how all major components of apomixis derived simultaneously several times independently. Recently, alternative concepts are discussed proposing the possibility of sexual reproduction and apomixis as ancient alternatives. In this view, conserved molecular machineries control the mode of reproduction dependent on nutritional state and environmental conditions. Irrespective of its evolutionary origin, apomixis is characterized by distinctive changes in the gene regulatory program as compared to sexual reproduction. Increasing evidence suggests that interrelated regulatory control mechanisms are involved, including epigenetic regulatory pathways, cell cycle control, regulation of protein turnover and degradation, signal transduction pathways, and hormonal regulatory pathways. Particularly for (apo)meiosis the assembly and regulation of ribosomes and associated factors emerge as novel important layer of regulation. Furthermore, precise coordination of repression and activation of gene activity appears to be involved in the transition from the female gametes to embryo and endosperm development. Thereby, BBM and BBML proteins have emerged as important players to promote embryogenesis. Taken together, recent insights into the molecular mechanisms and genetic basis underlying apomixis provide an important basis for future investigations directed on the detailed understanding of the regulatory programs involved. One important focus should be on fundamental and evolutionary conserved mechanisms to modify gene and protein activity and on the impact of environmental conditions on reproductive mode and success. This will not only allow us to gain fundamental new insights in the developmental processes of reproduction, but will also be an important basis for the harnessing of plant reproduction and apomixis for agricultural applications.

Funding: Work on the gene regulatory basis of apomixis and sexual reproduction is funded by the "Deutsche Forschungsgemeinschaft, DFG" grants SCHM2448/2-1 and SCHM2448/2-2 to A.S.

Acknowledgments: I thank Markus Kiefer (Centre for Organismal Studies, Heidelberg University, Germany) and David Ibberson (Deep Sequencing Core Facility, Heidelberg University, Germany) for critical reading of the manuscript and for helpful comments. I apologize to all colleagues whose work I could not reference due to space constrains.

Conflicts of Interest: The authors declare no conflict of interest. The funders had no role in the design of the study; in the collection, analyses, or interpretation of data; in the writing of the manuscript, or in the decision to publish the results.

References

1. Hörandl, E.; Hojsgaard, D. The evolution of apomixis in angiosperms: A reappraisal. *Plant Biosyst* **2012**, *146*, 681–693. [CrossRef]
2. Whitton, J.; Sears, C.; Baack, E.; Otto, S. The Dynamic Nature of Apomixis in the Angiosperms. *Int. J. Plant Sci.* **2008**, *169*, 169–182. [CrossRef]
3. Hojsgaard, D.; Klatt, S.; Baier, R.; Carman, J.G.; Hörandl, E. Taxonomy and Biogeography of Apomixis in Angiosperms and Associated Biodiversity Characteristics. *CRC Crit. Rev. Plant Sci.* **2014**, *33*, 414–427. [CrossRef]
4. Carman, J.G. Asynchronous expression of duplicate genes in angiosperms may cause apomixis, bispory, tetraspory, and polyembryony. *Biol. J. Linn. Soc.* **1997**, *61*, 51–94. [CrossRef]
5. Wang, X.; Xu, Y.; Zhang, S.; Cao, L.; Huang, Y.; Cheng, J. Genomic analyses of primitive, wild and cultivated citrus provide insights into asexual reproduction. *Nat. Genet.* **2017**, *49*, 765–772. [CrossRef] [PubMed]
6. Hojsgaard, D.; Horandl, E. The Rise of Apomixis in Natural Plant Populations. *Front. Plant Sci.* **2019**, *10*, 358. [CrossRef] [PubMed]
7. Fei, X.; Shi, J.; Liu, Y.; Jinshuang, N.; Wei, A. The steps from sexual reproduction to apomixis. *Planta* **2019**, *249*, 1715–1730. [CrossRef]
8. Spillane, C.; Steimer, A.; Grossniklaus, U. Apomixis in agriculture: The quest for clonal seeds. *Sex. Plant Reprod.* **2001**, *14*, 179–187. [CrossRef]
9. Koltunow, A.M.; Bicknell, R.A.; Chaudhury, A.M. Apomixis: Molecular Strategies for the Generation of Genetically Identical Seeds without Fertilization. *Plant Physiol.* **1995**, *108*, 1345–1352. [CrossRef]
10. Barcaccia, G.; Albertini, E. Apomixis in plant reproduction: A novel perspective on an old dilemma. *Plant Reprod.* **2013**, *26*, 159–179. [CrossRef]
11. Hisanaga, T.; Yamaoka, S.; Kawashima, T.; Higo, A.; Nakajima, K.; Araki, T.; Kohchi, T.; Berger, F. Building new insights in plant gametogenesis from an evolutionary perspective. *Nat. Plants* **2019**, *5*, 663–669. [CrossRef] [PubMed]
12. Nakajima, K. Be my baby: Patterning toward plant germ cells. *Curr. Opin. Plant Biol.* **2018**, *41*, 110–115. [CrossRef] [PubMed]
13. Erbasol Serbes, I.; Palovaara, J.; Gross-Hardt, R. Development and function of the flowering plant female gametophyte. *Curr. Top. Dev. Biol.* **2019**, *131*, 401–434. [CrossRef] [PubMed]
14. Schmid, M.W.; Schmidt, A.; Grossniklaus, U. The female gametophyte: An emerging model for cell type-specific systems biology in plant development. *Front. Plant Sci.* **2015**, *6*, 907. [CrossRef]
15. Koltunow, A.M.; Grossniklaus, U. Apomixis: A developmental perspective. *Annu. Rev. Plant Biol.* **2003**, *54*, 547–574. [CrossRef]
16. Nakano, M.; Shimada, T.; Endo, T.; Fujii, H.; Nesumi, H.; Kita, M.; Ebinac, M.; Shimizua, T.; Omurabd, M. Characterization of genomic sequence showing strong association with polyembryony among diverse Citrus species and cultivars, and its synteny with Vitis and Populus. *Plant Sci. Int. J. Exp. Plant Biol.* **2012**, *183*, 131–142. [CrossRef]
17. Bicknell, R.A.; Koltunow, A.M. Understanding apomixis: Recent advances and remaining conundrums. *Plant Cell* **2004**, *16* (Suppl. 1), S228–S245. [CrossRef]
18. Albertini, E.; Barcaccia, G.; Carman, J.G.; Pupilli, F. Did apomixis evolve from sex or was it the other way around? *J. Exp. Bot.* **2019**, *70*, 2951–2964. [CrossRef]
19. Nogler, G.A. Gametophytic Apomixis. In *Embryology of Angiosperms*; Johri, B.M., Ed.; Springer: Berlin/Heidelberg, Germany, 1984; pp. 475–518.

20. Leblanc, O.; Mazzucato, A. Screening procedures to identify and quantify apomixis. In *The Flowering of Apomixis: From Mechanisms to Genetic Engineering*; Savidan, Y., Carman, J.G., Dresselhaus, T., Eds.; CIMMYT: Texcoco, Mexico, 2001.
21. Van Dijk, P.J.; Bakx-Schotman, J.M.T. Formation of Unreduced Megaspores (Diplospory) in Apomictic Dandelions *(Taraxacum officinale, s.l.)* Is Controlled by a Sex-Specific Dominant Locus. *Genetics* **2004**, *166*, 483–492. [CrossRef]
22. Koltunow, A.M.; Johnson, S.D.; Rodrigues, J.C.; Okada, T.; Hu, Y.; Tsuchiya, T.; Wilson, S.; Fletcher, P.; Ito, K.; Suzuku, G. Sexual reproduction is the default mode in apomictic Hieracium subgenus Pilosella, in which two dominant loci function to enable apomixis. *Plant J.* **2011**, *66*, 890–902. [CrossRef]
23. Mandáková, T.; Ashby, K.; Price, B.; Windham, M.; Carman, J.; Lysak, M. Genome structure and apomixis in Phoenicaulis (Brassicaceae; Boechereae). *J. Syst. Evol.* **2019**. [CrossRef]
24. Van Dijk, P.; Vijverberg, K. The significance of apomixis in the evolution of the angiosperms: A reappraisal. In *Plant Species-Level Systematics. New Perspectives on Pattern and Process; Regnum Vegetabile 143*; Bakker, F.T., Chatrou, L.W., Gravendeel, B., Pelzer, P.B., Eds.; Koeltz Scientific Books: Ruggell, Liechtenstein, 2005; pp. 101–116.
25. Muller, H.J. The Relation of Recombination to Mutational Advance. *Mutat. Res.* **1964**, *106*, 2–9. [CrossRef]
26. R Marshall, D.; Brown, A. The evolution of apomixis. *Heredity* **1981**, *47*. [CrossRef]
27. Brukhin, V.; Osadtchiy, J.V.; Florez-Rueda, A.M.; Smetanin, D.; Bakin, E.; Nobre, M.S.; Bakin, E.; Grossniklaus, U. The Boechera Genus as a Resource for Apomixis Research. *Front. Plant Sci.* **2019**, *10*, 392. [CrossRef]
28. Lovell, J.T.; Aliyu, O.M.; Mau, M.; Schranz, M.E.; Koch, M.; Kiefer, C.; Song, B.-H.; Mitchell-Olds, T.; Sharbel, T.F. On the origin and evolution of apomixis in Boechera. *Plant Reprod.* **2013**, *26*, 309–315. [CrossRef]
29. Podio, M.; Cáceres, M.E.; Samoluk, S.S.; Seijo, J.G.; Pessino, S.C.; Ortiz, J.P.A.; Pupilli, F. A methylation status analysis of the apomixis-specific region in Paspalum spp. suggests an epigenetic control of parthenogenesis. *J. Exp. Bot.* **2014**, *65*, 6411–6424. [CrossRef]
30. Hojsgaard, D.H.; Martinez, E.J.; Quarin, C.L. Competition between meiotic and apomictic pathways during ovule and seed development results in clonality. *New Phytol.* **2013**, *197*, 336–347. [CrossRef]
31. Hojsgaard, D.; Hörandl, E. A little bit of sex matters for genome evolution in asexual plants. *Front. Plant Sci.* **2015**, *6*, 82. [CrossRef]
32. Hodač, L.; Klatt, S.; Hojsgaard, D.; Sharbel, T.F.; Hörandl, E. A little bit of sex prevents mutation accumulation even in apomictic polyploid plants. *BMC Evol. Biol.* **2019**, *19*, 170. [CrossRef]
33. Carman, J.G.; Mateo de Arias, M.; Gao, L.; Zhao, X.; Kowallis, B.M.; Sherwood, D.A.; Srivastava, M.K.; Dwivedi, K.K.; Price, B.J.; Watts, L. Apospory and Diplospory in Diploid Boechera (Brassicaceae) May Facilitate Speciation by Recombination-Driven Apomixis-to-Sex Reversals. *Front. Plant Sci.* **2019**, *10*, 724. [CrossRef]
34. Juranic, M.; Johnson, S.D.; Koltunow, A.M. Phenotypic plasticity of aposporous embryo sac development in Hieracium praealtum. *Plant Signal. Behav.* **2019**, *14*, 1622981. [CrossRef]
35. Hand, M.L.; Koltunow, A.M.G. The Genetic Control of Apomixis: Asexual Seed Formation. *Genetics* **2014**, *197*, 441–450. [CrossRef]
36. Ozias-Akins, P.; Roche, D.; Hanna, W.W. Tight clustering and hemizygosity of apomixis-linked molecular markers in Pennisetum squamulatum implies genetic control of apospory by a divergent locus that may have no allelic form in sexual genotypes. *Proc. Natl. Acad. Sci. USA* **1998**, *95*, 5127–5132. [CrossRef]
37. Grimanelli, D.; Leblanc, O.; Espinosa, E.; Perotti, E.; Gonzalez de Leon, D.; Savidan, Y. Mapping diplosporous apomixis in tetraploid Tripsacum: One gene or several genes? *Heredity* **1998**, *80*, 33–39. [CrossRef]
38. Grossniklaus, U.; Nogler, G.A.; van Dijk, P.J. How to Avoid Sex. *Genet. Control Gametophytic Apomixis* **2001**, *13*, 1491–1498. [CrossRef]
39. Grimanelli, D.; Leblanc, O.; Perotti, E.; Grossniklaus, U. Developmental genetics of gametophytic apomixis. *Trends Genet. TIG* **2001**, *17*, 597–604. [CrossRef]
40. Ozias-Akins, P.; Akiyama, Y.; Hanna, W.W. Molecular characterization of the genomic region linked with apomixis in Pennisetum/Cenchrus. *Funct. Integr. Genom.* **2003**, *3*, 94–104. [CrossRef]
41. Zappacosta, D.; Gallardo, J.; Carballo, J.; Meier, M.; Rodrigo, J.M.; Gallo, C.A.; Selva, J.P.; Stein, J.; Ortiz, J.P.A.; Albertini, E. A High-Density Linkage Map of the Forage Grass Eragrostis curvula and Localization of the Diplospory Locus. *Front. Plant Sci.* **2019**, *10*, 918. [CrossRef]

42. Ozias-Akins, P.; Dijk, P. Mendelian Genetics of Apomixis in Plants. *Annu. Rev. Genet.* **2007**, *41*, 509–537. [CrossRef]
43. Pupilli, F.; Barcaccia, G. Cloning plants by seeds: Inheritance models and candidate genes to increase fundamental knowledge for engineering apomixis in sexual crops. *J. Biotechnol.* **2012**, *159*, 291–311. [CrossRef]
44. Schallau, A.; Arzenton, F.; Johnston, A.J.; Hahnel, U.; Koszegi, D.; Blattner, F.R.; Altschmied, L.; Haberer, G.; Barcaccia, G.; Bäumlein, H. Identification and genetic analysis of the APOSPORY locus in Hypericum perforatum L. *Plant J.* **2010**, *62*, 773–784. [CrossRef]
45. Noyes, R.D.; Rieseberg, L.H. Two Independent Loci Control Agamospermy (Apomixis) in the Triploid Flowering Plant *Erigeron annuus*. *Genetics* **2000**, *155*, 379–390.
46. Vasut, R.J.; Vijverberg, K.; van Dijk, P.J.; de Jong, H. Fluorescent in situ hybridization shows DIPLOSPOROUS located on one of the NOR chromosomes in apomictic dandelions (Taraxacum) in the absence of a large hemizygous chromosomal region. *Genome* **2014**, *57*, 609–620. [CrossRef]
47. Vijverberg, K.; Milanovic-Ivanovic, S.; Bakx-Schotman, T.; van Dijk, P.J. Genetic fine-mapping of DIPLOSPOROUS in Taraxacum (dandelion; Asteraceae) indicates a duplicated DIP-gene. *BMC Plant Biol.* **2010**, *10*, 154. [CrossRef]
48. Akiyama, Y.; Hanna, W.W.; Ozias-Akins, P. High-resolution physical mapping reveals that the apospory-specific genomic region (ASGR) in Cenchrus ciliaris is located on a heterochromatic and hemizygous region of a single chromosome. *TAG Theor. Appl. Genet. Theor. Angew. Genet.* **2005**, *111*, 1042–1051. [CrossRef]
49. Conner, J.A.; Goel, S.; Gunawan, G.; Cordonnier-Pratt, M.M.; Johnson, V.E.; Liang, C.; Wang, H.; Pratt, L.H.; Mullet, J.E.; DeBarry, J. Sequence analysis of bacterial artificial chromosome clones from the apospory-specific genomic region of Pennisetum and Cenchrus. *Plant Physiol.* **2008**, *147*, 1396–1411. [CrossRef] [PubMed]
50. Goel, S.; Chen, Z.; Akiyama, Y.; Conner, J.A.; Basu, M.; Gualtieri, G.; Hanna, W.W.; Ozias-Akins, P. Comparative physical mapping of the apospory-specific genomic region in two apomictic grasses: Pennisetum squamulatum and Cenchrus ciliaris. *Genetics* **2006**, *173*, 389–400. [CrossRef]
51. Galla, G.; Siena, L.A.; Ortiz, J.P.A.; Baumlein, H.; Barcaccia, G.; Pessino, S.C.; Belluci, M.; Pupilli, F. A Portion of the Apomixis Locus of Paspalum Simplex is Microsyntenic with an Unstable Chromosome Segment Highly Conserved Among Poaceae. *Sci. Rep.* **2019**, *9*, 3271. [CrossRef]
52. Catanach, A.S.; Erasmuson, S.K.; Podivinsky, E.; Jordan, B.R.; Bicknell, R. Deletion mapping of genetic regions associated with apomixis in *Hieracium*. *Proc. Natl. Acad. Sci. USA* **2006**, *103*, 18650–18655. [CrossRef]
53. Kotani, Y.; Henderson, S.; Suzuki, G.; Johnson, S.; Okada, T.; Siddons, H.; Mukai, Y.; Koltunow, A.M. The LOSS OF APOMEIOSIS (LOA) locus in Hieracium praealtum can function independently of the associated large-scale repetitive chromosomal structure. *New Phytol.* **2014**, *201*, 973–981. [CrossRef]
54. Okada, T.; Ito, K.; Johnson, S.D.; Oelkers, K.; Suzuki, G.; Houben, A.; Mukai, Y.; Koltunow, A.M. Chromosomes carrying meiotic avoidance loci in three apomictic eudicot Hieracium subgenus Pilosella species share structural features with two monocot apomicts. *Plant Physiol.* **2011**, *157*, 1327–1341. [CrossRef]
55. Sharbel, T.F.; Voigt, M.L.; Mitchell-Olds, T.; Kantama, L.; de Jong, H. Is the aneuploid chromosome in an apomictic Boechera holboellii a genuine B chromosome? *Cytogenet. Genome Res.* **2004**, *106*, 173–183. [CrossRef] [PubMed]
56. Sharbel, T.F.; Mitchell-Olds, T.; Dobes, C.; Kantama, L.; de Jong, H. Biogeographic distribution of polyploidy and B chromosomes in the apomictic Boechera holboellii complex. *Cytogenet. Genome Res.* **2005**, *109*, 283–292. [CrossRef] [PubMed]
57. Mandakova, T.; Schranz, M.E.; Sharbel, T.F.; de Jong, H.; Lysak, M.A. Karyotype evolution in apomictic Boechera and the origin of the aberrant chromosomes. *Plant J.* **2015**, *82*, 785–793. [CrossRef] [PubMed]
58. Kantama, L.; Sharbel, T.F.; Schranz, M.E.; Mitchell-Olds, T.; de Vries, S.; de Jong, H. Diploid apomicts of the Boechera holboellii complex display large-scale chromosome substitutions and aberrant chromosomes. *Proc. Natl. Acad. Sci. USA* **2007**, *104*, 14026–14031. [CrossRef] [PubMed]
59. Schranz, M.E.; Kantama, L.; de Jong, H.; Mitchell-Olds, T. Asexual reproduction in a close relative of Arabidopsis: A genetic investigation of apomixis in Boechera (Brassicaceae). *New Phytol.* **2006**, *171*, 425–438. [CrossRef] [PubMed]
60. Dechaud, C.; Volff, J.-N.; Schartl, M.; Naville, M. Sex and the TEs: Transposable elements in sexual development and function in animals. *Mob. DNA* **2019**, *10*, 42. [CrossRef]
61. Taylor, D.; Chu, E.; Spektor, R.; Soloway, P. Long Non-Coding RNA Regulation of Reproduction and Development. *Mol. Reprod. Dev.* **2015**, *82*, 932–956. [CrossRef]

62. Galla, G.; Basso, A.; Grisan, S.; Bellucci, M.; Pupilli, F.; Barcaccia, G. Ovule Gene Expression Analysis in Sexual and Aposporous Apomictic Hypericum perforatum L. (Hypericaceae) Accessions. *Front. Plant Sci.* **2019**, *10*, 654. [CrossRef]
63. Corral, J.M.; Vogel, H.; Aliyu, O.M.; Hensel, G.; Thiel, T.; Kumlehn, J.; Sharbel, T.F. A Conserved Apomixis-Specific Polymorphism Is Correlated with Exclusive Exonuclease Expression in Premeiotic Ovules of Apomictic Boechera Species. *Plant Physiol.* **2013**, *163*, 1660–1672. [CrossRef]
64. Mau, M.; Lovell, J.T.; Corral, J.M.; Kiefer, C.; Koch, M.A.; Aliyu, O.M.; Sharbel, T.F. Hybrid apomicts trapped in the ecological niches of their sexual ancestors. *Proc. Natl. Acad. Sci. USA* **2015**, *112*, E2357–E2365. [CrossRef] [PubMed]
65. Mau, M.; Corral, J.M.; Vogel, H.; Melzer, M.; Fuchs, J.; Kuhlmann, M.; de Storme, N.; Geelen, D.; Sharbel, T.F. The conserved chimeric transcript UPGRADE2 is associated with unreduced pollen formation and is exclusively found in apomictic Boechera species. *Plant Physiol.* **2013**, *163*, 1640–1659. [CrossRef] [PubMed]
66. Koch, M.A. A new chromosome was born: Comparative chromosome painting in *Boechera*. *Trends Plant Sci.* **2015**, *20*, 533–535. [CrossRef] [PubMed]
67. Mancini, M.; Permingeat, H.; Colono, C.; Siena, L.; Pupilli, F.; Azzaro, C.; de Alencar Dusi, D.M.; de Campos Carneiro, V.T.; Podio, M.; Seijo, J.G. The MAP3K-Coding QUI-GON JINN (QGJ) Gene Is Essential to the Formation of Unreduced Embryo Sacs in Paspalum. *Front. Plant Sci.* **2018**, *9*, 1547. [CrossRef]
68. Siena, L.A.; Ortiz, J.P.; Calderini, O.; Paolocci, F.; Caceres, M.E.; Kaushal, P.; Grisan, S.; Pessino, S.C.; Pupilli, F. An apomixis-linked ORC3-like pseudogene is associated with silencing of its functional homolog in apomictic Paspalum simplex. *J. Exp. Bot.* **2016**, *67*, 1965–1978. [CrossRef]
69. Conner, J.A.; Gunawan, G.; Ozias-Akins, P. Recombination within the apospory specific genomic region leads to the uncoupling of apomixis components in Cenchrus ciliaris. *Planta* **2013**, *238*, 51–63. [CrossRef]
70. Conner, J.A.; Ozias-Akins, P. Apomixis: Engineering the Ability to Harness Hybrid Vigor in Crop Plants. *Methods Mol. Biol.* **2017**, *1669*, 17–34. [CrossRef]
71. Horstman, A.; Li, M.; Heidmann, I.; Weemen, M.; Chen, B.; Muino, J.M.; Angenent, G.C.; Boutilier, K. The BABY BOOM Transcription Factor Activates the LEC1-ABI3-FUS3-LEC2 Network to Induce Somatic Embryogenesis. *Plant Physiol.* **2017**, *175*, 848–857. [CrossRef]
72. Boutilier, K.; Offringa, R.; Sharma, V.K.; Kieft, H.; Ouellet, T.; Zhang, L.; Hattori, J.; Liu, C.-M.; van Lammeren, A.A.M. Ectopic Expression of BABY BOOM Triggers a Conversion from Vegetative to Embryonic Growth. *Plant Cell* **2002**, *14*, 1737–1749. [CrossRef] [PubMed]
73. Conner, J.A.; Mookkan, M.; Huo, H.; Chae, K.; Ozias-Akins, P. A parthenogenesis gene of apomict origin elicits embryo formation from unfertilized eggs in a sexual plant. *Proc. Natl. Acad. Sci. USA* **2015**, *112*, 11205–11210. [CrossRef]
74. Worthington, M.; Heffelfinger, C.; Bernal, D.; Quintero, C.; Zapata, Y.P.; Perez, J.G.; De Vega, J.; Miles, J.; Dellaporta, S.; Tohme, J. A Parthenogenesis Gene Candidate and Evidence for Segmental Allopolyploidy in Apomictic Brachiaria decumbens. *Genetics* **2016**, *203*, 1117–1132. [CrossRef] [PubMed]
75. Khanday, I.; Skinner, D.; Yang, B.; Mercier, R.; Sundaresan, V. A male-expressed rice embryogenic trigger redirected for asexual propagation through seeds. *Nature* **2019**, *565*, 91–95. [CrossRef]
76. Shimada, T.; Endo, T.; Fujii, H.; Nakano, M.; Sugiyama, A.; Daido, G.; Ohta, S.; Yoshioka, T.; Omura, M. MITE insertion-dependent expression of CitRKD1 with a RWP-RK domain regulates somatic embryogenesis in citrus nucellar tissues. *BMC Plant Biol.* **2018**, *18*, 166. [CrossRef] [PubMed]
77. Tedeschi, F.; Rizzo, P.; Rutten, T.; Altschmied, L.; Bäumlein, H. RWP-RK domain-containing transcription factors control cell differentiation during female gametophyte development in Arabidopsis. *New Phytol.* **2017**, *213*, 1909–1924. [CrossRef] [PubMed]
78. Koszegi, D.; Johnston, A.J.; Rutten, T.; Czihal, A.; Altschmied, L.; Kumlehn, J.; Wüst, S.E.; Kiroukhova, O.; Gheyselinck, J.; Grossniklaus, U. Members of the RKD transcription factor family induce an egg cell-like gene expression program. *Plant J.* **2011**, *67*, 280–291. [CrossRef]
79. Wuest, S.E.; Vijverberg, K.; Schmidt, A.; Weiss, M.; Gheyselinck, J.; Lohr, M.; Wellmer, F.; Rahnenführer, J.; von Mering, C.; Grossniklaus, U. *Arabidopsis* Female Gametophyte Gene Expression Map Reveals Similarities between Plant and Animal Gametes. *Curr. Biol.* **2010**, *20*, 506–512. [CrossRef]
80. Schmidt, A.; Schmid, M.W.; Klostermeier, U.C.; Qi, W.; Guthörl, D.; Sailer, C.; Waller, M.; Rosenstiel, P.; Grossniklaus, U. Apomictic and Sexual Germline Development Differ with Respect to Cell Cycle, Transcriptional, Hormonal and Epigenetic Regulation. *PLoS Genet.* **2014**, *10*, e1004476. [CrossRef]

81. Koi, S.; Hisanaga, T.; Sato, K.; Shimamura, M.; Yamato Katsuyuki, T.; Ishizaki, K.; Kohchi, T.; Nakajama, K. An Evolutionarily Conserved Plant RKD Factor Controls Germ Cell Differentiation. *Curr. Biol.* **2016**, *26*, 1775–1781. [CrossRef]
82. Rovekamp, M.; Bowman, J.L.; Grossniklaus, U. Marchantia MpRKD Regulates the Gametophyte-Sporophyte Transition by Keeping Egg Cells Quiescent in the Absence of Fertilization. *Curr. Biol.* **2016**, *26*, 1782–1789. [CrossRef]
83. Vielle-Calzada, J.P.; Nuccio, M.L.; Budiman, M.A.; Thomas, T.L.; Burson, B.L.; Hussey, M.A.; Wing, R.A. Comparative gene expression in sexual and apomictic ovaries of Pennisetum ciliare (L.) Link. *Plant Mol. Biol.* **1996**, *32*, 1085–1092. [CrossRef]
84. Sahu, P.P.; Gupta, S.; Malaviya, D.R.; Roy, A.K.; Kaushal, P.; Prasad, M. Transcriptome Analysis of Differentially Expressed Genes During Embryo Sac Development in Apomeiotic Non-Parthenogenetic Interspecific Hybrid of Pennisetum glaucum. *Mol. Biotechnol.* **2012**, *51*, 262–271. [CrossRef] [PubMed]
85. Chen, L.; Miyazaki, C.; Kojimai, A.; Saito, A.; Adachi, T. Isolation and Characterization of a Gene Expressed during Early Embryo Sac Development in Apomictic Guinea Grass (Panicum maximum). *J. Plant Physiol.* **1999**, *154*, 55–62. [CrossRef]
86. Radhakrishna, A.; Dwivedi, K.K.; Srivastava, M.K.; Roy, A.K.; Malaviya, D.R.; Kaushal, P. Transcriptomic data of pre-meiotic stage of floret development in apomictic and sexual types of guinea grass (Panicum maximum Jacq.). *Data Brief* **2018**, *18*, 590–593. [CrossRef]
87. Albertini, E.; Marconi, G.; Barcaccia, G.; Raggi, L.; Falcinelli, M. Isolation of candidate genes for apomixis in Poa pratensis L. *Plant Mol. Biol.* **2004**, *56*, 879–894. [CrossRef] [PubMed]
88. Albertini, E.; Marconi, G.; Reale, L.; Barcaccia, G.; Porceddu, A.; Ferranti, F.; Falcinelli, M. *SERK* and *APOSTART*. Candidate Genes for Apomixis in *Poa pratensis*. *Plant Physiol.* **2005**, *138*, 2185–2199. [CrossRef] [PubMed]
89. Leblanc, O.; Armstead, I.; Pessino, S.; Ortiz, J.P.A.; Evans, C.; do Valle, C.; Hayward, M.D. Non-radioactive mRNA fingerprinting to visualise gene expression in mature ovaries of Brachiaria hybrids derived from B. brizantha, an apomictic tropical forage. *Plant Sci.* **1997**, *126*, 49–58. [CrossRef]
90. Rodrigues, J.C.M.; Cabral, G.B.; Dusi, D.M.A.; de Mello, L.V.; Rigden, D.J.; Carneiro, V.T.C. Identification of differentially expressed cDNA sequences in ovaries of sexual and apomictic plants of Brachiaria brizantha. *Plant Mol. Biol.* **2003**, *53*, 745–757. [CrossRef]
91. Pessino, S.C.; Espinoza, F.; Martínez, E.J.; Ortiz, J.P.A.; Valle, E.M.; Quarín, C.L. Isolation of cDNA Clones Differentially Expressed in Flowers of Apomictic and Sexual Paspalum Notatum. *Hereditas* **2001**, *134*, 35–42. [CrossRef]
92. Polegri, L.; Calderini, O.; Arcioni, S.; Pupilli, F. Specific expression of apomixis-linked alleles revealed by comparative transcriptomic analysis of sexual and apomictic Paspalum simplex Morong flowers. *J. Exp. Bot.* **2010**, *61*, 1869–1883. [CrossRef]
93. Ochogavía, A.C.; Seijo, J.G.; González, A.M.; Podio, M.; Duarte Silveira, E.; Machado Lacerda, A.L.; de Campos Carneiro, V.T.; Ortiz, J.P.A.; Pessino, S.C. Characterization of retrotransposon sequences expressed in inflorescences of apomictic and sexual Paspalum notatum plants. *Sex. Plant Reprod.* **2011**, *24*, 231–246. [CrossRef]
94. Ortiz, J.P.A.; Revale, S.; Siena, L.A.; Podio, M.; Delgado, L.; Stein, J.; Leblanc, O.; Pessino, S.C. A reference floral transcriptome of sexual and apomictic Paspalum notatum. *BMC Genom.* **2017**, *18*, 318. [CrossRef]
95. Garbus, I.; Romero, J.R.; Selva, J.P.; Pasten, M.C.; Chinestra, C.; Carballo, J.; Zappacosta, D.C.; Echenique, V. De novo transcriptome sequencing and assembly from apomictic and sexual Eragrostis curvula genotypes. *PLoS ONE* **2017**, *12*, e0185595. [CrossRef] [PubMed]
96. Barcaccia, G.; Varotto, S.; Meneghetti, S.; Albertini, E.; Porceddu, A.; Parrini, P.; Lucchin, M. Analysis of gene expression during flowering in apomeiotic mutants of Medicago spp.: Cloning of ESTs and candidate genes for 2n eggs. *Sex. Plant Reprod.* **2001**, *14*, 233–238. [CrossRef] [PubMed]
97. Tang, Q.; Zang, G.; Cheng, C.; Luan, M.; Dai, Z.; Xu, Y.; Yang, Z.; Zhao, L.; Su, J. Diplosporous development in Boehmeria tricuspis: Insights from de novo transcriptome assembly and comprehensive expression profiling. *Sci. Rep.* **2017**, *7*, 46043. [CrossRef] [PubMed]
98. Galla, G.; Zenoni, S.; Avesani, L.; Altschmied, L.; Rizzo, P.; Sharbel, T.F.; Barcaccia, G. Pistil Transcriptome Analysis to Disclose Genes and Gene Products Related to Aposporous Apomixis in Hypericum perforatum L. *Front. Plant Sci.* **2017**, *8*, 79. [CrossRef]

99. Tucker, M.R.; Araujo, A.-C.G.; Paech, N.A.; Hecht, V.; Schmidt, E.D.L.; Rossell, J.-B.; de Vries, S.C.; Koltunow, A.M.G. Sexual and Apomictic Reproduction in *Hieracium* subgenus *Pilosella* Are Closely Interrelated Developmental Pathways. *Plant Cell* **2003**, *15*, 1524–1537. [CrossRef]
100. Brauning, S.; Catanach, A.; Lord, J.M.; Bicknell, R.; Macknight, R.C. Comparative transcriptome analysis of the wild-type model apomict Hieracium praealtum and its loss of parthenogenesis (lop) mutant. *BMC Plant Biol.* **2018**, *18*, 206. [CrossRef]
101. Sharbel, T.F.; Voigt, M.-L.; Corral, J.M.; Galla, G.; Kumlehn, J.; Klukas, C.; Schreiber, F.; Vogel, H.; Rotter, B. Apomictic and Sexual Ovules of Boechera Display Heterochronic Global Gene Expression Patterns. *Plant Cell* **2010**, *22*, 655–671. [CrossRef]
102. Sharbel, T.F.; Voigt, M.-L.; Corral, J.M.; Thiel, T.; Varshney, A.; Kumlehn, J.; Vogel, H.; Rotter, B. Molecular signatures of apomictic and sexual ovules in the Boechera holboellii complex. *Plant J.* **2009**, *58*, 870–882. [CrossRef]
103. Zühl, L.; Volkert, C.; Ibberson, D.; Schmidt, A. Differential activity of F-box genes and E3 ligases distinguishes sexual versus apomictic germline specification in Boechera. *J. Exp. Bot.* **2019**, *70*, 5643–5657. [CrossRef]
104. Florez Rueda, A.M.; Grossniklaus, U.; Schmidt, A. Laser-assisted Microdissection (LAM) as a Tool for Transcriptional Profiling of Individual Cell Types. *J. Vis. Exp. JoVE* **2016**, *111*, 53916. [CrossRef] [PubMed]
105. Schmidt, A.; Schmid, M.W.; Grossniklaus, U. Analysis of plant germline development by high-throughput RNA profiling: Technical advances and new insights. *Plant J.* **2012**, *70*, 18–29. [CrossRef]
106. Schmid, M.W.; Schmidt, A.; Klostermeier, U.C.; Barann, M.; Rosenstiel, P.; Grossniklaus, U. A Powerful Method for Transcriptional Profiling of Specific Cell Types in Eukaryotes: Laser-Assisted Microdissection and RNA Sequencing. *PLoS ONE* **2012**, *7*, e29685. [CrossRef] [PubMed]
107. Schmidt, A.; Wuest, S.E.; Vijverberg, K.; Baroux, C.; Kleen, D.; Grossniklaus, U. Transcriptome Analysis of the Arabidopsis Megaspore Mother Cell Uncovers the Importance of RNA Helicases for Plant Germline Development. *PLoS Biol.* **2011**, *9*, e1001155. [CrossRef] [PubMed]
108. Okada, T.; Hu, Y.; Tucker, M.R.; Taylor, J.M.; Johnson, S.D.; Spriggs, A.; Tsuchiya, T.; Oelkers, K.; Rodrigures, J.C.; Koltunow, A.M. Enlarging Cells Initiating Apomixis in Hieracium praealtum Transition to an Embryo Sac Program prior to Entering Mitosis. *Plant Physiol.* **2013**, *163*, 216–231. [CrossRef]
109. Juranic, M.; Tucker, M.R.; Schultz, C.J.; Shirley, N.J.; Taylor, J.M.; Spriggs, A.; Johnson, S.D.; Bulone, V.; Koltunow, A.M. Asexual Female Gametogenesis Involves Contact with a Sexually-Fated Megaspore in Apomictic Hieracium. *Plant Physiol.* **2018**, *177*, 1027–1049. [CrossRef]
110. Yamada-Akiyama, H.; Akiyama, Y.; Ebina, M.; Xu, Q.; Tsuruta, S.-I.; Yazaki, J.; Kishimoto, N.; Kikuchi, S.; Takahara, M.; Takamizo, T. Analysis of expressed sequence tags in apomictic guineagrass (Panicum maximum). *J. Plant Physiol.* **2008**, *166*, 750–761. [CrossRef]
111. Laspina, N.; Vega, T.; Seijo, G.; Gonzalez, A.; Martelotto, L.; Stein, J.; Podio, M.; Ortiz, J.P.A.; Echenique, V.C.; Quarin, C.L. Gene expression analysis at the onset of aposporous apomixis in Paspalum notatum. *Plant Mol. Biol.* **2008**, *67*, 615–628. [CrossRef]
112. Cervigni, G.; Paniego, N.; Pessino, S.; Selva, J.P.; Diaz, M.; Spangenberg, G.; Echenique, V. Gene expression in diplosporous and sexual Eragrostis curvula genotypes with differing ploidy levels. *Plant Mol. Biol.* **2008**, *67*, 11–23. [CrossRef]
113. Galla, G.; Vogel, H.; Sharbel, T.F.; Barcaccia, G. De novo sequencing of the Hypericum perforatum L. flower transcriptome to identify potential genes that are related to plant reproduction sensu lato. *BMC Genom.* **2015**, *16*, 254. [CrossRef]
114. Rabiger, D.S.; Taylor, J.M.; Spriggs, A.; Hand, M.L.; Henderson, S.T.; Johnson, S.D.; Oelkers, K.; Hrmova, M.; Saito, K.; Suzuki, G. Generation of an integrated Hieracium genomic and transcriptomic resource enables exploration of small RNA pathways during apomixis initiation. *BMC Biol.* **2016**, *14*, 86. [CrossRef] [PubMed]
115. Long, J.M.; Liu, Z.; Wu, X.M.; Fang, Y.N.; Jia, H.H.; Xie, Z.Z.; Deng, X.X.; Guo, W.W. Genome-scale mRNA and small RNA transcriptomic insights into initiation of citrus apomixis. *J. Exp. Bot.* **2016**, *67*, 5743–5756. [CrossRef] [PubMed]
116. Schmidt, A.; Schmid, M.W.; Grossniklaus, U. Plant germline formation: Common concepts and developmental flexibility in sexual and asexual reproduction. *Development* **2015**, *142*, 229–241. [CrossRef] [PubMed]
117. Mackenzie, A.; Heslop-Harrison, J.; Dickinson, H.G. Elimination of Ribosomes during Meiotic Prophase. *Nature* **1967**, *215*, 997–999. [CrossRef] [PubMed]

118. Zeng, Y.; Conner, J.; Ozias-Akins, P. Identification of ovule transcripts from the Apospory-Specific Genomic Region (ASGR)-carrier chromosome. *BMC Genom.* **2011**, *12*, 206. [CrossRef]
119. Xue, S.; Barna, M. Specialized ribosomes: A new frontier in gene regulation and organismal biology. *Nat. Rev. Mol. Cell Biol.* **2012**, *13*, 355–369. [CrossRef]
120. Yan, Q.; Zhu, C.; Guang, S.; Feng, X. The Functions of Non-coding RNAs in rRNA Regulation. *Front. Genet.* **2019**, *10*, 290. [CrossRef]
121. Dalla Venezia, N.; Vincent, A.; Marcel, V.; Catez, F.; Diaz, J.-J. Emerging Role of Eukaryote Ribosomes in Translational Control. *Int. J. Mol. Sci.* **2019**, *20*, 1226. [CrossRef]
122. Bazin, J.; Baerenfaller, K.; Gosai, S.J.; Gregory, B.D.; Crespi, M.; Bailey-Serres, J. Global analysis of ribosome-associated noncoding RNAs unveils new modes of translational regulation. *Proc. Natl. Acad. Sci. USA* **2017**, *114*, E10018–E10027. [CrossRef]
123. Ding, D.-Q.; Haraguchi, T.; Hiraoka, Y. The role of chromosomal retention of noncoding RNA in meiosis. *Chromosome Res.* **2013**, *21*, 665–672. [CrossRef]
124. Flórez-Zapata, N.M.V.; Reyes-Valdés, M.H.; Martínez, O. Long non-coding RNAs are major contributors to transcriptome changes in sunflower meiocytes with different recombination rates. *BMC Genom.* **2016**, *17*, 490. [CrossRef]
125. Hafidh, S.; Čapková, V.; Honys, D. Safe Keeping the Message: mRNP Complexes Tweaking after Transcription. In *RNA Infrastructure and Networks*; Collins, L.J., Ed.; Springer: New York, NY, USA, 2011; pp. 118–136.
126. Liu, Y.; Imai, R. Function of Plant DExD/H-Box RNA Helicases Associated with Ribosomal RNA Biogenesis. *Front. Plant Sci.* **2018**, *9*, 125. [CrossRef] [PubMed]
127. Owttrim, G. RNA helicases and abiotic stress. *Nucleic Acids Res.* **2006**, *34*, 3220–3230. [CrossRef] [PubMed]
128. Silveira, É.D.; Guimarães, L.A.; de Alencar Dusi, D.M.; da Silva, F.R.; Martins, N.F.; do Carmo Costa, M.M.; Alves-Ferreira, M.; de Campos Carneiro, V.T. Expressed sequence-tag analysis of ovaries of Brachiaria brizantha reveals genes associated with the early steps of embryo sac differentiation of apomictic plants. *Plant Cell Rep.* **2012**, *31*, 403–416. [CrossRef]
129. Gehring, M. Epigenetic dynamics during flowering plant reproduction: Evidence for reprogramming? *New Phytol.* **2019**, *224*, 91–96. [CrossRef] [PubMed]
130. Zhang, X.; Wang, W.; Zhu, W.; Dong, J.; Cheng, Y.; Yin, Z.; Shen, F. Mechanisms and Functions of Long Non-Coding RNAs at Multiple Regulatory Levels. *Int. J. Mol. Sci.* **2019**, *20*, 5573. [CrossRef] [PubMed]
131. Garcia-Aguilar, M.; Michaud, C.; Leblanc, O.; Grimanelli, D. Inactivation of a DNA Methylation Pathway in Maize Reproductive Organs Results in Apomixis-Like Phenotypes. *Plant Cell* **2010**, *22*, 3249–3267. [CrossRef]
132. Olmedo-Monfil, V.; Durán-Figueroa, N.; Arteaga-Vandázquez, M.; Demesa-Arévalo, E.; Autran, D.; Grimanelli, D.; Grimanelli, D.; Slotkin, K.; Martienssen, R.A.; Vielle-Calzada, J.-P. Control of female gamete formation by a small RNA pathway in Arabidopsis. *Nature* **2010**, *464*, 628–632. [CrossRef]
133. Singh, M.; Goel, S.; Meeley, R.B.; Dantec, C.; Parrinello, H.; Michaud, C.; Leblanc, O.; Grimanelli, D. Production of Viable Gametes without Meiosis in Maize Deficient for an ARGONAUTE Protein. *Plant Cell* **2011**, *23*, 443–458. [CrossRef]
134. Hernandez-Lagana, E.; Rodriguez-Leal, D.; Lua, J.; Vielle-Calzada, J.P. A Multigenic Network of ARGONAUTE4 Clade Members Controls Early Megaspore Formation in Arabidopsis. *Genetics* **2016**, *204*, 1045–1056. [CrossRef]
135. Meister, G. Argonaute proteins: Functional insights and emerging roles. *Nat. Rev. Genet.* **2013**, *14*, 447–459. [CrossRef] [PubMed]
136. Garbus, I.; Selva, J.P.; Pasten, M.C.; Bellido, A.M.; Carballo, J.; Albertini, E.; Echenique, V. Characterization and discovery of miRNA and miRNA targets from apomictic and sexual genotypes of Eragrostis curvula. *BMC Genom.* **2019**, *20*, 839. [CrossRef]
137. Ortiz, J.P.A.; Leblanc, O.; Rohr, C.; Grisolia, M.; Siena, L.A.; Podio, M.; Colono, C.; Azzaro, C.; Pessino, S.C. Small RNA-seq reveals novel regulatory components for apomixis in Paspalum notatum. *BMC Genom.* **2019**, *20*, 487. [CrossRef] [PubMed]
138. Amiteye, S.; Corral, J.M.; Vogel, H.; Kuhlmann, M.; Mette, M.F.; Sharbel, T.F. Novel microRNAs and microsatellite-like small RNAs in sexual and apomictic Boechera species. *MicroRNA* **2013**, *2*, 45–62. [CrossRef] [PubMed]
139. Amiteye, S.; Corral, J.M.; Vogel, H.; Sharbel, T.F. Analysis of conserved microRNAs in floral tissues of sexual and apomictic Boechera species. *BMC Genom.* **2011**, *12*, 500. [CrossRef] [PubMed]

140. Ravi, M.; Marimuthu, M.P.A.; Siddiqi, I. Gamete formation without meiosis in Arabidopsis. *Nature* **2008**, *451*, 1121. [CrossRef] [PubMed]
141. Pawlowski, W.P.; Wang, C.-J.R.; Golubovskaya, I.N.; Szymaniak, J.M.; Shi, L.; Hamant, O.; Zhu, T.; Harper, L.; Sheridan, W.F.; Cande, W.Z. Maize AMEIOTIC1 is essential for multiple early meiotic processes and likely required for the initiation of meiosis. *Proc. Natl. Acad. Sci. USA* **2009**, *106*, 3603–3608. [CrossRef]
142. D'Erfurth, I.; Cromer, L.; Jolivet, S.; Girard, C.; Horlow, C.; Sun, Y.; To, J.P.C.; Berchowitz, L.E.; Copenhaver, G.P.; Mercier, R. The CYCLIN-A CYCA1;2/TAM Is Required for the Meiosis I to Meiosis II Transition and Cooperates with OSD1 for the Prophase to First Meiotic Division Transition. *PLoS Genet.* **2010**, *6*, e1000989. [CrossRef]
143. D'Erfurth, I.; Jolivet, S.; Froger, N.; Catrice, O.; Novatchkova, M.; Mercier, R. Turning Meiosis into Mitosis. *PLoS Biol.* **2009**, *7*, e1000124. [CrossRef]
144. Mieulet, D.; Jolivet, S.; Rivard, M.; Cromer, L.; Vernet, A.; Mayonove, P.; Pereira, L.; Droc, G.; Courtois, B.; Guiderdoni, E. Turning rice meiosis into mitosis. *Cell Res.* **2016**, *26*, 1242–1254. [CrossRef]
145. Marimuthu, M.P.A.; Jolivet, S.; Ravi, M.; Pereira, L.; Davda, J.N.; Cromer, L.; Wang, L.; Nogue, F.; Chan, S.W.L.; Siddiqu, I. Synthetic Clonal Reproduction Through Seeds. *Science* **2011**, *331*, 876. [CrossRef]
146. Wang, C.; Liu, Q.; Shen, Y.; Hua, Y.; Wang, J.; Lin, J.; Wu, M.; Sun, T.; Cheng, Z.; Mercier, R. Clonal seeds from hybrid rice by simultaneous genome engineering of meiosis and fertilization genes. *Nat. Biotechnol.* **2019**, *37*, 283–286. [CrossRef] [PubMed]
147. Kelliher, T.; Starr, D.; Richbourg, L.; Chintamanani, S.; Delzer, B.; Nuccio, M.L.; Green, J.; Chen, Z.; McCuiston, J.; Wang, W. MATRILINEAL, a sperm-specific phospholipase, triggers maize haploid induction. *Nature* **2017**, *542*, 105–109. [CrossRef] [PubMed]
148. Pires, N.D. Seed evolution: Parental conflicts in a multi-generational household. *Biomol. Concepts* **2014**, *5*, 71–86. [CrossRef] [PubMed]
149. Sprunck, S. Twice the fun, double the trouble: Gamete interactions in flowering plants. *Curr. Opin. Plant Biol.* **2020**, *53*, 106–116. [CrossRef] [PubMed]
150. Hands, P.; Rabiger, D.S.; Koltunow, A. Mechanisms of endosperm initiation. *Plant Reprod.* **2016**, *29*, 215–225. [CrossRef]
151. Vijverberg, K.; Ozias-Akins, P.; Schranz, M.E. Identifying and Engineering Genes for Parthenogenesis in Plants. *Front. Plant Sci.* **2019**, *10*. [CrossRef] [PubMed]
152. Derkacheva, M.; Hennig, L. Variations on a theme: Polycomb group proteins in plants. *J. Exp. Bot.* **2013**, *65*, 2769–2784. [CrossRef]
153. Rodrigues, J.C.M.; Tucker, M.R.; Johnson, S.D.; Hrmova, M.; Koltunow, A.M.G. Sexual and apomictic seed formation in Hieracium requires the plant polycomb-group gene FERTILIZATION INDEPENDENT ENDOSPERM. *Plant Cell* **2008**, *20*, 2372–2386. [CrossRef]
154. Xie, Z.; Nolan, T.M.; Jiang, H.; Yin, Y. AP2/ERF Transcription Factor Regulatory Networks in Hormone and Abiotic Stress Responses in Arabidopsis. *Front. Plant Sci.* **2019**, *10*. [CrossRef]
155. Nic-Can, G.I.; López-Torres, A.; Barredo-Pool, F.; Wrobel, K.; Loyola-Vargas, V.M.; Rojas-Herrera, R.; De-la-Pena, C. New Insights into Somatic Embryogenesis: LEAFY COTYLEDON1, BABY BOOM1 and WUSCHEL-RELATED HOMEOBOX4 Are Epigenetically Regulated in Coffea canephora. *PLoS ONE* **2013**, *8*, e72160. [CrossRef]
156. Kirioukhova, O.; Shah, J.N.; Larsen, D.S.; Tayyab, M.; Mueller, N.E.; Govind, G.; Baroux, C.; Federer, M.; Gheyselinck, J.; Barrell, P.J. Aberrant imprinting may underlie evolution of parthenogenesis. *Sci. Rep.* **2018**, *8*, 10626. [CrossRef] [PubMed]
157. Aldrich, J.C.; Ferree, P.M. Genome Silencing and Elimination: Insights from a "Selfish" B Chromosome. *Front Genet.* **2017**, *8*, 50. [CrossRef]
158. Hörandl, E.; Speijer, D. How oxygen gave rise to eukaryotic sex. *Proc. R. Soc. B Biol. Sci.* **2018**, *285*, 20172706. [CrossRef]
159. Mirzaghaderi, G.; Horandl, E. The evolution of meiotic sex and its alternatives. *Proc. R. Soc. B Biol. Sci.* **2016**, *283*, 20161221. [CrossRef]
160. Horandl, E.; Hadacek, F. The oxidative damage initiation hypothesis for meiosis. *Plant Reprod.* **2013**, *26*, 351–367. [CrossRef]
161. Kelliher, T.; Walbot, V. Hypoxia Triggers Meiotic Fate Acquisition in Maize. *Science* **2012**, *337*, 345–348. [CrossRef]

162. Zhao, X.; Bramsiepe, J.; Van Durme, M.; Komaki, S.; Prusicki, M.A.; Maruyama, D.; Forner, J.; Medzihradszky, A.; Wijnker, E.; Harashima, H. RETINOBLASTOMA RELATED1 mediates germline entry in Arabidopsis. *Science* **2017**, *356*, eaaf6532. [CrossRef] [PubMed]
163. Huang, H.; Ullah, F.; Zhou, D.-X.; Yi, M.; Zhao, Y. Mechanisms of ROS Regulation of Plant Development and Stress Responses. *Front. Plant Sci.* **2019**, *10*. [CrossRef]
164. Ha, H.C.; Sirisoma, N.S.; Kuppusamy, P.; Zweier, J.L.; Woster, P.M.; Casero, R.A. The natural polyamine spermine functions directly as a free radical scavenger. *Proc. Natl. Acad. Sci. USA* **1998**, *95*, 11140–11145. [CrossRef]
165. Hong, Z.; Zhang, Z.; Olson, J.M.; Verma, D.P.S. A Novel UDP-Glucose Transferase Is Part of the Callose Synthase Complex and Interacts with Phragmoplastin at the Forming Cell Plate. *Plant Cell* **2001**, *13*, 769–779. [CrossRef]
166. Luna, E.; Pastor, V.; Robert, J.; Flors, V.; Mauch-Mani, B.; Ton, J. Callose deposition: A multifaceted plant defense response. *Mol. Plant Microbe Interact. MPMI* **2011**, *24*, 183–193. [CrossRef]
167. Rodkiewicz, B. Callose in cell walls during mega-sporogenesis in Angiosperms. *Planta* **1970**, *93*, 39–47. [CrossRef]
168. Koltunow, A.M. Apomixis: Embryo Sacs and Embryos Formed without Meiosis or Fertilization in Ovules. *Plant Cell* **1993**, *5*, 1425–1437. [CrossRef]
169. De Storme, N.; Geelen, D. The impact of environmental stress on male reproductive development in plants: Biological processes and molecular mechanisms. *Plant Cell Environ.* **2014**, *37*, 1–18. [CrossRef]
170. Otsubo, Y.; Nakashima, A.; Yamamoto, M.; Yamashita, A. TORC1-Dependent Phosphorylation Targets in Fission Yeast. *Biomolecules* **2017**, *7*, 50. [CrossRef]
171. Sherwood, D.A. A Simple Metabolic Switch May Activate Apomixis in *Arabidopsis thaliana*. Ph.D. Thesis, Utah University, Salt Lake City, UT, USA, 2018.
172. Pfeiffer, A.; Janocha, D.; Dong, Y.; Medzihradszky, A.; Schone, S.; Daum, G.; Suzaki, T.; Forner, J.; Langenecker, T.; Rempel, E. Integration of light and metabolic signals for stem cell activation at the shoot apical meristem. *Elife* **2016**, *5*, e17023. [CrossRef]
173. Mayer, C.; Grummt, I. Ribosome biogenesis and cell growth: mTOR coordinates transcription by all three classes of nuclear RNA polymerases. *Oncogene* **2006**, *25*, 6384–6391. [CrossRef]
174. Grossmann, J.; Fernández, H.; Chaubey, P.M.; Valdés, A.E.; Gagliardini, V.; Cañal, M.J.; Russo, G.; Grossniklaus, U. Proteogenomic Analysis Greatly Expands the Identification of Proteins Related to Reproduction in the Apogamous Fern Dryopteris affinis ssp. affinis. *Front. Plant Sci.* **2017**, *8*, 336. [CrossRef]

genes

MDPI

Review

How to Become an Apomixis Model: The Multifaceted Case of *Paspalum*

Juan Pablo A. Ortiz [1], Fulvio Pupilli [2], Carlos A. Acuña [3], Olivier Leblanc [4] and Silvina C. Pessino [1,*]

1 Instituto de Investigaciones en Ciencias Agrarias de Rosario (IICAR), CONICET, Facultad de Ciencias Agrarias, Universidad Nacional de Rosario, S2125ZAA Zavalla, Argentina; ortiz@iicar-conicet.gob.ar
2 Institute of Biosciences and Bioresources (IBBR-CNR), 06128 Perugia, Italy; fulvio.pupilli@ibbr.cnr.it
3 Instituto de Botánica del Nordeste (IBONE), CONICET, Facultad de Ciencias Agrarias, Universidad Nacional del Nordeste, 3400 Corrientes, Argentina; caalac77@gmail.com
4 UMR DIADE, IRD, Univ. Montpellier, 34090 Montpellier, France; olivier.leblanc@ird.fr
* Correspondence: pessino@iicar-conicet.gob.ar

Received: 13 July 2020; Accepted: 17 August 2020; Published: 21 August 2020

Abstract: In the past decades, the grasses of the *Paspalum* genus have emerged as a versatile model allowing evolutionary, genetic, molecular, and developmental studies on apomixis as well as successful breeding applications. The rise of such an archetypal system progressed through integrative phases, which were essential to draw conclusions based on solid standards. Here, we review the steps adopted in *Paspalum* to establish the current body of knowledge on apomixis and provide model breeding programs for other agronomically important apomictic crops. In particular, we discuss the need for previous detailed cytoembryological and cytogenetic germplasm characterization; the establishment of sexual and apomictic materials of identical ploidy level; the development of segregating populations useful for inheritance analysis, positional mapping, and epigenetic control studies; the development of omics data resources; the identification of key molecular pathways via comparative gene expression studies; the accurate molecular characterization of genomic loci governing apomixis; the in-depth functional analysis of selected candidate genes in apomictic and model species; the successful building of a sexual/apomictic combined breeding scheme.

Keywords: agamospermy; plant breeding; plant development; plant reproduction

1. Introduction

Apomixis (asexual reproduction through seeds) [1] has long been seen as an unprecedented natural tool to maximize plant breeding, with potential wide impact on global farming systems [2]. In close developmental connection with sexuality, it functions as either a digressed or a parallel pathway, ending in the generation of clonal embryos of maternal origin within viable seeds [3]. Besides the importance of understanding this puzzling reproductive mode to advance the basic knowledge on reproduction, the combined use of sexuality and apomixis could accelerate the generation of novel improved plant varieties while dramatically diminishing the cost of the process, turning the notion of customized, locally-adapted hybrid crops adapted to every farm plot into reality [2]. The potential benefits of harnessing apomixis vary from full exploitation and permanent fixing of heterosis to seed distribution for crops actually propagated vegetatively like potatoes or strawberries [4]. However, the effective use of this trait in plant breeding requires full knowledge of the molecular switch allowing the transition from sexuality to clonal seed reproduction.

In the last 30 years, coordinated international efforts have led to the elucidation of major molecular actors controlling apomictic development [5]. The integration of these genes into interlaced functional pathways is currently under investigation, in the prospect of generating optimized biotechnological tools. With this aim, modulating the expression of some critical genes has allowed the rewiring of apomixis components in sexual plants [5] and even induced the production of clonal seeds in rice [6–8]. However, the stability of synthetic apomictic rice remains to be tested by natural selection and validated in field conditions. In any case, the molecular triggers of apomixis in natural agamic complexes (i.e., species composed of diploid sexual cytotypes and polyploid apomictic counterparts) remain unknown. Moreover, the harnessing of the trait into plant breeding acquired an entire new dimension under the proposal that apomixis and sexuality might be ancient polyphenic phenotypes, with both pathways represented in all plant species, although many lineages have lost the capacity to shift from one to the other [9]. This hypothesis implies that sexual species (like major crops) could become apomictic by restoring the lost natural switch between both phenotypes, provided that the asexual route remained operational. In order to identify the natural triggers of apomixis, extensive reproductive characterization should be conducted in a high number of taxa, to extend our knowledge on sexual and asexual reproduction coevolution. Up to date, although apomixis was reported in at least 78 of the 460 angiosperm families [10–12], only a few species have been characterized from a molecular perspective. Moreover, less than 10% of the >350,000 flowering plant species have been examined using cytoembryological techniques, which suggests that new members previously assumed to be sexual might be added to the apomicts list soon [9]. Thus, there is a need to complete the information available on the cytoembryological, developmental, and molecular aspects of apomixis through the scrutiny of previously uncharted species, a task that should be ideally carried out by scientists from different countries, who have access and are familiarized with unique local materials collected from plant populations growing in their natural habitats.

Paspalum L. is one of the few genera in which sexual and asexual reproductive behaviors have been characterized side by side for more than 50 years. During this period, many approaches proved to be unsuccessful, while others offered significant advances. In this review, we aim at presenting the rationale supporting the work carried out in this particular genus, and, based on our previous experience, proposing a series of advisable steps that could help to explore the molecular control of the trait in other species, through genetic, genomic, and/or breeding approaches. We expect to favor the development of other research projects, in order to boost the investigation of this biologically amazing and complex field.

2. Cytoembryological and Cytogenetic Germplasm Characterization

While apomixis and sexuality frequently coexist within a species, sometimes within a single individual or even a single ovule (facultative apomicts), some genotypes reproduce mainly by apomixis (obligate apomicts) or only by sexuality (full sexuals). There is no possible way to carry out solid molecular comparisons of the sexual and apomictic developmental pathways in the absence of previous knowledge on the extent of apomictic reproduction in the particular species/genotype under study. Moreover, the cytological (e.g., the level of ploidy and meiotic behavior) and the embryological (e.g., the apomictic type and the temporal developmental progression of the trait) backgrounds must be carefully examined in order to select the appropriate plant material/time frame to be compared.

To provide materials for research and breeding purposes, an extensive living collection of *Paspalum* species was settled at IBONE CONICET, Corrientes, Argentina [13]. Cytoembryological and cytological analyses of these materials, covering 72 out of the total 370 species included in the genus, showed that approximately 75% of them are polyploid [14]. Besides, 68% showed some potential for apomixis [14]. Cytoembryological examination revealed that sexuality is represented by the double fertilization of reduced female gametophytes (FGs) of the *Polygonum* type, typically composed of an egg apparatus (one egg cell and two synergid cells), a large binucleated central cell, and a mass of proliferating antipodal cells at the chalazal pole [15]. Meanwhile, apomictic reproduction corresponds mainly to the aposporous type, with apospory initials (AIs) differentiating from companion nucellar cells and dividing mitotically to form non-reduced ESs, which occasionally coexist with a meiotic ES in the same ovule [16]. The *Paspalum* mature aposporous ESs usually contain an egg cell, one or two synergid cells, a binucleate central cell, and no antipodal cells [17–20]. *Hieracium*-type apospory (i.e., 8-nucleate non-reduced female gametophytes containing antipodal cells) was described in only two species: *Paspalum secans* Hitchc. and Chase [21] and *Paspalum simplex* Morong [22,23]. Moreover, *Taraxacum*-type diplospory (i.e., 8-nucleate non-reduced female gametophytes containing antipodals) was reported in three species: *Paspalum commersonii* Lam. (2n = 6x), *Paspalum longifolium* Roxb. (2n = 4x) and *Paspalum conjugatum* Berg. (2n = 4x) [24,25]. Finally, both aposporous and diplosporous types were detected together in *Paspalum minus* E. Fourn. [26] and *Paspalum scrobiculatum* L. [24].

In polyploid apomictic individuals, microsporogenesis is characterized by a statistically significant increase of asynapsis/desynapsis, lagged chromosomes, chromosome bridges, and micronuclei [16]. Such abnormalities were attributed to genetic rearrangements such as inversions or translocations affecting one particular chromosome [27,28]. Pagliarini et al. [29] reported the formation of non-reduced pollen in polyploid Brazilian accessions. Besides, the occurrence of restitution nuclei as a consequence of irregular or arrested meiosis has been associated with apomixis in *P. secans* [30], *P. conjugatum* [25], and *P. minus* [26].

All apomictic *Paspalum* species characterized so far are pseudogamous, i.e., they require the fertilization of the unreduced polar nuclei by a reduced male gamete to form the endosperm and a viable seed. The number of polar nuclei involved in pseudogamy is of practical importance in apomixis research, particularly when the mode of reproduction is determined by the flow cytometric seed screen (FCSS) method [31]. This methodology allows the identification of the origin of a seed (apomixis or sexuality) in large numbers of individuals by comparing embryo:endosperm DNA content ratios. While sexuality results in a 2:3 ratio, the outcome of apomixis usually differs. For instance, apomictic seeds of tetraploid pseudogamous *Paspalum notatum* show a 2:5 embryo:endosperm ratio (4x embryo and 4x + 4x + 2x = 10x endosperm). This method was successful during the last two decades for determining reproductive behaviors in individual or bulked seeds of several *Paspalum* species [23,32–38]. Another distinctive feature of seed development in *Paspalum* apomicts is the lack of deleterious response to deviations from the 2:1 maternal-to-paternal genomic ratio in the endosperm strictly required for sexual reproduction [39]. Typically, apomictic individuals form seeds with parental contributions showing a maternal genomic excess in the endosperm, e.g., 4:1 (8x:2x) in tetraploid apomicts [15].

Moreover, the study of the relationship between apomixis and ploidy is of importance when characterizing the biological materials to be used in apomixis research. For instance, according to the data reviewed by Ortiz et al. [14], 27 (37.5%) from 72 characterized *Paspalum* species showed multiple cytotypes with variable ploidy levels. Out of them, 22 (81.48%) displayed sexuality at lower ploidy levels and apomixis at higher ones. The exceptions were *Paspalum conspersum* Schrad. (only sexual individuals); *Paspalum distichum* L., *P. secans* (Itchc. & Chase), and *Paspalum proliferum* Arechav (only apomictic individuals); *P. scrobiculatum* L. (with sexuality reappearing at the highest ploidy level, i.e., 12x) [14]. Consequently, there is a positive yet not strict correlation between the increment of the ploidy level and the expression of apomixis. A flow cytometry estimation of the DNA content in the embryo and the endosperm in a seed-by-seed analysis of 77 *Paspalum* accessions allowed confirmation of the reproductive mode/ploidy level for several of the above-mentioned species and provided new information on the reproductive mode for 12 additional ones and one botanical variety [40]. Most apomictic *Paspalum* entities belong to multiploid species of autoploid origin. Each multiploid contains a sexual self-sterile diploid cytotype and a series of aposporous apomictic autopolyploid cytotypes, usually from 3x to 6x, with tetraploids as the most common cytotype. The two main *Paspalum* species used as models for apomixis research, *P. notatum* and *P. simplex*, form agamic complexes made up of diploid sexual and autopolyploid apomictic individuals [14].

Once a significant number of materials were subjected to cytogenetic and cytoembryological examinations, the next step to be followed is the selection of those species/genotypes in which sexual and apomictic development will be analyzed from the genetic and molecular points of view. In the case of *Paspalum*, the species that received most attention are *P. simplex*, *P. notatum*, *Paspalum rufum*, and *Paspalum guenoarum*. All of them grow as wild native populations in the argentine Litoral region, located within a limited radius around several local laboratories devoted to *Paspalum* apomixis research, which eases the collection and classification of novel germplasm. The genetic structure of some of these populations was carefully characterized, and sexual diploid individuals were proposed as the main genetic variability source, since they co-habit and hybridize with polyploid individuals [34,41,42]. Additionally, the selected species show sizes of ovaries and anthers that allow a relatively simple dissection with scalpels and forceps under a magnifying spectroscope when conducting experimental crosses. Moreover, they (as most *Paspalum* species) are perennial and flourish during the warm season, are well represented in rangelands used for cattle production systems, and cultivated within and without the natural distribution region to be used as forage, turf, and cereal [13]. Finally, besides the necessary considerations on germplasm availability, plant anatomy, life cycle, and agronomic potential, the species choice criteria might include the size of the genome. In this regard, the 1C DNA content values of the selected *Paspalum* species are among the smallest within the Poaceae tribe, ranging from 0.550 pg in *P. notatum* to 0.900 pg in *P. guenoarum* [40].

Chiefly among the points deserving consideration in the selection of the species for molecular analysis is the availability of natural or induced materials of identical ploidy and contrasting reproductive modes. These resources are essential for the subsequent establishment of useful populations segregating for apomixis, which will be used in inheritance analysis, genetic/epigenetic mapping, omics surveys, and breeding. Moreover, the comparison of sexual and apomictic developmental pathways using genotypes of different ploidy levels is directly not advised, simply because ploidy may affect expression levels independently of reproductive behavior at numerous loci [43,44]. Regarding this, a preliminary gene expression comparison between a *P. notatum* sexual diploid genotype and a newly formed sexual autotetraploid derivative revealed that at least 0.49% of the pre-meiotic floral transcripts changed their relative expression after an increment of ploidy [45]. According to these numbers and considering the existence of near 70,000 transcripts expressed in *Paspalum* flowers, as was revealed recently from RNAseq experiments [46], ploidy-related differential expression would involve around 3500 transcripts, a quantity equivalent to that detected in the sexual vs. apomictic reproduction comparisons [46]. These estimations agree with results reported by De Oliveira et al. [47], who selected 28,969 transcripts that were common to 2x sexual, 4x sexual, and 4x apomictic genotypes, and found 1173 of them differentially expressed between 2x sexual vs. 4x sexual plants, while 1317 were contrastingly represented between 4x apomictic and 4x sexual plants. This confirmed that ploidy-related polymorphic expression might severely interfere the unequivocal identification of transcripts related to the reproductive mode.

In several species of *Paspalum*, apomixis occurs mainly at polyploid levels and sexual counterparts of the same ploidy are not available in nature. Therefore, colchicine treatments were required to duplicate the chromosome content of sexual diploids [33,48–52]. Alternatively, colchicine-induced facultative apomictic polyploids were crossed to apomicts in order to produce full sexual descendants [53]. Comparative genetic characterization of these recent artificial polyploids and the original diploids in *Paspalum plicatulum* and *P. notatum* revealed that, immediately after polyploidization, genetic rearrangements affected 28–38% and 9.55% of the genome, respectively [54,55]. Moreover, ancestral alleles lost after polyploidization were spontaneously recovered in further generations, a phenomenon previously reported in species of other genera [55]. Once the alleged sexual polyploids were obtained at the laboratory, their reproductive phenotypes might be controlled periodically, since one cannot rule out that plants showing high levels of sexuality might express some capacity for apomixis in a different environmental/seasonal condition (and vice versa), as described for *Eragrostis curvula* [56] and *P. notatum* [36]. It should be noted that sexual mother plants used to produce families segregating for the reproductive mode are expected to lack the genomic region/s determining apomixis and, therefore, any capacity for asexual reproduction. Artificial sexual materials of the proper ploidy must be repeatedly and extensively checked by cytoembryological and/or molecular progeny tests [50,53,57].

Finally, comparative transcriptomic and epigenetic surveys require the establishment of detailed timelines on reproductive development (i.e., timeframes in which the subsequent reproductive developmental stages were unambiguously demarked), to maximize the discovery of true differential representation results. Several reproductive calendars were established by correlating morphological features, e.g., ovary/ovule size and morphology, as well as mega- and microgametophyte developmental stages in *P. simplex* [49], *P. notatum* [58], and *P. rufum* [59].

In summary, as schematized in Figure 1, preliminary germplasm characterization includes a series of successive obligatory steps leading to the selection/generation of plants with appropriate reproductive phenotype/genetic backgrounds to be used in genetic mapping, (epi)genomic and transcriptomic assays, together with the setting of precise time frames to standardize the collection of RNA/DNA samples.

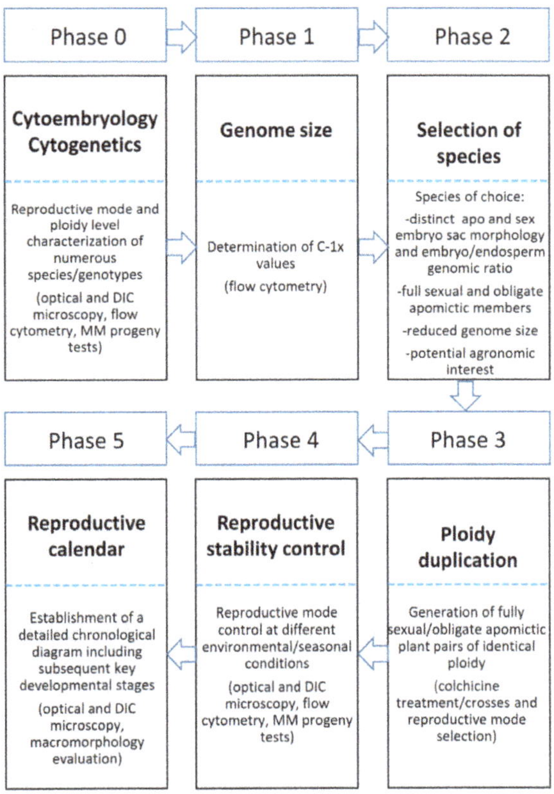

Figure 1. Consecutive phases of plant material preparation. In a clockwise sense, the series of phases that should be considered to produce germplasm suitable for apomixis molecular characterization. Methods/techniques are listed at the bottom of each column, within brackets. MM: molecular markers.

3. The Genetic Control of Apomixis in *Paspalum*

Understanding and exploiting the advantages apomixis represents for plant breeding requires detailed knowledge of the genetic and molecular bases that control both its inheritance and expressivity. Although genes governing specific reproductive steps have already been engineered into plants and artificial systems can mimic clonal reproduction [60–65], the whole apomictic program has not been reproduced in sexual models yet [9]. This is the consequence of a complex genetic and/or epigenetic control of the trait, far more intricate than initial inheritance studies had suggested.

In *Paspalum*, inheritance analyses were unfeasible until fully sexual tetraploid individuals were available [14]. Indeed, initial attempts suggesting a control by a few recessive genes [48] could not be further confirmed [66]. Towards the end of the 90s, however, sexual tetraploid individuals were experimentally generated for *P. notatum* [57], *P. simplex* [49] and, more recently, for *P. plicatulum* [33]. These genotypes became the female progenitors of segregating populations used in apomixis inheritance genetic dissection and breeding programs. Currently, a large number of well-characterized sexual tetraploid individuals, generated by chromosome doubling treatments (sometimes followed by experimental crosses) are available for the three species [33,50,51,53,67,68]. Moreover, several sexual genotypes of *P. simplex* and *P. plicatulum* were also used as female parents in interspecific crosses with different members of the *Anachyris* subgenus and the Plicatula group, respectively, widening the genetic studies to other species of the genus [60–70].

Genetic analyses of F_1 and BC_1 *P. notatum* progenies showed that apospory is inherited as a simple dominant trait (Aaaa). However, distorted segregations were often observed in favor of sexuality (aaaa) with apomixis:sexuality ratios varying from 1:1 (expected) up to 1:9 [27,71–75]. In the case of *P. simplex*, the segregation evidence indicates that apomixis is under the control of a single dominant allele, whose transmission is biased towards sexuality, while mapping reveals an association between the Apomixis Controlling Locus (ACL) and the telomeric region of the long arm of rice chromosome 12 [23,76]. Inheritance studies in species of the Plicatula group using F_1 and F_2 generations and three backcross populations derived from an artificial tetraploid sexual clone of *P. plicatulum* and an apomictic *P. guenoarum* cv. Rojas plant confirmed the previous genetic model, with a single Mendelian dominant factor with altered transmission (1 apo:1.6 sex) [67]. Interestingly, high rates of distortion against apomixis were also found in interspecific crosses involving *P. simplex* x *Paspalum malacophyllum* (1:5) [77] and *P. simplex* x *Paspalum procurrens* (1:15.7) [69]. Moreover, distortion of segregation has been reported in several grass and non-grass apomictic systems [78]. The large chromosome rearrangement (inversion or translocation) associated with the apomixis locus reported by Podio et al. [28] is to date the most relevant evidence to provide a mechanistic explanation for this phenomenon in *P. notatum*.

All genetic analyses in *Paspalum* showed a strict co-segregation between apospory, parthenogenesis, and unbalanced endosperm formation, suggesting that their relevant genetic determinants are located in the same low recombining chromosome area [23,67,79]. The parthenogenesis of reduced egg cells in the genus was never documented. Moreover, although sexual seeds require a strict 2:1 maternal:paternal genome contribution to form the endosperm, viable seeds from apomicts usually form despite unbalanced parental genomic contributions [15].

Another central topic demanding special attention is the transmission of the apomixis expressivity capacity. Usually, apomictic individuals are classified as obligate when almost all ovules (>90%) show one or several aposporous embryo sacs (AESs), or facultative, when at least some ovules bear meiotic or mixed (i.e., meiotic and aposporous) female gametophytes [14]. In *P. notatum*, the analysis of several segregating populations generated using an apomictic male progenitor revealed a high variability in the proportion of ovules bearing AESs among the hybrids [72,74]. In many cases, two well-differentiated groups of apomictic hybrids with low and high levels of apospory, respectively, were recovered from the progenies [74,75]. Moreover, only a small fraction (less than 10%) of them showed an apospory expressivity as high as that detected in the apomictic male progenitor, and this proportion decreased in subsequent crosses with sexual genotypes [72]. At least in *P. notatum*, apospory expressivity can also show seasonal variation, reaching its maximum during peak flowering (summer) and decreasing in fall [80]. A similar variation depending on the environmental conditions was reported in *Paspalum cromyorrhizon* [81]. Although the parental genetic distance positively correlates with the number of aposporous hybrids detected in the progeny (i.e., plants showing at least one ovule containing one or more AES, which also produce apomixis-associated molecular marker polymorphic bands), it does not display any association with the expressivity of apospory (i.e., the average proportion of ovules containing at least one non-reduced female gametophyte), suggesting a separate control for these two reproductive components [75]. The above-mentioned body of evidence suggests that genetic or epigenetic factors may be affecting the expressivity of the trait. Moreover, embryo parthenogenesis and endosperm development capacities are both variable: plants with well-developed AES in >90% of ovules cannot produce viable apomictic seeds [82]. Parthenogenesis seems to be under epigenetic control in *P. simplex*, as artificial wide-genome demethylation significantly reduced parthenogenesis but had no effect on apospory [83]. Therefore, the formation of viable apomictic seed relies not only on the presence of the ACL, but also on as-yet unknown modifiers potentially affecting apomeiosis, parthenogenesis, endosperm development, and germination. In this regard, a precise evaluation of the apomixis capacity for a given material entails the use of a combination of experimental approaches, including mature female gametophyte examination (apospory capacity), flow cytometry (seed development), and progeny tests (germination and establishment of apomixis-derived progenies) [13].

It is a well-known fact that many *Paspalum* species form multiploid complexes composed of self-sterile sexual diploid and self-fertile pseudogamous apomictic polyploid cytotypes [14]. However, this general rule is challenged by several observations in *P. notatum* and *P. rufum*, including aposporous-like embryo sacs (AES-like) reported at the diploid level [32,50,84]; seed formation from aposporous sacs of diploid cytotypes pollinated using tetraploid progenitors [32,52]; artificial apomictic tetraploid plants emerging from colchicine-induced chromosome duplication in sexual diploid individuals [50,52]; increased apospory expressivity at the diploid level after hybridization [52]. Moreover, a SCAR marker derived from the *P. notatum* ACL was detected in the genome of 10 out of 57 diploid plants from a natural population (Juan Pablo A. Ortiz, personal communication). These results support the hypothesis that the factor/s controlling apomixis might occasionally occur in diploid individuals, but they remain silent until they are activated in response to hybridization and/or polyploidization.

The *Paspalum* ACL, although quite simple in genetic terms, might be physically complex at the molecular level, as it was exposed by associating it to grass model species markers of defined map positions [69,76,77,85,86]. The apomixis-governing region is characterized by recombination restriction, hemizygosity, and heavy cytosine methylation (Figure 2) [27,77,79,83,86]. Comparative mapping showed syntenic relationships between the apomixis loci of at least four *Paspalum* species (*P. simplex*, *P. notatum*, *P. procurrens* and *P. malacophyllum*) and a 6–10 Mb region of rice chromosome 12, as well as a segment of 10 Mb of rice chromosome 2 (for *P. notatum* only) [77,86]. Although the ACL was conserved over generations within single species, small variations (short indels) were detected among species of the same section. More consistent changes occurred among species of different sections. In particular, members of the *Anachyris* subgenus show macrosynteny with the above-mentioned segment of rice chromosome 12 [69,77], whereas the ACL of *P. notatum* is a hybrid syntenic group resembling segments of chromosomes 12 and 2 of rice [77,86].

A deeper analysis of the ACL structure arose from the availability of apomixis-specific *P. simplex* bacterial artificial chromosomes (BACs) isolated with the aid of specific primers. Partial sequencing of one of them, corresponding to the ACL, revealed that although synteny at the marker level was conserved with respect to the rice genome, gene micro-collinearity was frequently interrupted with transposable elements and migrant genes from other rice chromosomes [87]. More recently, Galla et al. [88] sequenced two other apomixis-linked BACs and discovered that the region of synteny of the *P. simplex* ACL is conserved among five reference grass species, being located at a telomeric position in chromosomes 12, 8, 3, and 4 of rice, *Sorghum*, *Setaria*, and *Brachypodium*, respectively, and in a more centromeric region of maize chromosome 1 (Figure 3). Based on these findings, it was hypothesized that the ACL of *Paspalum* could have originated from an ancestral unstable genome region in which (i) sex-related genes were grouped by gene migration during speciation, (ii) a polyploidization event (through an intermediate triploid bridge) locally induced further small-scale rearrangements that, in turn, generated local sequence divergence, lack of chromosome pairing, and recombination blocking. This non-recombinant segment favored the accumulation of mutations, since they could not be discarded by meiosis. This kind of evolutionary and structural organization may have consequences on gene content and expression. Indeed, Polegri et al. [89] noticed that apomixis-linked genes tend to be expressed in a constitutive mode throughout reproductive development. Moreover, other apomixis linked genes are specifically expressed in germ line committed cells, i.e., nucellus cells originating apospory initials, polar nuclei, and egg cells [88,90].

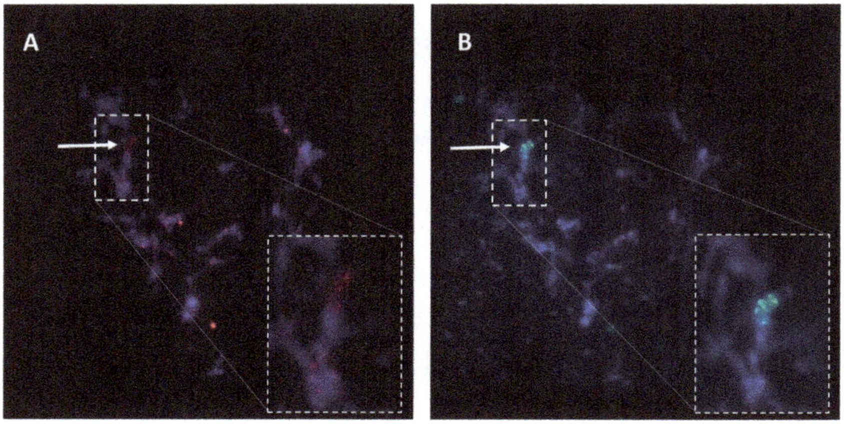

Figure 2. Methylcytosine immunodetection (**A**) and Apomixis Controlling Locus (ACL)-specific bacterial artificial chromosome fluorescence in situ hybridization (BAC-FISH) (**B**) on 4′,6-diamidino-2-phenylindole (DAPI)-stained pachytene chromosomes of apomictic *P. notatum* genotype Q4117. Arrows indicate an immunodetected heterochromatin knob (red dots) (A) and the ACL as revealed by BAC-FISH (green dots) (B) located at the same position. The region of interest was increased in order to show the details (bottom right corner). Maricel Podio (IICAR, CONICET-UNR, Argentina) kindly provided the images.

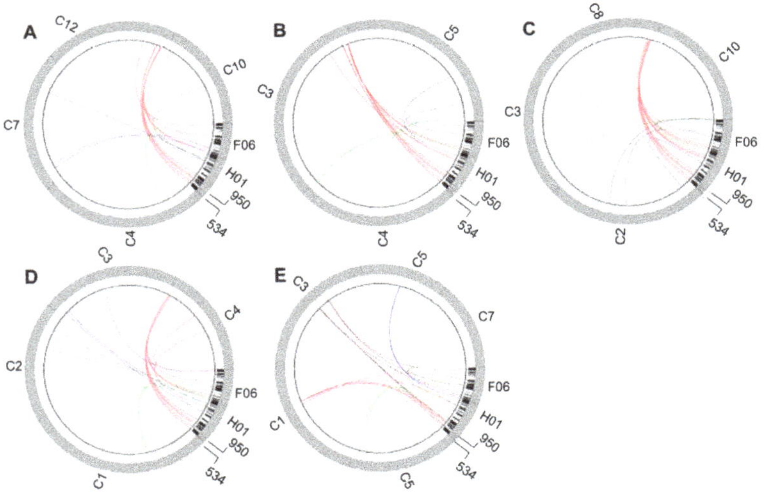

Figure 3. Mapping of genes located in four apomixis-linked BACs onto reference grass genomes, including *Oryza sativa* (**A**), *Setaria italica* (**B**), *Sorghum bicolor* (**C**), *Brachypodium distachyon* (**D**), and *Zea mays* (**E**). F06, H01, 950, and 534 correspond to contigs PS127F6_c1, PS366H1_c1, H10_950, and H10_534, respectively. C1-C12 represent chromosome numbers for each reference grass genome. Red lines link genes to the conserved chromosome area related to apomixis. Dr. Giulio Galla (DAFNAE, University of Padua, Italy) kindly provided the image.

The expression of the ACL genes seems to have common features with operon-like gene clusters, which are defined as a set of two or more non-homologous functionally related genes that share a close genomic neighborhood [91]. Operons in plants probably originated by an initial event of gene duplication followed by neo-functionalization [92]. Similarly to the ACL, the operon like gene clusters originated from subtelomeric dynamic regions and are characterized by high rates of gene rearrangements [93]. Another multigene complex that shows striking similarity with the ACL of *Paspalum* is that related to the Y-chromosome of dioecious plants, which originated from autosomal chromosomes by initial suppression of recombination in the regions containing the sex controlling genes and later on, by the migration of male determining genes [94]. From a functional point of view, the evolution of the Y-chromosome induces both the silencing of the female genes, and the development of male function by the action of specific genes [94]. This finding could support the hypothesis of apomixis silencing key genes of sexuality. Podio et al. [83] showed that parthenogenesis in *P. simplex* is superimposed over fertilization-mediated embryo development by a mechanism controlled by DNA methylation. Furthermore, several *P. simplex* apomixis-linked genes expressed sense and antisense transcripts in reproductively committed cells, and showed a putative silencing effect of the apomixis-linked alleles on their sexual-specific counterparts [88,90].

4. Identification of Candidate Genes through Transcriptome Comparisons

In addition to the identification of key triggering factors, our knowledge on apomixis should be complemented by the understanding of the molecular pathways involved in subsequent developmental steps. Although genetic analyses suggest that apomixis in *Paspalum* spp. is under the control of a single genomic region, the intrinsic characteristics of the ACL (i.e., absence of recombination within a large chromosomal segment and abundance of repetitive sequences) hampers the identification of the trait's key determinants through positional mapping strategies. In this scenario, a combination of transcriptomic and genetic/genomic information could provide hints on the primary or secondary role of DETs and be the method of choice for identifying genes and molecular routes involved in both activation and progression of the trait.

However, several concerns emerge for comparative transcriptomic analyses using apomictic/sexual systems: (i) the genetic nature of the materials, since comparisons usually involve polyploid, highly heterozygous individuals, which complicates the distinction between differential expression and genetic polymorphisms in PCR-based analyses as well as the classification of homologs (orthologs, paralogs) in RNAseq approaches, in particular if no genome sequence is available; (ii) the sample collection timing, because the rapid progression of the RNA landscape along with reproductive development might cause the frequent emergence of false DETs; (iii) the short- and long-term transcriptome responses after hybridization and polyploidization, since some of the materials used might have recently originated from colchicine treatments, while others emerged from natural whole genome duplication events in either ancient or relatively recent times [42]. Unraveling these major drawbacks requires consideration of way out strategies like, for instance, the use of bulked segregant analyses involving several apomictic and sexual segregating offsprings, differential expression validation in numerous individuals by qPCR and/or in situ hybridization, standardization of sample collection protocols, RNAseq technical and biological replication, as well as attenuation of circadian and environmental effects by collecting samples at defined daytimes and/or conditions. After identification of apomixis candidates, their genomic locations must be recognized, to determine if they are being transcribed from the ACL or anywhere else in the genome. In the first case, the candidate could potentially be one of the triggers of the apomictic pathway. Otherwise, it could be part of the downstream molecular cascade involved in asexual reproductive development. To unequivocally map the sequence of interest, relatively large family populations segregating for the mode of reproduction and/or a reference genome where the ACL has been located should be available.

Transcriptomic surveys carried out in *P. notatum* since the early 2000s produced a large list of genes whose expression correlate with the occurrence of key reproductive features. An initial comparative examination of bulked RNAs from spikelets of sexual and apomictic F_1 hybrids at anthesis was carried out by differential display and led to the identification of one transcript (*ARP1*) encoding an KSP consensus domain previously detected in several *cdc* 2-regulated cytoskeletal proteins [95]. A second transcriptome analysis using the same methodology revealed 65 new transcripts differentially expressed at the premeiosis/meiosis stage [58], including members of the *LORELEI* family (*GAP1*) [58,96], the S-adenosyl-L-methionine-dependent methyltransferase family (*TRIMETHYLGUANOSINE SYNTHASE1*) [58,97]; the *MAP3K YODA* family (*QUI-GON JINN* [58,98] as well as several retrotransposons carrying transduplicated gene segments [58,99], sometimes involving apomixis-associated genes like *SOMATIC EMBRYOGENESIS RECEPTOR-LIKE KINASE* (*SERK*) [58,99,100], and long-noncoding transcripts possibly involved in splicing regulation [101]. Furthermore, differential expression analyses during early seed development (3–24 h after pollination) revealed ≈100 DETs possibly associated with the unbalanced genomic contribution found in pseudogamous endosperms (4:1), including transcripts related with transcription regulation, signal transduction (e.g., lectin-like protein kinase and CK2 protein kinase α 1), growth/division, and response to changes in the extracellular ATP levels [102,103].

In *P. simplex* cv. Morong, the use of cDNA-AFLP on RNA extracted from flowers at several developmental stages, from premeiosis to 3–6 days after anthesis, rendered 202 DETs, the majority of which were present in apomictic florets only [89]. Near 20 of them, mostly related to signal transduction and nucleic acid binding, mapped within the ACL and showed constitutive expression in apomictic plants. Interestingly, the majority of these sequences displayed nonsense and frameshift mutations, revealing a probable pseudogene nature [89]. The remaining transcripts, transcribed from non-ACL genomic regions, mostly showed regulatory and seed storage functions. Several of the *P. simplex* DETs belong to the same annotation classes of those reported for *P. notatum*, including extensins, YODA-like MAP3Ks, LRR-like proteins, transferase proteins, and retrotransposon proteins.

In the last decade, progress in DNA and RNA sequencing have provided biologists powerful tools to study and understand gene functions and interactions. However, the establishment of genomes/transcriptomes from genetically poorly characterized, polyploid, and highly heterozygous species might easily result in chimeric assemblies and/or fragmented transcripts [104–106]. Therefore, robust reference floral transcriptomes were initially assembled from sexual and apomictic tetraploid *P. notatum* spikelets by using the long-read Roche 454-FLX + technology [46]. Out of these reference transcriptomes (48,842 genes identified), a preliminary list of 3732 sexual vs. apomictic DETs was generated, revealing several molecular networks putatively altered during apomixis, mainly related to ribonucleotide metabolic processes, protein complex biogenesis and assembly, monosaccharide catabolism, translation, gene expression, proteolysis, protein transport, DNA replication, and regulation of RAS activity [46]. Since these reference transcriptomes were constructed from long reads (mean sequence length around 500 bp), they provided a solid frame to establish future short reads-based assemblies and allowed recovering of putative alleles/paralogs full sequences for thousands of genes.

While long-read Roche 454-FLX + sequencing produced genuine transcript assemblies, the derived quantitative expression comparisons lacked statistical accuracy (e.g., no technical replicas were established). Recently, De Oliveira et al. [47] reported a global gene expression analysis in *P. notatum* using Illumina sequencing of RNA samples extracted from leave and florets of 2x sexual, 4x sexual, and 4x apomictic genotypes. The database contains 114,306 reference transcripts, 536 of which correspond to genes possibly associated with apomixis. Interestingly, 89 differentially expressed transcripts mapped onto rice chromosome regions syntenic to the ACL [47]. Moreover, to provide a more comprehensive view of the sense/antisense transcriptomic landscapes emerging during reproduction in *P. notatum*, we generated Illumina TruSeq floral RNA libraries in triplicate, from sexual and apomictic materials collected at four different developmental stages, from premeiosis to anthesis (NCBI repository SRA accession: PRJNA511813). Data analysis (read assembly, comparative quantification, and predictive network interactions) is on-going. Figure 4 shows an example of a network possibly involved in apomixis, identified using the String v11 software [107].

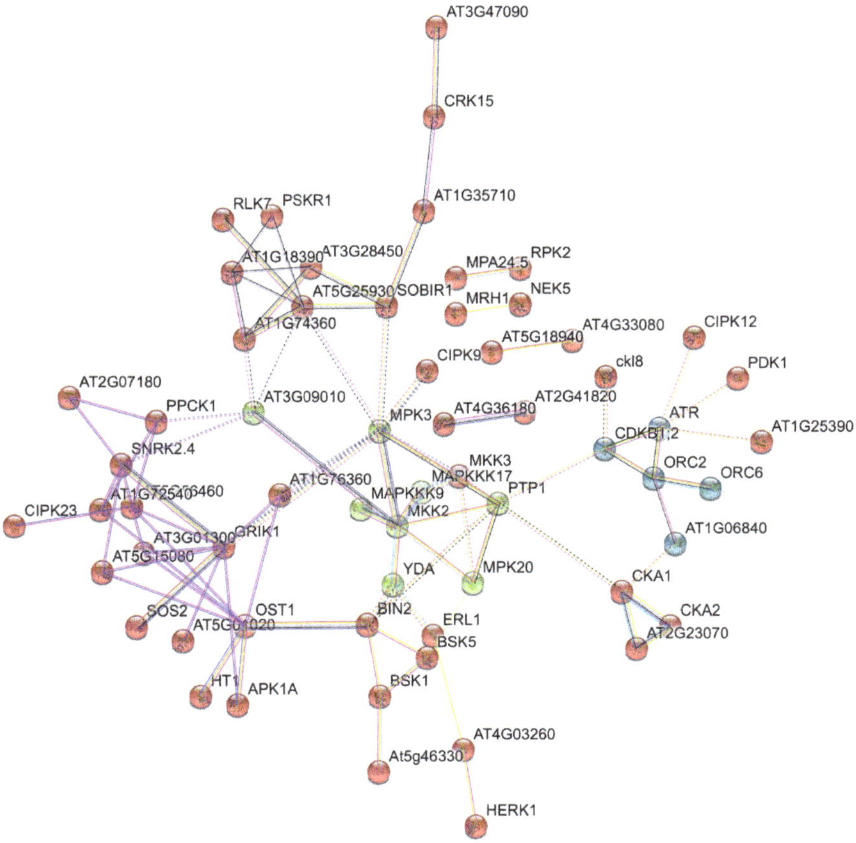

Figure 4. Gene network interactions controlling apomixis development. The scheme shows a particular group of apomixis-related predicted proteins and their functional interactions. The central node, MPK3 (mitogen-activated protein kinase 3), is encoded by the putative *Arabidopsis* ortholog to *QUI-GON JINN*, and was functionally related with apomixis by Mancini et al. [98] since its expression in the nucellus is necessary for aposporous embryo sacs (AESs) formation. MPK3 interacts with CIPK9 (CBL-interacting serine/threonine-protein kinase 9), a group of other MPK proteins (MKK2, MAPKKK17, MAPKKK9, and YDA), as well as to a homologous to AT1G76360 (a protein serine/threonine kinase) and AT5G25930 (protein kinase family protein with leucine-rich repeat domain). Nodes: query proteins and first shell of interactors with some known or predicted 3D structure. Edges: light blue: from curated databases; magenta: experimentally determined; green: gene neighborhood; red: gene fusions; blue: gene co-occurrence; yellow: text mining; black: coexpression; grey: protein homology.

Finally, the short-sequence components of the *P. notatum* sexual and apomictic floral transcriptomes (including siRNAs and miRNAs) and their possible target genes were characterized in detail [108]. A total of 1525 transcripts showed differential sRNA representation between sexual and apomictic plants, including genes related to meiosis, plant hormone signaling, biomolecules transport, transcription control, and cell cycle. Forty miRNAs precursors corresponding to conserved families and eight novel entities were identified. From them, 56 precursors showed miRNA differential representation between apomictic and sexual plants, always displaying upregulation in the apomictic sample. An analysis of the miRNA possible targets revealed 374 sequences, among which auxin-related genes were prevalent [108], opening a new line of research on the role of phytohormones in the switch from sexuality to apomixis.

5. Genomic Resources: The Lost Continent of Apomixis Research

As mentioned in previous sections, the mechanisms for establishing apomictic lineages from sexual ancestors and the evolutionary consequences of their emergence remain poorly understood and largely speculative. Mining the large body of knowledge gained in apomictic species, including *Paspalum*, points out hypotheses as diverse as polyploidization and hybridization [52,109,110], inactivation of epigenetic silencing pathways [111,112], oxidative stress during meiosis [113], functional trans-acting roles for a highly heterochromatic, hemizygous genomic region specific of apomictic plant genomes [90,98,114], miRNA deregulation [108,115], alterations in RNA splicing machinery [97], and hormonal signaling [116]. Although functional analyses in model species, including *Arabidopsis* and rice, have already provided valuable information regarding the role of some candidate genes, their positioning into a comprehensive genetic network controlling apomixis remains elusive to date.

Furthermore, apomictic plants usually have polyploid genomes, or possess aneuploid chromosome complements of highly heterozygous nature. Genetic approaches have revealed a singular genomic organization, characterized by a large, hemizygous and non-recombinant region of heterochromatic nature named Apomixis Controlling Region (ACR) in *Pennisetum* [78] or Apomixis Controlling Locus (ACL) in *Paspalum* [87]. Such genetic behavior and molecular features occur rarely in nature and, as mentioned before, are amazingly reminiscent of sex chromosome evolution, which involves the selection for suppression of recombination in a specific chromosomal region, followed by a massive rearrangement characterized by repetitive sequences accumulation, gene silencing, and loss of function [117–119]. Determining the genetic variation functionally relevant to build apomictic developments as well as the effects that in return may have shaped the architecture of apomictic genomes remain a key focus in apomixis research. This challenge requires the establishment of research platforms accommodating biological information and molecular resources for the genomes found in agamic complexes. Unfortunately, albeit advances in molecular and acid nucleic sequencing technologies and bioinformatics have widened genomics and transcriptomics approaches at an unprecedented scale [120,121], assembling the genome of an apomict remains a difficult task, because of complications rooted in their biology, as shown above.

To date, a few genomes of sexual relatives of apomictic plants have been sequenced, including *Boechera retrofracta* [122] and *E. curvula* [123], providing valuable information on genome evolution and relationships within agamic complexes. The vast amount of botanical, phylogenetical, cytoembryological, and transcriptomic information collected from numerous *Paspalum* species and agamic complexes [14,46,108] makes this genus a unique model system to explore the interplay between genome evolution and the emergence of apomictic reproduction in plants. *Paspalum* genomes are relatively small, as 2C values range from 0.5 to 6.5 pg [40]. Interestingly, in-depth characterization of the *Paspalum* ACL has revealed various extents of synteny with the subtelomeric part of the rice chromosome 12 long-arm [40,88], and shows structural features of heterochromatin [86,87]. Recently, we have initiated long-read sequencing and optical mapping approaches for generating genome assemblies of sexual and apomictic plants from several *Paspalum* species. Once assembled and annotated, we expect to molecularly resolve the ACL and determine its evolutionary trends; to re-analyze *Paspalum* transcriptomic resources, and to generate and compare epigenomes of sexual and apomictic plants. Finally, this effort will generate spillovers of great interest for forage grass breeding by providing critical genomic resources and allowing efficient molecular breeding, an important issue considering the need to mitigate both climatic change and anthropic pressures in economically important grassland and pasture agro-ecosystems.

6. Functional Analysis of Apomixis-Related Candidate Genes

After identifying candidate genes, the next rational step to advance apomixis research further consists of the establishment of functional analyses aimed at investigating their capacity to activate at least some steps of the trait. Those transcripts displaying differential regulation in apomictic/sexual backgrounds can be readily investigated through reverse genetic approaches, by using defective mutants or sense/antisense/RNAi transformants, since their down- or upregulation within the required genetic background would allegedly cause the emergence of apomixis-related phenotypes. Unfortunately, the absence of wide range germplasm banks for natural apomictic species forces either the prediction of orthology in model species or the adjustment of gene transformation techniques.

Based on these considerations, the steps to be followed when launching the functional characterization of an apomixis candidate gene in non-model apomictic/sexual systems like *Paspalum* are as follows: (1) previous spatio-temporal and sense/antisense expression validation of candidate genes with a protocol allowing cell-specific expression detection and distinction of the expressed RNA strand, like in situ hybridization; and (2) generation of overexpression/downregulation transformants with an efficient transformation platform, or identification of mutant germplasm involving model species orthologs. While the first step (spatio-temporal plus sense/antisense validation) will provide information useful for planning modulation (up- vs. downregulation), selecting the required reproductive background (sexual or apomictic) and guiding the promoter choice, the second one (mutant/transformant germplasm identification or generation) will facilitate the analysis of possible reproductive phenotype alterations.

A point that deserves special attention is that, at least for many candidate genes, wide-range expression is associated with different functions in a multiplicity of organs/developmental stages, and only a subtle, timely, and cell-specific activity change might produce a deviation in the reproductive mode without causing collateral, sometimes detrimental, consequences. Therefore, when intending transformation, gene promoters of choice might be preferably time and organ/cell-type specific. Even if basic research could take advantages from additional information derived from constitutive promoter-based functional analysis, due to its potential to reveal the candidate function in numerous organs/tissues, the identification and effective use of cell-specific promoters is no doubt a bottleneck for apomixis-based breeding, which requires the modification of a restricted function set. If the project aims at generating a methodology to induce apomixis for breeding purposes, it should contemplate the identification, isolation, and/or validation of appropriate cell-specific promoters in the species of interest. On the contrary, when dealing with functional characterization of the gene, a constitutive promoter with verified expression in the ovule could be of choice, unless it causes unviable phenotypes due to deleterious side effects.

In connection with this, in situ RNA hybridization protocols were developed for *Paspalum*, and gradually optimized to analyze the spatio-temporal plus sense/antisense expression distribution within the ovule [38,58,90,96–101]. An example of expression characterization in reproductive tissues at early megasporogenesis involving the apomixis candidate gene ORC3 is shown in Figure 5.

Regarding the development of protocols for modulating the expression of candidate genes in natural apomictic systems, Mancini et al. [124] considered several previously developed methodologies [125–127] as the starting point to examine alternative explants/conditions for biolistic transformation, and designed a platform best suited to a wide range of *Paspalum* genotypes. Such methodology is currently being used to produce *Paspalum* lines with up- or downregulated expression of apomixis candidates, which are later subjected to reproductive phenotype analyses. An example of the cytoembryological characterization of an antisense transformant line with downregulated expression of the apospory-inducer candidate gene *QGJ* is shown in Figure 6. Moreover, the potential of cell-type specific *Paspalum* promoters identified from genome sequencing is currently under analysis (S. Pessino, unpublished). All these methods have allowed the functional characterization of three reproductive candidate genes (*ORC3*, *QGJ*, and *TGS1*) in species of *Paspalum* (see descriptions below), and several others are undergoing the same process.

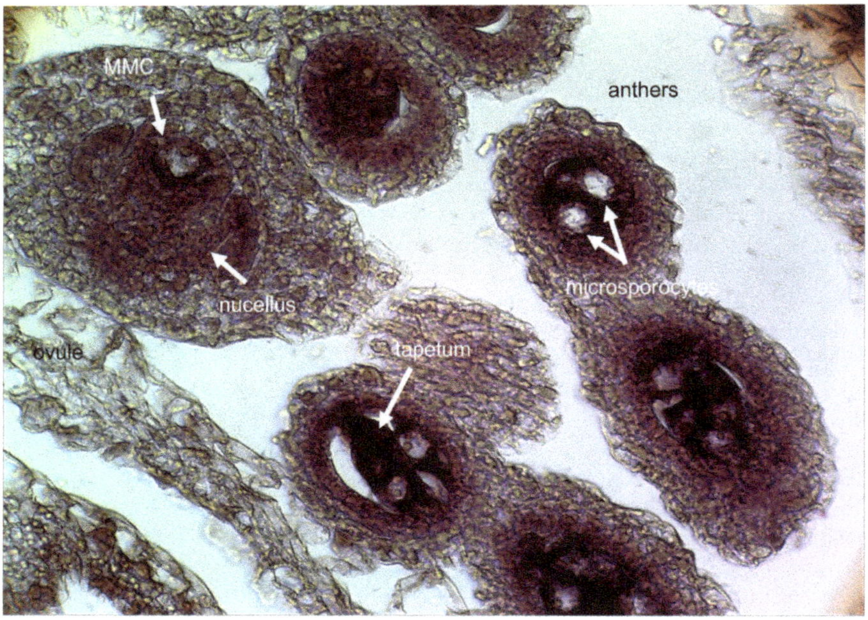

Figure 5. In situ hybridization of gene *ORC3* in reproductive tissues of *P. simplex*. Hybridization pattern of *PsORC3* (sense probe) in ovules and anthers of an apomictic genotype before the onset of meiosis. A strong hybridization signal is visible in nucellar and tapetum cells. The megaspore mother cell (MMC) does not express the target transcript. Dr. Lorena Siena (IICAR, CONICET-UNR, Argentina) kindly provided the image.

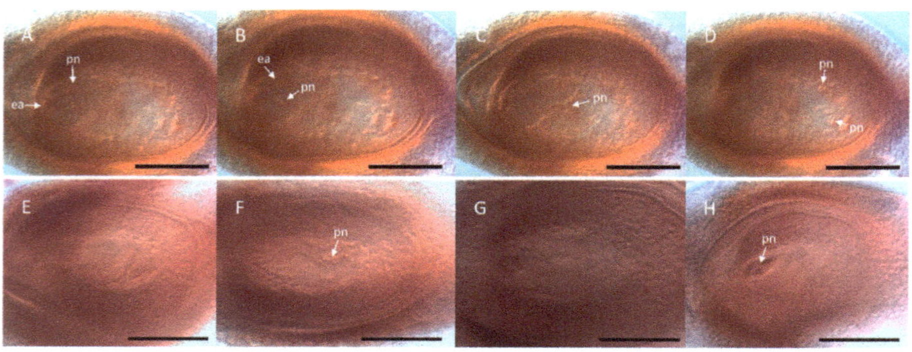

Figure 6. Cytoembryological analysis of RNAi lines with defective expression of the apospory candidate gene *QUI-GON JINN* (*QGJ*). The *QGJ* downregulation inhibits the formation of AESs in obligate apomictic plants [98]. (**A**–**D**) DIC consecutive focal planes of a single ovule originated from an obligate apomictic control, at anthesis, showing at least five typical supernumerary mature aposporous embryo sacs. (**E**–**H**) DIC single focal planes of a *QUI-GON JINN* defective RNAi line. Each image corresponds to a different ovary of the same plant, at anthesis. All of them (**E**–**H**) display abnormal morphology. pn: polar nuclei. ea: egg apparatus. Black bars: 100 μm.

6.1. PsORC3

A particular copy of the gene *ORC3* (*ORIGIN OF RECOGNITION COMPLEX 3*), identified as *PsORC3a*, resulted genetically linked to apomixis in all *Paspalum* spp. for which segregating populations were available [90,128]. In *P. simplex*, *ORC3* exists as three different copies, of which *PsORC3a* is a pseudogene specific for apomicts expressing an RNA transcript unlikely to be translated in a functional protein, whereas *PsORC3b*, probably coding for a highly conserved functional protein, together with *PsORC3c*, coding for a truncated protein, are common to both apomictic and sexual plants [90]. *PsORC3a* is poorly and constitutively expressed at all developmental stages in apomictic flowers only. In situ analysis showed that, in apomictic plants, sense and antisense strands of *PsORC3* are represented in cells and nuclei committed to reproduction (i.e., polar nuclei and egg cells), whereas both transcripts are silenced in the endosperm. Conversely, in sexual plants this gene is expressed as a sense transcript in the egg cell, polar nuclei, and endosperm, but not in the embryo [90]. Reverse genetics in both *Arabidopsis* and rice showed that *ORC3* defective genotypes display normal gametophyte development, but endosperm/embryo arrest at early stages of development [90]. Based on these considerations, we argued that the effect of the regulation of this gene on apomixis should be related to endosperm development. Particularly, we hypothesized that the apomixis-linked copy of *PsORC3* (*PsORC3a*) could be involved in a relaxation of control mechanisms, which allow endosperm development even facing a maternal genome contribution excess [90].

6.2. *QGJ*

The transcriptomic surveys carried out by Laspina et al. [58] identified a DET homologous to a mitogen-activated protein kinase kinase kinase gene (N46). Apomictic and sexual *P. notatum* Roche 454-FLX + floral transcriptomes were used to recover N46 full cDNA sequences and carry out a molecular phylogenetic analysis [98]. N46 was classified as a member of the *YODA* MAP3K family and renamed *QUI-GON JINN* (*QGJ*). At meiosis, in situ hybridization analysis revealed an altered pattern of expression in *P. notatum* apomictic plants, which could be analyzed at even more detail in apomictic *Brachiaria brizantha* [98]. While in sexual plants *QGJ* was strongly expressed at micropylar degenerating megaspores, in apomictic ones it showed activity within the enlarged meiocyte and the distal ovule nucellus, but was absent from the cell layer surrounding the meiocyte, from which apospory initials (AIs) originate. The effect of a *QGJ* diminished expression in an apomictic background was further investigated by producing RNAi lines [98]. Relatively high proportions of aborted ovaries, defects in both initiation and completion of AES formation, as well as a substantially lower number of AES per ovule were detected in two independent RNAi lines in comparison to both wild type and transformation control plants [98]. The proportion of ovules containing meiotic ES (MES) showed no statistical difference among the obligate apomictic wild type (3–8% MES according to Ortiz et al. [57]), the control, and the RNAi lines [98]. The conclusion emerging from these results was that the significant reduction of *QGJ* expression in an obligate apomictic background impaired the formation of AESs and, therefore, the expression of *QGJ* in distal nucellar cells is needed for aposporous development [98]. Genetic mapping analysis showed no evidence of a genetic link between *QGJ* and the ACL, but a long non-coding RNA partially related in sequence to *QGJ* (*LNC_QGJ*) cosegregated strictly with apomixis when mapped in a *P. notatum* segregating population [98]. Moreover, reverse-transcribed PCR experiments using *LNC_QGJ* specific primers, which were conducted in several apomictic and sexual *P. notatum* individuals, showed that *LNC_QGJ* is expressed only in apomictic plants. The existence of a functional link between the particular *QGJ* expression pattern detected in apomictic plants and a putative regulatory long non-coding *LNC_QGJ* activity operating from the ACL should be further investigated [98].

6.3. *TGS1*

After identification of an apomixis-associated DET (N69) homologous to a PRIP-interacting methyltransferase S-adenosyl domain protein by Laspina et al. [58], this candidate sequence was extended and confirmed to be a plant-specific *TRIMETHYLGUANOSINE SYNTHASE 1* (*TGS1*) gene, which was named *PN_TGS1*-like [97]. In facultative apomictic plants, *PN_TGS1*-like showed expression levels positively correlated with sexuality rates. Moreover, it displayed contrasting in situ hybridization patterns in apomictic and sexual plant ovules from premeiosis to anthesis [97], with higher expression in ovules (including nucellus, integuments, and reproductive lineage) of sexual plants throughout development, from premeiosis to maturity. Since the nucellus is the site of aposporous initials (AIs) differentiation, we proposed that *PN_TGS1*-like might be preventing the differentiation of apospory initials in sexual *P. notatum* plants [97]. Thereafter, a full-sexual *P. notatum* genotype was transformed with a *TGS1*-like antisense construction under a constitutive promoter, to obtain lines with a reduced transcript representation [38]. Antisense plants developed prominent trichomes on the adaxial leaf surface, occasionally formed twin ovules, and showed around 15% of ovules bearing what looked like supernumerary aposporous-like gametophytes (i.e., numerous female gametophytes with a typical *Paspalum* unreduced megagametophyte morphology, including an egg cell, one-two synergids, two polar nuclei, and no antipodal cells). Moreover, around 9% of ovules showed a combination of meiotic and aposporous-like sacs. At early developmental stages, 32% of ovules displayed nucellar cells with prominent nuclei resembling apospory initials (AIs) surrounding the megaspore mother cell (MMC) or the MMC-derived meiotic products. Occasionally, immature binucleated (FG2) female gametophytes of the aposporous type (i.e., the two nuclei located at the same side of the central vacuole) were detected [38]. Neither multiple meiosis nor early proembryos were registered, which suggested a non-reduced nature for the extra nuclei observed in the mature ovules and an absence of parthenogenesis, respectively. The antisense lines produced viable pollen and formed an equivalent full seed set after self-pollination. Flow cytometry analyses of caryopses revealed that all full seeds had originated from meiotic female gametophytes (i.e., by sexuality) supporting the hypothesis that parthenogenesis might not be operative [38]. Moreover, antisense lines showed a significant reduction of the germination percentage, indicating that *PN_TGS1*-like might also be involved in either embryogenesis or endosperm development. These results suggest that *PN_TGS1*-like is a developmental repressor, whose expression in leaves blocks the formation of trichomes, while in ovules inhibit the onset of apospory initials and/or the progression of gametophytes. However, it does not influence parthenogenesis, even when it might play an unknown role during embryogenesis.

7. Advances in Methods for Improving Apomictic *Paspalum* Species

The identification of molecular markers (MM) cosegregating with apomixis, the generation of artificial sexual polyploids after colchicine duplication, the construction of transcriptome databases and, in a close future, genomic assemblies, the identification of genes controlling the reproductive mode, and the establishment of biolistic transformation platforms offer good and innovative prospects for harnessing reproductive and non-reproductive traits of interest in *Paspalum* forage grasses, which have been selected and improved by classical and molecular methods over 80 years. The diversity present in the genus *Paspalum* for the modes of reproduction and ploidy levels is directly linked to adaptation to different environmental conditions and variation for a large group of traits of agronomic interest. Since the 1940s, near 30 apomictic cultivars belonging to the genus *Paspalum* have been released mainly for forage or turf [13]. All of these cultivars, except for one recently developed, are ecotypes collected in South America, evaluated in their target area, and used in different parts of the world, mainly in Australia, the United States, Japan, Thailand, Brazil, and Argentina. The success of this approach lies in the large diversity available in the genus and even within individual species, as reported for *P. simplex* [129]. However, a large number of desirable traits remain dispersed in the apomictic germplasm, among ecotypes and species, since genetic recombination is locked by apomixis [130].

The majority of the *Paspalum* species form agamic complexes [14]. The release of the genetic variability present in the apomictic germplasm was made possible after sexual tetraploid plants were generated by diploid plant chromosome duplication [33,48–51]. Since then, large segregating progenies have been generated by crossing these artificially induced sexual tetraploid genotypes as female parents and apomictic ecotypes as pollen donors. This procedure has been repeatedly used over the last years attempting to improve *P. notatum* [72–75] and, more recently, *P. simplex* [131] and species of the Plicatula group [70]. The experience with *P. notatum* indicates that only a reduced fraction of the progeny (around 10%) inherits the full capacity to express apomixis [72–74] and variable degrees of apospory expressivity are observed within hybrids [73–75]. Moreover, the self-incompatibility present in the diploid germplasm [132] and the induced sexual tetraploids is not transmitted to the sexual or apomictic progeny [72]. There is also evidence indicating that segregation for apomixis is independent of the segregation for traits of agronomic interest in *P. simplex* [131]. In the Plicatula group, several intra and interspecific crosses have been performed and different degrees of crossability and fertility of the resulting hybrids were detected [70]. Nevertheless, the possibility of hybridization is restricted because of flowering asynchrony among ecotypes and the sexual tetraploid germplasm. This issue may be overcome by storing the pollen of the apomictic ecotypes [133] or by creating new sexual tetraploid genotypes with different flowering times [70]. As a general rule for the genus, a large diversity is observed for agronomic traits as a result of crossing sexual and apomictic genotypes [13]. Heterosis for traits of interest, such as forage yield and cold tolerance, have been repeatedly reported in *P. notatum* [72,73,75,134] and for interspecific hybrids between *P. plicatulum* and *P. guenoarum* [70,135]. An apomictic hybrid of *P. notatum* identified as upright and fast-growing was recently released as a forage cultivar named Boyero UNNE, resulting from a collaborative research between the University of Florida and the National University of the North-East, Argentina [136].

MMs have been used to monitor the process of hybridization between sexual and apomictic genotypes in *Paspalum*, particularly for the identification of true hybrids within progeny [35,70,74,131]. The general idea is to test the progeny for the presence of male-specific MMs and for the segregation of female-specific markers indicative of recombination events in the female side. This procedure is particularly useful, since sexual tetraploid hybrids exhibit high levels of seed fertility when self-pollinated [72]. Moreover, several markers 100% linked to apospory have also been developed for the genus [77,85,137]. Some of these markers have been successfully applied to identify hybrids exhibiting apospory in *P. notatum* [74,75] and *P. simplex* [131]. Usually, they are useful for an initial evaluation that allows separating all the highly sexual progeny from the aposporous, but further analyses are needed to determine the different levels of apospory expressivity. This can be achieved by assessing phenotypic homogeneity in the progeny or by female gametophyte observations [74,75]. Moreover, random MMs are used in progeny tests, to determine the apomixis expressivity, and the genetic stability of an apomictic cultivar [136]. This is an important point to consider before going forward with the breeding process, because a relatively high expression of sexuality will reduce the stability of the new cultivar in a few years. An original alternative is to carry out gene expression analysis, by targeting those candidates whose activity is positively or negatively correlated with apomixis [97]. Although this approach can result much more complicated than cytoembryological analyses and progeny tests, it is expected to provide a good estimation of apomixis capacity and can be extended to high numbers of individuals. Finally, the lack of available apomixis-linked markers in many *Paspalum* species, including the Plicatula group, makes the use of cytometric seed analysis an attractive screening technique for determining the reproductive mode [138].

MMs have also been used to identify heterotic groups in *Paspalum*. Since the main objective of hybridization in apomictic species is to combine high levels of apomixis with heterosis for traits of agronomic interest, Marcón et al. [75] used ISSR and SSR markers to determine the genetic distances between a group of sexual and apomictic genotypes of *P. notatum*. Crosses were made between pairs of sexual and apomictic parents with low, intermediate, and high genetic distances. As expected, higher genetic distances between parents increased the proportion of hybrids exhibiting heterosis for forage yield. Moreover, the same positive relationship was observed between genetic distances and the proportion of aposporous hybrids identified within the progeny. However, apospory expressivity was not significantly related to the genetic distance. These results suggest that random MMs may be successfully used to determine heterotic groups for forage yield in *P. notatum*, but also to predict a high occurrence of apomictic hybrids within segregating families. Furthermore, as part of a hybridization project in the Plicatula group, the genetic distances between an induced sexual genotype of *P. plicatulum* and 22 accessions belonging to 12 different species were determined using AFLP markers [70]. In this case, the large variation for seed fertility and crossability was not correlated to the genetic distances.

A synthetic sexual tetraploid population (SSTP) was established in *P. notatum* to increase the sexual tetraploid gene pool, as follows: (1) crossing the few available sexual tetraploid genotypes with a group of ecotypes collected across the Americas, and (2) polycrossing the resulting highly sexual hybrids to create the SSTP. This procedure allowed the transfer of the genetic diversity present among geographically and genetically distant ecotypes into a single sexual and cross-pollinated population [68,139]. More recently, a similar approach has been used to create a sexual tetraploid population in the Plicatula group [138]. In this case, nine accessions from six species with contrasting morphological characteristics were crossed to a single sexual induced genotype. The sexual hybrids were set apart using a cytometric seed screening and a group representing the different origins was polycrossed to obtain a large sexual population containing alleles from this diverse germplasm. This approach may also be used to generate sexual tetraploid populations for other species or groups of species in the genus sharing the same genetic characteristics, such as *P. simplex* or other species of the subgenus *Anachyris*.

The availability of sexual tetraploid populations has allowed the application of breeding methods originally developed for cross-pollinated crops, such as maize or alfalfa, as was originally suggested for apomictic tetraploid *Brachiaria* spp. [140]. Marcón et al. [141] evaluated the use of recurrent phenotypic selection (RPS) and recurrent selection based on combining ability (RSCA) in tetraploid *P. notatum*. Both methods proved to be efficient for improving forage yield in *P. notatum*. Although RSCA was expected to be more appropriate for exploiting heterosis since it was developed to accumulate heterotic effects over cycles, no difference was observed between the RPS and RSCA procedures for increasing forage yield [141]. A modified version of RSCA was used to identify superior tetraploid hybrids within the sexual population of Plicatula [138]. In this case, an initial phase of phenotypic selection was used to reduce the number of superior sexual hybrids, which were test-crossed to two elite apomictic clones belonging to *P. guenoarum*. This approach identified a few sexual hybrids as possible progenitors to continue tapping into genetic diversity of apomictic individuals or to generate improved sexual populations.

As was previously stated, molecular techniques may play an important role in the different phases of the selection procedures for generating superior apomictic hybrids. A theoretical scheme of RSCA developed for tetraploid *P. notatum* is represented in Figure 7 to show how MMs may assist the breeding process.

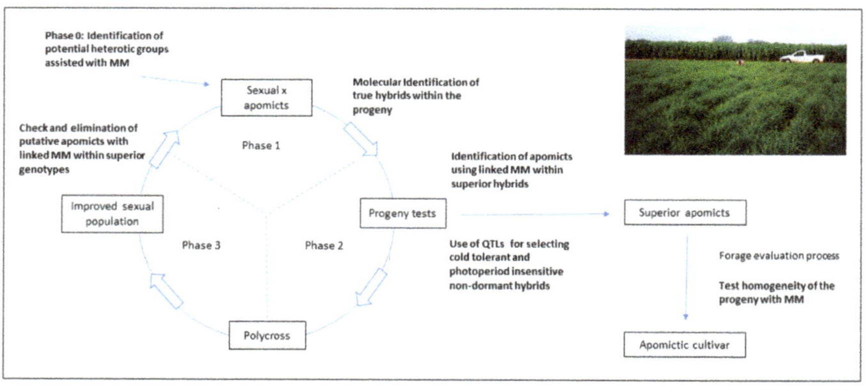

Figure 7. Theoretical scheme of recurrent phenotypic selection based on a combination of classical and molecular techniques in *P. notatum*. The molecular markers (MM) use scenarios are indicated at each phase of the process. An image of an experimental/seed production plot of an improved apomictic hybrid cultivar of *P. notatum* (cv. Boyero UNNE) is shown at top right.

Random molecular markers may aid the identification of heterotic groups for the initial phase of the process (indicated as Phase 0 in Figure 7), as it was suggested by Marcón et al. [75]. Phase 1 consists of test-crossing a group of sexual genotypes to a single apomictic tester or a group of testers. When genetic data is expected to be collected, the true hybrids can be identified by using male-specific markers. The evaluation of the resulting progeny is indicated as Phase 2. The use of QTLs associated with traits of agronomic interest may be applied for the rapid identification of superior progeny. For instance, development of QTLs for cold tolerance and photoperiod insensitivity is underway for tetraploid *P. notatum,* since growth and forage production is markedly reduced during the winter in the subtropics. Once phenotypic and molecular data is available for selecting the best progeny, the apomictic genotypes may be separated using apomixis-linked markers. Superior apomictic hybrids usually need to go through a multiphase forage evaluation procedure [142]. The genetic homogeneity of the progeny coming from each selected line can be tested with random markers [136], to ensure the high expressivity of apomixis among the best lines. Phase 2 allows the selection of the best sexual genotypes that were crossed in phase 1, and these are polycrossed to generate an improved sexual population (phase 3 in Figure 7). This population is evaluated under field conditions, and only a few genotypes enter phase 1 of a new cycle. It is very important to check the mode of reproduction of this group of selected genotypes, and this can be done using apomixis-linked markers.

8. Conclusions

Given the dramatic increase in the world population, which is expected to reach 9.7 billion by 2050, and the concomitant expanding food demand, élite resilient crops should be readily developed on a massive scale in order to address the environmental, climate and overcrowding challenges [143]. Current methods used to produce new plant varieties, which rely entirely in the exploitation of sexuality, involve 7–20 years to bring them to the market, cost even millions of euros, and are limited regarding the harnessing of favorable gene interactions. Apomixis breeding could rapidly overcome the drawbacks associated with sexuality-based programs, facilitating a better adaptation to local environments, the avoidance of monocultures, and substantial increases in crop resilience [143]. However, specific methodological hitches complicate the generation and interpretation of the apomixis-related available molecular data and large areas of knowledge have barely been touched upon. To begin with, appropriate criteria should be established in order to reduce the number of candidate genes emerging from genomic and comparative transcriptomic surveys, based on both the combination of positional evidence with expression data and the establishment of reliable network predictions. Moreover, until now, no comprehensive study on the reproductive developmental impact of novel apomixis-related mRNA splice variants is available and protein post-translational regulatory mechanisms possibly influencing the trait remain uncharted. Finally, the regulatory role of transposons and lncRNAs was only poorly outlined and needs further clarification. The functional analysis of these aspects, once implemented, in combination with data body already available (Nanopore-derived genome and Roche 454-FLX +/Illumina transcriptome sequence databases) will lead to the application of accurate genetic engineering approaches by using sense or antisense transgenesis, RNA-directed DNA methylation, CRISPR-Cas9 editing and other techniques and allow the complete harnessing of the trait into plant breeding by inducing apomixis in sexual plants.

From the *Paspalum* example we can derive the complex series of steps necessary to produce useful and applicable knowledge in apomictic species, consisting of (1) germplasm groundwork, including collection of variable materials, reproductive phenotyping, ploidy level assessment, genome content characterization, ploidy manipulation and reproductive calendar construction; (2) inheritance analysis, dense genetic mapping, and identification of markers cosegregating with apomixis; (3) genome sequencing; (4) transcriptome sequencing; (5) functional analysis in sexual and apomictic species; (6) traditional and molecular marker-assisted breeding to release élite apomictic hybrids. The molecular elucidation of the sexual–asexual transition would require the use of this methodological scheme in as many species as possible, in order to expose subtle variants that might have emerged during evolution and provide a comprehensive perspective of alternative pathways, disclosing optimal tools for harnessing apomixis into the breeding of major crops by genetic engineering. With this review, we hope to contribute to the design of the necessary experimental phases, which have been laid down in *Paspalum* after decades of effort, in a process that often involved trial and error. At best, the ordered management of the flowchart presented here could save both time and costs and stimulate apomixis research/breeding in other species of agronomic interest.

Author Contributions: Conceptualization: S.C.P. and J.P.A.O.; original draft compilation: S.C.P.; writing and editing: S.C.P., J.P.A.O., F.P., O.L. and C.A.A. All authors have read and agreed to the published version of the manuscript.

Funding: Agencia Nacional de Promoción Científica y Tecnológica (ANPCyT), Argentina (Project PICT-2017-1956); Consejo Nacional de Investigaciones Científicas y Técnicas, Argentina (CONICET) (Projects: PIP 11220150100702CO; PUE 22920160100043CO); Universidad Nacional de Rosario (UNR), Argentina, Project: AGR271; Ufficio Relazioni Internazionali del Consiglio Nazionale delle Ricerche, Italy (Laboratori Congiunti Bilaterali Internazionali CNR, Prot. 0005651); Ministero degli Affari Esteri, e della Cooperazione Internazionale, Direzione Generale per la Promozione del Sistema Paese (Italy).

Acknowledgments: We thank Maricel Podio and Lorena Siena (both from IICAR CONICET-UNR, Rosario, Argentina) as well as Giulio Galla (DAFNAE, University of Padua, Italy) for providing some of the images used to illustrate the text.

Conflicts of Interest: The authors declare no conflict of interest.

References

1. Nogler, G.A. Gametophytic apomixis. In *Embryology of Angiosperms*; Johri, B.M., Ed.; Springer: Berlin, Germany, 1984; pp. 475–518.
2. Toenniessen, G.H. Feeding the world in the 21st century: Plant breeding, biotechnology, and the potential role of apomixis. In *The Flowering of Apomixis: From Mechanisms to Genetic Engineering*; Savidan, Y., Carman, J.G., Dresselhaus, T., Eds.; CIMMYT: Mexico DF, Mexico; IRD: Marseille, France; European Commission OC VI (FAIR): Brussels, Belgium, 2001; pp. 1–7.
3. Hand, M.; Koltunow, A. The genetic control of apomixis: Asexual seed formation. *Genetics* **2014**, *197*, 441–450. [CrossRef] [PubMed]
4. Barcaccia, G.; Albertini, E. Apomixis in plant reproduction: A novel perspective on an old dilemma. *Plant Reprod.* **2013**, *26*, 159–179. [CrossRef] [PubMed]
5. Schmidt, A. Controlling apomixis: Shared features and distinct characteristics of gene regulation. *Genes* **2020**, *11*, 329. [CrossRef] [PubMed]
6. Fayos, I.; Mieulet, D.; Petit, J.; Meunier, A.C.; Périn, C.; Nicolas, A.; Guiderdoni, E. Engineering meiotic recombination pathways in rice. *Plant Biotechnol. J.* **2019**, *17*, 2062–2077. [CrossRef] [PubMed]
7. Wang, K. Fixation of hybrid vigor in rice: Synthetic apomixis generated by genome editing. *aBIOTECH* **2020**, *1*, 15–20. [CrossRef]
8. Kaushal, P.; Malaviya, D.R.; Roy, A.K. Prospects for breeding apomictic rice: A reassessment. *Curr. Sci.* **2004**, *87*, 292–296.
9. Albertini, E.; Barcaccia, G.; Carman, J.G.; Pupilli, F. Did apomixis evolve from sex or was it the other way around? *J. Exp. Bot.* **2019**, *70*, 2951–2964. [CrossRef]
10. Carman, J.G. Asynchronous expression of duplicate genes in angiosperms may cause apomixis, bispory, tetraspory, and polyembryony. *Biol. J. Linn. Soc.* **1997**, *61*, 51–94. [CrossRef]
11. Hörandl, E.; Hojsgaard, D.H. The evolution of apomixis in angiosperms: A reappraisal. *Plant Biosyst.* **2012**, *146*, 681–693.
12. Hojsgaard, D.; Klatt, S.; Baier, R.; Carman, J.G.; Hörandl, E. Taxonomy and biogeography of apomixis in angiosperms and associated biodiversity characteristics. *Crit. Rev. Plant Sci.* **2014**, *33*, 414–427. [CrossRef]
13. Acuña, C.A.; Martínez, E.J.; Zilli, A.L.; Brugnoli, E.A.; Espinoza, F.; Marcón, F.; Urbani, M.H.; Quarin, C.L. Reproductive systems in *Paspalum*: Relevance for germplasm collection and conservation, breeding techniques, and adoption of released cultivars. *Front. Plant Sci.* **2019**, *10*, 1377. [CrossRef] [PubMed]

14. Ortiz, J.P.A.; Quarin, C.L.; Pessino, S.C.; Acuña, C.; Martínez, E.J.; Espinoza, F.; Hojsgaard, D.H.; Sartor, M.E.; Cáceres, M.E.; Pupilli, F. Harnessing apomictic reproduction in grasses: What we have learned from *Paspalum*. *Ann. Bot.* **2013**, *112*, 767–787. [CrossRef] [PubMed]
15. Quarin, C.L. Effect of pollen source and pollen ploidy on endosperm formation and seed set in pseudogamous apomictic *Paspalum notatum*. *Sex. Plant Reprod.* **1999**, *11*, 331–335. [CrossRef]
16. Quarin, C.L. The nature of apomixis and its origin in Panicoid grasses. *Apomixis Newsl.* **1992**, *5*, 8–15.
17. Burson, B.L.; Bennett, H.W. Chromosome numbers, microsporogenesis, and mode of reproduction of seven *Paspalum* species. *Crop Sci.* **1971**, *11*, 292–294. [CrossRef]
18. Quarin, C.L.; Pozzobon, M.T.; Valls, J.F.M. Cytology and reproductive behavior of diploid, tetraploid and hexaploid germplasm accessions of a wild forage grass: *Paspalum compressifolium*. *Euphytica* **1996**, *9*, 345–349. [CrossRef]
19. Espinoza, F.; Urbani, M.H.; Martínez, E.J.; Quarin, C.L. The breeding system of three *Paspalum* species. *Trop. Grassl.* **2001**, *35*, 211–217.
20. Ma, G.; Huang, X.; Zhao, N.; Xu, Q. Apospory in *Paspalum thunbergii*. *Aust. J. Bot.* **2004**, *52*, 81–86. [CrossRef]
21. Snyder, L.A. Apomixis in *Paspalum secans*. *Am. J. Bot.* **1957**, *44*, 318–324. [CrossRef]
22. Caponio, I.; Quarin, C.L. El sistema genético de *Paspalum simplex* y de un híbrido interespecífico con *P. dilatatum*. *Kurtziana* **1987**, *19*, 35–45.
23. Cáceres, M.E.; Matzk, F.; Busti, A.; Pupilli, F.; Arcioni, S. Apomixis and sexuality in *Paspalum simplex*: Characterization of the mode of reproduction in segregating progenies by different methods. *Sex. Plant Reprod.* **2001**, *14*, 201–206. [CrossRef] [PubMed]
24. Chao, C.Y. Megasporogenesis and megagametogenesis in *Paspalum commersonii* and *Paspalum longifolium* at two polyploid levels. *Bot. Not.* **1974**, *127*, 267–275.
25. Chao, C.Y. Autonomous development of embryo in *Paspalum conjugatum* Berg. *Bot. Not.* **1980**, *133*, 215–222.
26. Bonilla, J.R.; Quarin, C.L. Diplosporous and aposporous apomixis in pentaploid *Paspalum minus*. *Plant Sci.* **1997**, *127*, 97–104. [CrossRef]
27. Stein, J.; Quarin, C.L.; Martínez, E.J.; Pessino, S.C.; Ortiz, J.P.A. Tetraploid races of *Paspalum notatum* show polysomic inheritance and preferential chromosome pairing around the apospory-controlling locus. *Theor. Appl. Genet.* **2004**, *109*, 186–191. [CrossRef]
28. Podio, M.; Siena, L.A.; Hojsgaard, D.; Stein, J.; Quarin, C.L.; Ortiz, J.P.A. Evaluation of meiotic abnormalities and pollen viability in aposporous and sexual tetraploid *Paspalum notatum* (Poaceae). *Plant Syst. Evol.* **2012**, *298*, 1625–1633. [CrossRef]
29. Pagliarini, M.S.; Takayama, S.Y.; de Freitas, P.M.; Carraro, L.R.; Adamowski, E.V.; Silva, N.; Batista, L.A.R. Failure of cytokinesis and 2n gamete formation in Brazilian accessions of *Paspalum*. *Euphytica* **1999**, *108*, 129–135. [CrossRef]
30. Snyder, L.A. Asyndesis and meiotic non-reduction in microsporogenesis of apomitic *Paspalum secans*. *Cytologia* **1961**, *26*, 50–61. [CrossRef]
31. Matzk, F.; Meister, A.; Schubert, I. An efficient screen for reproductive pathways using mature seeds of monocots and dicots. *Plant J.* **2000**, *21*, 97–108. [CrossRef]
32. Siena, L.A.; Sartor, M.E.; Espinoza, F.; Quarin, C.L.; Ortiz, J.P.A. Genetic and embryological evidences of apomixis at the diploid level in *Paspalum rufum* support recurrent auto-polyploidization in the species. *Sex. Plant Reprod.* **2008**, *21*, 205–215. [CrossRef]
33. Sartor, M.E.; Quarin, C.L.; Espinoza, F. Mode of reproduction of colchicine-induced *Paspalum plicatulum* tetraploids. *Crop Sci.* **2009**, *49*, 1270–1276. [CrossRef]
34. Sartor, M.E.; Quarin, C.L.; Urbani, M.H.; Espinoza, F. Ploidy levels and reproductive behaviour in natural populations of five *Paspalum* species. *Plant Syst. Evol.* **2011**, *293*, 31–41. [CrossRef]
35. Aguilera, P.M.; Sartor, M.E.; Galdeano, F.; Espinoza, F.; Quarin, C.L. Interspecific tetraploid hybrids between two forage grass species: Sexual *Paspalum plicatulum* and apomictic *Paspalum guenoarum*. *Crop Sci.* **2011**, *51*, 1544–1550. [CrossRef]
36. Rebozzio, R.N.; Sartor, M.E.; Quarin, C.L.; Espinoza, F. Residual sexuality and it seasonal variation in natural apomictic *Paspalum notatum* accessions. *Biol. Plant.* **2011**, *55*, 391–395. [CrossRef]
37. Hojsgaard, D.H.; Martínez, E.J.; Quarin, C.L. Competition between meiotic and apomictic pathways during ovule and seed development results in clonality. *New Phytol.* **2013**, *197*, 336–347. [CrossRef] [PubMed]

38. Colono, C.M.; Ortiz, J.P.A.; Permingeat, H.; Souza Canada, D.; Siena, L.A.; Spoto, N.; Galdeano, F.; Espinoza, F.; Leblanc, O.; Pessino, S.C. A plant-specific *TGS1* homolog influences gametophyte development in sexual tetraploid *Paspalum notatum* ovules. *Front. Plant Sci.* **2019**, *10*, 1566. [CrossRef]
39. Haig, D.; Westoby, M. Genomic imprinting in endosperm: Its effect on seed development in crosses between species, and between different ploidies of the same species, and its implication for the evolution of apomixis. *Philos. Trans. R. Soc. Lond. B* **1991**, *333*, 1–13.
40. Galdeano, F.; Urbani, M.H.; Sartor, M.E.; Honfi, A.I.; Espinoza, F.; Quarin, C.L. Relative DNA content in diploid, polyploid, and multiploid species of *Paspalum* (Poaceae) with relation to reproductive mode and taxonomy. *J. Plant Res.* **2016**, *129*, 697–710. [CrossRef]
41. Urbani, M.H.; Quarin, C.L.; Espinoza, F.; Penteado, M.I.O.; Rodrigues, I.F. Cytogeography and reproduction of the *Paspalum simplex* polyploid complex. *Plant Syst. Evol.* **2002**, *236*, 99–105. [CrossRef]
42. Daurelio, L.D.; Espinoza, F.; Quarin, C.L.; Pessino, S.C. Genetic diversity in sexual diploid and apomictic tetraploid populations of *Paspalum notatum* situated in sympatry or allopatry. *Plant Syst. Evol.* **2004**, *244*, 189–199. [CrossRef]
43. Adams, K.L.; Wendel, J.F. Novel patterns of gene expression in polyploid plants. *Trends Genet.* **2005**, *21*, 539–543. [CrossRef] [PubMed]
44. Cheng, F.; Wu, J.; Cai, X.; Liang, J.; Freeling, M.; Wang, X. Gene retention, fractionation and subgenome differences in polyploid plants. *Nat. Plants* **2018**, *4*, 258–268. [CrossRef] [PubMed]
45. Martelotto, L.G.; Ortiz, J.P.A.; Stein, J.; Espinoza, F.; Quarin, C.L.; Pessino, S.C. A comprehensive analysis of gene expression alterations in a newly synthesized *Paspalum notatum* autotetraploid. *Plant Sci.* **2005**, *169*, 211–220. [CrossRef]
46. Ortiz, J.P.A.; Revale, S.; Siena, L.A.; Podio, M.; Delgado, L.; Stein, J.; Leblanc, O.; Pessino, S.C. A reference floral transcriptome of sexual and apomictic *Paspalum notatum*. *BMC Genom.* **2017**, *18*, 318. [CrossRef]
47. De Oliveira, F.A.; Vigna, B.B.Z.; Da Silva, C.C.; Fávero, A.P.; De Matta, F.P.; Azevedo, A.L.S.; De Souza, A.P. Coexpression and transcriptome analyses identify active apomixis-related genes in *Paspalum notatum* leaves. *BMC Genom.* **2020**, *21*, 78. [CrossRef]
48. Burton, G.W.M.; Forbes, I. The genetics and manipulation of obligate apomixis in common Bahia grass (*Paspalum notatum* Flugge). In Proceedings of the 8th International Grassland Congress, Reading, UK, 11–21 July 1960; pp. 66–71.
49. Cáceres, M.E.; Pupilli, F.; Quarin, C.L.; Arcioni, S. Feulgen-DNA densitometry of embryo sacs permits discrimination between sexual and apomictic plants in *Paspalum simplex*. *Euphytica* **1999**, *110*, 161–167. [CrossRef]
50. Quarin, C.L.; Espinoza, F.; Martínez, E.J.; Pessino, S.C.; Bovo, O.A. A rise of ploidy level induces the expression of apomixis in *Paspalum notatum*. *Sex. Plant Reprod.* **2001**, *13*, 243–249. [CrossRef]
51. Quesenberry, K.H.; Dampier, J.M.; Lee, Y.Y.; Smith, R.L.; Acuña, C.A. Doubling the chromosome number of bahiagrass via tissue culture. *Euphytica* **2010**, *175*, 43–50. [CrossRef]
52. Delgado, L.; Sartor, M.E.; Espinoza, F.; Soliman, M.; Galdeano, F.; Ortiz, J.P.A. Hybridity and autopolyploidy increase the expressivity of apospory in diploid *Paspalum rufum*. *Plant Syst. Evol.* **2016**, *302*, 1471–1481. [CrossRef]
53. Quarin, C.L.; Urbani, M.H.; Blount, A.R.; Martínez, E.J.; Hack, C.M.; Burton, G.W.; Quesenberry, K. Registration of Q4188 and Q4205, sexual tetraploid germplasm lines of bahiagrass. *Crop Sci.* **2003**, *43*, 745–746. [CrossRef]
54. Martelotto, L.G.; Ortiz, J.P.A.; Stein, J.; Espinoza, F.; Quarin, C.L.; Pessino, S.C. Genome rearrangements derived from autopolyploidization in *Paspalum* sp. *Plant Sci.* **2007**, *172*, 970–977. [CrossRef]
55. Weihmüller, E.; Beltrán, C.; Sartor, M.E.; Espinoza, F.; Spampinato, C.; Pessino, S. Genomic response of *Paspalum plicatulum* to genome duplication. *Genetica (GENE)* **2014**, *142*, 227–234. [CrossRef] [PubMed]
56. Zappacosta, D.; Ochogavía, A.; Rodrigo, J.M.; Romero, J.; Meier, M.; Garbus, I.; Pessino, S.; Echenique, V. Increased apomixis expression concurrent with genetic and epigenetic variation in a newly synthesized *Eragrostis curvula* polyploid. *Sci. Rep.* **2014**, *4*, 4223. [CrossRef] [PubMed]
57. Ortiz, J.P.A.; Pessino, S.C.; Leblanc, O.; Hayward, M.D.; Quarin, C.L. Genetic fingerprint for determinig the mode of reproduction in *Paspalum notatum*, a subtropical apomictic forage grass. *Theor. Appl. Genet.* **1997**, *95*, 850–856. [CrossRef]

58. Laspina, N.V.; Vega, T.; Seijo, G.; González, A.M.; Martelotto, L.G.; Stein, J.; Podio, M.; Ortiz, J.P.A.; Echenique, V.C.; Quarin, C.L.; et al. Gene expression analysis at the onset of aposporous apomixis in *Paspalum notatum*. *Plant Mol. Biol.* **2008**, *67*, 615–628. [CrossRef]
59. Soliman, M.; Espinoza, F.; Ortiz, J.P.A.; Delgado, L. Heterochronic reproductive developmental processes between diploid and tetraploid cytotypes of *Paspalum rufum*. *Ann. Bot.* **2019**, *123*, 901–915. [CrossRef]
60. Ravi, M.; Marimuthu, M.P.A.; Siddiqi, I. Gamete formation without meiosis in *Arabidopsis*. *Nature* **2008**, *451*, 1121–1124. [CrossRef]
61. D'Erfurth, I.; Jolivet, S.; Froger, N.; Catrice, O.; Novatchkova, M.; Mercier, R. Turning meiosis into mitosis. *PLoS Biol.* **2009**, *7*, e1000124. [CrossRef]
62. Khanday, I.; Skinner, D.; Yang, B.; Mercier, R.; Sundaresan, V. A male-expressed rice embryogenic trigger redirected for asexual propagation through seeds. *Nature* **2019**, *565*, 91–95. [CrossRef]
63. Wang, C.; Liu, Q.; Shen, Y.; Hua, Y.; Wang, J.; Lin, J.; Wu, M.; Sun, T.; Cheng, Z.; Mercier, R.; et al. Clonal seeds from hybrid rice by simultaneous genome engineering of meiosis and fertilization genes. *Nat. Biotechnol.* **2019**, *37*, 283–286. [CrossRef]
64. Xie, E.; Li, Y.; Tang, D.; Lv, Y.; Shen, Y.; Cheng, Z. A strategy for generating rice apomixis by gene editing. *J. Integr. Plant Biol.* **2019**, *61*, 911–916. [CrossRef] [PubMed]
65. Mieulet, D.; Jolivet, S.; Rivard, M.; Cromer, L.; Vernet, A.; Mayonove, P.; Pereira, L.; Droc, G.; Courtois, B.; Guiderdoni, E.; et al. Turning rice meiosis into mitosis. *Cell Res.* **2016**, *26*, 1242–1254. [CrossRef] [PubMed]
66. Gates, R.N.; Quarin, C.L.; Pedreira, C.G. Bahiagrass. In *Warm-Season (C4) Grasses*; American Society of Agronomy: Madison, WI, USA; Crop Science Society of America: Madison, WI, USA; Soil Science Society of America: Madison, WI, USA, 2004; Volume 45, pp. 651–680.
67. Aguilera, P.M.; Galdeano, F.; Quarin, C.L.; Ortiz, J.P.A.; Espinoza, F. Inheritance of aposporous apomixis in inter-specific hybrids derived from sexual *Paspalum plicatulum* and apomictic *Paspalum guenoarum*. *Crop Sci.* **2015**, *55*, 1947–1956. [CrossRef]
68. Zilli, A.L.; Acuña, C.A.; Schulz, R.R.; Brugnoli, E.A.; Guidalevich, V.; Quarin, C.L.; Martínez, E.J. Widening the gene pool of sexual tetraploid bahiagrass: Generation and reproductive characterization of a sexual synthetic tetraploid population. *Crop Sci.* **2018**, *58*, 762–772. [CrossRef]
69. Hojsgaard, D.; Martínez, E.J.; Acuña, C.; Quarin, C.L.; Pupilli, F. A molecular map of the apomixis-control locus in *Paspalum procurrens* and its comparative analysis with other species of *Paspalum*. *Theor. Appl. Genet.* **2011**, *123*, 959–971. [CrossRef]
70. Novo, P.E.; Acuña, C.A.; Quarin, C.L.; Urbani, M.H.; Marcón, F.; Espinoza, F. Hybridization and heterosis in the *Plicatula* group of *Paspalum*. *Euphytica* **2017**, *213*, 198. [CrossRef]
71. Martínez, E.J.; Urbani, M.H.; Quarin, C.L.; Ortiz, J.P. Inheritance of apospory in bahiagrass, *Paspalum notatum*. *Hereditas* **2001**, *135*, 19–25. [CrossRef]
72. Acuña, C.A.; Blount, A.R.; Quesenberry, K.H.; Kenworthy, K.E.; Hanna, W.W. Bahiagrass tetraploid germplasm: Reproductive and agronomic characterization of segregating progeny. *Crop Sci.* **2009**, *49*, 581–588. [CrossRef]
73. Acuña, C.A.; Blount, A.R.; Quesenberry, K.H.; Kenworthy, K.E.; Hanna, W.W. Tetraploid bahiagrass hybrids: Breeding technique, genetic variability and proportion of heterotic hybrids. *Euphytica* **2011**, *179*, 227–235. [CrossRef]
74. Zilli, A.L.; Brugnoli, E.A.; Marcón, F.; Billa, M.B.; Rios, E.F.; Martínez, E.J.; Acuña, C.A. Heterosis and expressivity of apospory in tetraploid bahiagrass hybrids. *Crop Sci.* **2015**, *55*, 1189–1201. [CrossRef]
75. Marcón, F.; Martínez, E.J.; Rodríguez, G.R.; Zilli, A.L.; Brugnoli, E.A.; Acuña, C.A. Genetic distance and the relationship with heterosis and reproductive behavior in tetraploid bahiagrass hybrids. *Mol. Breed.* **2019**, *39*, 89. [CrossRef]
76. Pupilli, F.; Labombarda, P.; Cáceres, M.E.; Quarin, C.L.; Arcioni, S. The chromosome segment related to apomixis in *Paspalum simplex* is homoeologous to the telomeric region of the long arm of rice chromosome 12. *Mol. Breed.* **2001**, *8*, 53–61. [CrossRef]
77. Pupilli, F.; Martinez, E.J.; Busti, A.; Calderini, O.; Quarin, C.L.; Arcioni, S. Comparative mapping reveals partial conservation of synteny at the apomixis locus in *Paspalum* spp. *Mol. Genet. Genom.* **2004**, *270*, 539–548. [CrossRef] [PubMed]
78. Ozias-Akins, P.; Van Dijk, P.J. Mendelian genetics of apomixis in plants. *Annu. Rev. Genet.* **2007**, *41*, 509–537. [CrossRef] [PubMed]

79. Stein, J.; Pessino, S.C.; Martínez, E.J.; Rodríguez, M.P.; Siena, L.A.; Quarin, C.L.; Ortiz, J.P.A. A genetic map of tetraploid *Paspalum notatum* Flügge (bahiagrass) based on single-dose molecular markers. *Mol. Breed.* **2007**, *20*, 153–166. [CrossRef]
80. Rios, E.F.; Blount, A.; Kenworthy, K.E.; Acuña, C.A.; Quesenberry, K.H. Seasonal expression of apospory in bahiagrass. *Trop. Grassl. Forrajes Trop.* **2013**, *1*, 116–118. [CrossRef]
81. Quarin, C.L. Seasonal changes in the incidence of apomixis of diploid, triploid, and tetraploid plants of *Paspalum cromyorrhizon*. *Euphytica* **1986**, *35*, 512–522. [CrossRef]
82. Acuña, C.A.; Martínez, E.J.; Quarin, C.L. Apospory followed by sterility in a hypotriploid hybrid (2x X 4x) of *Paspalum*. *Caryologia* **2004**, *57*, 373–378. [CrossRef]
83. Podio, M.; Cáceres, M.E.; Samoluk, S.S.; Seijo, J.G.; Pessino, S.C.; Ortiz, J.P.; Pupilli, F. A methylation status analysis of the apomixis–specific region in *Paspalum* spp. suggests an epigenetic control of parthenogenesis. *J. Exp. Bot.* **2014**, *65*, 6411–6424. [CrossRef]
84. Delgado, L.; Galdeano, F.; Sartor, M.E.; Quarin, C.L.; Espinoza, F.; Ortiz, J.P.A. Analysis of variation for apomictic reproduction in diploid *Paspalum rufum*. *Ann. Bot.* **2014**, *113*, 1211–1218. [CrossRef]
85. Martínez, E.J.; Hopp, H.E.; Stein, J.; Ortiz, J.P.A.; Quarin, C.L. Genetic characterization of apospory in tetraploid *Paspalum notatum* based on the identification of linked molecular markers. *Mol. Breed.* **2003**, *12*, 319–327. [CrossRef]
86. Podio, M.; Rodríguez, M.P.; Felitti, S.; Stein, J.; Martínez, E.J.; Siena, L.A.; Quarin, C.L.; Pessino, S.C.; Ortiz, J.P.A. Sequence characterization, in silico mapping and cytosine methylation analysis of markers linked to apospory in *Paspalum notatum*. *Genet. Mol. Biol.* **2012**, *35*, 827–837. [CrossRef] [PubMed]
87. Calderini, O.; Chang, S.B.; De Jong, H.; Busti, A.; Paolocci, F.; Arcioni, S.; De Vries, S.C.; Abma-Henkens, M.H.; Lankhorst, R.M.; Donnison, I.S.; et al. Molecular cytogenetics and DNA sequence analysis of an apomixis-linked BAC in *Paspalum simplex* reveal a non pericentromere location and partial microcolinearity with rice. *Theor. Appl. Genet.* **2006**, *112*, 1179–1191. [CrossRef] [PubMed]
88. Galla, G.; Siena, L.A.; Ortiz, J.P.A.; Baumlein, H.; Barcaccia, G.; Pessino, S.C.; Bellucci, M.; Pupilli, F. A portion of the apomixis locus of *Paspalum simplex* is microsyntenic with an unstable chromosome segment highly conserved among Poaceae. *Sci. Rep.* **2019**, *9*, 1–12. [CrossRef] [PubMed]
89. Polegri, L.; Calderini, O.; Arcioni, S.; Pupilli, F. Specific expression of apomixis-linked alleles revealed by comparative transcriptomic analysis of sexual and apomictic *Paspalum simplex* Morong flowers. *J. Exp. Bot.* **2010**, *61*, 1869–1883. [CrossRef]
90. Siena, L.A.; Ortiz, J.P.A.; Calderini, O.; Paolocci, F.; Cáceres, M.E.; Kaushal, P.; Grisan, S.; Pessino, S.C.; Pupilli, F. An apomixis-linked *ORC3*-like pseudogene is associated with silencing of its functional homolog in apomictic *Paspalum simplex*. *J. Exp. Bot.* **2016**, *67*, 1965–1978. [CrossRef]
91. Nützmann, H.W.; Scazzocchio, C.; Osbourn, A. Metabolic gene clusters in eukaryotes. *Annu. Rev. Genet.* **2018**, *52*, 159–183. [CrossRef]
92. Chu, H.Y.; Wegel, E.; Osbourn, A. From hormones to secondary metabolism: The emergence of metabolic gene clusters in plants. *Plant J.* **2011**, *66*, 66–79. [CrossRef]
93. Boycheva, S.; Daviet, L.; Wolfender, J.L.; Fitzpatrick, T.B. The rise of operon-like gene clusters in plants. *Trends Plant Sci.* **2014**, *19*, 447–459. [CrossRef]
94. Bergero, R.; Charlesworth, D. The evolution of restricted recombination in sex chromosomes. *Trends Ecol. Evol.* **2008**, *24*, 94–102. [CrossRef]
95. Pessino, S.C.; Espinoza, F.; Martínez, E.J.; Ortiz, J.P.A.; Valle, E.M.; Quarin, C.L. Isolation of cDNA clones differentially expressed in flowers of apomictic and sexual *Paspalum notatum*. *Hereditas* **2001**, *134*, 35–42. [CrossRef] [PubMed]
96. Felitti, S.A.; Seijo, J.G.; González, A.M.; Podio, M.; Laspina, N.V.; Siena, L.; Ortiz, J.P.A.; Pessino, S.C. Expression of *LORELEI*-like genes in aposporous and sexual *Paspalum notatum* plants. *Plant Mol. Biol.* **2011**, *77*, 337–354. [CrossRef] [PubMed]
97. Siena, L.A.; Ortiz, J.P.A.; Leblanc, O.; Pessino, S. *PNTGS1*-like expression during reproductive development supports a role for RNA methyltransferases in the aposporous pathway. *BMC Plant Biol.* **2014**, *14*, 297. [CrossRef] [PubMed]
98. Mancini, M.; Permingeat, H.; Colono, C.; Siena, L.; Pupilli, F.; Azzaro, C.; Dusi, D.M.; Tavares de Campos Carneiro, V.; Podio, M.; Seijo, J.G.; et al. The MAP3K-coding *QUI-GON JINN* (*QGJ*) gene is essential to the formation of unreduced embryo sacs in *Paspalum*. *Front. Plant Sci.* **2018**, *9*, 1547. [CrossRef] [PubMed]

99. Ochogavía, A.C.; Seijo, J.G.; González, A.M.; Podio, M.; Laspina, N.V.; Duarte Silveira, E.; Machado Lacerda, A.L.; Tavares de Campos Carneiro, V.; Pessino, S. Characterization of retrotransposon sequences expressed in inflorescences of apomictic and sexual *Paspalum notatum* plants. *Sex. Plant Reprod.* **2011**, *24*, 231–246. [CrossRef] [PubMed]
100. Podio, M.; Felitti, S.A.; Siena, L.A.; Delgado, L.; Mancini, M.; Seijo, G.; González, A.M.; Pessino, S.C.; Ortiz, J.P.A. Characterization and expression analysis of *SOMATIC EMBRYOGENESIS RECEPTOR KINASE* (*SERK*) genes in sexual and apomictic *Paspalum notatum*. *Plant Mol. Biol.* **2013**, *84*, 479–495. [CrossRef]
101. Ochogavía, A.; Galla, G.; Seijo, J.G.; González, A.M.; Bellucci, M.; Pupilli, F.; Barcaccia, G.; Albertini, E.; Pessino, S.C. Structure, target-specificity and expression of *PN_LNC_N13*, a lncRNA differentially expressed in apomictic and sexual *Paspalum notatum*. *Plant Mol. Biol.* **2018**, *96*, 53–67. [CrossRef]
102. Felitti, S.A.; Acuña, C.A.; Ortiz, J.P.A.; Quarin, C.L. Transcriptome analysis of seed development in apomictic *Paspalum notatum*. *Ann. Appl. Biol.* **2015**, *167*, 36–54. [CrossRef]
103. Depetris, M.B.; Acuña, C.A.; Pozzi, F.I.; Quarin, C.L.; Felitti, S.A. Identification of genes related to endosperm balance number insensitivity in *Paspalum notatum*. *Crop Sci.* **2018**, *58*, 813–822. [CrossRef]
104. Martin, J.A.; Wang, Z. Next-Generation transcriptome assembly. *Nat. Rev. Genet.* **2011**, *12*, 671–682. [CrossRef]
105. Gutiérrez-Gonzalez, J.J.; Garvin, D.F. De Novo transcriptome assembly in polyploid species. In *Methods in Molecular Biology: Oat*; Gasparis, S., Ed.; Humana Press: New York, NY, USA, 2017; Volume 1536, pp. 209–221.
106. Surget-Groba, Y.; Montoya-Burgos, J.I. Optimization of de novo transcriptome assembly from next generation sequencing data. *Genome Res.* **2010**, *20*, 1432–1440. [CrossRef] [PubMed]
107. Szklarczyk, D.; Gable, A.L.; Lyon, D.; Junge, A.; Wyder, S.; Huerta-Cepas, J.; Simonovic, M.; Doncheva, N.T.; Morris, J.H.; Bork, P.; et al. STRING v11: Protein-Protein association networks with increased coverage, supporting functional discovery in genome-wide experimental datasets. *Nucleic Acids Res.* **2019**, *47*, 607–613. [CrossRef] [PubMed]
108. Ortiz, J.P.A.; Leblanc, O.; Rohr, C.; Grisolia, M.; Siena, L.A.; Podio, M.; Colono, C.; Azzaro, C.; Pessino, S.C. Small RNA-seq reveals novel regulatory components for apomixis in *Paspalum notatum*. *BMC Genom.* **2019**, *20*, 487. [CrossRef] [PubMed]
109. Neiman, M.; Sharbel, T.F.; Schwander, T. Genetic causes of transitions from sexual reproduction to asexuality in plants and animals. *J. Evol. Biol.* **2014**, *27*, 1346–1359. [CrossRef] [PubMed]
110. León-Martínez, G.; Vielle-Calzada, J.-P. Apomixis in flowering plants: Developmental and evolutionary considerations. *Curr. Top. Dev. Biol.* **2019**, *131*, 565–604. [CrossRef]
111. Grimanelli, D. Epigenetic regulation of reproductive development and the emergence of apomixis in angiosperms. *Curr. Opin. Plant Biol.* **2012**, *15*, 57–62. [CrossRef]
112. Selva, J.P.; Siena, L.; Rodrigo, J.M.; Garbus, I.; Zappacosta, D.; Romero, J.R.; Ortiz, J.P.A.; Pessino, S.C.; Leblanc, O.; Echenique, V. Temporal and spatial expression of genes involved in DNA methylation during reproductive development of sexual and apomictic *Eragrostis curvula*. *Sci. Rep.* **2017**, *7*, 15092. [CrossRef]
113. Mirzaghaderi, G.; Horandl, E. The evolution of meiotic sex and its alternatives. *Proc. Biol. Sci.* **2016**, *283*, 1838. [CrossRef]
114. Conner, J.A.; Mookkan, M.; Huo, H.; Chae, K.; Ozias-Akins, P. A parthenogenesis gene of apomict origin elicits embryo formation from unfertilized eggs in a sexual plant. *Proc. Natl. Acad. Sci. USA* **2015**, *112*, 11205–11210. [CrossRef]
115. Amiteye, S.; Corral, J.M.; Vogel, H.; Kuhlmann, M.; Mette, M.F.; Sharbel, T.F. Novel microRNAs and microsatellite-like small RNAs in sexual and apomictic *Boechera* species. *MicroRNA* **2013**, *2*, 45–62. [CrossRef]
116. Tucker, M.R.; Okada, T.; Johnson, S.D.; Takaiwa, F.; Koltunow, A.M. Sporophytic ovule tissues modulate the initiation and progression of apomixis in *Hieracium*. *J. Exp. Bot.* **2012**, *63*, 3229–3241. [CrossRef] [PubMed]
117. Charlesworth, D. Plant contributions to our understanding of sex chromosome evolution. *New Phytol.* **2015**, *208*, 52–65. [CrossRef] [PubMed]
118. Graves, J.A. Evolution of vertebrate sex chromosomes and dosage compensation. *Nat. Rev. Genet.* **2016**, *17*, 33–46. [CrossRef] [PubMed]
119. Muyle, A.; Zemp, N.; Deschamps, C.; Mousset, S.; Widmer, A.; Marais, G.A. Rapid de novo evolution of X chromosome dosage compensation in *Silene latifolia*, a plant with young sex chromosomes. *PLoS Biol.* **2012**, *10*, e1001308. [CrossRef] [PubMed]

120. De Lannoy, C.; de Ridder, D.; Risse, J. The long reads ahead: De novo genome assembly using the MinION. *F1000Research* **2017**, *6*, 1083. [CrossRef]
121. Ho, S.S.; Urban, A.E.; Mills, R.E. Structural variation in the sequencing era. *Nat. Rev. Genet.* **2020**, *21*, 171–189. [CrossRef]
122. Kliver, S.; Rayko, M.; Komissarov, A.; Bakin, E.; Zhernakova, D.; Prasad, K.; Rushworth, C.; Baskar, R.; Smetanin, D.; Schmutz, J.; et al. Assembly of the *Boechera retrofracta* genome and evolutionary analysis of apomixis-associated genes. *Genes* **2018**, *9*, 185. [CrossRef]
123. Carballo, J.; Santos, B.A.C.M.; Zappacosta, D.; Garbus, I.; Selva, J.P.; Gallo, C.A.; Díaz, A.; Albertini, E.; Caccamo, M.; Echenique, V. A high-quality genome of *Eragrostis curvula* grass provides insights into Poaceae evolution and supports new strategies to enhance forage quality. *Sci. Rep.* **2019**, *9*, 10250. [CrossRef]
124. Mancini, M.; Woitovich, N.; Permingeat, H.; Podio, M.; Siena, L.A.; Ortiz, J.P.A.; Pessino, S.C.; Felitti, S.A. Development of a modified transformation platform for apomixis candidate genes research in *Paspalum notatum* (Bahiagrass). *In Vitro Cell. Dev. Plant* **2014**, *50*, 412–424. [CrossRef]
125. Altpeter, F.; James, V.A. Genetic transformation of turf-type bahiagrass (*Paspalum notatum* Flugge) by biolistic gene transfer. *Int. Turfgrass Soc. Res. J.* **2005**, *10*, 485–489.
126. Gondo, T.; Tsuruta, S.I.; Akashi, R.; Kawamura, O.; Hoffmann, F. Green, herbicide-resistant plants by particle inflow gun-mediated gene transfer to diploid bahiagrass (*Paspalum notatum*). *J. Plant Physiol.* **2005**, *162*, 1367–1375. [CrossRef] [PubMed]
127. Himuro, Y.; Gondo, T.; Yamakawa, K.; Akashi, R. Genetic transformation of bahiagrass (*Paspalum notatum* Flügge) by visually screening cells expressing green fluorescent protein. *Grassl. Sci.* **2009**, *55*, 216–220. [CrossRef]
128. Calderini, O.; Donnison, I.; Polegri, L.; Panara, F.; Thomas, A.; Arcioni, S.; Pupilli, F. Partial isolation of the genomic region linked with apomixis in *Paspalum simplex*. *Mol. Breed.* **2011**, *28*, 265–276. [CrossRef]
129. Brugnoli, E.A.; Urbani, M.H.; Quarin, C.L.; Zilli, A.L.; Martínez, E.J.; Acuña, C.A. Diversity in apomictic populations of *Paspalum simplex* Morong. *Crop Sci.* **2014**, *54*, 1656–1664. [CrossRef]
130. Vogel, K.P.; Burson, B.L. Breeding and genetics. In *Warm-Season (C4) Grasses*; Moser, L.E., Burson, B.L., Sollenberger, L.E., Eds.; John Wiley & Sons, Ltd.: Hoboken, NJ, USA, 2004; pp. 51–94.
131. Brugnoli, E.A.; Martínez, E.J.; Ferrari Usandizaga, S.C.; Zilli, A.L.; Urbani, M.H.; Acuña, C.A. Breeding tetraploid *Paspalum simplex*: Hybridization, early identification of apomicts, and impact of apomixis on hybrid performance. *Crop Sci.* **2019**, *59*, 1–8. [CrossRef]
132. Burton, G.W. Breeding Pensacola bahiagrass, *Paspalum notatum*: Method of reproduction. *Agron. J.* **1955**, *47*, 311–314. [CrossRef]
133. Dinato, N.B.; Santos, I.R.I.; Leonardecz, E.; Burson, B.L.; Quarín, C.L.; de Paula, A.F.; Fávero, A.P. Storage of bahiagrass pollen at different temperatures. *Crop Sci.* **2018**, *58*, 2391–2398. [CrossRef]
134. Weiler, R.L.; Dall'Agnol, M.; Simioni, C.; Krycki, K.C.; Pereira, E.A.; Medianeira Machado, J.; Minski da Motta, E.A. Intraspecific tetraploid hybrids of *Paspalum notatum*: Agronomic evaluation of segregating progeny. *Sci. Agric.* **2018**, *75*, 36–42. [CrossRef]
135. Da Motta, E.A.M.; Dall'Agnol, M.; Nascimento, F.L.; Pereira, E.A.; Machado, J.M.; Barbosa, M.R.; Simioni, C.; Ferreira, P.B. Forage performance of *Paspalum* hybrids from an interspecific cross. *Ciênc. Rural* **2016**, *46*, 1025–1031. [CrossRef]
136. Urbani, M.H.; Acuña, C.A.; Doval, D.W.; Sartor, M.E.; Galdeano, F.; Blount, A.R.; Quesenberry, K.H.; Mackowiak, C.L.; Quarin, C.L. Registration of 'Boyero UNNE' Bahiagrass. *J. Plant Regist.* **2017**, *11*, 26–32. [CrossRef]
137. Rebozzio, R.N.; Rodríguez, M.P.; Stein, J.; Ortiz, J.P.A.; Quarin, C.L.; Espinoza, F. Validation of molecular markers linked to apospory in tetraploid races of bahiagrass, *Paspalum notatum* Flüggé. *Mol. Breed.* **2012**, *29*, 189–198. [CrossRef]
138. Novo, P.E.; Acuña, C.A.; Urbani, M.H.; Galdeano, F.; Espinoza, F.; Quarin, C.L. Genetic transfer from several apomictic tetraploid *Paspalum* species to an elite group of sexual plants. *Crop Sci.* **2020**, *60*. [CrossRef]
139. Zilli, A.L.; Acuña, C.A.; Schulz, R.R.; Marcón, F.; Brugnoli, E.A.; Novo, S.F.; Quarin, C.L.; Martínez, E.J. Transference of natural diversity from the apomictic germplasm of *Paspalum notatum* to a sexual synthetic population. *Ann. Appl. Biol.* **2019**, *175*. [CrossRef]
140. Miles, J.W. Apomixis for cultivar development in tropical forage grasses. *Crop Sci.* **2007**, *47*, 238–249. [CrossRef]

141. Marcón, F.; Martínez, E.J.; Zilli, A.L.; Rodríguez, G.R.; Brugnoli, E.A.; Acuña, C.A. Recurrent phenotypic selection and recurrent selection based on combining ability in tetraploid bahiagrass. *Crop Sci.* **2020**, *60*. [CrossRef]

142. Sollenberger, L.E.; Cherney, D.J.R. Evaluating forage production and quality. In *Forages*; Barnes, R.F., Miller, D.A., Nelson, C.J., Eds.; Iowa State University Press: Ames, IA, USA, 1995; Volume 2, pp. 97–110.

143. Hojsgaard, D. Apomixis technology: Separating the wheat from the chaff. *Genes* **2020**, *11*, 411. [CrossRef]

genes

MDPI

Article

Whether Gametophytes Are Reduced or Unreduced in Angiosperms Might Be Determined Metabolically

Mayelyn Mateo de Arias [1,2,†], Lei Gao [1,3,†], David A. Sherwood [1,4], Krishna K. Dwivedi [5,6], Bo J. Price [1,7], Michelle Jamison [5,8], Becky M. Kowallis [5,9] and John G. Carman [1,*]

1 Plants, Soils, and Climate Department, Utah State University, Logan, UT 84322-4820, USA; mayelyn.mateo@intec.edu.do (M.M.d.A.); gao.lei@jju.edu.cn (L.G.); david@sherwoodpethealth.com (D.A.S.); bojprice57@gmail.com (B.J.P.)
2 Instituto Tecnológico de Santo Domingo, 10103 Santo Domingo, Dominican Republic
3 College of Pharmacy and Life Science, Jiujiang University, Jiujiang 332000, China
4 Sherwood Pet Health, Logan, UT 84321, USA
5 Caisson Laboratories, Inc., Smithfield, UT 84335, USA; dwivedi1976@gmail.com (K.K.D.); m.jamison@elitechgroup.com (M.J.); becky.kowallis@cytiva.com (B.M.K.)
6 Crop Improvement Division, Indian Grassland and Fodder Research Institute, 284003 Jhansi, India
7 Molecular Biology Program, University of Utah, Salt Lake City, UT 84112-5750, USA
8 Wescor, Inc. An Elitech Company, Logan, UT 84321, USA
9 Cytiva, Inc., Logan, UT 84321, USA
* Correspondence: john.carman@usu.edu; Tel.: +1-435-512-4913
† These authors contributed equally.

Received: 19 October 2020; Accepted: 27 November 2020; Published: 2 December 2020

check for updates

Abstract: In angiosperms, meiotic failure coupled with the formation of genetically unreduced gametophytes in ovules (apomeiosis) constitute major components of gametophytic apomixis. These aberrant developmental events are generally thought to be caused by mutation. However, efforts to locate the responsible mutations have failed. Herein, we tested a fundamentally different hypothesis: apomeiosis is a polyphenism of meiosis, with meiosis and apomeiosis being maintained by different states of metabolic homeostasis. Microarray analyses of ovules and pistils were used to differentiate meiotic from apomeiotic processes in *Boechera* (Brassicaceae). Genes associated with translation, cell division, epigenetic silencing, flowering, and meiosis characterized sexual *Boechera* (meiotic). In contrast, genes associated with stress responses, abscisic acid signaling, reactive oxygen species production, and stress attenuation mechanisms characterized apomictic *Boechera* (apomeiotic). We next tested whether these metabolic differences regulate reproductive mode. Apomeiosis switched to meiosis when premeiotic ovules of apomicts were cultured on media that increased oxidative stress. These treatments included drought, starvation, and H_2O_2 applications. In contrast, meiosis switched to apomeiosis when premeiotic pistils of sexual plants were cultured on media that relieved oxidative stress. These treatments included antioxidants, glucose, abscisic acid, fluridone, and 5-azacytidine. High-frequency apomeiosis was initiated in all sexual species tested: Brassicaceae, *Boechera stricta, Boechera exilis,* and *Arabidopsis thaliana*; Fabaceae, *Vigna unguiculata*; Asteraceae, *Antennaria dioica*. Unreduced gametophytes formed from ameiotic female and male sporocytes, first division restitution dyads, and nucellar cells. These results are consistent with modes of reproduction and types of apomixis, in natural apomicts, being regulated metabolically.

Keywords: 5-azacytidine; abscisic acid; apomixis; apospory; diplospory; expression profiling; fluridone; metabolic homeostasis; oxidative stress; sucrose non-fermenting-related protein kinase

1. Introduction

Apomixis renews life cycles by producing unreduced, mitotically active (parthenogenetic) gametes or gamete like cells [1], and it occurs in all eukaryote kingdoms [2–5]. Once discovered in angiosperms [6], it was soon documented that the timing of apomixis induction during ovule development and the types of ovule cells involved vary among taxa. This temporal and spatial variation became the focal point for defining multiple types of gametophytic apomixis in angiosperms, and within the first decade of the 20th century, three morphologically distinct types had been described [4,7]. Subsequent Mendelian analyses suggested that these types, and their elements, i.e., apomeiosis (unreduced gametophyte formation), parthenogenesis (embryo formation without fertilization), and endosperm formation without the normal 2M:1P (maternal paternal) genome ratio, are under separate genetic controls [4,8–11]. Accordingly, it became widely viewed that the different apomixis types are caused by different mutations that in mysterious ways destabilize meiosis (megasporogenesis), gametophyte (embryo sac) formation, egg formation, and syngamy [8–13].

In the Antennaria and Eragrostis types of diplospory (gonial apospory), the unreduced gametophyte, with its parthenogenetic egg, forms from a meiosis-aborted megasporocyte (megaspore mother cell, MMC). In Taraxacum type diplospory, meiosis aborts slightly later, during the meiotic prophase stage of the MMC. The heterotypic division (reductional) then fails and a first division restitution replaces it. The equational mitotic-like second division then occurs producing two unreduced spores. The unreduced gametophyte, with its parthenogenetic egg, then forms from one of the unreduced spores. In Hieracium type apospory, sex aborts during meiosis, and the unreduced gametophyte, with its parthenogenetic egg, forms from a cell of the ovule wall (nucellus). In sporophytic apomixis (adventitious embryony), sex—if it aborts at all—aborts after a genetically reduced egg forms. The clonal embryo then develops adventitiously, from an unreduced egg-like cell that forms in the ovule wall, and this embryo is nourished by the sexually derived endosperm [8].

Based on numerous reports of apomixis switching to sex in response to stress, reviewed in [14,15], we hypothesized that (i) sexual development is induced and maintained by a stress-like state of metabolic homeostasis, and (ii) ovules in apomictic plants maintain a state of metabolic homeostasis that detoxifies oxidative stress and induces apomixis. To test this, we profiled the reproductive tissues of sexual and apomictic *Boechera* (Brassicaceae). Our goal was to identify differentially expressed genes (DEGs) indicative of differences in metabolic states. Upon observing such differences, we designed and conducted whole-plant stress physiology experiments that tested the effects of drought and heat on frequencies of sexual and apomictic development in facultatively apomictic *Boechera*. We also excised immature pistils, containing ovules in the MMC stage, of sexual and apomictic *Boechera*, sexual *Arabidopsis thaliana* (L.) Heynh. (Brassicaceae), and sexual *Vigna unguiculata* (L.) Walp. (Fabaceae). These were then cultured on media amended with chemicals selected to alter stress associated metabolic pathways. Additionally, we placed cut stems of male dioecious *Antennaria dioica* (L.) Gaertn. (Asteraceae), which contained premeiotic microsporocytes (pollen mother cells, PMCs), in chemically amended liquid media. Through these experiments, we demonstrated that apomeiotic spore and gametophyte formation, of various types (Taraxacum, Antennaria, and Hieracium), are inducible in sexual species of *Boechera*, *Arabidopsis*, *Vigna*, and *Antennaria* by altering homeostasis-based processes of stress perception and attenuation.

2. Materials and Methods

2.1. Plant Materials and Growth Conditions

Plants of the following taxa were grown from seed: diploid aposporous *Boechera microphylla* (Nuttall) Dorn from Millard County, UT USA (*Boechera imnahaensis* × *yellowstonensis*, UT10003), diploid diplosporous and aposporous *B. microphylla* from Cache County, UT USA (UT05001), diploid diplosporous and aposporous *Boechera retrofracta* × *stricta* (CO11010), diploid sexual *A. thaliana* (Col O), diploid sexual *Boechera exilis* (A. Nelson) Dorn (NV14003), diploid sexual *Boechera stricta* (Graham)

Al-Shehbaz (UT10007), diploid sexual *V. unguiculata* (cowpea), diploid diplosporous *B. exilis* × *retrofracta* (UT11004), triploid diplosporous *Boechera* cf. *gunnisoniana* (Rollins) W.A. Weber (CO11005), and diploid diplosporous *Boechera lignifera* (A. Nelson) W. A. Weber (*B. exilis* × *thompsonii*, WY05001). A population of diplosporous hybrids of mixed parentage, which resemble sexual *B. formosa* (Greene) Windham and Al-Shehbaz (UT10006), are referred to herein as *B.* × *formosa* and were transplanted from native habitats, six from Duchesne County, UT USA (40.169 N, −110.328 W) and 58 from Carbon County, UT USA (39.547 N, −110.649 W). Male plants of diploid *A. dioica* were transplanted from landscape plantings at Utah State University, Logan, UT USA. Unless otherwise indicated, field collection information, ploidy, and voucher specimen numbers of *Boechera* taxa are from [16,17]. Seed germination, establishment, vernalization, and growing conditions were as previously reported [16].

2.2. Microarray Analyses

Based on the close phylogenic relationship between *Boechera* and *Arabidopsis* [18,19], Affymetrix (Santa Clara, CA, USA) ATH1 gene chips were chosen in 2005 for *Boechera* profiling, and this profiling system was retained for consistency through 2013. Ovules and pistils were collected from 9:00 A.M. to 12:00 noon, placed in Nuclease-Free water (Ambion, Life Technologies Corporation, Carlsbad, CA, USA), and measured. Needles from 1.0 mL syringes were used to excise ovules. Instruments and work surfaces were cleaned with RNAse ZAP (Ambion) prior to dissection and at 20–30 min intervals. Numbers of ovules per replicate were ca. 1600, 1400, 1200, and 800 for the MMC, active meiocyte (meiotic MMC, AM), young 1–4 nucleate gametophyte (YG), and mature gametophyte (MG) stages, respectively. Eighty pistils were obtained per replicate for *B. stricta*. Ovules and pistils were immediately placed in RNAlater (Ambion) and stored at −80 °C. RNA was extracted and purified from ovules using TRIzol reagent (Invitrogen, Life Technologies Corporation) and RNeasy columns (Qiagen, Germantown, MD, USA) and from pistils using PureLink® RNA Mini Kits (Ambion, Austin, TX, USA) and TURBO DNA-free™ Kits (Ambion, Austin, TX, USA). NanoDrop (Thermo Scientific, Wilmington, DE, USA) and Bioanalyzer (Agilent, Santa Clara, CA, USA) instruments were used to determine RNA yield and integrity. Affymetrix two-cycle cDNA synthesis kits (P/N 900494) were used for cDNA synthesis and RNA amplification of ovule RNA. The MEGAscript T7 kit (Ambion) with unlabeled nucleotides was used for first-cycle cRNA amplifications. For in vitro transcription (IVT), 20 μL cDNA was mixed with 30 μL first-cycle IVT master mix, and reactions were incubated at 37 °C for 24 h. Second-cycle first and second strand cDNA syntheses were performed as above. The Affymetrix GeneChip IVT labeling kit (P/N 900449) was used for the second IVT. First and second cycle cRNA was purified using Affymetrix sample cleanup modules P/N 900371, and cRNA was fragmented at 94 °C for 35 min. Pistil RNA (500 ng) was amplified, labeled, and fragmented using GeneChip® 3′ IVT Express Kits (Affymetrix). ATH1 arrays were hybridized for 16 h at 45 °C using 15 μg of labeled and fragmented cRNA (Affymetrix Technical Analysis Manual). Arrays were stained with streptavidin-phycoerythrin in an Affymetrix GeneChip® Fluidics Workstation 400 and scanned with a GeneChip® Scanner 3000.

2.3. Expression Profiling

Cel files (ncbi.nlm.nih.gov/geo/query/acc.cgi?acc=GSE156684) were deleted if RMAExpress (rmaexpress.bmbolstad.com/) values exceeded 99% of the RLE-NUSE T2 multivariate statistic. The remaining files were RMA normalized using BRB-ArrayTools (4.6.0-beta-2, Dr. Richard Simon, and the BRB-ArrayTools Development Team, linus.nci.nih.gov/BRB-ArrayTools.html). Low variability genes (<10% of expression values exceeding 1.5-fold change from the median) or genes missing >50% of their expression data were removed. The BRB-ArrayTools class comparison option (random variance model with a normalized significance level of 0.001 for univariate tests) was used to identify DEGs. Where multiple ATH1 probe sets identified the same gene (~0.8% of probe sets), probe sets producing the greatest fold change were retained. Each comparison used to identify DEGs involved detectable expression data from ca. 16,000 (59%) of 27,048 *Boechera* genes [19]. Tests for DEGs were performed within stages for all pairwise cross taxa comparisons and within taxa but across stages. Multi-stage comparisons, across taxa, were performed when comparisons of adjoining stages within taxa produced few or no DEGs.

2.4. Gene Ontology Analyses

Two overrepresentation gene ontology (OGO) analyses (Panther Version 15, pantherdb.org/, 2020–10-9 release) were performed for all comparisons that produced ≥10 DEGs, one for DEGs upregulated in one taxon and one for DEGs upregulated in the other. For Panther enrichment GO (EGO) analyses, gene trimming criteria were relaxed. Here, genes were removed if 0–40% of sample values differed from the median by ≥8%. The sample value threshold was varied so as to retain 10–16 thousand genes. Since EGO analyses are based on all expression data, rather than just DEGs, it was possible to perform these analyses for all within stage cross taxa comparisons. All OGO and EGO analyses were corrected for false discovery rate (FDR, $p \leq 0.05$). GO terms (all GOs of DEGs, OGOs, and EGOs) were partitioned into higher-order groups, e.g., [20–22], that were defined by the broader biological concepts being investigated. Partitioning was performed manually by assigning each GO term to one of four groups and to one of 13 subgroups (Tables S1 and S2). Subgroup frequencies were then compared across taxa by chi-square tests for independence [23], and subgroups responsible for significant pooled chi-square values were identified based on p values of adjusted residuals [24] (Table S3). Molecular pathway components were identified by searching DEG narratives (Araport11, 1/1/2020 release) and additional literature as referenced for DEGs in TAIR (www.arabidopsis.org).

2.5. Expression Verification

Fourteen DEGs identified between the YG stages of *B. microphylla* and *B.* ×*formosa* were selected for quantitative reverse transcription PCR (qRT-PCR). RNA was converted to cDNA using Superscript III First-Strand Synthesis kits (Invitrogen). Primers were designed from *A. thaliana* data using Primer3 (bioinfo.ut.ee/primer3/), quality checked by PrimerSelect (dnastar.com), and searched (blast.ncbi.nlm.nih.gov/Blast.cgi) for like sequences. Sequences predicted to form internal loops or dimers or to be homologous to other genes were discarded. Primers were tested using *Boechera* gDNA. qRT-PCR reactions (25 μL) contained QuantiTect SYBR Green (Qiagen), 200–250 nM forward and reverse primers and 3–5 μL of dilute cDNA. Reaction cycles were: 50 °C, 2 min; 95 °C, 15 min; 40 cycles of 94 °C for 15 s, 58 °C for 30 s, and 72 °C for 30 s. Melt curve analyses were performed from 65 °C to 95 °C in 5 sec 0.5 °C increments. A DNA Engine (Opticon 2, Continuous Fluorescence Detection System) with MicroAmp Optical 96-well plates (Bio-Rad, Hercules, CA, USA) was used. Three reps with two technical reps each were conducted. Expression values and standard deviations were calculated using the comparative C_T method using separate sets of housekeeping replicates for each gene. Delta Ct values were normalized (delta-delta Ct method) and subjected to Student's *t*-tests ($p \leq 0.05$).

2.6. Drought and Heat-Stress Experiments

Seedlings (ca. 60 days post-planting) were randomly assigned to three groups, well-watered, drought-stressed, and heat plus drought-stressed (8–12 plants per taxon per group) and vernalized (60 days, 4 °C). After vernalization, plants targeted for the well-watered and droughted treatments were grown in the greenhouse, and plants targeted for heat plus drought were grown in a growth chamber (32 °C, 500 μmol m^{-2} s^{-1} photosynthetic photon flux, 16/8 h day/night photoperiod). To minimize surface evaporation, perlite was added to the surface of pots, and field capacity weights were obtained (weighed 2 h after drenching). Enough water was added each day to the well-watered plants to bring them to field capacity (based on pot weight). Water was added to droughted plants each day to equal 50% of the transpiration rate of well-watered plants. Upon flowering, pistils were collected for cytology and expression profiling. Ovule parameters were compared by chi-square tests [25].

2.7. In Vitro Pistil Culture Experiments

Chemicals were from MilliporeSigma (St. Louis, MO, USA) unless otherwise indicated. The basal tissue culture medium was MS [26] salts and vitamins (Caisson Laboratories, Smithfield, UT, USA) with 20 g L^{-1} sucrose and solidified with 1.5 g L^{-1} Phytogel (pH 5.7 prior to autoclaving). For osmotic stress, basal media were amended with 0, 20, 30, 40, or 60 g L^{-1} polyethylene glycol (PEG) 6000. Water potentials (Ψ) were estimated by adding the water potential of basal MS medium (ca. −0.5 MPa [27]) to the water potential of PEG in water (ca. 0.0, −0.013, −0.022, −0.034, −0.065 MPa, respectively [28]) and multiplying the sums by 1.075 [28,29]. Brassinozole (BZR, dissolved in minimal DMSO), epibrassinolide (epiBL, dissolved in 80% ethanol), ABA (dissolved in ethanol), fluridone (dissolved in DMSO), 5-azacytidine (5-azaC, dissolved in water), and (S)-2-aminobutane-1,4-dithiol hydrochloride (DTBA, dissolved in water) were added by filter sterilization after media were autoclaved and before solidification. Two factorial experiments were conducted to determine the effects of multiple treatments on ovule development in *B. stricta* and *A. thaliana*. Variables tested were: sucrose (20 or 30 g L^{-1}), glucose (20 g L^{-1}); MS or B5 vitamins; 0.5 mmol L^{-1} 5-azaC; 0.5 or 1.0 µmol L^{-1} epiBL; 0.5 or 1.0 µmol L^{-1} DTBA; and 0.5 and 1.0 µmol L^{-1} epiBL and DTBA. Results were combined to show the main effects. A third factorial experiment was conducted to determine the effects of multiple treatments on ovule development in *V. unguiculata*. The basal medium here was amended with B5 vitamins, 0.5 µmol L^{-1} naphthalene acetic acid (NAA), and 5.0 µmol L^{-1} 6-benzylaminopurine (BAP). Variables tested were: 60 g L^{-1} sucrose or 30 g L^{-1} glucose; 1.0 µmol L^{-1} epiBL; 1.0 µmol L^{-1} DTBA; and 1.0 µmol L^{-1} epiBL and DTBA. Results were combined to show main effects. Pistils for individual and factorial experiments were staged by length (based on predetermined ovule development stages), aseptically excised, and immediately transferred without disinfestation to media (horizontally, ca. 30% of pistil bodies pressed into the medium). For experiments involving a pretreatment, excised pistils were soaked in liquid media followed by transfer to solidified media. Ovule parameters were tabulated 24–72 h after culture initiation and analyzed by chi-square tests [25].

2.8. Cytoembryological Analyses

Pistils and anthers were fixed, cleared, and observed by differential interference contrast microscopy as previously described [16]. The following characteristics were recorded: (i) MMCs (without a large vacuole), (ii) Antennaria type diplosporous 1–2 nucleate gametophytes (with one or more large vacuoles; ADGs), (iii) sexual or Taraxacum type diplosporous dyads, (iv) sexual tetrads (no large vacuoles), (v) sexual 1–2 nucleate gametophytes (with one or more large vacuoles and degenerating tetrad remnants still visible); (vi) Taraxacum type diplosporous 1–2 nucleate gametophytes (with one or more large vacuoles; TDGs), (vii) aposporous initials (enlarged non-vacuolate nucellar cells), (viii) aposporous gametophytes (nucellar cells with one or more large vacuoles and one or more nuclei), (ix) microspore dyads and tetrads, and (x) 1–3 nucleate unreduced male gametophytes. Developmental stages of ovules per pistil were recorded, and pistil lengths corresponding to the following *A. thaliana* stages [30] were recorded: 1-I, late pre-MMC; 1-II, ovule meristems protruding from placentae (precedes MMC formation); 2-II, enlarging MMC; 2-III, mature MMC; 2-IV, meiosis; 3-III, large vacuole present in immature gametophyte.

3. Results

3.1. Expression Profiling

We recently documented reproduction in 44 *Boechera* taxa. These ranged from 100% sexual in species to nearly 100% apomictic, by TDG formation, in hybrids. Between these extremes were facultative apomicts that expressed combinations of meiosis and TDG and/or Hieracium-type gametophyte (HAG) formation [16]. Here, we report transcriptome analyses for four of these taxa: sexual *B. stricta*, strongly diplosporous *B.* × *formosa*, facultatively aposporous, and diplosporous *B. microphylla* (Cache County, UT), and facultatively diplosporous *B. lignifera*. Pistil lengths, array normalization results, array vs. qRT-PCR comparisons, and primers used for qRT-PCR are shown in Figures S1–S3 and Table S4, respectively.

Comparisons among the MMC (megasporocyte), AM (active meiocyte), and YG (young gametophyte) stages within diplosporous *B. lignifera* and aposporous *B. microphylla* produced few DEGs (Table S5). This near absence of DEGs may have occurred because pistil length intervals for the respective samples differed only minimally (Figure S1). However, few DEGs were detected at these stages even when comparisons were made across taxa. Hence, we combined the MMC, AM, and YG stages, within taxa, to make a single *B. lignifera* vs. *B. microphylla* comparison, which yielded 283 DEGs. At the MG stage (mature gametophyte), 609 DEGs were detected (Figure 1a, Table S5).

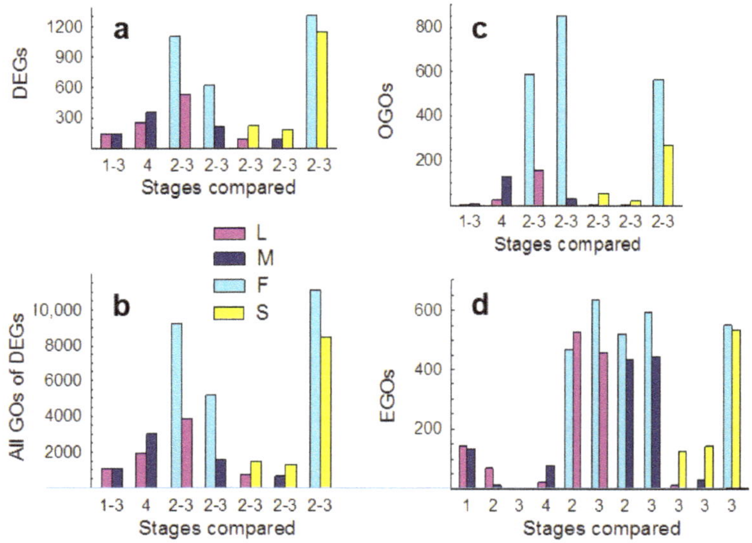

Figure 1. Differentially expressed genes (DEGs) and associated gene ontology categories (GOs) by comparison: (**a**) DEGs, (**b**) all GOs of all DEGs, (**c**) overrepresented GOs (OGOs), (**d**) enriched GOs (EGOs). L (*Boechera lignifera*), M (*Boechera microphylla*), F (*Boechera* × *formosa*), S (*Boechera stricta*), 1 (megasporocyte (MMC) stage), 2 (active meiocyte stage), 3 (young gametophyte stage), 4 (mature gametophyte stage).

Across all comparisons, 3723 distinct DEGs were detected (Figure 1a; Table S5), which are linked to 3980 distinct GO terms (Figure 1b; Table S6). Overrepresented GOs (OGOs) occurs when more DEGs are linked to a GO term than expected by chance (pantherdb.org/). Among our OGO comparisons (Figure 1c), 930 distinct OGOs were detected (Table S7). EGOs occur when more genes of a GO term are up or down-regulated (significant or not) than expected by chance (pantherdb.org/). Since all expression data are used, EGO analyses can detect shifts in gene expression that may be missed by OGO analyses [31,32]. Among our EGO comparisons (Figure 1d), 1773 distinct EGOs were detected (Table S8). Collectively, 4822 distinct GO terms were identified. To more readily visualize large metabolism-related shifts in gene expression, we partitioned the 4822 GO terms into 13 user-defined subgroups (Tables S1 and S2). Subgroup frequencies per taxon were then compared among taxa by chi-square analyses (Figures 2 and 3; Table S3; see Materials and Methods).

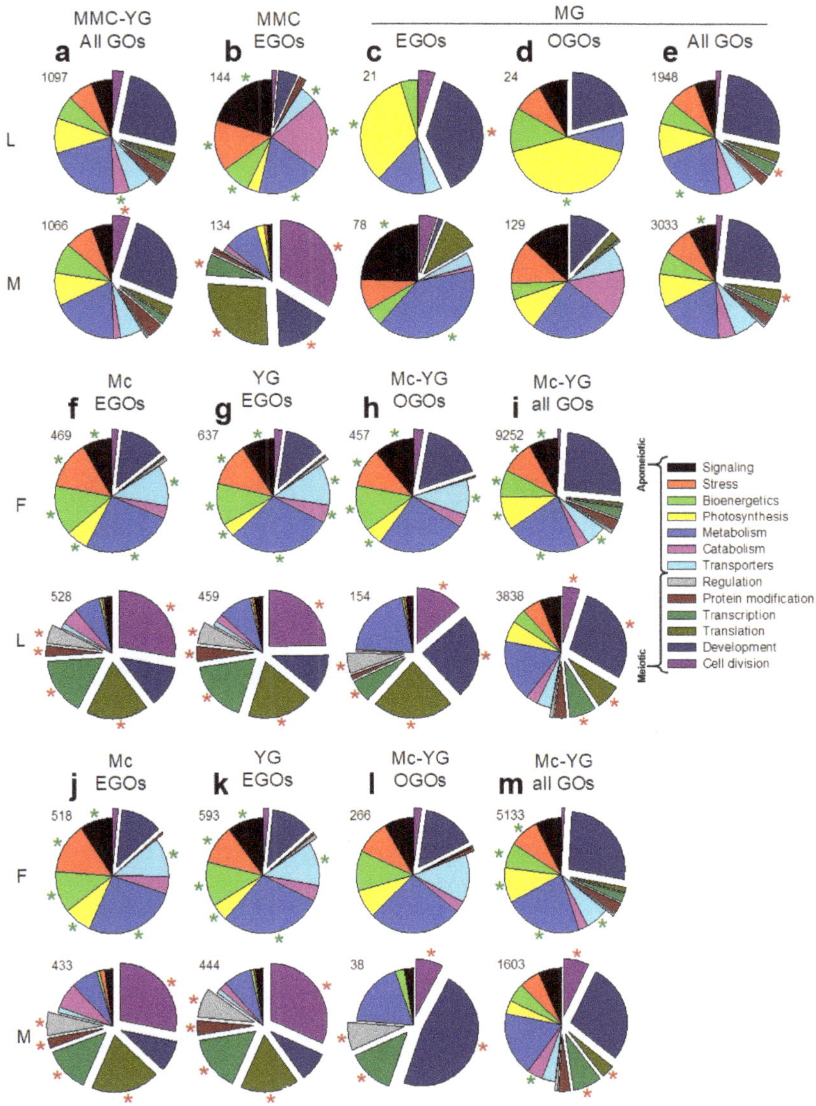

Figure 2. Pie charts of enriched gene ontology categories (EGOs), overexpressed GOs (OGOs), and GOs of differentially expressed genes (all GOs) by GO group and by comparison: (**a–e**), L vs. M; (**f–i**), F vs. L; (**j–m**), F vs. M. F (*Boechera* × *formosa*), L (*Boechera lignifera*), M (*Boechera microphylla*), MMC (megasporocyte stage), Mc (meiocyte stage), YG (young gametophyte stage), MG (mature gametophyte stage). Normal and exploded sections tended to be more frequently observed in less and more meiotic taxa, respectively. Starred GO groups differed significantly in frequency ($p \leq 0.05$): green stars, normal; red stars, exploded; values next to pies are total numbers of GOs observed.

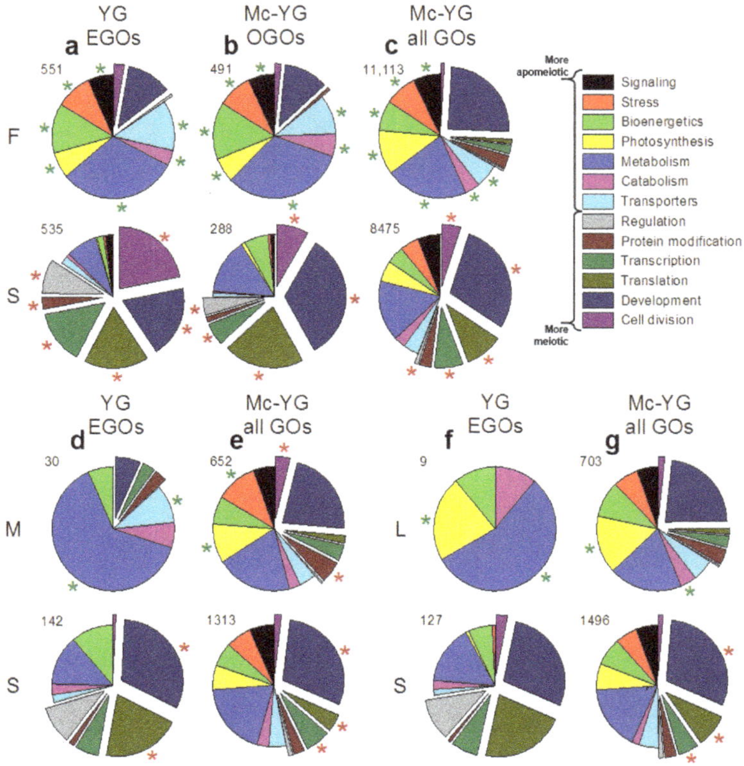

Figure 3. Pie charts of enriched gene ontology categories (EGOs), overexpressed GOs (OGOs), and GOs of differentially expressed genes (all GOs) by GO group and by comparison: (**a–c**), F vs. S; (**d**,**e**), M vs. S; (**f**,**g**), L vs. S. F (*Boechera × formosa*), L (*Boechera lignifera*), M (*Boechera microphylla*), S (*Boechera stricta*), Mc (meiocyte stage), YG (young gametophyte stage). Normal and exploded subgroups tended to be more frequently observed in less and more meiotic taxa, respectively. Starred subgroups differed significantly in frequency ($p \leq 0.05$): green stars, normal; red stars, exploded; values next to pies are total numbers of GOs observed.

Only 5–20% of ovules of the TDG forming *B. lignifera* were facultatively meiotic [16]. Hence, we tested whether fewer meiosis related GO terms (all GOs of DEGs) occurred in *B. lignifera* compared to the TDG and HAG forming *B. microphylla*, where ca. 40% of ovules were meiotic [16]. For the combined MMC to YG comparison, meiosis and gametophyte related GOs occurred nearly exclusively in the more meiotic (and aposporous) *B. microphylla* (Table S6, lines 1–35 vs. 1098–1151, see also lines 2164–92 vs. 4112–62). These meiosis and gametophyte related GOs belong to the 'cell division' subgroup, which collectively represented a significantly larger percentage of GO terms in the aposporous *B. microphylla* compared to the diplosporous *B. lignifera* (Figure 2a, red star).

We used EGO analyses to compare *B. lignifera* to *B. microphylla* at each stage individually. The MMC stage produced the most EGOs (Figure 1d). Here, signaling, stress, bioenergetics, metabolism, catabolism, and transporter EGOs occurred more frequently in ovules of the TDG forming *B. lignifera* (least meiotic) (Figure 2b). These included carbohydrate, fatty acid, and protein metabolism processes, responses to organic and inorganic stimuli, activation of homeostasis processes, regulation of ROS and other stresses, and ABA signaling (Table S8, lines 11–79, 84–91, 95–144). In contrast, transcription, translation, development, and cell division EGOs were more frequent in ovules of the more meiotic HAG forming *B. microphylla* (Figure 2b). These included 13 meiosis associated terms and numerous ribosome-associated terms (Table S8, lines 145–210, 232–75). Fewer differences were observed at the AM, YG, and MG stages (Table S8, lines 290–471; Figure 2a,b). Collectively, these findings revealed major metabolism-related differences. Gene expression in the TDG forming *B. lignifera* favored signaling, stress, and catabolism. In the meiotic HAG forming *B. microphylla*, gene expression favored transcription, translation, development, and cell division.

Meiosis associated EGO subgroups at the MMC stage (exploded pie sections, Figure 2b) comprised ca. 10% and 80% of the diplosporous *B. lignifera* and the more meiotic but aposporous *B. microphylla*, respectively. Interestingly, nearly the same percentages were observed when diplosporous *B. lignifera* and aposporous *B. microphylla* were compared individually to the more diplosporous *B. ×formosa* (Figure 2f,j). These results reveal a continuum where meiosis associated gene expression was highest in *B. microphylla* and lowest in *B. ×formosa* with *B. lignifera* in between (compare Figure 2b,f,j). One of our objectives was to determine if these distinct shifts in gene expression are responsible for reproductive mode and apomixis type or are simply coincidental (discussed below). Analyses involving EGOs, OGOs, and all GOs of DEGs at the YG or AM/YG stages generally mirrored those at the AM stage for comparisons between the nearly obligate apomictic *B. ×formosa* and the facultatively apomictic *B. lignifera* and *B. microphylla* (Figure 2f–i, j–m).

For this study, we collected *Boechera* species from native habitats in the Great Basin and Rocky Mountain Cordillera of North America [16]. However, embryological analyses revealed apomixis for taxa we thought, at the time of collection, were sexual. We eventually found sexual species, but resources for ovule dissections were no longer available. Instead, we created expression profiles from whole pistils of sexual *B. stricta*, each pistil of which contained ca. 160–180 ovules. We then compared ovule profiles of the apomicts to the same pistil profile from sexual *B. stricta* (Figure 3). We first considered differential gene expression between the nearly obligate TDG forming *B. ×formosa* versus the completely sexual *B. stricta*. This comparison produced the most DEGs, OGOs, all GOs of DEGs and EGOs (Figure 1). Importantly, it produced highly similar frequency patterns for meiosis-associated vs. apomeiosis-associated EGO subgroups (Figure 3a) as had been observed for the more vs. the less meiotic apomicts (Figure 2b,f,j). Similar GO subgroup patterns were observed for OGOs and all GOs of DEGs (Figure 3a–c) and for the facultative apomicts compared to sexual *B. stricta* (Figure 3d–g).

In the three most meiotic taxa, *B. stricta* (100%), *B. microphylla* (40%), and *B. lignifera* (5–20%), 17, 14, and 10 meiosis associated EGOs per comparison were detected compared to the nearly obligate *B. x formosa*. Meiosis associated EGOs in these comparisons were not enriched in *B. × formosa* (Table S8: 458–65 vs. 927–1073, 1455–69 vs. 2092–205, 2551–8 vs. 3072–193, 3505–14 vs. 4098–237, 4851–65 vs. 5402–518). The more meiotic taxon of each comparison was also enriched for EGOs associated with transcriptional regulators involving DNA conformation changes, histone and DNA modifications, chromatin remodeling and silencing, epigenetic regulation, and posttranscriptional gene silencing by RNA (Table S8: 92–4 vs. 232–9; 816–23 vs. 1259–345; 1946–55 vs. 2369–449; 2942–8 vs. 3348–403; 3962–72 vs. 4386–4455; 5306–11 vs. 5756–831). In contrast, the more diplosporous taxon of each comparison tended to be enriched for EGOs associated with carbohydrate, fatty acid, and lipid metabolism; photosynthesis; responses to biotic and abiotic stimuli; kinase and phosphatase activities; calcium signaling; hormone metabolic pathways; second messenger signaling; ABA, salt, drought, hypoxia and ROS responses; regulation of ABA signaling; detoxification processes; homeostasis processes; pH regulation; peroxidase activities; and the biosynthesis of ascorbate, flavonoid and glutathione antioxidants (Table S8, 11–91 and 95–144 vs. 211–29 and 276–8; 306–42 vs. 348–57; 525–815 and 824–926 vs. 1137–213 and 1440–54; 1545–945 and 1956–2091 vs. 2257–319 and 2537–50; 2620–941 and 2949–3071 vs. 3232–3306 and 3489–504; 3586–961 and 3973–4097 vs. 4276–330 and 4530–4541; 4542–51 vs. 4592–625 and 4678; 4681–706 vs. 4755–96; 4931–5303 and 5312–401 vs. 5620–90 and 5923–36).

3.2. Molecular Pathways Upregulated in Meiotic Taxa

To better understand large scale shifts in gene expression between meiotic and apomeiotic taxa (Figures 2 and 3), we searched gene narratives (Araport11 release, 1 January 2020) of our DEGs (Table S5) using the following terms: ABA, AGL, AGO, AOX, ARF, ascorbate, BR, Ca, calmodulin, catalase, chromatin modification, detoxification, epigenetic, ERF, ethylene, flavonol, gametophyte, glutathione, H_2O_2, H3K4, HDAC, histone, homeostasis, HSP, imprinting, LEA, MAPK, meiosis, miRNA, NaCl, oxidative damage, peroxidase, PP2C, RBOH, RDDM, ribosome, RNAi, ROS, salt responsive, silencing, siRNA, SnRK, SPO, stress, superoxide dismutase, thioredoxin, translation, and transport. After manual screening, 514 DEGs were identified for further analyses across all comparisons (Table S9).

Carbon starvation downregulates translation by inactivating TARGET OF RAPAMYCIN (TOR) [33]. DEGs in the red and green boxes (Figure 4a) are translation machinery genes (Table S9). Most of these were upregulated in *B. stricta* (red box), which suggests that diplosporous *B.* × *formosa* was experiencing carbon starvation. Further evidence for this involved upregulated levels of ARABIDOPSIS NAC DOMAIN CONTAINING PROTEIN 2 (ATAF1), a transcription factor (TF) that upregulates TREHALASE1 (TRE1). TRE1 promotes the formation of a sugar starvation metabolome by catabolizing the signaling molecule trehalose-6-phosphate, thus further reducing TOR signaling by increasing SUCROSE-NON-FERMENTING-RELATED PROTEIN KINASE-1 (SnRK1) activity [34–36]. Carbon starvation also upregulates the vacuolar glucose exporter EARLY RESPONSIVE TO DEHYDRATION-LIKE 6 gene (*ERDL6*) [37] and the DUAL-SPECIFICITY PROTEIN PHOSPHATASE 4 gene (*SEX4*) [38], the products of which replenish monosaccharide levels to maintain primary metabolism. SEX4 might also inactivate SnRK1, thus reinitiating TOR activity in the presence of restored energy levels [39]. These three-carbon starvation genes were upregulated in diplosporous *B.* × *formosa* (Figure 4a). In contrast, SNF1-RELATED PROTEIN KINASE REGULATORY β SUBUNIT 1 (*AKINβ1*), which encodes an SnRK1 subunit that directs SnRK1 to the nucleus for gene regulation but also silences SnRK1 upon *N*-myristoylation [40], was downregulated in *B.* × *formosa* (Figure 4a), possibly as a feedback mechanism responding to TOR inactivation.

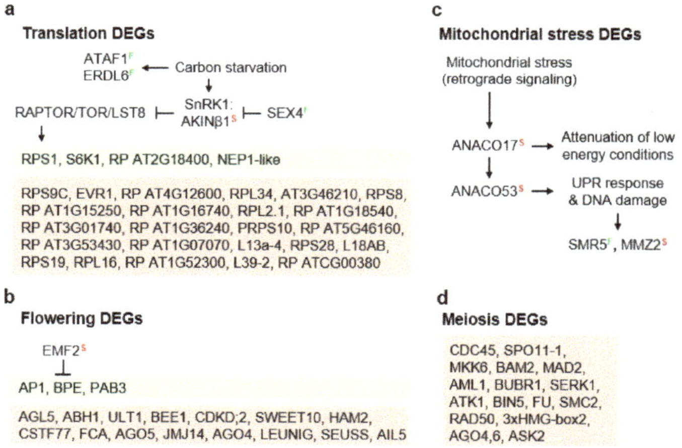

Figure 4. Differentially expressed genes (DEGs) observed between diplosporous *Boechera* × *formosa* and sexual *Boechera stricta* (Tables S5 and S9) as associated with: (**a**), translation; (**b**), flowering; (**c**), mitochondrial stress; and (**d**), meiosis. Superscripts identify taxa in which DEGs are upregulated; F, *B.* × *formosa*; S, *B. stricta*; red and green highlights, up or downregulation in *B. stricta*, respectively; arrows indicate upregulation or activation; blunt end lines represent downregulation or suppression; RP, ribosomal protein; UPR, unfolded protein response.

The early flowering gene EMBRYONIC FLOWER 2 (*EMF2*) encodes a polycomb repressive complex member that suppresses flowering [41]. It was downregulated in *B.* ×*formosa* (Figure 4b), suggesting that this apomict may exhibit an early flowering phenotype, which is common for apomicts in nature [4]. Additionally, genes responsible for flower formation were mostly downregulated in *B.* ×*formosa* (Figure 4b, red box). Taken together, these results suggest that the earliness of flowering was favored in apomictic *B.* ×*formosa* but that several epigenetically regulated programs of reproduction were only weakly expressed in this taxon.

In apomictic strains of *Saccharomyces cerevisiae* (brewer's yeast), heat stress switches apomeiotic divisions, which normally produce dyads of unreduced spores, to meiotic divisions, which produce tetrads of reduced spores. However, when mitochondrial transcription in apomictic yeast is interrupted by erythromycin, the heat-stress switch from apomeiosis to meiosis no longer occurs, i.e., the yeast remains apomictic. This finding, plus research in other organisms, suggests that meiosis depends on stress-induced nucleomitochondrial communication [3,42,43]. In the present study, genes for two nuclear TFs, NAC DOMAIN CONTAINING PROTEIN 17 (ANACO17) and 53 (ANACO53), which participate in drought-induced mitochondrial retrograde signaling, were upregulated in sexual *B. stricta* (downregulated in the TDG-forming 1st division restitution *B.* ×*formosa*, Figure 4c). This is consistent with a mitochondrial role in the initiation and completion of meiosis I, a role that is downregulated during 1st division restitution apomeioses. ANAC017 normally induces senescence, suppresses auxin and chloroplast function, reprograms mitochondrial metabolism for lower energy conditions, and possibly activates SnRK1 [44]. It also upregulates ANACO53, which upregulates ROS production genes and initiates unfolded protein responses (UPRs) that attenuate proteotoxic conditions in the ER [45]. MMS ZWEI HOMOLOGUE 2 (MMZ2), which was strongly upregulated in *B. stricta*, is upregulated by DNA damage and UPRs, and it normally participates in post replicative DNA repair. Since these genes were upregulated in sexual *B. stricta*, their expression is consistent with the view that meiosis evolved as a DNA repair mechanism [15,46–48] at about the same time that early eukaryotes were domesticating their oxygen respiring endocytically obtained organelles [46]. The downregulation of meiosis genes and mitochondrial stress response genes in apomictic *B.* ×*formosa* (Figure 4c,d) is further evidence that reproductive processes, from flower formation through meiosis, were only weakly supported at the molecular level in apomictic *B.* ×*formosa*.

DEGs of two additional signaling related categories, gene silencing and BR signaling, also provide clues concerning sex apomixis switching (Figure 5). Chromatin remodeling by methylation and demethylation occurs throughout plant development. In sexual cell lineages, RNA-directed DNA methylation (RdDM) is required to accomplish meiosis-specific gene splicing [49]. RdDM may also occur in tissues that produce sex cells because it silences potentially harmful transposons [49]. In the present study, silencing processes by RdDM and by polycomb repressive complex activity were upregulated in the ovules of sexual *B. stricta* vs. apomictic *B.* ×*formosa* (Figure 5a–c). This again is consistent with other upregulated reproductive processes in *B. stricta* as described above.

ALFIN-LIKE 7 (AL7) is a nuclear-localized H3K4me3 binding protein that directs polycomb repressive complex (PcG) 1s (PCR1s) to transcriptionally active chromatin where the H3K4me3 expression marks are converted to H2Aub1 repression marks. These are then stabilized by PCR2 H3K27me3 silencing marks [50]. ULTRAPETALA1 (ULT1) is a trithorax factor that functions as an anti-repressor to counteract PcG silencing at 1000s of loci, including many genes where specific up or downregulation is required for normal reproductive development [51,52]. The upregulation in *B. stricta* of AL7 and ULT1 genes (Figure 5a–c) is further evidence that floral development at the molecular level was strongly supported in sexual *B. stricta* but weakly supported in apomictic *B.* ×*formosa*. Overall, it appears that *B. stricta* ovules embarked on epigenetic paths of sexual development that apomeiotic ovules were reluctant to follow.

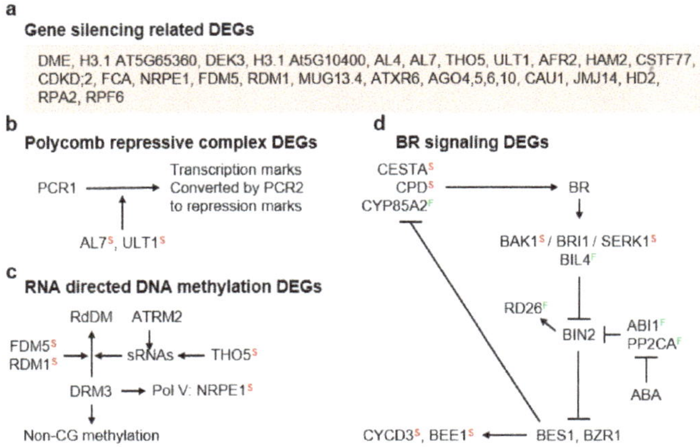

Figure 5. Differentially expressed genes (DEGs) observed between diplosporous *Boechera* ×*formosa* and sexual *Boechera stricta* (see Tables S5 and S9) as associated with: (**a**), gene silencing; (**b**), polycomb repressive complex formation, (**c**), RNA directed DNA methylation, and (**d**), brassinosteroid (BR) signaling. Superscripts identify taxa in which DEGs are upregulated; F, *B.* ×*formosa*; S, *B. stricta*; red highlights, upregulation in sexual *B. stricta*; arrows indicate upregulation or activation; blunt end lines represent downregulation or suppression.

BR interacts with other processes to regulate growth and development as well as abiotic and biotic stresses [53,54]. The major hub of BR signaling is the kinase BRASSINOSTEROID-INSENSITIVE 2 (BIN2), which phosphorylation inactivates the major BR TFs BRASSINAZOLE-RESISTANT 1 (BZR1) and BRI1-EMS-SUPPRESSOR 1 (BES1). When BIN2 is inhibited, BZR1, and BES1 are activated, and this activation upregulates cascades of gene expression involving multiple downstream TFs and ca. 2000 downstream genes [55]. BR is a growth-promoting hormone. Hence, it is generally upregulated by TOR and downregulated by ABA. Consistent with this pattern, the two BR regulated signaling genes identified in our *B.* ×*formosa* vs. *B. stricta* comparison (Table S9) were upregulated in *B. stricta* (Figure 5d): CYCLIN D3;1 (CYCD3) is a cyclin that initiates cell division [56], and BR-ENHANCED EXPRESSION 1 (BEE1) combines with a blue light photoreceptor, CRYPTOCHROME 2 (CRY2), to initiate photoperiod induced flowering [57]. Several additional BR associated genes were upregulated in *B. stricta*. These included two BR synthesis genes (Figure 5d), which, like TOR (Figure 4a), are upregulated under favorable conditions: HALF FILLED (CESTA) is a TF that upregulates multiple BR biosynthesis genes [58], including CONSTITUTIVE PHOTOMORPHOGENIC DWARF (*CPD*), which is critical for BR biosynthesis [59]. Additionally upregulated in *B. stricta* were two BR receptor-associated proteins, including BRI1-ASSOCIATED RECEPTOR KINASE (BAK1) [54] and SOMATIC EMBRYOGENESIS RECEPTOR-LIKE KINASE 1 (SERK1) [60]. Both of these proteins enhance BR signal transduction across membranes by forming protective complexes with the cell surface BR receptor kinase BR INSENSITIVE 1 (BRI1). One of these, SERK1, also binds EXCESS MICROSPOROCYTES 1 (EMS1) during reproductive development in processes that convert somatic cells to reproductive cells (microsporocytes) in anthers [61]. The downregulation of these BR genes in *B.* × *formosa* is consistent with other downregulated reproduction processes in *B.* ×*formosa* (Figures 4 and 5).

3.3. Molecular Pathways Upregulated in Apomeiotic Taxa

Several BR associated genes were upregulated in *B.* × *formosa* (Figure 5d). One of these, BRASSINOSTEROID-6-OXIDASE 2 (*CYP85A2*), is a circadian rhythm regulated gene that encodes an enzyme that catalyzes the last step in BR biosynthesis. When upregulated, CYP85A2 increases BR but decreases ABA signaling [62]. BRZ-INSENSITIVE-LONG HYPOCOTYLS 4 (*BIL4*) was also upregulated in *B.* × *formosa*. It encodes a BRI1-interacting transmembrane protein that is involved in cell elongation. BIL4 suppresses BRI1 degradation and physically maintains BRI1 close to the plasma membrane [63]. Two protein phosphatase 2C (PP2C) genes, ABA INSENSITIVE 1 (*ABI1*) and PROTEIN PHOSPHATASE 2CA (*PP2CA*) were also upregulated in *B.* × *formosa*. These genes upregulate BZR1 and BES1 by phosphatase inactivating BIN2 (Figure 5d). They are also upregulated by ABA as part of a feedback inhibition mechanism where (i) stress induces ABA biosynthesis, (ii) ABA inactivates ABI1 and PP2CA, (iii) ABA signaling upregulates ABI1 and PP2CA biosynthesis, and (iv) additional accumulation of ABI1 and PP2CA suppresses ABA signaling (Figure 6). These variable findings are not surprising given the abundant crosstalk that occurs among tissues and cells during hormonal and environmental signaling [54].

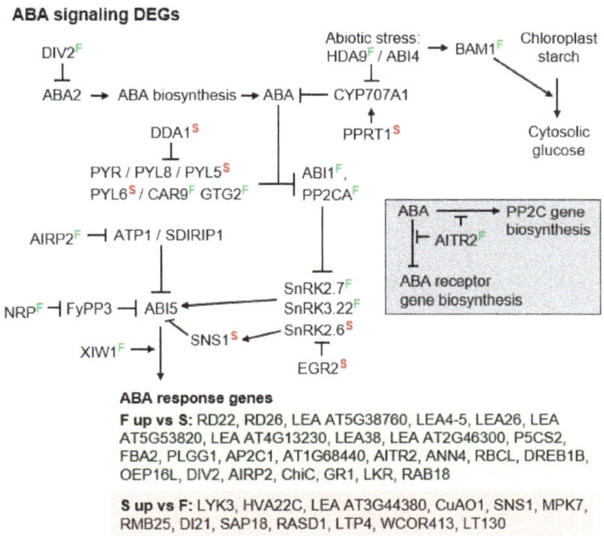

Figure 6. ABA signaling associated differentially expressed genes (DEGs) observed between diplosporous *Boechera* × *formosa* and sexual *Boechera stricta* (see Tables S5 and S9). Superscripts identify taxa in which DEGs are upregulated; F, *B.* × *formosa*; S, *B. stricta*; red and green highlights, up and down-regulation in *B. stricta*, respectively; arrows indicate upregulation or activation; blunt end lines represent downregulation or suppression.

A very different state of cellular homeostasis exists during stress than during favorable conditions. During stress, ABA-inactivated PP2Cs permit the self-activation by autophosphorylation of SnRK2,3s as well as BIN2. Activated BIN2 phosphorylation inactivates the growth-enhancing BR TFs but phosphorylation activates the NAC TF RESPONSIVE TO DESICCATION 26 (RD26), which upregulates multiple stress response genes [53,64]. *RD26* was upregulated 4.2 fold in *B.* × *formosa* vs. *B. stricta* (Figure 5d, Table S9). Activated SnRK2,3s upregulate ABA specific stress response genes [65] (Figure 6). These data suggest that apomictic ovules experienced states of cellular homeostasis that were maintained by enhanced stress perception with feedback suppression of ABA signaling occurring through *CYP85A2*, *ABI1*, and *PP2CA* [53,62,64,65].

The molecular pathways presented to this point suggest that apomeiosis is caused by carbon starvation (Figure 4a, low-level TOR activity), ABA (Figure 6), and upregulated drought response genes (Figure 5d). However, these pathways do not explain how a stress-like state of metabolic homeostasis induces apomixis, especially when stressed apomicts in many cases revert to sex [15,66]. Recognizing this paradox, we hypothesized that the stress-like state of metabolic homeostasis observed in ovules of apomicts is a chronic condition that is maintained by imbalances in wild type gene expression that occur as a result of hybridization, polyploidization, or other chromosome aberrations [13,67,68]. We further hypothesized that this chronic stress response state enhances stress tolerance, and it is this enhanced stress tolerance that simultaneously suppresses sex while inducing apomixis. Accordingly, we searched our DEGs for candidates that might be responsible for shifting metabolic homeostasis toward a chronic dual state of stress-response but even stronger stress-attenuation (Figures 6 and 7).

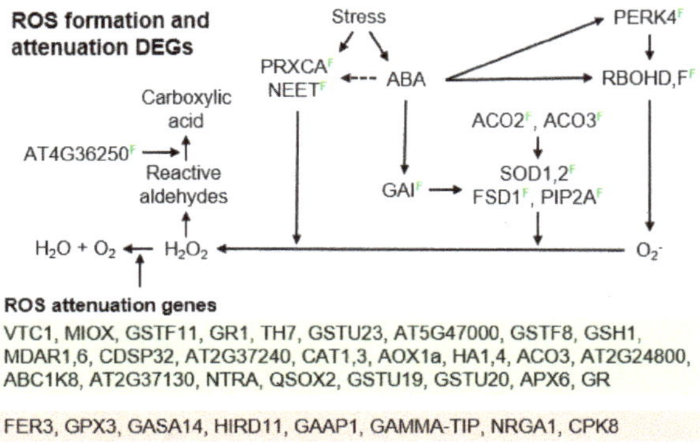

Figure 7. Reactive oxygen species (ROS) formation and attenuation pathways and associated differentially expressed genes (DEGs) as observed between diplosporous *Boechera* × *formosa* and sexual *Boechera stricta* (see Tables S5 and S9). Superscripts identify taxa in which DEGs were upregulated; F, *B.* × *formosa*; S, *B. stricta*; red and green highlights indicate up and down-regulation in *B. stricta*, respectively; arrows indicate upregulation or activation; blunt end lines represent downregulation or suppression.

We first considered the possibility that the biosynthesis of ABA itself, which is upregulated by stress [69], might be chronically upregulated in the apomicts. ABA DEFICIENT 2 (*ABA2*), an ABA biosynthesis gene, was upregulated in aposporous *B. microphylla* vs. sexual *B. stricta*. However, this gene was also upregulated in *B. microphylla* vs. *B.* × *formosa* (Table S9), which is inconsistent with ABA synthesis being a major contributor. Furthermore, the ABA and salt stress-induced TF DIVARICATA 2 (DIV2), which negatively regulates ABA biosynthesis by downregulating *ABA1,2* [70], was upregulated in *B.* × *formosa* vs. *B. stricta* (Table S9). Hence, a chronic stress-like phenotype being caused by upregulated ABA biosynthesis is unlikely.

Decreased catabolism of ABA is another possibility. Here, a drought-upregulated histone deacetylase gene, HISTONE DEACETYLASE 9 (*HDA9*), was upregulated 1.9 fold in *B.* × *formosa* vs. *B. stricta* (Figure 6). HDA9 forms a complex with ABA INSENSITIVE 4 (ABI4) that downregulates the major ABA catabolism genes CYTOCHROME P450, FAMILY 707, SUBFAMILY A, POLYPEPTIDE 1,2 (*CYP707A1,2*) [71]. Moreover, the salt stress-induced PROTEIN WITH THE RING DOMAIN AND TMEMB_185A 1 (*PPRT1*) gene [72], a C3HC4 zinc-finger ubiquitin E3 ligase that upregulates the ABA catabolism genes *CYP707A1,3* [73], was downregulated in *B.* × *formosa* vs. *B. stricta*. Hence, the up and downregulation of *HDA9* and *PPRT1*, respectively, in *B.* × *formosa* likely contributed to its stronger ABA signaling. Hence, from our expression profiles, suppressing ABA catabolism is a reasonable candidate for upregulating ABA signaling and inducing a homeostatic state of stress attenuation.

Upregulating the biosynthesis of ABA receptor genes is another way in which ABA signaling in *B.* × *formosa* might have been enhanced. Normally, stress-induced ABA feedback inhibits ABA receptor gene biosynthesis, and this may have occurred for the ABA receptor genes PYRABACTIN RESISTANCE 1-LIKE 5,6 (*PYL5,6*) (downregulated 67% in *B.* × *formosa* vs. *B. stricta*; PYL5 was also downregulated 67% in *B. lignifera* and *B. microphylla* vs. *B. stricta*; Table S9). However, the downregulation of ABA receptor genes by feedback mechanisms is suppressed by the ABA-INDUCED TRANSCRIPTION REPRESSOR 2 gene (*AITR2*) [74,75], which was upregulated 2.1-fold in *B.* × *formosa* vs. *B. stricta*.

Additionally, the C2-DOMAIN ABA-RELATED 9 gene (*CAR9*), an ABA receptor (Figure 6), was upregulated 4.3-fold in *B.* × *formosa* vs. *B. stricta*. CAR9 positively regulates ABA signaling by interacting with ABA and various PYRABACTIN RESISTANCE (PYR) and PYL ABA receptors at the plasma membrane [76]. Furthermore, the ABA receptor gene GPCR-TYPE G PROTEIN 2 (*GTG2*), which encodes a G-protein with nine transmembrane domains [77], was also upregulated 2.1-fold in *B.* × *formosa* vs. *B. stricta*. Hence, ABA receptor upregulation in *B.* × *formosa* likely enhanced ABA signaling in this apomict.

Downregulating proteins that target ABA receptors for degradation, could also enhance ABA signaling. Here, the DET1 AND DDB1 ASSOCIATED 1 gene (*DDA1*), which downregulates ABA signaling by targeting *PYL4,8,9* for degradation [78], was downregulated in all three apomicts (Table S9). Hence, DDA1 is a reasonable candidate for shifting metabolic homeostasis toward ABA signaling. In general, the majority of ABA receptors were, for various reasons, upregulated in apomictic *B.* × *formosa* (Figure 6).

Under favorable conditions, PP2Cs promote growth by promoting BR signaling (Figure 5d, by inactivating BIN2), promoting TOR activity (Figure 4a, by inactivating SnRK1), and preventing ABA signaling (Figure 6, by inactivating SnRK2,3s). During stress, ABA synthesis is upregulated, ABA and ABA receptors bind, and the receptor complexes inactivate PP2Cs, thus suppressing growth. In addition to downregulating ABA receptors, stress-induced ABA signaling also upregulates PP2C biosynthesis (Figure 6), thus reducing ABA signaling by feedback inhibition. Hence, the upregulated PP2Cs (ABI1 and PP2CA) in *B.* × *formosa* (Table S9) is further evidence of upregulated ABA signaling. Interestingly, AITR2, which suppresses ABA-induced suppression of ABA receptor gene biosynthesis, also suppresses ABA-induced upregulation of PP2C biosynthesis [75] (Figure 6). AITR2 was upregulated in *B. lignifera* vs. *B. stricta* 29.5 fold and in *B.* × *formosa* vs. *B. stricta* 2.1 fold (Table S9). Hence, *AITR2* is an interesting candidate gene for upregulating ABA signaling.

Next in ABA signaling are certain SnRK2,3s that phosphorylation activate the major ABA TF ABA INSENSITIVE 5 (ABI5) and other ABA TFs (Figure 6). During favorable conditions, SnRK2,3s are inactivated by PP2Cs. Hence, by upregulating the biosynthesis of SnRK2,3s, ABA signaling might be upregulated simply by overwhelming PP2C suppression. Here, *SnRK2.6* is an interesting case. It was upregulated in the opposite direction, i.e., 3-fold in sexual *B. stricta* vs. *B.* × *formosa* (Table S9). However, an even greater upregulation of *SnRK2.6* was likely prevented by the EARLY GROWTH RESPONSE 2 gene *EGR2*, which also was upregulated in *B. stricta* (2.4-fold). In warm temperatures, the EGR2 phosphatase inactivates SnRK2.6. In cold weather, EGR2 is inactivated, and SnRK2.6 upregulates cold stress genes [79]. Furthermore, SnRK2.6 activates SNRK2-SUBSTRATE 1 (SNS1), which suppresses ABA signaling [80]. Compared to diplosporous *B.* × *formosa*, *SNS1* was upregulated 5.3, 3.4, and 3.2-fold in the more meiotic *B. stricta*, *B. microphylla*, and *B. lignifera*, respectively (Table S9). Hence, SnRK upregulation in *B.* × *formosa* may have also contributed to a stress-attenuation-based metabolic homeostasis.

Several SnRK2,3s upregulate stress response genes independently of ABA. SnRK2.7 is activated by salt and osmotic stress [81], and its synthesis was upregulated in *B.* × *formosa* vs. *B. stricta* 2.5-fold. The *SnRK3.22* gene (PROTEIN KINASE SOS2-LIKE 5, *PKS5*), which activates ABI5 [82], is also upregulated by salt stress, and it was upregulated 3-fold in *B.* × *formosa* vs. *B. stricta*. In the absence of stress, SnRK3.22 inactivates the H^+ ATPase driven SALT OVERLY SENSITIVE 2 Na^+/H^+ antiporter gene (*SOS2*). During salt stress, this antiporter is activated, by Ca^{2+} signaling, whereupon it exports Na^+ from cells [83].

Other targets for altering ABA signaling are the ABA TFs themselves (Figure 6). XPO1-INTERACTING WD40 PROTEIN 1 (*XIW1*) was upregulated in *B.* × *formosa* 3.5, 2.8, and 2.1-fold vs. *B. stricta*, *B. microphylla*, and *B. lignifera*, respectively. It upregulates ABA signaling by stabilizing ABI5 [84]. ASPARAGINE-RICH PROTEIN (*NRP*) was upregulated in *B.* × *formosa* 3.0, 5.6, and 4.1-fold vs. *B. stricta*, *B. microphylla*, and *B. lignifera*, respectively. It upregulates ABA signaling by targeting FLOWER-SPECIFIC, PHYTOCHROME-ASSOCIATED PROTEIN PHOSPHATASE 3 (FyPP3) for degradation [85]. FyPP3 decreases ABA signaling by inactivating ABI5. Likewise, ABA INSENSITIVE RING PROTEIN 2 (*AIRP2*) was upregulated in *B. ×formosa* 2.2-fold vs. *B. stricta*. It encodes a cytosolic RING-type E3 ubiquitin ligase that upregulates ABA signaling, in response to ABA or salt stress, by targeting ATAIRP2 TARGET PROTEIN 1 (ATP1) for ubiquitination [86]. In the absence of salt stress, ATP1 decreases ABA signaling by silencing ABI5. Accordingly, enhancing ABI5 stability represents another means whereby ABA signaling was enhanced in *B. ×formosa*. Other heat, salt, and osmotic stress induced DEGs that were upregulated in *B. ×formosa* vs. sexual *B. stricta* included *SHE, HSP70 AT3G12580, HSP AT3G47940, AT1G49670, MYC2, CML9, CALS5, ZAT10, AKR4C8, ADC2, DOB1, ChlAKR, DREB A-2, HSP70 AT3G09440, FBP7, TPR4, DJC77, ALDA10A9, HMAD1, AT4G29780, RDUF1, CYS-3A, SRP3, HDA9, SnRK2.7, MPK4, AT4G37530, AT2G37130, AT2G24800, GSTU19, APG6, EGR3, DJA7, ERDL6, DNAJ, RD21, SAT32, MKK9, NHX1, PCaP1, TYDC, AT2G25940,* and *GSTF11* (Tables S5 and S9). In most of these cases, it is not known whether ABA alone or other stresses activate their expression.

Many of the ABA response DEGs (Figure 6) are also upregulated by heat, salt, or osmotic stress (Table S9), suggesting that hypersensitivity to these conditions may have also contributed to the chronic stress induction/tolerance homeostasis observed in *B. ×formosa*. Many of these were late embryogenesis abundant (LEA) genes, the proteins of which protect cells and proteins from the deleterious effects of ROS, drought, and freezing [87]. Others included: DELTA 1-PYRROLINE-5-CARBOXYLATE SYNTHASE 2 (*P5CS2*), upregulated 9.9, 17.1 and 6.6-fold in *B. ×formosa* vs. *B. stricta, B. microphylla* and *B. lignifera*, respectively; FRUCTOSE-BISPHOSPHATE ALDOLASE 2 (*FBA2*) [88], upregulated 8.8, 8.8 and 7.3-fold in *B. ×formosa* vs. *B. stricta, B. microphylla* and *B. lignifera*, respectively; RAB GTPASE HOMOLOG B18 (*RAB18*), upregulated 4.7-fold in *B. ×formosa* vs. *B. stricta, B. microphylla* and *B. lignifera*, which is a dehydrin that interacts with aquaporins in responses to drought stress [89]; RESPONSIVE TO DESICCATION 22 (*RD22*) [90], a NAC TF upregulated in *B. ×formosa* 11.9, 9.4 and 6.3-fold vs. *B. stricta, B. microphylla* and *B. lignifera*, respectively; RD26 [64], a NAC TF upregulated 4.2-fold in *B. ×formosa* vs. *B. stricta*, which upregulates drought response genes; GLUTATHIONE-DISULFIDE REDUCTASE (*GR1*), upregulated 1.9-fold in *B. ×formosa* vs. *B. stricta*, which is a major peroxisomal glutathione reductase that functions in the H_2O_2 detoxification ascorbate glutathione pathway [91]; and the protein-serine/threonine phosphatase gene *AP2C1*, upregulated 12.1 and 6.1-fold in *B. ×formosa* vs. *B. stricta* and *B. lignifera*, respectively. AP2C1 modulates K^+ homeostasis by dephosphorylation inactivating CBL-INTERACTING PROTEIN KINASE 9 (CIPK9), which normally enhances K^+ ion uptake through K^+ channels, particularly under low K^+ conditions [92]. It also moderates responses to pathogens by dephosphorylation inactivating MAP KINASE 4,6 (MPK4,6) in the cytoplasm and nucleus. These MAPKs affect the expression of hundreds of genes by modifying the activity of 88 TFs from 21 TF families [93].

ABA signaling upregulates ROS synthesis, ROS signaling, and large suites of peroxisomal and other antioxidant network genes [91,94]. The first step in ROS signaling is an upregulation of ROS biosynthesis by peroxidases or RBOHs. Five ROS biosynthesis genes were highly upregulated in the TDG forming *B. ×formosa*. PEROXIDASE CA (*PRXCA*) (Figure 7) encodes a cell wall peroxidase that produces H_2O_2 in the apoplast [95]. It was upregulated in *B. ×formosa* 11.4, 10.0, and 8.5-fold vs. the meiosis exhibiting *B. stricta, B. microphylla,* and *B. lignifera*, respectively (Table S9). NEET is a ROS homeostasis regulator that was upregulated in *B. ×formosa* 7.8, 4.9, and 10.7-fold vs. *B. stricta, B. microphylla,* and *B. lignifera*. AT-NEET (*NEET*) encodes a protein that regulates Fe and ROS homeostasis by controlling the biogenesis of Fe-S clusters that function in electron transport. In the absence of NEET, excess Fe-S clusters form and react with oxygen to produce ROS [96]. PROLINE-RICH RECEPTOR-LIKE PROTEIN KINASE 4 (PERK4) enhances ROS accumulation by enabling ABA-induced expression of RBOH genes, including *RBOHC* [97]. PERK4 was upregulated in *B. ×formosa* 6.7, 5.6, and 6.5-fold vs. *B. stricta, B. microphylla,* and *B. lignifera*. RBOHD produces superoxide in response to heat and wounding and its gene was upregulated in *B. ×formosa* 2.9, 5.1, and 8.2-fold vs. *B. stricta, B. microphylla,* and *B. lignifera*. Both *RBOHD* and *RBOHF* are upregulated by ABA, and their expression, through ROS signaling, is critical to cell survival under ER stress [98]. *RBOHF* was upregulated in *B.* ×*formosa* 2.9, 4.2, and 3.5-fold vs. *B. stricta*, *B. microphylla*, and *B. lignifera*.

Conversion of superoxides to H_2O_2 by superoxide dismutases (SODs) is the next step in ROS signaling. Five genes responsible for this conversion were upregulated in *B.* ×*formosa*. ACONITASE 2,3 (ACO2,3) upregulate SUPEROXIDE DISMUTASE 2 (*SOD2*) (Figure 7). *ACO2* was upregulated 3.1, 4.1, and 5.0-fold in *B.* ×*formosa* vs. *B. stricta*, *B. microphylla*, and *B. lignifera*, respectively, and *ACO3* was upregulated 2.5-fold in *B.* ×*formosa* vs. *B. stricta* (Table S9). SOD1 is a cytosolic copper/zinc SOD. Under favorable conditions, its transcripts are cleaved by miR398, which decreases SOD1 levels. Under stress conditions, miR398 expression is suppressed, and SOD1 levels increase [99]. SOD1 was upregulated 3.2-fold in *B.* ×*formosa* vs. *B. lignifera* and 2.9-fold in *B. microphylla* vs. *B. lignifera*. SOD2 is a chloroplastic copper/zinc SOD. Its transcripts are also cleaved by miR398 under favorable conditions. Under stress conditions, miR398 expression is suppressed, and SOD2 levels increase [99,100]. SOD2 was upregulated 1.8-fold in *B.* ×*formosa* vs. *B. microphylla*, 2.8-fold in *B.* ×*formosa* vs. *B. lignifera*, and 2.0-fold in *B. microphylla* vs. *B. lignifera*. FE SUPEROXIDE DISMUTASE 1 (FSD1) is a salt stress-induced chloroplastic SOD and is also an osmoprotectant [101]. It was upregulated 5.6-fold in *B.* ×*formosa* vs. *B. stricta*, 12.2-fold in *B.* ×*formosa* vs. *B. microphylla*, and 5.8-fold in *B. lignifera* vs. *B. microphylla*.

Several other DEGs increase H_2O_2 levels in cells. PLASMA MEMBRANE INTRINSIC PROTEIN 2A (PIP2A) is a water and H_2O_2 transmembrane aquaporin that is downregulated by salt stress. Apoplastic RBOHF-generated ROS cause PIP2A clustering, and subsequent endocytosis of the PIP2A clusters appears to regulate ROS levels [102]. *PIP2A* transcripts were upregulated 2.8-fold in *B.* ×*formosa* vs. *B. stricta* and 5.3-fold in *B.* ×*formosa* vs. *B. lignifera* (Table S9). GIBBERELLIC ACID INSENSITIVE (GAI) is a transcriptional regulator that interacts with TFs. It is upregulated by ABA, ethylene, and salt stress and increases drought tolerance by restraining cell proliferation and expansion. It possibly increases H_2O_2 levels by upregulating SODs [103]. GAI was upregulated 2.4-fold in *B.* ×*formosa* vs. *B. stricta* and 2.2-fold in *B.* ×*formosa* vs. *B. lignifera*.

Several H_2O_2 attenuation processes were upregulated in *B.* ×*formosa*. The most prominent of these, based on numbers of DEGs, was the classic glutathione ascorbate H_2O_2 attenuation cycle. Here, VITAMIN C DEFECTIVE 1 (*VTC1*), which encodes an essential enzyme in ascorbate biosynthesis, the levels of which are correlated with ascorbate levels [104], was upregulated 2.1, 4.5, and 2.1-fold in *B.* ×*formosa* vs. *B. stricta*, *B. microphylla* and *B. lignifera*, respectively (Table S9). MYO-INOSITOL OXYGENASE 2 (MIOX) also functions in ascorbate biosynthesis [105]. *MIOX* was upregulated 4.7, 3.7, and 3.3-fold in *B.* ×*formosa* vs. *B. stricta*, *B. microphylla*, and *B. lignifera*, respectively. ALTERNATIVE OXIDASE 1A (AOX1a) is an ascorbate peroxidase that is highly suppressed at the transcriptional and post-transcriptional levels under favorable growing conditions. During stress, *AOX1a* is upregulated by multiple stress-related TFs including SnRK1 [106]. It was upregulated 1.7 and 2.1-fold in *B.* ×*formosa* vs. *B. stricta* and *B. lignifera*. MONODEHYDROASCORBATE REDUCTASE 1 (MDAR1) is a peroxisomal monodehydroascorbate reductase that functions in the ascorbate-glutathione cycle. *MDAR1* was upregulated 6.2 and 2.6-fold in *B.* ×*formosa* vs. *B. stricta* and *B. lignifera*. ASCORBATE PEROXIDASE 6 (APX6) is a major cytosolic ascorbate peroxidase [107]. *APX6* was upregulated 3.9 and 1.9-fold in *B.* ×*formosa* vs. *B. stricta* and *B. microphylla*. It was also upregulated 2.1-fold in *B. microphylla* vs. *B. stricta* and 2.0-fold in *B. lignifera* vs. *B. microphylla*. GLUTAMATE-CYSTEINE LIGASE (GSH1) catalyzes the rate-limiting step in glutathione biosynthesis. *GSH1* was upregulated 2.0 and 4.3-fold in *B.* ×*formosa* vs. *B. stricta* and *B. lignifera*. GLUTATHIONE REDUCTASE (GR) is a plastid glutathione reductase that maintains a highly reduced glutathione pool that balances ROS and enables the redox buffering required to maintain efficient photosynthesis [108]. *GR* was upregulated 2.9, 7.7, and 7.3-fold in *B.* ×*formosa* vs. *B. stricta*, *B. microphylla*, and *B. lignifera*, respectively. It was also upregulated 2.9-fold in *B. stricta* vs. *B. microphylla*. GLUTATHIONE-DISULFIDE REDUCTASE (GR1) is a glutathione reductase that is upregulated by drought, dehydration, and ABA and is a major peroxisomal glutathione reductase in the glutathione ascorbate cycle [91]. *GR1* was upregulated in *B.* ×*formosa* 1.9-fold vs. *B. stricta*. Five glutathione transferases that are involved in ROS attenuation were also upregulated. GLUTATHIONE S-TRANSFERASE U23 (GSTU23) detoxifies ROS under high oxidative stress conditions [109]. *GSTU23* was upregulated 3.8, 4.0, and 3.9-fold in *B. ×formosa* vs. *B. stricta*, *B. microphylla*, and *B. lignifera*, respectively. GLUTATHIONE S-TRANSFERASE F11 (*GSTF11*) was upregulated 7.7, 12.3, and 20.9-fold in *B. ×formosa* vs. *B. stricta*, *B. microphylla*, and *B. lignifera*, respectively. *GSTF11* was also upregulated 3.3-fold in *B. stricta* vs. *B. lignifera*. GSTF8 reduces oxidative damage by binding to heavy metals. *GSTF8* was upregulated 2.0 and 2.5-fold in *B. ×formosa* vs. *B. stricta* and *B. microphylla*, respectively. *GSTU19* is plant-specific and is upregulated by salt, drought, and oxidative stress. Its protein functions redundantly with other GSTs to maintain redox homeostasis [110]. It was upregulated 1.7-fold in *B. ×formosa* vs. *B. stricta*. GSTU20 is required for plant responses to far-red light [111]. *GSTU20* is upregulated by stress and was upregulated 10.6 and 3.2-fold in *B. ×formosa* vs. *B. stricta* and *B. lignifera*, respectively. It was also upregulated 5.9-fold in *B. microphylla* vs. *B. stricta* and 3.6-fold in *B. lignifera* vs. *B. stricta*.

Several other classes of ROS attenuation genes were upregulated in *B. × formosa*, including catalases, peroxidases, and thioredoxins. CATALASE 1 (CAT1) catalyzes the reduction of H_2O_2. *CAT1* is upregulated by ROS and was upregulated 3.1-fold in *B. ×formosa* vs. *B. stricta* (Table S9). CAT3 also catalyzes the reduction of H_2O_2 but also catalyzes the transnitrosylation of REPRESSOR OF GSNOR1 (GSNOR1) to activate H_2O_2 reduction in the presence of NO [112]. *CAT3* was upregulated 3.2 and 4.4-fold in *B. × formosa* vs. *B. stricta* and *B. lignifera*, respectively. Three peroxidases, AT5G47000, AT2G24800, AT2G37130, which are upregulated in response to oxidative stress, were upregulated 3.0, 2.1, and 1.9-fold in *B. × formosa* vs. *B. stricta*, respectively. The former was also upregulated 2.6-fold in *B. × formosa* vs. *B. microphylla* and *B. lignifera*. The respiratory burst oxidase homolog (RBOH) NADPH-DEPENDENT THIOREDOXIN REDUCTASE A (NTRA) is a major cytosolic isoform of NADPH-dependent thioredoxin reductases. It redundantly regulates enzymes involved in photorespiration, ATP synthesis, stress-related reactions, redox homeostasis processes, and TCA cycle enzymes [113]. *NTRA* was upregulated 1.9-fold in *B. ×formosa* vs. *B. stricta*. CHLOROPLASTIC DROUGHT-INDUCED STRESS PROTEIN OF 32 KD (CDSP32), thioredoxin upregulated by oxidative stress and drought, participates in maintaining cell redox homeostasis. *CDSP32* was upregulated 1.8 and 2.2-fold in *B. ×formosa* vs. *B. stricta* and *B. lignifera*, respectively. AT2G37240, thioredoxin that participates in maintaining cell redox homeostasis, was upregulated 1.9 and 1.6-fold in *B. ×formosa* vs. *B. stricta* and *B. lignifera*, respectively. QUIESCIN-SULFHYDRYL OXIDASE 2 (QSOX2) is a thioredoxin involved in protein folding and in maintaining cell redox homeostasis. *QSOX2* was upregulated 1.8-fold in *B. ×formosa* vs. *B. stricta*.

Three additional classes of ROS attenuation genes were upregulated in *B. × formosa*. THIOREDOXIN H-TYPE 7 (TH7) functions as a redox sensor/transmitter in the ER where it participates in protein folding and transport of NADPH to peroxisomes [114]. *TH7* was upregulated 5.0, 4.5, and 4.9-fold in *B. × formosa* vs. *B. stricta*, *B. microphylla*, and *B. lignifera*, respectively (Table S9). H(+)-ATPase 1,4 (HA1,4) are plasma membrane proton ATPases that acidify the apoplast while increasing cytosolic pH and decreasing ROS levels [115]. *HA1* and *HA4* were upregulated 3.7 and 2.4-fold in *B. ×formosa* vs. *B. stricta*, respectively. ABC2 HOMOLOG 13 (ABC1K8) is a chloroplast localized protein kinase that is upregulated by heavy metals and mediates iron distribution and lipid membrane changes during oxidative stress [116]. It was upregulated 2-fold in *B. ×formosa* vs. *B. stricta*.

3.4. Metabolically Induced Switching from Apomeiosis to Meiosis

We hypothesized above that apomicts express chronic stress responses coupled with chronic, overcompensating stress attenuation mechanisms that suppress meiosis and induce apomeiosis. If this is correct, meiosis should be inducible in apomicts by applying stresses that exceed the apomict's stress attenuation capabilities. In 1951, Bocher [117] reported shifts from 1st division restitutions in microsporocytes of apomictic *Boechera*, which produced unreduced pollen, to complete meioses in weak plants that flowered sporadically late in the season. He noted that the processes that produce unreduced male and female spores, 1st division restitution, occurred in anthers and ovules. He also suggested that stress had induced the observed shifts from apomeiosis to meiosis, late in the season, and he speculated that stress might affect ovules in the same way. We viewed Bocher's speculations as an opportunity to gain further insights into these processes of sex/apomixis switching in facultative apomicts. Accordingly, we exposed randomly selected groups of *Boechera* plants to three growth conditions: well-watered, droughted, and droughted plus heat. Taxa studied were sexual *B. stricta*, facultative TDG forming *B. lignifera*, triploid, and highly TDG forming *B.* cf. *gunnisoniana*, HAG forming *B. microphylla* from Millard County Utah (meiotic and aposporous, but not diplosporous), and facultative TDG forming *B. exilis* × *retrofracta* [16]. The combination of drought and heat increased meiosis frequencies (tetrad formation) from 15 to 75% in *B. lignifera* and from 2 to 77% in *B.* cf. *gunnisoniana* (Figure 8a). Since most dyads in these droughted and heat-stressed plants (23 to 25% of ovules, Figure 8a) were possibly produced by meiosis, actual frequencies of sexual meioses in these stressed TDG-forming apomicts may have been closer to 90%. Drought alone in TDG forming *B. lignifera* and *B.* cf. *gunnisoniana* was less effective in switching apomeiosis to meiosis (Figure 8a). Disturbed meioses that produced pentads and hexads of imbalanced megaspores also occurred among the stress-induced meioses of triploid *B.* cf. *gunnisoniana* (Figure 8c,d). Likewise, Bocher also observed microspore pentads and hexads in the anthers of late-flowering *Boechera* triploids [117].

Due to lethality, data were not obtained for the double stressed (drought and heat) *B. stricta*, *B. exilis* × *retrofracta*, and *B. microphylla*. Drought alone had minor effects on the reproductive mode in sexual *B. stricta* and facultative *B. exilis* × *retrofracta*. The sexual dyad stage of meiosis is brief and difficult to catch cytologically. Accordingly, it is possible that drought accelerated the dyad stage (e.g., more rapid onset of M_{II}) such that fewer dyads were observed in ovules of *B. stricta* (Figure 8a). Drought also shifted frequencies of gametophyte formation from aposporous to sexual in the Millard County *B. microphylla* (Figure 8b). Figure 8e–g provides examples of sexual tetrads, diplosporous dyads, and early onset stages of aposporous gametophyte formation as observed in this experiment.

Our greenhouse experiment exposed plants to chronic drought or drought plus heat for the 10–40 days required for flowering to occur following vernalization. Accordingly, we asked whether long-term chronic stress is required to switch apomeiosis to meiosis or whether this switch might occur rapidly, i.e., following acute stresses applied to pistils within days or even hours of apomeiosis. To answer this question, we developed a simple MS-based [26] pistil culture protocol that supported ovule development in cultured pistils from the late pre-MMC stage (1-I) through the enlarging MMC (2-II), mature MMC (2-III), meiosis (2-IV), late meiosis to functional megaspore (2-V), and early vacuolate multinucleate embryo sac (3-III) stages. Ovules of pistils cultured on basal medium (without pharmacological additives) progressed through these developmental stages within 48–72 h of culture initiation. However, development beyond the early gametophyte stage was irregular, in controls and treatments, which prevented us from evaluating in vitro the effects of metabolic modifications on later stages of gametophyte development and seed formation.

The TDG forming triploid *B.* cf. *gunnisoniana* responded strongly to chronic drought (Figure 8a). Hence, we chose it for acute drought experiments that involved culturing 1.5–1.9 mm long pistils, each containing 40–50 ovules at the early to late MMC stage, on low water potential media (Figure 9a). Even on control media (−0.50 MPa), frequencies of meiotic tetrads were elevated above the 2–3% background level. Apparently, excising and culturing these young desiccation sensitive pistils was itself stressful enough to switch apomeiosis to meiosis in some ovules, presumably by overwhelming stress attenuation processes in ovules of the cultured pistils. Meiotic tetrad frequencies were highest (~50%) in the −0.56 MPa treatment, and the highest tetrad frequencies occurred in pistils that contained late MMC staged ovules at the time of culture initiation. Hence, the duration of drought exposure may be less important than drought exposure occurring at the late MMC stage when meiosis or apomeiosis fates are determined. Ovules with less mature MMCs at the time of culture initiation possibly had more time in culture to restore metabolic homeostasis. Importantly, this experiment revealed that extended periods of stress are not required to reprogram genomes for meiosis. Instead, this reprogramming occurred within hours of stress induction.

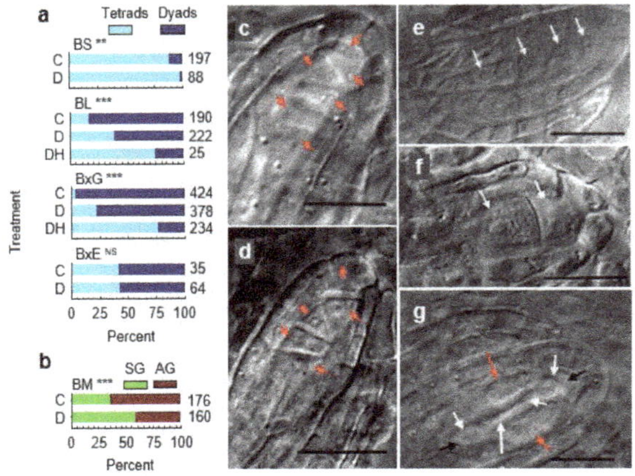

Figure 8. Effects of no stress (C), drought (D) and drought plus heat (DH) on (**a**) percentages of meiotic or apomeiotic dyads and meiotic tetrads in sexual *Boechera stricta* (BS), facultatively diplosporous *Boechera lignifera* (BL), triploid and nearly obligately diplosporous *Boechera* cf. *gunnisoniana* (BxG), and facultatively diplosporous *Boechera exilis* × *retrofracta* (BxE); (**b**) percentages of aposporous (AG) and sexual (SG) gametophyte formation in aposporous *Boechera microphylla* (BM); (**c**,**d**) meiotic hexad and pentad of megaspores from stress-induced meioses in triploid diplosporous *B.* cf. *gunnisoniana*; (**e**) meiotic tetrad of megaspores from sexual *B. stricta*; (**f**) diplosporous dyad from diplosporous *B. lignifera*; (**g**) degenerating tetrad (between black arrows) of Millard County *B. microphylla* with early vacuolate, 1-2 nucleate aposporous gametophytes on each side (red arrows). Numbers of ovules examined in (**a**) and (**b**) that contained scorable dyads, tetrads, SGs, and AGs per treatment are indicated at the right of the colored bars; cytology bars, 20 μm. Chi-square *p* values: **, *** are 0.01, 0.001; NS, not significant.

Comparisons of the more apomeiotic vs. the more meiotic taxa revealed reduced molecular support for translation in ovules of apomicts (Figure 2b,f,j and Figure 4a). Moreover, these ovules were more metabolically active, were responding to ABA, and were more actively attenuating stress (Figures 6 and 7). Hence, it appears that our chronic and acute drought stresses overwhelmed the stress attenuation capabilities of these apomicts and switched apomeiosis to meiosis. We next asked whether severe carbon starvation, which generates oxidative stress and terminates translation [118], might also increase apomeiosis to meiosis conversions. To address this question, we cultured pistils of *B.* cf. *gunnisoniana* (containing early to late-stage MMCs) on basal medium either with or without sucrose. Again, younger pistils on the control medium (with sucrose) produced fewer tetrads than more mature pistils (Figure 9b, compare 2-III to 2-IV results). Presumably, this occurred because cells of the younger 2-III staged pistils had more time to adjust their stress shocked metabolism to pre-excision levels prior to reaching the point of stress tolerance that defines facultativeness (apomeiosis/meiosis switching) in apomicts. In the no sucrose treatment, TDG (1st division restitution dyad) frequencies decreased by ca. 50%. This was matched by corresponding increases in meiotic tetrads, but it was also matched by HAG formation onset (Figure 9g). Frequencies of both phenomena approached 50% in ovules of the more mature pistils (Figure 9b). Hence, under severe carbon starvation, apomeiosis switched to meiosis, but the meioses in these ovules were often accompanied by HAG formation (apospory) with little or no net decrease in apomixis. We next tested whether oxidative stress (exogenously applying H_2O_2) would induce apomeiosis to meiosis conversions as well as diplospory to apospory conversions, i.e., switching nucellar cell programmed cell death (the norm) to HAG formation. For this experiment, we used pistils (with ovules at the MMC stage) from three apomicts, the facultative TDG forming *B. lignifera* (1.0–1.4 mm long pistils), the nearly obligate TDG forming *B.* cf. *gunnisoniana* (1.5–1.9 mm long pistils), and the facultative TDG forming *B. retrofracta* × *stricta*

(1.6–2.1 mm long pistils). Pistils were excised and soaked for 5 min in H_2O_2 solutions prior to culture (Figure 9c). These H_2O_2 pretreatments induced up to 60% apomeiosis to meiosis conversions. However, all megaspores from many of the induced meiotic tetrads degenerated, and in many cases, this degeneration was accompanied by HAG formation (Figure 9h–i). These conversions were again most prominent in ovules of more mature pistils, suggesting that younger ovules had more time (ca. 12–24 h) to detoxify the exogenously added H_2O_2 before committing themselves to very different developmental paths.

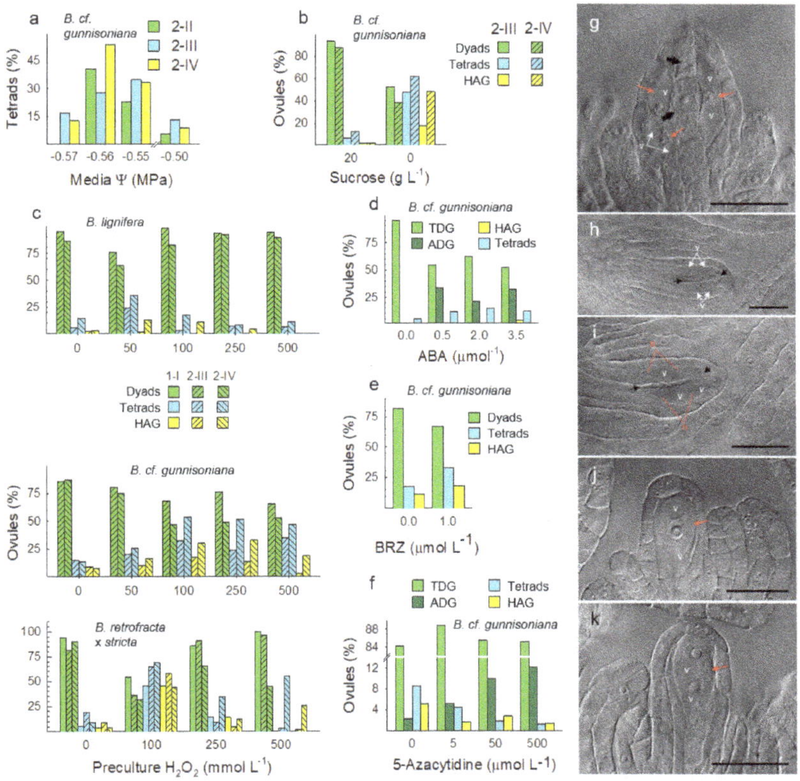

Figure 9. Effects of pharmacological treatments and pistil maturity on conversion from Taraxacum-type gametophyte formation (TDG) to (i) meiosis, (ii) Hieracium-type gametophyte formation (HAG), and (iii) Antennaria-type gametophyte formation (ADG): (**a**) drought (media water potential); (**b**) energy starvation; (**c**) reactive oxygen species stress (H_2O_2); (**d**) abscisic acid (ABA); (**e**) brassinazole (BRZ, brassinosteroid synthesis inhibitor); (**f**) 5-azacytidine (DNA methylation inhibitor); (**g**) sucrose starved *Boechera* cf. *gunnisoniana* ovule with a Taraxacum-type diplosporous dyad (black arrows) and three 1 nucleate vacuolate HAGs (red arrows; v, vacuoles); (**h–i**) H_2O_2 induced conversions from TDG formation to meiosis, but with all four megaspores degenerating (between black arrows), and from programmed cell death of nucellar cells to apospory (n, nuclei); (**j**,**k**) ABA induced conversions from TDGs to ADGs. Numbers of informative ovules per treatment (average, range): (**a**), 222, 103–289; (**b**), 81, 22-164; (**c**), *Boechera lignifera* (70, 29–168), *B.* cf. *gunnisoniana* (217, 78–465), *Boechera retrofracta* × *stricta* (77, 6–138); (**d**), 183, 160–201; (**e**), 102, 84–119; (**f**), 382, 178–487. Pooled chi-squares were significant ($p < 0.001$). 1-I, late pre-MMC; 2-II, enlarging MMC; 2-III, mature MMC; 2-IV, meiosis; cytology bars, 20 µm.

The experiments reported so far illustrate the importance of plants maintaining highly stable states of metabolic homeostasis that restrict reproduction to a single sexual or apomictic type, and they suggest that apomixis occurs when oxidative stress is overly attenuated (Figure 7). Continuing with this theme, we asked whether upregulating stress attenuation by increasing ABA signaling (Figures 6 and 7) might prevent meiotic prophase and shift TDG formation to ADG formation in *B.* cf. *gunnisoniana* (Figure 9d). Such treatments induced ADG formation in 25–35% of ovules (gametophyte formation directly from MMCs; Figure 9j–k). These findings place oxidative stress and oxidative stress attenuation at the center of reproductive decisions in *Boechera*, with strong and moderate stress attenuation causing ADG and TDG formation and oxidative stress causing nucellar cell destabilization followed by HAG formation.

Our expression profiling studies suggested that ABA signaling in the TDG forming *B.* × *formosa* reduces BR signaling at BIN2 (Figure 5d). To further evaluate this, we cultured TDG forming *B.* cf. *gunnisoniana* pistils, with ovules at the MMC stage, on media containing the BR biosynthesis inhibitor brassinazole (BRZ). This more thorough reduction in BR signaling increased frequencies of meiosis and apospory (Figure 9e), possibly by further suppressing mechanisms of oxidative stress attenuation.

Meiosis is a cell-lineage-specific process that requires a meiosis-specific transcriptome that includes meiosis-specific *de novo* RNA-directed DNA methylation (RdDM) [49]. Here, we asked whether disrupting RdDM using 5-azaC, a methyltransferase inhibitor that reduces DNA methylation [119], might prevent meiotic prophase in TDG forming *B.* cf. *gunnisoniana* but still permit gametophyte formation directly from MMCs (ADG formation). Though not as efficient as ABA, 5-azaC induced ADG formation at frequencies that were correlated with 5-azaC concentrations (Figure 9f). That both ABA and 5-azaC induce ADG formation (Figure 9d,f,j–k) is consistent with ABA-induced oxidative stress attenuation silencing meiosis-specific RdDM in germline cells (Figures 4–7).

3.5. Metabolically Induced Switching from Meiosis to Apomeiosis

We demonstrated above that the apomicts studied herein chronically express stress attenuation mechanisms, but when these mechanisms are overwhelmed, apomeiosis reverts to meiosis. Here, we show that the reverse also occurs, i.e., when stress attenuation mechanisms are enhanced in sexual plants, meiosis reverts to apomeiosis. For these experiments, we chose four sexual diploid species: *A. thaliana*, *B. stricta*, and *B. exilis* of the Brassicaceae family and *V. unguiculata* of the Fabaceae family. Pistils containing MMC staged ovules were used. Pistil lengths were 0.5–1.2, 1.3–1.9, 1.5–2.3, and 1.5–1.9 mm, respectively. Dioecious *A. dioica* (Asteraceae) was used to study microsporogenesis.

Treatments applied to excised *A. thaliana* pistils included a 60 min pretreatment emersion. In the controls (Figure 10a), percentages of tetrads to dyads were as expected. However, 17% of control ovules (presoaked for 60 min) contained vacuolate HAGs. This response may have been caused by anoxia induced ROS (e.g., Figure 9c). To test this, we pre-soaked sets of pistils for 7 min in water or 100 mmol L^{-1} H_2O_2 and then cultured them on basal medium or medium amended with 1.0 μmol L^{-1} epiBL. Water alone did not induce HAG formation, but such formation did occur in 12–15% of ovules pretreated with H_2O_2 (Figure 10b). Diplospory was absent in these treatments. Hence, exogenous H_2O_2 may have reinforced meiosis while destabilizing nucellar cells and either inducing or allowing aposporous development. The TF BZR1, which is activated by endogenous H_2O_2 [120], may have also been involved. These experiments indicate that nucellar cells in ovules of diplosporous and sexual plants respond similarly to H_2O_2 treatments in terms of inducing apospory.

To determine whether apomeiosis is inducible by reducing oxidative stress, we excised early MMC staged *A. thaliana* pistils, soaked them in the antioxidant (S)-2-aminobutane-1,4-dithiol hydrochloride (DTBA) for 60 min, and cultured them on DTBA containing medium for 4 days. In 15 of 41 scored ovules, a first division meiotic restitution occurred, which produced a dyad of megaspores (Figure 10a). The micropylar dyad member in some of these was degenerating and the chalazal dyad member was becoming vacuolate and undergoing endomitosis to form a multinucleate gametophyte (Figure 10c,d).

An additional 7 of 41 scored ovules completed meiosis, but all four megaspores in these ovules were degenerating and being replaced by HAGs (Figure 10e–f).

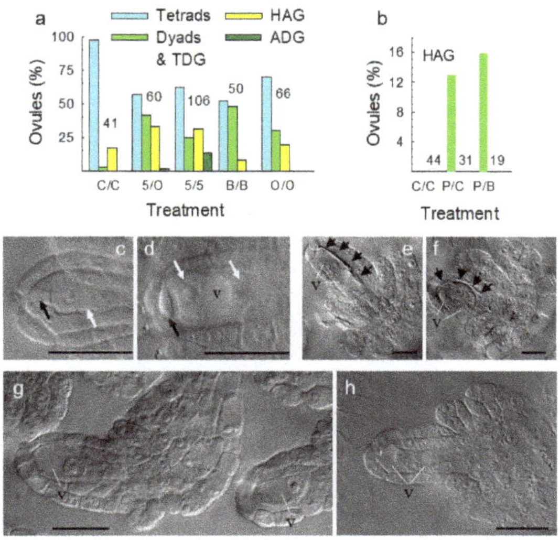

Figure 10. Taraxacum (TDG), Antennaria (ADG), and Hieracium (HAG) types of gametophyte formation induced in cultured *Arabidopsis thaliana* pistils: (**a**) percentages of ovules that produced sexual tetrads, sexual or TDG dyads or TDGs, HAGs and ADGs by treatment; (**b**) percentages of ovules that produced HAGs by treatment; (**c**–**d**) 1 and 2-nucleate TDGs induced by DTBA; (**e**–**f**) 1 and 2-nucleate HAGs induced by DTBA; (**g**–**h**) 1 and 2 nucleate ADGs induced by 5-azaC. 5, 5-azaC; B, brassinolide (BR); C, control; O, other (B or DTBA); P, 7 min H_2O_2 pretreatment. Numbers of informative ovules analyzed appear next to treatment bars in (**a**) and (**b**) (values are sums of four and two replicates, respectively). Pooled chi-squares were significant ($p < 0.001$). v, vacuole; cytology bars, 20 μm.

We also asked whether BR alone would induce apomeiosis. In 24 of 50 ovules, BR induced a 1st division restitution and subsequent TDG formation. Low-level HAG formation (8%) also occurred. Adding an H_2O_2 pretreatment to BR-treated pistils prevented TDG formation possibly by reinforced meiosis (Figure 10b). Nevertheless, the combination of BR and H_2O_2 supported post-meiotic abortion coupled with HAG formation. We next tested whether 5-azaC would terminate meiosis and induce ADG formation in *A. thaliana* as it had done in *B.* cf. *gunnisoniana* (Figure 9d,j,k). Here, we soaked pre-MMC staged *A. thaliana* pistils in 5-azaC for 60 min and cultured them on control or 5-azaC medium. In 5-azaC treated pistils, 14 of 106 appropriately staged ovules were forming ADGs (Figure 10a,g,h). Of all treated pistils (Figure 10a), 64% (180 of 282) initiated either TDG, ADG, or HAG formation, i.e., all three major types of apomeiosis were induced in *A. thaliana* by metabolic modifications.

To determine if apomeiosis is inducible in other angiosperms, we exposed pistils of sexual *B. stricta* and sexual *V. unguiculata* to treatments similar to those used for *A. thaliana* (Figure 11a,b). The *A. thaliana* and *B. stricta* treatments were performed in duplicate, but with glucose and sucrose used individually as energy sources. The sugar main effect was not significant in *A. thaliana*, but TDG formation (Figure 11c–d) increased significantly in *B. stricta* when glucose was used (38 of 152 ovules vs. 26 of 164 ovules). ADG formation was only rarely observed in *B. stricta* but was induced by all treatments in *V. unguiculata* (Figure 11b,j–k). The *V. unguiculata* study included 74 pistils, but only 34 ovules were at the MMC stage at culture initiation. Of these 34, 33 had been exposed to apomeiosis-inducing treatments, of which 24 produced ADGs (73%). One MMC staged ovule had been exposed to the sucrose control, and it was meiotic.

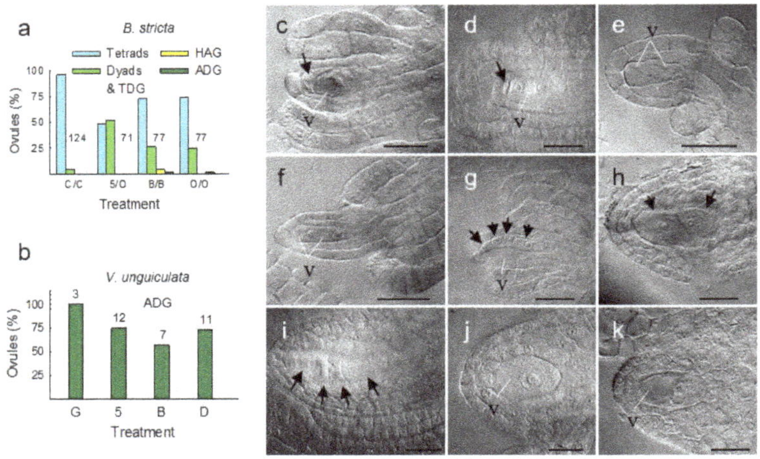

Figure 11. Taraxacum (TDG), Antennaria (ADG) and Hieracium (HAG) types of gametophyte formation induced in sexual *Boechera stricta* and ADG formation induced in sexual *Vigna unguiculata*: (**a**) percentages of ovules that produced sexual tetrads, TDGs, ADGs and HAGs in *B. stricta* by treatment; (**b**) percentages of ovules that produced ADGs in *V. unguiculata* by treatment; (**c**–**d**) 1 and 2 nucleate TDGs in *B. stricta*; (**e**–**f**) 1 and 2 nucleate ADG formation in *B. stricta*; (**g**) 1 nucleate HAG in *B. stricta*; (**h**–**i**) sexual dyad and tetrad formation in *V. unguiculata*; (**j**–**k**) vacuolate 1 nucleate ADG formation in *V. unguiculata*. Pretreatment/treatment acronyms: 5, 5-azaC; B, brassinolide (BR); C, control; D, DTBA; G, glucose; O, other (B or D). Numbers of informative ovules analyzed appear next to treatment bars in (**a**) and (**b**) (values are sums of two replicates). The *B. stricta* pooled chi-square was significant ($p < 0.001$). v, vacuole; cytology bars, 20 μm.

ABA signaling includes oxidative stress induction by respiratory burst oxidase homologs and superoxide dismutases and subsequent oxidative stress attenuation by antioxidants. The evolution of this compensatory stress tolerance process facilitated the evolution of land plants. Here, we wondered if terminating ABA synthesis might induce apomeiosis by downregulating PP2C production, hence upregulating SnRK2,3 activities and the subsequent production of stress attenuation mechanisms (Figures 6 and 7). To evaluate this, we exposed pistils of sexual *B. exilis* to 5.0 μmol L^{-1} fluridone treatments for 24–48 h before meiosis initiation (early to the mid-MMC stage). Fluridone, which inhibits ABA synthesis [121,122], induced 17%, 22%, and 2% of ovules to undergo ADG, HAG, and TDG formation, respectively (Figure 12a–g). In some ovules, ADG and HAG formation occurred simultaneously (Figure 12d–f). Evolutionarily, the nucellus is sporangial, like fern sporangia [5]. By altering ABA signaling, nucellar cells may have been released from programmed cell death, their normal fate, to pursue an ancestral fate, that of gametophyte formation.

We also tested whether chemicals that induce apomeiosis in megasporocytes might also induce apomeiosis in microsporocytes. Here we placed cut stems of male dioecious *A. dioica* (Asteraceae) in liquid MS media amended with fluridone. As was observed for MMCs, the ameiotic PMCs developed precociously into unreduced male gametophytes (Figure 12h–j). Note that the four-fold meiotically replicated and expanding tetrads of microspores crowded the anther locule in the controls (Figure 12i). In contrast, the meiotically skipped, precociously forming unreduced pollen grains of the fluridone treated plants were widely spaced throughout the locule (Figure 12j).

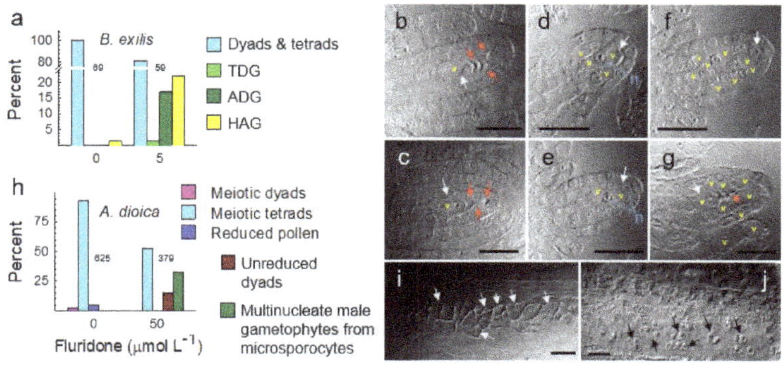

Figure 12. Fluridone induced unreduced gametophyte formation in ovules of *Boechera exilis* and anthers of *Antennaria dioica*: (**a**) percentages of ovules that produced Taraxacum (TDG), Antennaria (ADG) and Hieracium (HAG) types of gametophyte formation in ovules of sexual *B. exilis* when treated with fluridone; (**b**,**c**) tetrad and 2-nucleate sexual gametophyte in control pistils; (**d**,**e**) 1 and 2-nucleate ADGs in ovules of fluridone treated pistils (apospory is also occurring in these ovules); (**f**) 2-nucleate ADG and HAG formation in the same ovule of a fluridone treated pistil; (**g**) formation of multiple 1–2 nucleate HAGs in the same ovule of a fluridone treated pistil (this pistil had also successfully completed meiosis and was forming a vacuolate sexual gametophyte (white arrow)); (**h**) fluridone induced ameiotic male gametophyte formation from microsporocytes in young anthers of stem-cultured dioecious *A. dioica*; (**i**) normal microspore tetrads in anthers of control *A. dioica* plants; (**j**) multinucleate and vacuolate ameiotic 1–3 nucleate pollen grains that formed directly from microsporocytes in *A. dioica* anthers from stems that were cultured in fluridone amended medium. Numbers of informative ovules or anther meiocytes analyzed appear next to treatment bars in (**a**) and (**h**). Pooled chi-squares were significant ($p < 0.001$). red arrows, degenerating haploid megaspores; n, nucleus; v, vacuole; cytology bars, 20 μm.

4. Discussion

4.1. Sucrose Non-Fermenting-Related Protein Kinases May Be Central to Meiosis Apomeiosis Switching

The evolution of land plants required new water retaining and transporting morphologies, new stress detection and signaling pathways, and new processes for catabolizing toxins induced by terrestrial stresses [123–125]. Here, the evolution of ABA and SnRK2,3 stress response pathways was critical for increasing ROS attenuation capabilities [124]. ROS form constantly as byproducts of respiration, photosynthesis, and the activities of peroxidases and peroxisomes. However, in land plants, ROS are also produced to propagate waves of ROS signals from stressed cells to more distant cells [126]. Two ABA-induced pathways accomplish this: (i) propagation of ROS signaling through tissues via RBOHs and SODs (Figure 7) and (ii) upregulation of stress attenuation mechanisms through SnRK2,3 ABI5 signaling (Figure 6). During such signaling, H_2O_2 activates or inactivates redox-regulated protein phosphatases and other enzymes [127,128]. It is critical that excess H_2O_2 is catabolized by ROS attenuation mechanisms [94,129,130]. Metabolic processes that establish such redox homeostasis function by balancing ROS generation, from multiple sources, with ROS catabolism. We discovered, and report herein, that apomeiosis occurs when this redox balance is chronically tilted toward H_2O_2 catabolism, i.e., when SnRK2,3s and RD26 are activated (Figure 13). We further show that meiosis apomeiosis interconversions are inducible by applying chemicals that acutely disturb these taxa specific states of redox homeostasis.

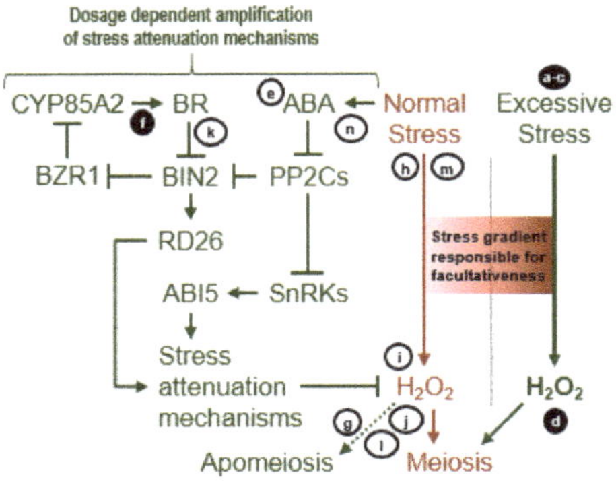

Figure 13. Simplified model of apomixis sex switching in plants. Green text to the left represents enhanced stress attenuation pathways in apomeiotic plants that suppress meiosis under non-stress conditions. The red text represents normal levels of stress attenuation in sexual plants, which are meiosis permissive. Green text to the right represents apomictic plants exposed to excessive stress, which overwhelms stress attenuation mechanisms and induces meiosis. White letters represent experiments that tested metabolic processes of apomeiosis to meiosis switching: **a**, heat and drought; **b**, osmotic stress; **c**, carbon starvation; **d**, H_2O_2; **f**, inhibition of brassinosteroid (BR) synthesis by brassinazole. Black numbers represent experiments that tested metabolic processes of meiosis to apomeiosis switching: **e**, abscisic acid (ABA); **g** and **l**, 5-azacytidine (DNA methylation suppression); **h**, anoxia (60 min emersion prior to pistil culture); **i**, H_2O_2; **j**, (S)-2-aminobutane-1,4-dithiol hydrochloride (antioxidant); **k**, BR; **m**, glucose; **n**, fluridone (ABA synthesis suppression). Arrows indicate upregulation or activation; blunt end lines represent downregulation or suppression; ABI5, ABA INSENSITIVE 5; BZR1, BRASSINAZOLE-RESISTANT 1; BIN2, BRASSINOSTEROID-INSENSITIVE 2; CYP85A2, BRASSINOSTEROID-6-OXIDASE 2; PP2Cs, protein phosphatase 2Cs; RD26, RESPONSIVE TO DESICCATION 26; SnRKs, sucrose non-fermenting related protein kinases.

In addition to the plant-specific SnRK2,3s (Figure 6), SnRK1 was also more active in the apomicts (Figure 4a). In eukaryotes, SnRK1 regulates growth no-growth decisions. When phosphorylated, SnRK1 dissociates the TORC1 complex [33,36,124,131–135], and this suppresses translation and BR regulated cell proliferation [33,120,124,136–139]. Translation downregulation in the more apomeiotic taxa (14 of 16 comparisons during early ovule development, Figure 2a,e–i and Figure 3a–g) indicates that TOR was less active in the apomicts due to increased SnRK1 activity. That stress or low energy conditions in the apomicts activated SnRK1, which suppresses TOR, is further shown by upregulated genes in the apomicts for fatty acid, amino acid, carbohydrate, and other macromolecule catabolism (Figure 2a,f,g,l, and Figure 3a–d,g).

In the apomicts, profiling revealed upregulated genes associated with signaling, stress responses, metabolism, photosynthesis, and bioenergetics (Figure 1a,e–j,l, and Figure 3a–g). Here, the *B.* ×*formosa* population may have contained triploids, which is the common polyploid level among *Boechera* apomicts [17]. Polyploids often show distinct patterns of gene expression [140] as well as increased stress attenuation capabilities [15,141]. Increased metabolic activity seems counterintuitive given the low energy induced suppression of TOR in this taxon. However, the evolution of SnRK2,3s enabled land plants to avoid toxic levels of ROS, due to enhanced ROS attenuation processes, while maintaining growth and other metabolic processes [124,125]. Under favorable conditions, all three categories of SnRKs (SnRK1,2,3s) were largely phosphatase inactivated in the more meiotic plants by ABI1,2 and the HIGHLY ABA-INDUCED PP2C GENE 2 (AIP1). In contrast, these SnRKs appeared to remain partially active in the apomicts, which likely explains their greater ROS attenuation capacity. Switching from apomeiosis to meiosis (facultativeness) in response to stress (Figures 8 and 9) suggests that ROS accumulation, due to stress treatments, exceeded ROS catabolism (Figure 13).

4.2. Metabolically Inducing Meiosis Apomeiosis Interconversions

Stress attenuation in plants involves the activities of ascorbic acid, ascorbate peroxidases, glutathiones, thioredoxins, catalases, superoxide dismutases (SODs), tocopherols, flavonoids, carotenoids, and ion homeostasis regulators. These are upregulated by oxidative stress, ABA-activated SnRKs, and ABA insensitive SnRKs (activated by salt, osmotic, or other stress) [124,128]. Genes encoding many of these were upregulated in ovules of the more apomeiotic taxa (Tables S2 and S5). In this respect, Shah et al. [142] provided evidence that chronic stress tolerances in apomictic *Boechera* might persist throughout the life cycle. They further suggested that this increased tolerance contributes to geographical parthenogenesis (superior stress tolerance in apomicts). We provide evidence herein that the chronic states of stress tolerance in apomicts suppress meiosis and induce apomeiosis. They might also suppress syngamy and induce parthenogenesis. This notion recently gained traction by the demonstration that parthenogenesis in rice is inducible by altering the location of expression of a wild type *APETALA2/ERF* gene [143], a member of a TF family that regulates development and stress by integrating redox, hormone, and environmental signals [144].

Using pharmacological approaches, we tested whether the differences in molecular pathways observed between apomeiotic and meiotic taxa (Figures 2–7) are responsible for inducing meiosis or apomeiosis or are simply coincidental. To do this, we identified chemicals that perturb specific elements of metabolic homeostasis. Here, apomeiosis switched to meiosis when the chosen treatments increased oxidative stress. These included drought, energy starvation, topical H_2O_2 applications, and BR synthesis inhibition (Figure 9a–c,e; Figure 13a–d,f). In contrast, meiosis switched to apomeiosis when the chosen treatments relieved oxidative stress. These included glucose, BR, antioxidants, ABA, and the ABA synthesis inhibitor fluridone (Figures 10–12 and Figure 13e,g–n). Additionally, 5-azaC terminated meiosis in sexual plants (Figures 10 and 11), terminated meiotic prophase in TDG forming plants (Figure 9f), and permitted or induced ADG formation, directly from ameiotic MMCs, in sexual and TDG forming plants. This suggests that while RdDM may be required for meiosis [49,145], it is not required for the early stages of gametophyte formation.

When used as the sole carbon source, glucose blocked meiosis, and induced TDG formation in *B. stricta* and ADG formation in *V. unguiculata* (Figure 11b–d,j,k). In *S. cerevisiae*, glucose, which inhibits Snf1 kinase (SnRK1 in plants) and possibly activates TOR [146], also blocks meiosis [147]. Thus, in plants and yeast, meiosis is correlated with Snf1/SnRK1 inactivity and possibly TOR activity (Figure 4a). In plants, meiosis was further correlated with SnRK2,3 inactivity (reduced ABA signaling). In contrast, the more apomeiotic taxa expressed elevated levels of ABA signaling and correspondingly higher levels of antioxidant production (Figures 6 and 7). Since glucose in plants also upregulates ABA synthesis [148], glucose upregulated apomeiosis in *B. stricta* and *V. unguiculata* (Figure 11) is consistent with upregulated ABA signaling followed by increased stress attenuation (Figure 13).

Surprisingly, carefully timed H_2O_2 treatments not only converted apomeiosis to meiosis in TDG forming *Boechera* (apomixis to sex) but also induced HAG formation (sex to apomixis) in many of the same ovules (Figure 9c). H_2O_2 also induced HAG formation in sexual *A. thaliana* (Figure 10b). We suggested that H_2O_2 induces HAG formation directly. However, treating pistils with the antioxidant DTBA also induced HAG formation (Figure 10a,e,f). Hence, it is also possible that the exogenous H_2O_2 treatments triggered an amplification of ROS attenuation mechanisms by upregulating ABA signaling (Figures 6 and 7). Upregulated ROS catabolism may have then freed nucellar cells from redox induced programmed cell death, thus allowing them to pursue gametophyte formation.

In addition to inducing ADG and HAG formation in sexual *B. exilis* (Figure 12a–g), fluridone also induced gametophyte formation directly from PMCs (Figure 12h–j). Since ABA upregulates PP2C formation, the suppression of ABA synthesis may have indirectly increased antioxidant production downstream of SnRK2,3s (Figure 6). PMCs, MMCs, and nucellar cells share an evolutionary origin that includes homosporous meioses, e.g., as occur in sporangia of many ferns [12,149,150]. Hence, it is not surprising that metabolic disturbances that induce ADG formation from MMCs and HAG formation from nucellar cells (Figure 12a–g) can also induce gametophyte formation from PMCs (Figure 12h–j).

The effects of fluridone persisted throughout the 3–4 days duration of the experiment. Accordingly, most nucellar cells and some integument cells initiated HAG formation (Figure 12d–g).

4.3. Apomixis Genes: Horses of a Different Color

From the early 1900s to the 1940s, common views concerning the inheritance of apomixis stemmed from Strasburger's work, which suggested that certain sexual plants possess propensities for apomixis that are released by hybridization and polyploidization [4,151]. Starting in the late 1930s, genetic studies suggested that apomixis genes exist, which were later found to be isolated in nonrecombining genomic regions [10]. Aside from these more common views, a short-lived hypothesis (in the 1920s) was introduced by Ernst [152] who concluded that apomixis was caused by hybridization. Importantly, Ernst emphasized that apomixis propensities in parent plants are not required. In the 1990's, Carman [13] comprehensively reviewed the genomics of known apomicts and reported highly significant correlations between apomixis and hybridization, polyploidization, and a relative absence of recent paleopolyploidization (low chromosome base numbers). Carman also concluded that hybridization was singularly important to apomixis evolution and that apomixis specific genes are not necessary. Ernst thought that phylogenetic distance between parental species determined apomixis [152]. In contrast, Carman cited physiological and biochemical examples of asynchronously expressed duplicate genes in hybrids and hypothesized that apomixis is caused by asynchronous expression of germline development genes in the ovules of certain hybrids.

The asynchrony hypothesis suggests that ecotype-driven divergences in wild type genes occurred during speciation along ecological gradients and that these new alleles modified ovule development timing. Following secondary contact hybridization, germline asynchronies were resolved in certain hybrids by eliminating meiosis and syngamy, thus producing apomixis [13]. The present paper updates this hypothesis. Here, asynchronies in germline development are replaced by dosage-dependent disparities in metabolic homeostasis (Figure 13). In terms of genetic mapping, this revision simply shifts the types of genes expected to be found in apomixis associated genomic regions from meiosis/syngamy genes to metabolic homeostasis genes. Interestingly, apomixis is much more common in land plants than in many other eukaryotes. The addition of a major ROS attenuation mechanism (ABA/SnRK2,3 signaling) possibly made apomixis-conducive states of redox homeostasis more readily attainable through hybridization and polyploidization.

It is important to address the question as to why "occasional apomixis" in sexual plants—sensu Asker and Jerling [4]—is rare. The simplest explanation is that the chronic states of stress attenuation responsible for apomixis (Figure 13, left side) are not environmentally inducible in sexual plants. Instead, such states of homeostasis are achieved only rarely through genomic aberrations associated with hybridization, recombination, or polyploidization. In contrast, states of stress attenuation in natural apomicts, induced by genomic aberrations, are variable (Figure 13, right side) such that some plants are weakly apomictic while others remain nearly obligately apomictic even when stressed.

Several interesting examples of homeostatic states of chronic stress attenuation, which possibly enforce apomixis, exist in apomicts of other kingdoms. Apomeiosis in bdelloid and monogonont rotifers (Metazoa) is mitotic-like, and the 2n eggs develop into embryos parthenogenetically. Bdelloid rotifers are obligately apomictic, and they exhibit constitutive anti-oxidant protection provided by extensively duplicated anti-oxidant genes [153,154]. Generally, in cyclical animal apomicts, it is the onset of overwhelming stress, from various factors, that cause apomictic females to apomictically produce clonal but sexually functional males and females (genders induced environmentally). These breed to produce resting eggs that hatch into apomictic females. This pattern is observed in monogonont rotifers, water fleas (*Daphnia*), aphids (Arthropoda, Metazoa) [155,156], diatoms (Chromista) [157], and green alga (Plantae) from the genera *Volvox* (Plantae). As in bdelloid rotifers, a major class of oxidative stress detoxifying enzymes, glutathione S-transferases, are upregulated during the apomictic female phase in the pea aphid *Acyrthosiphon pisum* [156]. Major oxidative stresses are also required to induce the switch from apomictic to sexual reproduction in *Daphnia* [155] and *Volvox* [158–160], and to induce apomixis to sex conversions in facultatively apomictic slime molds (Protozoa) [161–163] and brewer's yeast (Fungi) [3]. In our experiments, attenuating oxidative stress in sexual sporocytes of *Boechera*, *Arabidopsis*, *Vigna*, and *Antennaria* caused meiosis to switch to apomeiosis. In contrast, when we overwhelmed ovules of facultatively to nearly obligately apomictic *Boechera* with oxidative stress, apomeiosis switched to meiosis. If our interpretations are correct, states of metabolic homeostasis that induce apomeiosis in extant eukaryotes are generally rare, are characterized by strong anti-oxidant properties, and are usually caused by genomic aberrations that grossly disturb the expression of wild type genes.

5. Conclusions

Switching from apomeiosis to meiosis in apomictic plants and from meiosis to apomeiosis in sexual plants was readily achieved through appropriate treatments. However, the treatment effects dissipated rapidly, with taxon-specific levels of cellular homeostasis apparently returning quickly to pretreatment levels. For example, the effectiveness of treatments applied 24–48 h before apomeiosis were diminished by up to 90% compared to treatments applied only a few hours prior to apomeiosis (Figure 9). Accordingly, achieving levels of cellular homeostasis that assure complete penetrance of apomeiosis (and parthenogenesis) will likely require genetic engineering modifications that acutely alter cellular levels of metabolic homeostasis throughout ovule and early seed development. Candidate genes and pathways for achieving such modifications are described herein.

Our studies shed light on two important barriers that have frustrated the development of apomixis as a technology for stabilizing hybrid vigor in crops: (i) a correct understanding of what apomixis is, and (ii) a correct understanding of what biological properties trigger meiosis/apomeiosis interconversions. Concerning the first, we have demonstrated that apomeiosis does not require mutations, meiotic or otherwise, but is instead a polyphenism of meiosis, a conserved polyphenism that may be as ancient as meiosis itself [5]. Concerning the second, we have demonstrated that the polyphenic shifts from apomeiosis to meiosis or vice versa are metabolically regulated. Here, MMCs of apomictic TDG forming taxa reverted to meiosis at high frequencies when exposed to meiosis promoting metabolic conditions. Likewise, MMCs, nucellar cells, and/or microsporocytes from five sexual taxa—representing three angiospermous families—reverted to apomeiosis at high frequencies when exposed to apomeiosis promoting metabolic conditions.

The experiments reported herein can be easily modified to further elucidate metabolic and molecular details of meiosis apomeiosis switching using florets from a single genotype, or even from a single plant. Through such studies, it may be possible to identify wild type genes that faithfully induce apomixis in crops when their levels and locations of expression are transgenically manipulated.

Supplementary Materials: The following are available online at http://www.mdpi.com/2073-4425/11/12/1449/s1, Figure S1: Mean (±SD) lengths of pistils by ovule development stage, Figure S2: Microarrays that passed quality control analyses, Figure S3: qRT-PCR verification of microarray expression profiling, Table S1: Rules used to assign gene ontology (GO) terms to groups and subgroups, Table S2: Gene ontology (GO) groups and subgroups for all GO terms of differentially expressed genes, Table S3: Calculations for chi-square tests of independence and for the determination of standard and adjusted residuals, Table S4: Primers used for qRT-PCR verification of microarray data, Table S5: Differentially expressed genes for 17 microarray comparisons, Table S6: All gene ontology categories (GOs) of all differentially expressed genes (DEGs) by comparison, Table S7: GO terms by comparison using DEGs that were overrepresented or underrepresented, Table S8: Enriched gene ontology terms (EGOs) of comparisons, Table S9: Genes by comparison identified by keyword searches of DEG descriptions.

Author Contributions: Conceptualization: transcriptomics, J.G.C., K.K.D., M.J., B.M.K.; gene expression, pathway analyses, and plant experimentation, J.G.C., D.A.S., M.M.d.A., L.G., B.J.P. Data curation, J.G.C.; Formal analysis: J.G.C., M.M.d.A., L.G., D.A.S., K.K.D., B.J.P., M.J., B.M.K.; Funding acquisition, J.G.C.; Investigation: J.G.C., M.M.d.A., L.G., D.A.S., K.K.D., B.J.P., M.J., B.M.K. Methodology: J.G.C., M.M.d.A., L.G., D.A.S., K.K.D., B.J.P., M.J., B.M.K. Project administration, J.G.C. Resources, J.G.C. Supervision, J.G.C., L.G., D.A.S., K.K.D., B.J.P., M.J.; Validation: J.G.C., K.K.D., M.J., B.M.K.; Writing: original draft, J.G.C., M.M.d.A., L.G., D.A.S., K.K.D.; review and editing, J.G.C., M.M.d.A., L.G., D.A.S., K.K.D. All authors have read and agreed to the published version of the manuscript.

Funding: This research was funded by the United States Department of Commerce, National Institute of Standards and Technology, Advanced Technology Program, Cooperative Agreement Numbers 70NANB4H3039, and 70NANB7H7022; Caisson Laboratories, Smithfield, UT; Pioneer, Corteva Agriscience, Johnston, IA; and the Utah Agricultural Experiment Station (UAES), Utah State University, Logan, UT 84322–4845 (approved as UAES journal paper no. 9375).

Acknowledgments: We thank Michael Windham for assistance in identifying *Boechera* taxa, Gradimir Georgevich for encouraging molecular aspects of this study, Bruce Bugbee for guidance in developing drought stress treatments, and John Stevens for guidance concerning statistical analyses. We also acknowledge Aaron Lawyer, Teri Manwaring, Estella Elliott, Jayasree Pattanayak, John Carman Jr., Jonathan Cardwell, Devin Wright, and Landon Watts for technical assistance.

Conflicts of Interest: A U.S. patent related to this work has been filed by Utah State University and is pending. The funders had no role in the design of the study; in the collection, analyses, or interpretation of data; in the writing of the manuscript, or in the decision to publish the results.

References

1. Burnett, J.H. *Fundamentals of Mycology*, 2nd ed.; Edward Arnold: London, UK, 1976.
2. Suomalainen, E.; Saura, A.; Lokki, J. *Cytology and Evolution in Parthenogenesis*; CRC Press, Inc.: Boca Raton, FL, USA, 1987; p. 216.
3. Bilinski, C.A.; Marmiroli, N.; Miller, J.J. Apomixis in *Saccharomyces cerevisiae* and other eukaryotic micro-organisms. *Adv. Microb. Physiol.* **1989**, *30*, 23–52.
4. Asker, S.E.; Jerling, L. *Apomixis in Plants*; CRC Press, Inc.: Boca Raton, FL, USA, 1992; p. 298.
5. Albertini, E.; Barcaccia, G.; Carman, J.G.; Pupilli, F. Did apomixis evolve from sex or was it the other way around? *J. Exp. Bot.* **2019**, *70*, 2951–2964. [CrossRef]
6. Juel, H.O. Vergleichende untersuchungen uber typische und parthenogenetische fortpflanzung bei der gattung *Antennaria Kongl Sven. Vetenskars Akad. Handl.* **1900**, *33*, 1–59.
7. Gustafsson, A. Apomixis in higher plants. I. The mechanisms of apomixis. *Lunds Univ. Arsskrift* **1946**, *42*, 1–67.
8. Hand, M.L.; Koltunow, A.M. The genetic control of apomixis: Asexual seed formation. *Genetics* **2014**, *197*, 441–450. [CrossRef]
9. Nogler, G.A. Gametophytic apomixis. In *Embryology of Angiosperms*; Johri, B.M., Ed.; Springer-Verlag: New York, NY, USA, 1984; pp. 475–518.
10. Ozias-Akins, P.; van Dijk, P.J. Mendelian genetics of apomixis in plants. *Ann. Rev. Genet.* **2007**, *41*, 509–537. [CrossRef]
11. Savidan, Y. Apomixis: Genetics and breeding. In *Plant Breeding Reviews*; Janick, J., Ed.; John Wiley & Sons, Inc.: New York, NY, USA, 2000; Volume 18, pp. 13–86.
12. Mogie, M. *The Evolution of Asexual Reproduction in Plants*; Chapman & Hall: London, UK, 1992; p. 276.
13. Carman, J.G. Asynchronous expression of duplicate genes in angiosperms may cause apomixis, bispory, tetraspory, and polyembryony. *Biol. J. Linn. Soc.* **1997**, *61*, 51–94. [CrossRef]
14. Carman, J.G.; Jamison, M.; Elliott, E.; Dwivedi, K.K.; Naumova, T.N. Apospory appears to accelerate onset of meiosis and sexual embryo sac formation in sorghum ovules. *BMC Plant Biol.* **2011**, *11*, 9. [CrossRef]
15. Horandl, E.; Hadacek, F. The oxidative damage initiation hypothesis for meiosis. *Plant Reprod.* **2013**, *26*, 351–367. [CrossRef]
16. Carman, J.G.; Mateo de Arias, M.; Gao, L.; Zhao, X.; Kowallis, B.M.; Sherwood, D.A.; Srivastava, M.K.; Dwivedi, K.K.; Price, B.J.; Watts, L.; et al. Apospory and diplospory in diploid *Boechera* (Brassicaceae) may facilitate speciation by recombination-driven apomixis-to-sex reversals. *Front. Plant Sci.* **2019**, *10*, 724. [CrossRef]
17. Li, F.W.; Rushworth, C.A.; Beck, J.B.; Windham, M.D. Boechera microsatellite website: An online portal for species identification and determination of hybrid parentage. *Database (Oxford)* **2017**, *2017*. [CrossRef] [PubMed]
18. Windsor, A.J.; Schranz, M.E.; Formanova, N.; Gebauer-Jung, S.; Bishop, J.G.; Schnabelrauch, D.; Kroymann, J.; Mitchell-Olds, T. Partial shotgun sequencing of the Boechera stricta genome reveals extensive microsynteny and promoter conservation with *Arabidopsis*. *Plant Physiol.* **2006**, *140*, 1169–1182. [CrossRef]
19. Brukhin, V.; Osadtchiy, J.V.; Florez-Rueda, A.M.; Smetanin, D.; Bakin, E.; Nobre, M.S.; Grossniklaus, U. The *Boechera* genus as a resource for apomixis research. *Front. Plant Sci.* **2019**, *10*, 392. [CrossRef]

20. Na, D.; Son, H.; Gsponer, J. Categorizer: A tool to categorize genes into user-defined biological groups based on semantic similarity. *BMC Genom.* **2014**, *15*, 1091. Available online: http://www.biomedcentral.com/1471-2164/15/1091 (accessed on 22 October 2020). [CrossRef]
21. Hinderer, E.W., 3rd; Flight, R.M.; Dubey, R.; MacLeod, J.N.; Moseley, H.N.B. Advances in gene ontology utilization improve statistical power of annotation enrichment. *PLoS ONE* **2019**, *14*, e0220728. [CrossRef]
22. Hinderer, E.W., 3rd; Moseley, H.N.B. GOcats: A tool for categorizing Gene Ontology into subgraphs of user-defined concepts. *PLoS ONE* **2020**, *15*, e0233311. [CrossRef]
23. Ostle, B.; Mensing, R.W. *Statistics in Research*, 3rd ed.; The Iowa State University Press: Ames, UK, 1975; p. 596.
24. Agresti, A. *An Introduction to Categorical Data Analysis*, 3rd ed.; John Wiley & Sons, Inc.: New York, NY, USA, 2019.
25. SYSTAT. *SYSTAT for Windows*; SYSTAT Software, Inc.: San Jose, CA, USA, 2004.
26. Murashige, T.; Skoog, F. A revised medium for rapid growth and bio assays with tobacco tissue cultures. *Phsiologia Plant* **1962**, *15*, 473–497. [CrossRef]
27. Buah, J.N.; Kawamitsu, U.; Sato, S.; Murahama, S. Effects of different types and concentrations of gelling agents on the physical and chemical properties of media and the growth of banana (*Musa* spp.) in vitro. *Plant Prod. Sci.* **1999**, *2*, 138–145. [CrossRef]
28. Michel, B.E. Evaluation of the water potentials of solutions of polyethylene glycol 8000 both in the absence and presence of other solutes. *Plant Physiol.* **1983**, *72*, 66–70. [CrossRef]
29. Michel, B.E.; Kaufmann, M.R. The osmotic potential of polyethylene glycol 6000. *Plant Physiol.* **1973**, *51*, 914–916. [CrossRef]
30. Schneitz, K.; Hulskamp, M.; Pruitt, R.E. Wild-type ovule development in *Arabidopsis thaliana*: A light microscope study of cleared whole-mount tissue. *Plant J.* **1995**, *7*, 731–749. [CrossRef]
31. Clark, A.G.; Glanowski, S.; Nielsen, R.; Thomas, P.D.; Kejariwal, A.; Todd, M.A.; Tanenbaum, D.M.; Civello, D.; Lu, F.; Murphy, B.; et al. Inferring nonneutral evolution from human-chimp-mouse orthologous gene trios. *Science* **2003**, *302*, 1960–1963. [CrossRef] [PubMed]
32. Mootha, V.K.; Lindgren, C.M.; Eriksson, K.F.; Subramanian, A.; Sihag, S.; Lehar, J.; Puigserver, P.; Carlsson, E.; Ridderstrale, M.; Laurila, E.; et al. PGC-1alpha-responsive genes involved in oxidative phosphorylation are coordinately downregulated in human diabetes. *Nat. Genet.* **2003**, *34*, 267–273. [CrossRef] [PubMed]
33. Broeckx, T.; Hulsmans, S.; Rolland, F. The plant energy sensor: Evolutionary conservation and divergence of SnRK1 structure, regulation, and function. *J. Exp. Bot* **2016**, *67*, 6215–6252. [CrossRef]
34. Garapati, P.; Feil, R.; Lunn, J.E.; Van Dijck, P.; Balazadeh, S.; Mueller-Roeber, B. Transcription factor arabidopsis activating factor1 integrates carbon starvation responses with trehalose metabolism. *Plant Physiol.* **2015**, *169*, 379–390. [CrossRef]
35. Sun, L.; Zhang, P.; Wang, R.; Wan, J.; Ju, Q.; Rothstein, S.J.; Xu, J. The SNAC-A transcription factor ANAC032 reprograms metabolism in *Arabidopsis*. *Plant Cell Physiol.* **2019**, *60*, 999–1010. [CrossRef]
36. Carvalho, R.F.; Szakonyi, D.; Simpson, C.G.; Barbosa, I.C.; Brown, J.W.; Baena-Gonzalez, E.; Duque, P. The *Arabidopsis* SR45 splicing factor, a negative regulator of sugar signaling, modulates SNF1-related protein kinase 1 stability. *Plant Cell* **2016**, *28*, 1910–1925. [CrossRef]
37. Poschet, G.; Hannich, B.; Raab, S.; Jungkunz, I.; Klemens, P.A.; Krueger, S.; Wic, S.; Neuhaus, H.E.; Buttner, M. A novel *Arabidopsis* vacuolar glucose exporter is involved in cellular sugar homeostasis and affects the composition of seed storage compounds. *Plant Physiol.* **2011**, *157*, 1664–1676. [CrossRef]
38. Ma, J.; Jiang, Q.T.; Wei, L.; Yang, Q.; Zhang, X.W.; Peng, Y.Y.; Chen, G.Y.; Wei, Y.M.; Liu, C.; Zheng, Y.L. Conserved structure and varied expression reveal key roles of phosphoglucan phosphatase gene starch excess 4 in barley. *Planta* **2014**, *240*, 1179–1190. [CrossRef]
39. Crozet, P.; Margalha, L.; Confraria, A.; Rodrigues, A.; Martinho, C.; Adamo, M.; Elias, C.A.; Baena-Gonzalez, E. Mechanisms of regulation of SNF1/AMPK/SnRK1 protein kinases. *Front. Plant Sci.* **2014**, *5*, 190. [CrossRef]
40. Wang, Y.; Wang, L.; Micallef, B.J.; Tetlow, I.J.; Mullen, R.T.; Feil, R.; Lunn, J.E.; Emes, M.J. AKINbeta1, a subunit of SnRK1, regulates organic acid metabolism and acts as a global modulator of genes involved in carbon, lipid, and nitrogen metabolism. *J. Exp. Bot.* **2020**, *71*, 1010–1028. [CrossRef]
41. Liu, M.S.; Chen, L.F.; Lin, C.H.; Lai, Y.M.; Huang, J.Y.; Sung, Z.R. Molecular and functional characterization of broccoli EMBRYONIC FLOWER 2 genes. *Plant Cell Physiol.* **2012**, *53*, 1217–1231. [CrossRef]
42. Zhao, H.; Wang, Q.; Liu, C.; Shang, Y.; Wen, F.; Wang, F.; Liu, W.; Xiao, W.; Li, W. A role for the respiratory chain in regulating meiosis initiation in *Saccharomyces cerevisiae*. *Genetics* **2018**, *208*, 1181–1194. [CrossRef]

43. Beermann, F.; Hansmann, I. Follicular maturation, luteinization and first meiotic division in oocytes after inhibiting mitochondrial function in mice with chloramphenicol. *Mutat. Res.* **1986**, *160*, 47–54. [CrossRef]
44. Meng, X.; Li, L.; De Clercq, I.; Narsai, R.; Xu, Y.; Hartmann, A.; Claros, D.L.; Custovic, E.; Lewsey, M.G.; Whelan, J.; et al. ANAC017 coordinates organellar functions and stress responses by reprogramming retrograde signaling. *Plant Physiol.* **2019**, *180*, 634–653. [CrossRef]
45. Gladman, N.P.; Marshall, R.S.; Lee, K.H.; Vierstra, R.D. The proteasome stress regulon is controlled by a pair of NAC transcription factors in *Arabidopsis*. *Plant Cell* **2016**, *28*, 1279–1296. [CrossRef]
46. Cavalier-Smith, T. Origin of the cell nucleus, mitosis and sex: Roles of intracellular coevolution. *Biol. Direct.* **2010**, *5*, 7. [CrossRef]
47. Bernstein, H.; Bernstein, C. Evolutionary origin and adaptive function of meiosis. In *Meiosis*; Bernstein, H., Bernstein, C., Eds.; InTech: Rijeka, Croatia, 2013; p. 123.
48. Mirzaghaderi, G.; Horandl, E. The evolution of meiotic sex and its alternatives. *Proc. Biol. Sci.* **2016**, *283*. [CrossRef]
49. Walker, J.; Gao, H.; Zhang, J.; Aldridge, B.; Vickers, M.; Higgins, J.D.; Feng, X. Sexual-lineage-specific DNA methylation regulates meiosis in *Arabidopsis*. *Nat. Genet.* **2018**, *50*, 130–137. [CrossRef]
50. Peng, L.; Wang, L.; Zhang, Y.; Dong, A.; Shen, W.H.; Huang, Y. Structural analysis of the *Arabidopsis* AL2-PAL and PRC1 complex provides mechanistic insight into active-to-repressive chromatin state switch. *J. Mol. Biol.* **2018**, *430*, 4245–4259. [CrossRef]
51. Tyler, L.; Miller, M.J.; Fletcher, J.C. The trithorax group factor ULTRAPETALA1 regulates developmental as well as biotic and abiotic stress response genes in *Arabidopsis*. *G3 (Bethesda)* **2019**, *9*, 4029–4043. [CrossRef]
52. Wils, C.R.; Kaufmann, K. Gene-regulatory networks controlling inflorescence and flower development in *Arabidopsis thaliana*. *Biochim. Biophys. Acta Gene. Regul. Mech.* **2017**, *1860*, 95–105. [CrossRef]
53. Planas-Riverola, A.; Gupta, A.; Betegon-Putze, I.; Bosch, N.; Ibanes, M.; Cano-Delgado, A.I. Brassinosteroid signaling in plant development and adaptation to stress. *Development* **2019**, *146*. [CrossRef]
54. Lv, M.; Li, J. Molecular mechanisms of brassinosteroid-mediated responses to changing environments in *Arabidopsis*. *Int. J. Mol. Sci.* **2020**, *21*, 2737. [CrossRef]
55. Oh, E.; Zhu, J.Y.; Wang, Z.Y. Interaction between BZR1 and PIF4 integrates brassinosteroid and environmental responses. *Nat. Cell Biol.* **2012**, *14*, 802–809. [CrossRef]
56. Xu, J.; Lee, Y.J.; Liu, B. Establishment of a mitotic model system by transient expression of the D-type cyclin in differentiated leaf cells of tobacco (*Nicotiana benthamiana*). *New Phytol.* **2020**, *226*, 1213–1220. [CrossRef]
57. Wang, F.; Gao, Y.; Liu, Y.; Zhang, X.; Gu, X.; Ma, D.; Zhao, Z.; Yuan, Z.; Xue, H.; Liu, H. BES1-regulated BEE1 controls photoperiodic flowering downstream of blue light signaling pathway in *Arabidopsis*. *New Phytol.* **2019**, *223*, 1407–1419. [CrossRef]
58. Poppenberger, B.; Rozhon, W.; Khan, M.; Husar, S.; Adam, G.; Luschnig, C.; Fujioka, S.; Sieberer, T. CESTA, a positive regulator of brassinosteroid biosynthesis. *EMBO J.* **2011**, *30*, 1149–1161. [CrossRef]
59. Zhiponova, M.K.; Vanhoutte, I.; Boudolf, V.; Betti, C.; Dhondt, S.; Coppens, F.; Mylle, E.; Maes, S.; Gonzalez-Garcia, M.P.; Cano-Delgado, A.I.; et al. Brassinosteroid production and signaling differentially control cell division and expansion in the leaf. *New Phytol.* **2013**, *197*, 490–502. [CrossRef]
60. Santiago, J.; Henzler, C.; Hothorn, M. Molecular mechanism for plant steroid receptor activation by somatic embryogenesis co-receptor kinases. *Science* **2013**, *341*, 889–892. [CrossRef]
61. Li, Z.; Wang, Y.; Huang, J.; Ahsan, N.; Biener, G.; Paprocki, J.; Thelen, J.J.; Raicu, V.; Zhao, D. Two SERK receptor-like kinases interact with EMS1 to control anther cell fate determination. *Plant Physiol.* **2017**, *173*, 326–337. [CrossRef]
62. Northey, J.G.; Liang, S.; Jamshed, M.; Deb, S.; Foo, E.; Reid, J.B.; McCourt, P.; Samuel, M.A. Farnesylation mediates brassinosteroid biosynthesis to regulate abscisic acid responses. *Nat. Plants* **2016**, *2*, 16114. [CrossRef]
63. Yamagami, A.; Saito, C.; Nakazawa, M.; Fujioka, S.; Uemura, T.; Matsui, M.; Sakuta, M.; Shinozaki, K.; Osada, H.; Nakano, A.; et al. Evolutionarily conserved BIL4 suppresses the degradation of brassinosteroid receptor BRI1 and regulates cell elongation. *Sci. Rep.* **2017**, *7*, 5739. [CrossRef]
64. Jiang, H.; Tang, B.; Xie, Z.; Nolan, T.; Ye, H.; Song, G.Y.; Walley, J.; Yin, Y. GSK3-like kinase BIN2 phosphorylates RD26 to potentiate drought signaling in *Arabidopsis*. *Plant J.* **2019**, *100*, 923–937. [CrossRef]

65. Yoshida, T.; Nishimura, N.; Kitahata, N.; Kuromori, T.; Ito, T.; Asami, T.; Shinozaki, K.; Hirayama, T. ABA-hypersensitive germination3 encodes a protein phosphatase 2C (AtPP2CA) that strongly regulates abscisic acid signaling during germination among *Arabidopsis* protein phosphatase 2Cs. *Plant Physiol.* **2006**, *140*, 115–126. [CrossRef]
66. Carman, J.G. The evolution of gametophytic apomixis. In *Embryology of Flowering Plants, Vol. 3: The Systems of Reproduction*; Batygina, T.B., Ed.; Russian Academy of Sciences: St. Petersburg, Russia, 2000; pp. 218–245.
67. Bar-Zvi, D.; Lupo, O.; Levy, A.A.; Barkai, N. Hybrid vigor: The best of both parents, or a genomic clash? *Curr. Opin. Syst. Biol.* **2017**, *6*, 22–27. [CrossRef]
68. Kirk, H.; Choi, Y.H.; Kim, H.K.; Verpoorte, R.; van der Meijden, E. Comparing metabolomes: The chemical consequences of hybridization in plants. *New Phytol.* **2005**, *167*, 613–622. [CrossRef]
69. Hauser, F.; Li, Z.; Waadt, R.; Schroeder, J.I. SnapShot: Abscisic acid signaling. *Cell* **2017**, *171*, 1708–1708 e1700. [CrossRef]
70. Fang, Q.; Wang, Q.; Mao, H.; Xu, J.; Wang, Y.; Hu, H.; He, S.; Tu, J.; Cheng, C.; Tian, G.; et al. AtDIV2, an R-R-type MYB transcription factor of *Arabidopsis*, negatively regulates salt stress by modulating ABA signaling. *Plant Cell Rep.* **2018**, *37*, 1499–1511. [CrossRef]
71. Baek, D.; Shin, G.; Kim, M.C.; Shen, M.; Lee, S.Y.; Yun, D.J. Histone deacetylase HDA9 with ABI4 contributes to abscisic acid homeostasis in drought stress response. *Front. Plant Sci.* **2020**, *11*, 143. [CrossRef]
72. Liu, Y.; Pei, L.; Xiao, S.; Peng, L.; Liu, Z.; Li, X.; Yang, Y.; Wang, J. AtPPRT1 negatively regulates salt stress response in *Arabidopsis* seedlings. *Plant Signal. Behav.* **2020**, *15*, 1732103. [CrossRef]
73. Pei, L.; Peng, L.; Wan, X.; Xiong, J.; Liu, Z.; Li, X.; Yang, Y.; Wang, J. Expression pattern and function analysis of AtPPRT1, a novel negative regulator in ABA and drought stress responses in *Arabidopsis*. *Int. J. Mol. Sci.* **2019**, *20*, 394. [CrossRef]
74. Song, L.; Huang, S.C.; Wise, A.; Castanon, R.; Nery, J.R.; Chen, H.; Watanabe, M.; Thomas, J.; Bar-Joseph, Z.; Ecker, J.R. A transcription factor hierarchy defines an environmental stress response network. *Science* **2016**, *354*. [CrossRef]
75. Tian, H.; Chen, S.; Yang, W.; Wang, T.; Zheng, K.; Wang, Y.; Cheng, Y.; Zhang, N.; Liu, S.; Li, D.; et al. A novel family of transcription factors conserved in angiosperms is required for ABA signalling. *Plant Cell Environ.* **2017**, *40*, 2958–2971. [CrossRef]
76. Rodriguez, L.; Gonzalez-Guzman, M.; Diaz, M.; Rodrigues, A.; Izquierdo-Garcia, A.C.; Peirats-Llobet, M.; Fernandez, M.A.; Antoni, R.; Fernandez, D.; Marquez, J.A.; et al. C2-domain abscisic acid-related proteins mediate the interaction of PYR/PYL/RCAR abscisic acid receptors with the plasma membrane and regulate abscisic acid sensitivity in *Arabidopsis*. *Plant Cell* **2014**, *26*, 4802–4820. [CrossRef]
77. Alvarez, S.; Roy Choudhury, S.; Hicks, L.M.; Pandey, S. Quantitative proteomics-based analysis supports a significant role of GTG proteins in regulation of ABA response in *Arabidopsis* roots. *J. Proteome Res.* **2013**, *12*, 1487–1501. [CrossRef]
78. Irigoyen, M.L.; Iniesto, E.; Rodriguez, L.; Puga, M.I.; Yanagawa, Y.; Pick, E.; Strickland, E.; Paz-Ares, J.; Wei, N.; De Jaeger, G.; et al. Targeted degradation of abscisic acid receptors is mediated by the ubiquitin ligase substrate adaptor DDA1 in *Arabidopsis*. *Plant Cell* **2014**, *26*, 712–728. [CrossRef]
79. Ding, Y.; Lv, J.; Shi, Y.; Gao, J.; Hua, J.; Song, C.; Gong, Z.; Yang, S. EGR2 phosphatase regulates OST1 kinase activity and freezing tolerance in *Arabidopsis*. *EMBO J.* **2019**, *38*. [CrossRef]
80. Umezawa, T.; Sugiyama, N.; Takahashi, F.; Anderson, J.C.; Ishihama, Y.; Peck, S.C.; Shinozaki, K. Genetics and phosphoproteomics reveal a protein phosphorylation network in the abscisic acid signaling pathway in *Arabidopsis thaliana*. *Sci. Signal.* **2013**, *6*, rs8. [CrossRef]
81. Fujii, H.; Verslues, P.E.; Zhu, J.K. *Arabidopsis* decuple mutant reveals the importance of SnRK2 kinases in osmotic stress responses in vivo. *Proc. Natl. Acad. Sci. USA* **2011**, *108*, 1717–1722. [CrossRef]
82. Zhou, X.; Hao, H.; Zhang, Y.; Bai, Y.; Zhu, W.; Qin, Y.; Yuan, F.; Zhao, F.; Wang, M.; Hu, J.; et al. SOS2-LIKE PROTEIN KINASE5, an SNF1-RELATED PROTEIN KINASE3-type protein kinase, is iImportant for abscisic acid responses in *Arabidopsis* through phosphorylation of ABSCISIC ACID-INSENSITIVE5. *Plant Physiol.* **2015**, *168*, 659–676. [CrossRef]
83. Yang, Z.; Wang, C.; Xue, Y.; Liu, X.; Chen, S.; Song, C.; Yang, Y.; Guo, Y. Calcium-activated 14-3-3 proteins as a molecular switch in salt stress tolerance. *Nat. Commun.* **2019**, *10*, 1199. [CrossRef] [PubMed]

84. Xu, X.; Wan, W.; Jiang, G.; Xi, Y.; Huang, H.; Cai, J.; Chang, Y.; Duan, C.G.; Mangrauthia, S.K.; Peng, X.; et al. Nucleocytoplasmic trafficking of the *Arabidopsis* WD40 repeat protein XIW1 regulates ABI5 stability and abscisic acid responses. *Mol. Plant* **2019**, *12*, 1598–1611. [CrossRef]
85. Zhu, T.; Wu, Y.; Yang, X.; Chen, W.; Gong, Q.; Liu, X. The asparagine-rich protein NRP facilitates the degradation of the PP6-type phosphatase FyPP3 to promote ABA response in *Arabidopsis*. *Mol. Plant* **2018**, *11*, 257–268. [CrossRef]
86. Oh, T.R.; Kim, J.H.; Cho, S.K.; Ryu, M.Y.; Yang, S.W.; Kim, W.T. AtAIRP2 E3 ligase affects ABA and high-salinity responses by stimulating Its ATP1/SDIRIP1 substrate turnover. *Plant Physiol.* **2017**, *174*, 2515–2531. [CrossRef]
87. Liang, Y.; Kang, K.; Gan, L.; Ning, S.; Xiong, J.; Song, S.; Xi, L.; Lai, S.; Yin, Y.; Gu, J.; et al. Drought-responsive genes, late embryogenesis abundant group3 (LEA3) and vicinal oxygen chelate, function in lipid accumulation in *Brassica napus* and *Arabidopsis* mainly via enhancing photosynthetic efficiency and reducing ROS. *Plant Biotechnol. J.* **2019**, *17*, 2123–2142. [CrossRef]
88. Lu, W.; Tang, X.; Huo, Y.; Xu, R.; Qi, S.; Huang, J.; Zheng, C.; Wu, C.A. Identification and characterization of fructose 1,6-bisphosphate aldolase genes in *Arabidopsis* reveal a gene family with diverse responses to abiotic stresses. *Gene* **2012**, *503*, 65–74. [CrossRef]
89. Hernandez-Sanchez, I.E.; Maruri-Lopez, I.; Molphe-Balch, E.P.; Becerra-Flora, A.; Jaimes-Miranda, F.; Jimenez-Bremont, J.F. Evidence for in vivo interactions between dehydrins and the aquaporin AtPIP2B. *Biochem. Biophys. Res. Commun.* **2019**, *510*, 545–550. [CrossRef]
90. Seok, H.Y.; Nguyen, L.V.; Park, H.Y.; Tarte, V.N.; Ha, J.; Lee, S.Y.; Moon, Y.H. *Arabidopsis* non-TZF gene AtC3H17 functions as a positive regulator in salt stress response. *Biochem. Biophys. Res. Commun.* **2018**, *498*, 954–959. [CrossRef]
91. Ebeed, H.T.; Stevenson, S.R.; Cuming, A.C.; Baker, A. Conserved and differential transcriptional responses of peroxisome associated pathways to drought, dehydration and ABA. *J. Exp. Bot.* **2018**, *69*, 4971–4985. [CrossRef]
92. Singh, A.; Yadav, A.K.; Kaur, K.; Sanyal, S.K.; Jha, S.K.; Fernandes, J.L.; Sharma, P.; Tokas, I.; Pandey, A.; Luan, S.; et al. A protein phosphatase 2C, AP2C1, interacts with and negatively regulates the function of CIPK9 under potassium-deficient conditions in *Arabidopsis*. *J. Exp. Bot.* **2018**, *69*, 4003–4015. [CrossRef] [PubMed]
93. Shubchynskyy, V.; Boniecka, J.; Schweighofer, A.; Simulis, J.; Kvederaviciute, K.; Stumpe, M.; Mauch, F.; Balazadeh, S.; Mueller-Roeber, B.; Boutrot, F.; et al. Protein phosphatase AP2C1 negatively regulates basal resistance and defense responses to *Pseudomonas syringae*. *J. Exp. Bot.* **2017**, *68*, 1169–1183. [CrossRef] [PubMed]
94. Chapman, J.M.; Muhlemann, J.K.; Gayomba, S.R.; Muday, G.K. RBOH-dependent ROS synthesis and ROS scavenging by plant specialized metabolites to modulate plant development and stress responses. *Chem. Res. Toxicol.* **2019**, *32*, 370–396. [CrossRef] [PubMed]
95. Kaman-Toth, E.; Danko, T.; Gullner, G.; Bozso, Z.; Palkovics, L.; Pogany, M. Contribution of cell wall peroxidase- and NADPH oxidase-derived reactive oxygen species to *Alternaria brassicicola*-induced oxidative burst in *Arabidopsis*. *Mol. Plant Pathol.* **2019**, *20*, 485–499. [CrossRef]
96. Zandalinas, S.I.; Song, L.; Sengupta, S.; McInturf, S.A.; Grant, D.G.; Marjault, H.B.; Castro-Guerrero, N.A.; Burks, D.; Azad, R.K.; Mendoza-Cozatl, D.G.; et al. Expression of a dominant-negative AtNEET-H89C protein disrupts iron-sulfur metabolism and iron homeostasis in *Arabidopsis*. *Plant J.* **2020**, *101*, 1152–1169. [CrossRef]
97. Ma, X.; Zhang, X.; Yang, L.; Tang, M.; Wang, K.; Wang, L.; Bai, L.; Song, C. Hydrogen peroxide plays an important role in PERK4-mediated abscisic acid-regulated root growth in *Arabidopsis*. *Funct. Plant Biol.* **2019**, *46*, 165–174. [CrossRef]
98. Angelos, E.; Brandizzi, F. NADPH oxidase activity is required for ER stress survival in plants. *Plant J.* **2018**, *96*, 1106–1120. [CrossRef]
99. Li, L.; Yi, H.; Xue, M.; Yi, M. miR398 and miR395 are involved in response to SO2 stress in *Arabidopsis thaliana*. *Ecotoxicology* **2017**, *26*, 1181–1187. [CrossRef]
100. Panda, S.K.; Sunkar, R. Nutrient- and other stress-responsive microRNAs in plants: Role for thiol-based redox signaling. *Plant Signal. Behav.* **2015**, *10*, e1010916. [CrossRef]

101. Dvorak, P.; Krasylenko, Y.; Ovecka, M.; Basheer, J.; Zapletalova, V.; Samaj, J.; Takac, T. FSD1: Developmentally-regulated plastidial, nuclear and cytoplasmic enzyme with anti-oxidative and osmoprotective role. *Plant Cell Environ.* **2020**. [CrossRef]
102. Martiniere, A.; Fiche, J.B.; Smokvarska, M.; Mari, S.; Alcon, C.; Dumont, X.; Hematy, K.; Jaillais, Y.; Nollmann, M.; Maurel, C. Osmotic stress activates two reactive oxygen species pathways with distinct effects on protein nanodomains and diffusion. *Plant Physiol.* **2019**, *179*, 1581–1593. [CrossRef]
103. Achard, P.; Renou, J.P.; Berthome, R.; Harberd, N.P.; Genschik, P. Plant DELLAs restrain growth and promote survival of adversity by reducing the levels of reactive oxygen species. *Curr. Biol.* **2008**, *18*, 656–660. [CrossRef] [PubMed]
104. Fujiwara, A.; Togawa, S.; Hikawa, T.; Matsuura, H.; Masuta, C.; Inukai, T. Ascorbic acid accumulates as a defense response to Turnip mosaic virus in resistant *Brassica rapa* cultivars. *J. Exp. Bot.* **2016**, *67*, 4391–4402. [CrossRef]
105. Nepal, N.; Yactayo-Chang, J.P.; Medina-Jimenez, K.; Acosta-Gamboa, L.M.; Gonzalez-Romero, M.E.; Arteaga-Vazquez, M.A.; Lorence, A. Mechanisms underlying the enhanced biomass and abiotic stress tolerance phenotype of an *Arabidopsis* MIOX over-expresser. *Plant Direct.* **2019**, *3*, e00165. [CrossRef] [PubMed]
106. Selinski, J.; Scheibe, R.; Day, D.A.; Whelan, J. Alternative oxidase is positive for plant performance. *Trends Plant Sci.* **2018**, *23*, 588–597. [CrossRef]
107. Chen, C.; Letnik, I.; Hacham, Y.; Dobrev, P.; Ben-Daniel, B.H.; Vankova, R.; Amir, R.; Miller, G. ASCORBATE PEROXIDASE6 protects *Arabidopsis* desiccating and germinating seeds from stress and mediates cross talk between reactive oxygen species, abscisic acid, and auxin. *Plant Physiol.* **2014**, *166*, 370–383. [CrossRef]
108. Muller-Schussele, S.J.; Wang, R.; Gutle, D.D.; Romer, J.; Rodriguez-Franco, M.; Scholz, M.; Buchert, F.; Luth, V.M.; Kopriva, S.; Dormann, P.; et al. Chloroplasts require glutathione reductase to balance reactive oxygen species and maintain efficient photosynthesis. *Plant J.* **2020**. [CrossRef]
109. Tossounian, M.A.; Van Molle, I.; Wahni, K.; Jacques, S.; Gevaert, K.; Van Breusegem, F.; Vertommen, D.; Young, D.; Rosado, L.A.; Messens, J. Disulfide bond formation protects *Arabidopsis thaliana* glutathione transferase tau 23 from oxidative damage. *Biochim. Biophys. Acta Gen. Subj.* **2018**, *1862*, 775–789. [CrossRef]
110. Horvath, E.; Bela, K.; Galle, A.; Riyazuddin, R.; Csomor, G.; Csenki, D.; Csiszar, J. Compensation of mutation in *Arabidopsis* glutathione transferase (AtGSTU) genes under control or salt stress conditions. *Int. J. Mol. Sci.* **2020**, *21*, 2349. [CrossRef]
111. Dixon, D.P.; Edwards, R. Glutathione transferases. *Arab. Book* **2010**, *8*, e0131. [CrossRef]
112. Chen, L.; Wu, R.; Feng, J.; Feng, T.; Wang, C.; Hu, J.; Zhan, N.; Li, Y.; Ma, X.; Ren, B.; et al. Transnitrosylation mediated by the non-canonical catalase ROG1 regulates nitric oxide signaling in plants. *Dev. Cell* **2020**, *53*, 444–457.e445. [CrossRef]
113. da Fonseca-Pereira, P.; Daloso, D.M.; Gago, J.; de Oliveira Silva, F.M.; Condori-Apfata, J.A.; Florez-Sarasa, I.; Tohge, T.; Reichheld, J.P.; Nunes-Nesi, A.; Fernie, A.R.; et al. The mitochondrial thioredoxin system contributes to the metabolic responses under drought episodes in *Arabidopsis*. *Plant Cell Physiol.* **2019**, *60*, 213–229. [CrossRef]
114. Baune, M.C.; Lansing, H.; Fischer, K.; Meyer, T.; Charton, L.; Linka, N.; von Schaewen, A. The *Arabidopsis* plastidial glucose-6-phosphate transporter GPT1 is dually targeted to peroxisomes via the endoplasmic reticulum. *Plant Cell* **2020**, *32*, 1703–1726. [CrossRef]
115. Bissoli, G.; Munoz-Bertomeu, J.; Bueso, E.; Sayas, E.; Vilcara, E.A.; Felipo, A.; Ninoles, R.; Rubio, L.; Fernandez, J.A.; Serrano, R. An *Arabidopsis* mutant over-expressing subtilase SBT4.13 uncovers the role of oxidative stress in the inhibition of growth by intracellular acidification. *Int. J. Mol. Sci.* **2020**, *21*, 1173. [CrossRef]
116. Manara, A.; DalCorso, G.; Leister, D.; Jahns, P.; Baldan, B.; Furini, A. AtSIA1 AND AtOSA1: Two Abc1 proteins involved in oxidative stress responses and iron distribution within chloroplasts. *New Phytol.* **2014**, *201*, 452–465. [CrossRef]
117. Bocher, T.W. Cytological and embryological studies in the amphi-apomictic *Arabis holboellii* complex. *KongelDanske Vidensk-SelskabBiolSkr* **1951**, *6*, 1–59.
118. Contento, A.L.; Kim, S.J.; Bassham, D.C. Transcriptome profiling of the response of *Arabidopsis* suspension culture cells to Suc starvation. *Plant Physiol.* **2004**, *135*, 2330–2347. [CrossRef]

119. Griffin, P.T.; Niederhuth, C.E.; Schmitz, R.J. A comparative analysis of 5-azacytidine- and zebularine-induced DNA demethylation. *G3 (Bethesda)* **2016**, *6*, 2773–2780. [CrossRef]
120. Tian, Y.; Fan, M.; Qin, Z.; Lv, H.; Wang, M.; Zhang, Z.; Zhou, W.; Zhao, N.; Li, X.; Han, C.; et al. Hydrogen peroxide positively regulates brassinosteroid signaling through oxidation of the BRASSINAZOLE-RESISTANT1 transcription factor. *Nat. Commun.* **2018**, *9*, 1063. [CrossRef]
121. Rasmussen, R.D.; Hole, D.; Hess, J.R.; Carman, J.G. Wheat kernel dormancy and +abscisic acid level following exposure to fluridone. *J. Plant Physiol.* **1997**, *150*, 440–445. [CrossRef]
122. Bartels, P.G.; Watson, C.W. Inhibition of carotenoid synthesis by fluridone and norflurazon. *Weed Sci.* **1978**, *26*, 198–203. [CrossRef]
123. Bowman, J.L.; Kohchi, T.; Yamato, K.T.; Jenkins, J.; Shu, S.; Ishizaki, K.; Yamaoka, S.; Nishihama, R.; Nakamura, Y.; Berger, F.; et al. Insights into land plant evolution garnered from the *Marchantia polymorpha* genome. *Cell* **2017**, *171*, 287–304e215. [CrossRef] [PubMed]
124. Jamsheer, K.M.; Jindal, S.; Laxmi, A. Evolution of TOR-SnRK dynamics in green plants and its integration with phytohormone signaling networks. *J. Exp. Bot.* **2019**, *70*, 2239–2259. [CrossRef]
125. Szymanska, K.P.; Polkowska-Kowalczyk, L.; Lichocka, M.; Maszkowska, J.; Dobrowolska, G. SNF1-related protein kinases SnRK2.4 and SnRK2.10 modulate ROS homeostasis in plant response to salt stress. *Int. J. Mol. Sci.* **2019**, *20*, 143. [CrossRef] [PubMed]
126. Gilroy, S.; Bialasek, M.; Suzuki, N.; Gorecka, M.; Devireddy, A.R.; Karpinski, S.; Mittler, R. ROS, calcium, and electric signals: Key mediators of rapid systemic signaling in plants. *Plant Physiol.* **2016**, *171*, 1606–1615. [CrossRef]
127. Bheri, M.; Pandey, G.K. Protein phosphatases meet reactive oxygen species in plant signaling networks. *Environ. Exp. Bot.* **2019**, *161*, 26–40. [CrossRef]
128. Mhamdi, A.; Van Breusegem, F. Reactive oxygen species in plant development. *Development* **2018**, *145*. [CrossRef]
129. Uzilday, B.; Ozgur, R.; Sekmen, A.H.; Turkan, I. Endoplasmic reticulum stress regulates glutathione metabolism and activities of glutathione related enzymes in *Arabidopsis*. *Funct. Plant Biol.* **2018**, *45*, 284–296. [CrossRef]
130. Mittler, R. ROS are good. *Trends Plant Sci.* **2017**, *22*, 11–19. [CrossRef]
131. Nukarinen, E.; Nagele, T.; Pedrotti, L.; Wurzinger, B.; Mair, A.; Landgraf, R.; Bornke, F.; Hanson, J.; Teige, M.; Baena-Gonzalez, E.; et al. Quantitative phosphoproteomics reveals the role of the AMPK plant ortholog SnRK1 as a metabolic master regulator under energy deprivation. *Sci. Rep.* **2016**, *6*, 31697. [CrossRef]
132. Wang, P.; Zhao, Y.; Li, Z.; Hsu, C.C.; Liu, X.; Fu, L.; Hou, Y.J.; Du, Y.; Xie, S.; Zhang, C.; et al. Reciprocal regulation of the TOR kinase and ABA receptor balances plant growth and stress response. *Mol. Cell* **2018**, *69*, 100–112e106. [CrossRef]
133. Jamsheer, K.M.; Sharma, M.; Singh, D.; Mannully, C.T.; Jindal, S.; Shukla, B.N.; Laxmi, A. FCS-like zinc finger 6 and 10 repress SnRK1 signalling in *Arabidopsis*. *Plant J.* **2018**, *94*, 232–245. [CrossRef] [PubMed]
134. Soto-Burgos, J.; Bassham, D.C. SnRK1 activates autophagy via the TOR signaling pathway in *Arabidopsis thaliana*. *PLoS ONE* **2017**, *12*, e0182591. [CrossRef] [PubMed]
135. Rosenberger, C.L.; Chen, J. To grow or not to grow: TOR and SnRK2 coordinate growth and stress response in *Arabidopsis*. *Mol. Cell* **2018**, *69*, 3–4. [CrossRef] [PubMed]
136. Tang, W.; Yuan, M.; Wang, R.; Yang, Y.; Wang, C.; Oses-Prieto, J.A.; Kim, T.W.; Zhou, H.W.; Deng, Z.; Gampala, S.S.; et al. PP2A activates brassinosteroid-responsive gene expression and plant growth by dephosphorylating BZR1. *Nat. Cell Biol.* **2011**, *13*, 124–131. [CrossRef] [PubMed]
137. Zhang, Z.; Zhu, J.Y.; Roh, J.; Marchive, C.; Kim, S.K.; Meyer, C.; Sun, Y.; Wang, W.; Wang, Z.Y. TOR Signaling promotes accumulation of BZR1 to balance growth with carbon availability in *Arabidopsis*. *Curr. Biol.* **2016**, *26*, 1854–1860. [CrossRef]
138. Caldana, C.; Martins, M.C.M.; Mubeen, U.; Urrea-Castellanos, R. The magic 'hammer' of TOR: The multiple faces of a single pathway in the metabolic regulation of plant growth and development. *J. Exp. Bot.* **2019**, *70*, 2217–2225. [CrossRef]
139. Rodriguez, M.; Parola, R.; Andreola, S.; Pereyra, C.; Martinez-Noel, G. TOR and SnRK1 signaling pathways in plant response to abiotic stresses: Do they always act according to the "yin-yang" model? *Plant Sci.* **2019**, *288*, 110220. [CrossRef]

140. Pellino, M.; Hojsgaard, D.; Horandl, E.; Sharbel, T.F. Chasing the apomictic factors in the *Ranunculus auricomus* complex: Exploring gene expression patterns in microdissected sexual and apomictic ovules. *Genes (Basel)* **2020**, *11*, 728. [CrossRef]
141. Ulum, F.B.; Costa Castro, C.; Horandl, E. Ploidy-dependent effects of light stress on the mode of reproduction in the *Ranunculus auricomus* complex (Ranunculaceae). *Front. Plant Sci.* **2020**, *11*, 104. [CrossRef]
142. Shah, J.N.; Kirioukhova, O.; Pawar, P.; Tayyab, M.; Mateo, J.L.; Johnston, A.J. Depletion of key meiotic genes and transcriptome-wide abiotic stress reprogramming mark early preparatory events ahead of apomeiotic transition. *Front. Plant Sci.* **2016**, *7*, 1539. [CrossRef]
143. Khanday, I.; Skinner, D.; Yang, B.; Mercier, R.; Sundaresan, V. A male-expressed rice embryogenic trigger redirected for asexual propagation through seeds. *Nature* **2019**, *565*, 91–95. [CrossRef]
144. Muller, M.; Munne-Bosch, S. Ethylene response factors: A key regulatory hub in hormone and stress signaling. *Plant Physiol.* **2015**, *169*, 32–41. [CrossRef]
145. Oliver, C.; Santos, J.L.; Pradillo, M. Accurate chromosome segregation at first meiotic division requires AGO4, a protein involved in RNA-dependent DNA methylation in *Arabidopsis thaliana*. *Genetics* **2016**, *204*, 543–553. [CrossRef]
146. Shashkova, S.; Welkenhuysen, N.; Hohmann, S. Molecular communication: Crosstalk between the Snf1 and other signaling pathways. *FEMS Yeast Res.* **2015**, *15*, fov026. [CrossRef] [PubMed]
147. Honigberg, S.M.; Lee, R.H. Snf1 kinase connects nutritional pathways controlling meiosis in *Saccharomyces cerevisiae*. *Mol. Cell Biol.* **1998**, *18*, 4548–4555. [CrossRef]
148. Chen, Y.S.; Chao, Y.C.; Tseng, T.W.; Huang, C.K.; Lo, P.C.; Lu, C.A. Two MYB-related transcription factors play opposite roles in sugar signaling in *Arabidopsis*. *Plant Mol. Biol.* **2017**, *93*, 299–311. [CrossRef]
149. Haufler, C.H.; Pryer, K.M.; Schuettpelz, E.; Sessa, E.B.; Farrar, D.R.; Moran, R.; Schneller, J.J.; Watkins, J.E.; Windham, M.D. Sex and the single gametophyte: Revising the homosporous vascular plant life cycle in light of contemporary research. *BioScience* **2016**, *66*, 928–937. [CrossRef]
150. Herr, J.M.J. The origin of the ovule. *Am. J. Bot.* **1995**, *82*, 547–564. [CrossRef]
151. Stebbins, G.L. Apomixis in the angiosperms. *Bot. Rev.* **1941**, *7*, 507–542. [CrossRef]
152. Ernst, A. *Bastardierung als Ursache der Apogamie im Pflanzenreich*; Fischer: Jena, Germany, 1918.
153. Krisko, A.; Leroy, M.; Radman, M.; Meselson, M. Extreme anti-oxidant protection against ionizing radiation in bdelloid rotifers. *Proc. Natl. Acad. Sci. USA* **2012**, *109*, 2354–2357. [CrossRef] [PubMed]
154. Flot, J.F.; Hespeels, B.; Li, X.; Noel, B.; Arkhipova, I.; Danchin, E.G.; Hejnol, A.; Henrissat, B.; Koszul, R.; Aury, J.M.; et al. Genomic evidence for ameiotic evolution in the bdelloid rotifer Adineta vaga. *Nature* **2013**, *500*, 453–457. [CrossRef]
155. Pattiwael, M.R.; Mangindaan, R.E.P.; Prabowo, R.; Rumengan, I.F.M. Acute and chronic effect of cyaniade on survival rate, behavior, and reproduction of *Daphnia* sp. *Aquat. Sci. Manag.* **2013**, *1*, 52. [CrossRef]
156. Jedlicka, P.; Jedlickova, V.; Lee, H.J. Expression of stress-related genes in the parthenogenetic forms of the pea aphid, *Acyrthosiphon pisum*. *Comp. Biochem. Physiol. A Mol. Integr. Physiol.* **2015**, *180*, 32–37. [CrossRef] [PubMed]
157. Sabbe, K.; Chepurnov, V.A.; Vyverman, W.; Mann, D.G. Apomixis in *Achnanthes* (Bacillariophyceae); development of a model system for diatom reproductive biology. *Eur. J. Phycol.* **2004**, *39*, 327–341. [CrossRef]
158. Nedelcu, A.M. Sex as a response to oxidative stress: Stress genes co-opted for sex. *P Roy. Soc. B-Biol. Sci.* **2005**, *272*, 1935–1940. [CrossRef]
159. Nedelcu, A.M.; Marcu, O.; Michod, R.E. Sex as a response to oxidative stress: A twofold increase in cellular reactive oxygen species activates sex genes. *P Roy. Soc. B-Biol. Sci.* **2004**, *271*, 1591–1596. [CrossRef]
160. Nedelcu, A.M.; Michod, R.E. Sex as a response to oxidative stress: The effect of antioxidants on sexual induction in a facultatively sexual lineage. *P Roy. Soc. B-Biol. Sci.* **2003**, *270*, S136–S139. [CrossRef]
161. Fiore-Donno, A.M.; Novozhilov, Y.K.; Meyer, M.; Schnittler, M. Genetic structure of two protist species (Myxogastria, Amoebozoa) suggests asexual reproduction in sexual Amoebae. *PLoS ONE* **2011**, *6*, e22872. [CrossRef]
162. Collins, O.R. Myxomycete biosystematics: Some recent developments and future research opportunities. *Bot. Rev.* **1979**, *45*, 145–201. [CrossRef]

163. Collins, O.N.R.; Gong, T. Genetical relatedness of a former apomict and a heterothallic isolate in *Didymium Iridis* (Myxomycetes). *Mycologia* **1985**, *77*, 300–307. [CrossRef]

Publisher's Note: MDPI stays neutral with regard to jurisdictional claims in published maps and institutional affiliations.

genes

MDPI

Article

Genes Modulating the Increase in Sexuality in the Facultative Diplosporous Grass *Eragrostis curvula* under Water Stress Conditions

Juan Pablo Selva [1,†], Diego Zappacosta [1,2,†], José Carballo [1,2,†], Juan Manuel Rodrigo [1,2], Andrés Bellido [1], Cristian Andrés Gallo [1], Jimena Gallardo [1,2] and Viviana Echenique [1,2,*]

1 Centro de Recursos Naturales Renovables de la Zona Semiárida (CERZOS—CCT—CONICET Bahía Blanca), Camino de la Carrindanga km 7, Bahía Blanca 8000, Argentina; jpselva@criba.edu.ar (J.P.S.); dczappa@criba.edu.ar (D.Z.); jcarballo@cerzos-conicet.gob.ar (J.C.); juanmanuelrodrigo@outlook.com (J.M.R.); andresmbellido@gmail.com (A.B.); gallo.cristian.andres@gmail.com (C.A.G.); jgallardo@cerzos-conicet.gob.ar (J.G.)

2 Departamento de Agronomía, Universidad Nacional del Sur (UNS), San Andrés 800, Bahía Blanca 8000, Argentina

* Correspondence: echeniq@criba.edu.ar; Tel.: +54-291-486-1124

† These authors contributed equally to this work.

Received: 15 July 2020; Accepted: 17 August 2020; Published: 21 August 2020

Abstract: *Eragrostis curvula* presents mainly facultative genotypes that reproduce by diplosporous apomixis, retaining a percentage of sexual pistils that increase under drought and other stressful situations, indicating that some regulators activated by stress could be affecting the apomixis/sexual switch. Water stress experiments were performed in order to associate the increase in sexual embryo sacs with the differential expression of genes in a facultative apomictic cultivar using cytoembryology and RNA sequencing. The percentage of sexual embryo sacs increased from 4 to 24% and 501 out of the 201,011 transcripts were differentially expressed (DE) between control and stressed plants. DE transcripts were compared with previous transcriptomes where apomictic and sexual genotypes were contrasted. The results point as candidates to transcripts related to methylation, ubiquitination, hormone and signal transduction pathways, transcription regulation and cell wall biosynthesis, some acting as a general response to stress and some that are specific to the reproductive mode. We suggest that a DNA glycosylase EcROS1-like could be demethylating, thus de-repressing a gene or genes involved in the sexuality pathways. Many of the other DE transcripts could be part of a complex mechanism that regulates apomixis and sexuality in this grass, the ones in the intersection between control/stress and apo/sex being the strongest candidates.

Keywords: weeping lovegrass; apomixis; drought stress; RNA-seq; differentially expressed genes; sexuality

1. Introduction

Apomixis refers to asexual propagation by seeds and it is a process composed of three components: apomeiosis, parthenogenesis and autonomous endosperm development or pseudogamy. Three hundred out of more than 400 species of angiosperms that reproduce by apomixis occur in the Poaceae, Asteraceae or Rosaceae [1]. However, some capacity for sexuality is usually maintained; thus, they benefit from using a very sophisticated combination of reproductive strategies, generating diversity and, concurrently, allowing the best fitted individuals to propagate clonally [2]. Due to its polyphyletic origin, there are numerous forms of apomixis: diplospory, apospory and adventitious embryogenesis [3]. The elucidation of the mechanism of apomixis is important not only for biological interest, but also for agricultural technology. In the agricultural industry, it is predicted that apomixis would decrease the cost of hybrid seed production significantly and increase the yield of existing inbred crops by converting them into high-yielding hybrids [4].

Although several comparative transcriptomic studies have already been performed in apomictic species, such as *Pennisetum ciliare* [5], *Brachiaria brizantha* [6], *Poa pratensis* [7], *Panicum maximum* [8], *Paspalum simplex* [9], *Hieracium praealtum* [10], *Ranunculus auricomus* [11], *Boechera gunnisoniana* [12] and *Hypericum perforatum* [13,14], and several candidate genes triggering specific components of apomixis are known, the function of their proteins is still not clear. Among the genes associated with the components of apomixis, found both in studies of apomictic species and in mutants that resemble apomixis, the following can be mentioned: *SERK* and *APOSTART* [15], *DIF1* [16], *BABY BOOM* [17], *APOLLO* [18], *DEMETER* [19], *MSII* [20], *RDR6* and *SGS3* [21], *DYAD/SWITCH1* [22,23], *AGO9* [24], *AGO104* [25], *ORC* [26], *GID1* [27], *FIE* [28], *AGAMOUS-LIKE 62* [29], *PnTgs1-like* [30] and *DMC1* [31].

In organisms that can reproduce both sexually and asexually (facultative apomictic), stress plays an important role in determining which reproductive mode is used [32]. For example, in species with cyclical apomixis (like *Daphnia pulex*), reproduction is apomictic in one season, generally under favorable conditions, and sexual during stressful conditions [33]. Different species of fungi, algae and insects induce sexual reproduction under unfavorable conditions or in response to abiotic stress [34–36]. In plants, abiotic stress, such as drought or heat, can induce megaspore mother cells (MMCs) to undergo meiosis in the ovules of apomictic plants and produce genetically-reduced (sexual) embryo sacs [37–40]. Drought- and heat-stressed *Boechera lignifera* and *B. gunnisoniana* achieved major shifts from apomeiosis to meiosis in MMCs, whereas the non-stressed control plants exhibited 87–98% apomictic dyad formation, the heat- and water-stressed plants exhibited 75–80% sexual tetrad formation [38]. Differences in the photoperiod also induce increases in sexuality, such as in *Themeda australis* [41] and *Ranunculus auricomus* under long photoperiods [39] and in *Paspalum cromyorrhizon* [42] and *Brachiaria brizantha* [43] under short day conditions. The reproductive modes of other species also respond to environmental conditions; for instance, Gounaris et al. [44] detected a greater number of reduced embryo sacs under salt stress in *Cenchrus ciliaris*. Based on this, it is reasonable to expect that facultative apomicts tend to switch to sexual reproduction more often under stress conditions, and that such a stress-dependent switch facilitates the organism's adaptation to a stressful environment [40].

Few studies associate the change in the frequency of apomictic/sexual embryo sacs under stress conditions with changes in gene expression. RNA-Seq studies conducted with immature pistils taken from drought-stressed and well-watered sexual and apomictic *Boechera* spp. plants show that this stress-induced switching includes global epigenetic-based changes in gene expression [45]. Gene ontology (GO) analyses of these differences in gene expression indicate that oxidative stress induces meiosis to occur instead of apomeiosis in apomictic *Boechera* [45]. Several authors, in different model plants, found that the genes that participate in stress pathways are related to the determination or regulation of apomixis [13,39,40].

Eragrostis curvula (weeping lovegrass), an African grass with cytotypes of different ploidy levels (e.g., 2x–8x) and displaying obligate and facultative apomixis and sexual reproduction [46], has become a model for the analysis of apomixis mechanisms, due to its particular diplosporous development (meiotic diplospory maintaining the same embryo: endosperm ploidy ratio as in sexual seeds). In recent years, the reproductive mode of this grass was studied extensively, providing information about the cytoembryological aspects of its apomictic–sexual development [47], differentially expressed (DE) transcripts [48–51], epigenetic aspects of apomictic regulation [52,53], mapping of the apomixis locus [54] and a high quality genome assembly [55]. Our group also demonstrated that under different internal and external stressful situations, including a change in ploidy, water stress, in vitro culture and intraspecific hybridization, the number of sexual embryo sacs increased in facultative apomictic plants of this grass [40,56]. Our group was able to observe that plants of the Tanganyika INTA cultivar produced fewer than 2% of sexual embryo sacs when growing in optimal conditions, but under different stress situations these plants showed an increase in the number of sexual embryo sacs [40]. This increase in the level of sexuality was associated with genetic and epigenetic changes, like methylation and molecular markers [40,56]. Evidence of epigenetic mechanisms controlling apomixis was also observed since differentially expressed patterns of RNA-directed DNA methylation (RdDM) genes [52] and microRNA between sexual and apomictic genotypes [53] were recently reported in this grass.

The aim of the present study was to identify genes that are differentially expressed in weeping lovegrass inflorescences of control and water-stressed facultative apomictic plants and to compare them with the differentially expressed genes between apomictic and sexual plants previously reported by our group [51]. This approach was taken based on previous findings about increases of sexual processes under stress conditions [40], in order to look for common pathways between stress and apomixis. It could give clues about the regulation of this intriguing reproductive mode.

2. Materials and Methods

2.1. Plant Materials

Plants of the tetraploid (2n = 4x = 40) facultative apomictic Don Walter cultivar were grown in 10 l pots in the greenhouse. Three plants coming from the same seed set (apomictic background) were divided asexually (two tillers each) and one tiller was assigned to the control treatment and the other one to the stress condition, totaling six plants. To avoid the noise represented by the genotype effect in a high heterozygous grass, we did not include a sexual or a full apomictic genotype as a control. For this reason we worked with clonal plants.

2.2. Stress Treatments

Plants were exposed to water stress conditions by water deprivation from three months before the onset of flowering until the end of the flowering season (September to March, 2016–2017). In order to maintain the biological functions of the plants, they were watered weekly with 50–80 mL per pot and a supplementary irrigation of an extra 100 mL per pot was carried out to induce the flowering close to the flowering season. Plants grown under normal conditions (300–500 mL water/week) were used as controls. As an indicator of the plant water status, the relative water content (RWC) was determined at the inflorescence collection time in leaves using the following formula: RWC (%) = (FM−DM)/(TM−DM) × 100, where FM, DM and TM are the fresh, dry, and turgid tissue weights, respectively [57].

2.3. Embryo Sacs Analyses

To analyze the effect of water stress on the reproductive mode, the different stages of megasporogenesis and megagametogenesis were observed under an optical microscope. Inflorescences from the control and treated plants were collected at the beginning of anthesis, when all embryo sac developmental stages were observable [47] and they were fixed in FAA (50% ethanol, 5% acetic acid, 10% formaldehyde in distilled water). Then, individual spikelets were dehydrated in a tertiary butyl alcohol series and embedded in Paraplast [58]. Samples were sectioned at 10 µm and stained with safranin-fast green. Observations were carried out with a Nikon Eclipse TE300 light transmission microscope. To assess the reproductive mode, the presence of meiosis or the number and position of nuclei in the embryo sac were observed according to Meier et al. [47]. More than three hundred spikelets were observed (41 from control plants and 271 from stressed ones).

2.4. RNA Extraction and Sequencing

Spikelets with basal flowers at the beginning of anthesis, containing embryo sacs at all developmental stages, were collected from control (DWC1, DWC2 and DWC3) and treated plants (DWS1, DWS2 and DWS3). In total, 30 mg of fresh tissue from each sample were ground to a fine powder using liquid nitrogen. The total RNA was extracted from the plant tissue as two fractions, small and large RNA, including RNA sequences smaller and larger than 200 bp, respectively, using a commercial NucleoSpin® miRNA kit (Macherey-Nagel) according to the manufacturer's instructions. The large RNA fraction was sequenced in 150 bp reads in pair-end through an Illumina HiSeq1500 platform at INDEAR (Rosario, Argentina).

2.5. Bioinformatics Analyses

Quality assessments of the reads were performed using the FastQC software. Subsequently, the reads were filtered using the Trimmomatic software [59] with the following parameters: ILLUMINACLIP = TruSeq3-PE-2.fa:2:30:7:4:false, LEADING = 3, TRAILING = 3, SLIDINGWINDOW = 4 = 20, CROP = 150, HEADCROP = 13 and MINLEN = 36. The resulting paired-end reads were assembled using the Trinity software [60] with a KMER_SIZE:32. In order to remove redundant transcripts, a clustering with a sequence identity threshold of 0.9 was performed using the CD-HIT software [61,62]. A quality assessment of the assembly was made using the standard metrics provided by the downstream analysis of Trinity [63] and BUSCO [64,65].

Regarding the differential expression analysis, the transcript quantification (estimation of the abundance of each assembled transcript) was performed with RSEM software [66] using the trimmed paired and unpaired reads aligned by Bowtie2 as an input [67] with the parameters: fragment_length = 137 and fragment_std = 23. The differential expression analysis was carried out using the EDGE R-Cran package [68,69] and the DE transcripts were selected using a fold change (FC) = 2 and an *e*-value of 1e-3.

A functional annotation of the DE genes was performed with Blast2GO [70]. The distribution of level 2 and 3 GO terms—including biological process, cellular component and molecular function among the DE annotated transcripts—were plotted with Blast2GO. A comparative analysis of the GO terms containing the annotated down- and upregulated transcripts under stress conditions was performed and plotted on a bar chart. The KEGG pathways (Kyoto Encyclopedia of Genes and Genomes [71]) were also compared to detect differentially enriched pathways between the control and treated plants.

The Heatmap analysis was performed using the R Package pheatmap [68] using as an input the expression matrix used for the differential expression analysis.

2.6. Comparison with Previously Sequenced E. Curvula Transcriptomes

The DE transcripts between the control and stressed plants and the DE transcripts between sexual and apomictic plants previously obtained [51] were compared in order to find a relationship between the genes/pathways related to stress/increase in sexuality and the genes/pathways involved in apomixis/sexuality using a unidirectional exonerate [72] alignment with a minimum identity of 90% and minimum coverage of 50% to match the sequences.

2.7. Search for Long Noncoding RNAs

Detection of long noncoding RNA (lncRNA) was carried out using the DE transcripts that could not be annotated with the Blast2GO software using the Magnoliopsida nonredundant protein database. The Coding Potential Calculator (CPC) software [73] was run using these transcripts as input, with the default parameters to detect the potential coding and lncRNA sequences. Finally, the lncRNA sequences annotated with the CPC software were searched in the dataset of DE transcripts between sexual and apomictic plants [51] in order to identify common transcript sequences.

2.8. Validation of Gene Expression by Quantitative Real-Time PCR (qRT-PCR)

RNA was extracted as detailed above, under the same conditions and using the same RNA extraction kit used for the Illumina sequencing. The cDNA synthesis was performed using the ImProm II Reverse Transcription System (Promega) following the supplier's instructions. The cDNA amplification was performed using specific primers designed according to the Integrated DNA technology (IDT) webpage (https://www.idtdna.com/scitools/Applications/RealTimePCR/). The primer pairs used in the qRT-PCR experiments are shown in Table S1. Real-time PCR reactions included 50 pmol of forward and reverse primers, 5 μL of cDNA diluted 100-fold and 10 μL of Real Mix (Biodynamics, Argentina). The amplification was carried out in a Rotor Gene 6000 thermocycler (Corbett Research, Australia). The expression level was normalized against the *E. curvula* UBICE gene (Table S1). The thermal cycling used for amplifications was as follows: 95 °C for 2 min, followed by 45 cycles at 94 °C for 10 s, then 15 s at the optimal annealing temperature for each primer pair and finally at 72 °C for 20 s using three biological and three technical replicates. The specificity of each reaction was verified through the dissociation curve profiles. To calculate the relative expression level and primer efficiency estimation, background-corrected raw fluorescence data were imported into LinRegPCR software version 11.0 [74,75]. The program uses a linear regression analysis to fit a straight line and estimate the PCR efficiency of each individual sample based on the slope of this line. The statistical analysis of the qRT-PCR fold change in the expression of genes among different treatments was analyzed using a Student's t Test. A *p*-value of 0.05 was considered to be significant.

2.9. Data Availability

The Transcriptome Shotgun Assembly project has been deposited at DDBJ/EMBL/GenBank under the accession GIQX00000000. The version described in this paper is the first version GIQX01000000.

3. Results

3.1. Percentage of Sexual vs. Apomictic Processes in Control and Water-Stressed Plants

The number of pistils and the percentages of sexual and apomictic embryo sacs from the control and water-stressed plants are shown in Table 1. The cytoembryology analyses showed that 4.97% of the embryo sacs were sexual (n = 161) in control plants, with an average RWC of 81.9%. Under water-stressed conditions, with an average RWC of 49.7%, the percentage of sexual embryo sacs increased to 23.84% (n = 172). The spikelets from stressed plants were smaller and had fewer flowers than the control ones. They also had a higher number of aborted seeds, so it was necessary to observe more spikelets to get an n = 172.

Table 1. Percentage of sexual and apomictic embryo sacs in the control and water-stressed *E. curvula* plants (Don Walter cultivar) and the relative water content (RWC) for each treatment.

Embryo Sacs (%)			
Control (n = 161)		Stress (n = 172)	
Apo	Sex	Apo	Sex
95.03%	4.97%	76.16%	23.84%
(RWC = 81.9 ± 4.1)		(RWC = 49.7 ± 0.6)	

3.2. Sequencing and Assembly

Six RNA TruSeq HiSeq1500 Illumina libraries, corresponding to three biological replicates of control plants and three from water-stressed plants, were sequenced producing a total of 172,128,258 paired-end reads (2 × 150 bp). Then, after the quality analysis, performed with FastQC [76], the reads were trimmed resulting in 95,127,754 paired-end reads. These paired-end reads were de novo assembled (Trinity software) resulting in 305,798 transcripts. Finally, the redundant transcripts were removed and the remaining transcripts were clustered using the CDHIT software, resulting in a final number of 201,011.

Although an *E. curvula* genome assembly is available [55] and we could have used it as a reference genome, for this study we decided to de novo assemble the reads since the sequenced genome belongs to a sexual diploid genotype and the sequences corresponding to the region that determine apomixis might not be present, as was observed by Zappacosta et al. [54] using molecular markers linked to the trait. The assembly quality analysis, performed with Trinity stats, indicated an N50 of 1553 bp, an average contig length of 919.09 bp and a GC content of 46.38%. The percentage of aligned reads, determined by the Bowtie2 software, was 96.45% and the percentage of complete BUSCO genes in the final assembly was 87.2% (S = 54.6%, D = 32.6%), with 9.6% of fragmented and 3.2% of missing genes.

3.3. Differentially Expressed Transcript Analysis

A total of 501 DE transcripts were obtained with a fold change of two and an *e*-value of 1e-3. Out of these, 350 were downregulated while 151 were upregulated in the stressed plants. Table S2 shows all the information about the DE transcripts (ID and Blast, Blast2GO and KEGG analyses).

A principal component analysis (PCA) computed on the DE transcripts effectively separated the control from the treated samples, with the two first principal components explaining approximately 90% of the overall variance (Figure S1).

3.4. Gene Ontology Analysis

The gene ontology classification made in order to identify the pathways potentially associated with the increase in sexual pistils under stress conditions gave a result of 380 out of 501 DE annotated transcripts. This classification (Figure 1) shows that the transcripts upregulated and downregulated under stress treatments were included in the same main categories (except the rhythmic process and growth). However, the number of GO terms with downregulated transcripts was higher (264 downregulated vs. 116 upregulated under stress). This effect could be part of the general decrease in gene expression that happens under stress conditions due the lack of resources.

To carry out a more specific analysis, we searched for the presence of DE transcripts in the GO terms reproduction and reproductive processes. We found two upregulated DE transcripts with homology to a G-type lectin S-receptor-like serine/threonine-protein kinase SD2-5 GsSRK (DN35954_c0_g1_i7) and *grassy tillers1* (DN35816_c1_g1_i1) and five downregulated DE transcripts under stress with homology to stromal processing peptidase (DN36893_c0_g1_i3), β-expansin (DN35576_c3_g4_i12, DN35576_c3_g4_i9), B3 domain-containing protein LFL1 (DN37977_c0_g1_i3) and indole-3-pyruvate monooxygenase YUCCA2 (DN36728_c1_g2_i3). Although some of these genes are related to stress responses, such as GsSRK, the overexpression of this gene in *Arabidopsis* exhibited more siliques at the adult developmental stage, among other traits [77]. All these DE transcripts are present in pathways that could participate in reproduction regulation, since they are involved in auxin pathways (YUCCA2, [9]), transcription regulation (*grassy tillers1*, [78]), embryogenesis (stromal processing peptidase, [79]), flowering time regulation (LFL1, [80]) and cell wall biosynthesis and pollen tube penetration through the stigma (β-expansin, [81]).

The GO terms only represented by upregulated transcripts at level 3 (Figure S2) were: chemical response, cell wall organization or biogenesis, response to biotic stimulus, response to external stimulus, response to other organisms, drug binding, cofactor binding, ligase activity, catalytic activity (acting on RNA), oxygen carrier activity, carbohydrate binding, envelope, RNA polymerase complex, thylakoid and Sm-like protein family complex. Although several of these GO terms may play a role in determining or regulating apomixis, we would like to highlight the ligase activity, where one of the differential transcripts shows homology with a putative E3 ubiquitin-protein ligase RING1a, an enzyme cited by other authors as a candidate to be involved in apomixis [82].

The GO level 3 terms that were only represented by downregulated transcripts (Figure S2) were: regulation of biological quality, regulation of molecular function, cell communication, signal transduction, response to abiotic stimulus, lyase activity, lipid binding, signaling receptor activity, isomerase activity, quaternary ammonium group binding, structural constituent of ribosome, membrane protein complex, extracellular space, external encapsulating structure and the microtubule associated complex. Among these GO terms, it can be highlighted that the membrane protein complex term contains two transcripts with homology to the AP-1 complex subunit sigma-1 and ENTH domain-containing protein, which were associated with SCD1-mediated vesicle transport and protein ubiquitination, a pathway cited by other authors as related to apomixis [83].

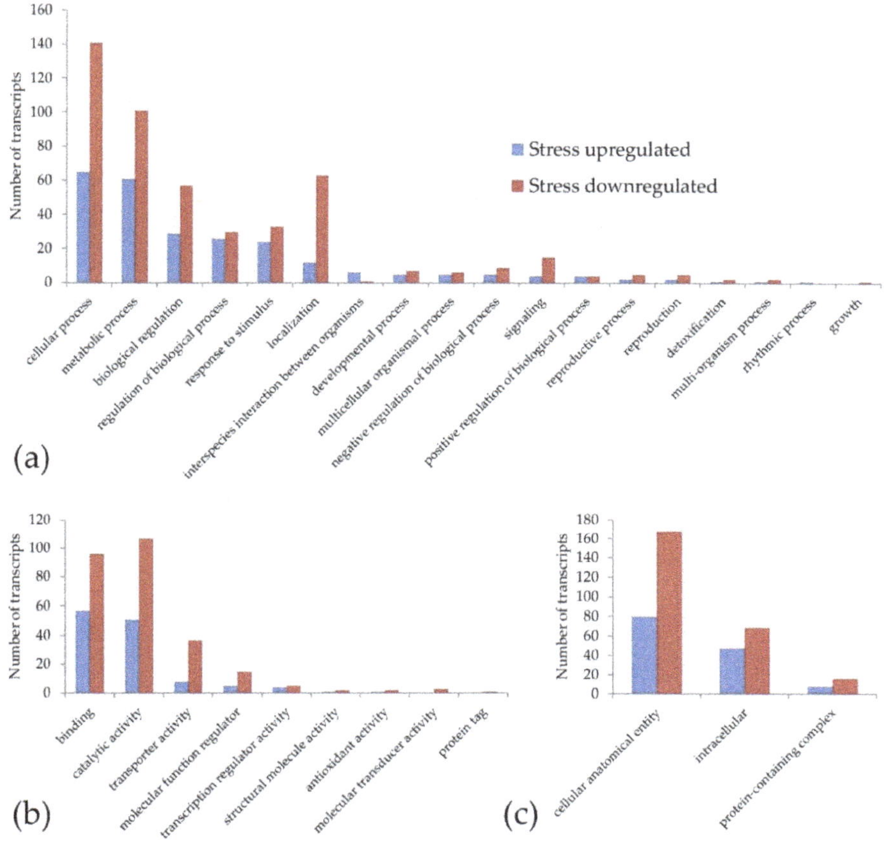

Figure 1. Classification of the weeping lovegrass differentially expressed transcripts according to gene ontology: (**a**) biological process (**b**) molecular function and (**c**) cellular component. Each main category was classified at level 2. Blue bars show the transcripts that are upregulated and red bars represent the transcripts that are downregulated under stress conditions.

3.5. KEGG Pathway Classification

To identify the additional levels of regulation acting on the apomixis/sexuality switch under stress, the DE transcripts were analyzed using the KEGG database. Using the Blast2GO software, 13 transcripts that were upregulated and 37 that were downregulated under stress conditions were assigned to 48 different pathways (Figure 2).

In eleven pathways, DE transcripts from both categories, up- and downregulated, were present while 16 pathways were composed only by upregulated and 21 by downregulated transcripts under stress conditions. The pathways with the highest number of downregulated transcripts were in thiamine metabolism and purine metabolism. On the other hand, the pathways with the highest number of upregulated transcripts were the alanine, aspartate and glutamate metabolisms, arginine biosynthesis and the amino sugar and nucleotide sugar metabolisms. The upregulated transcripts included in more pathways were the enzymes aspartate aminotransferase cytoplasmic (ec:2.6.1.1—transaminase, DN38803_c1_g1_i6) and glutamine synthetase cytosolic isozyme 1–3 (ec:6.3.1.2—synthetase, DN38913_c0_g1_i4). The downregulated ones included the enzymes NADPH-dependent aldo–keto reductase (ec:1.1.1.2—dehydrogenase NADP+, DN38835_c1_g1_i3) and phosphoenolpyruvate carboxylase 4 (ec:4.1.1.31—carboxylase, DN19844_c0_g1_i1).

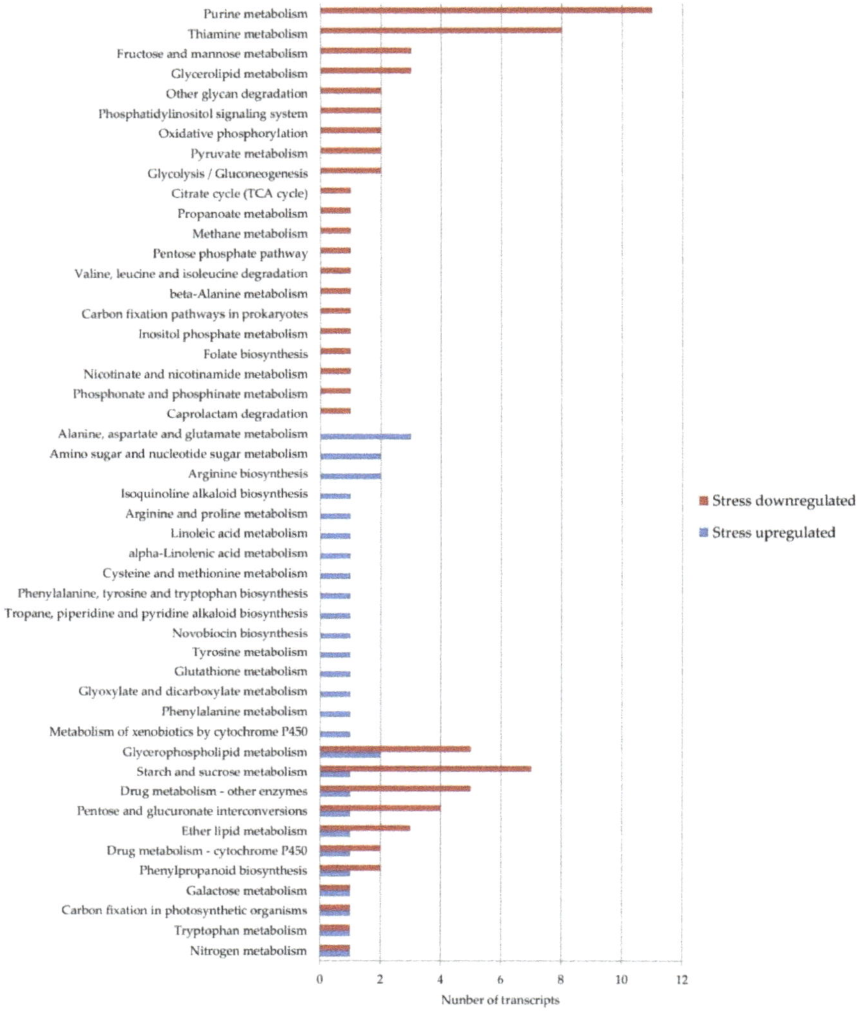

Figure 2. Weeping lovegrass differentially expressed the transcripts present in Kyoto Encyclopedia of Genes and Genomes (KEGG) pathways. The blue bars show the number of transcripts upregulated and the red bars show the number of transcripts downregulated under stress conditions in each pathway.

3.6. Comparison with Previous E. curvula Transcriptomes

In order to look for an association between the DE transcripts in plants under stress conditions and the increase in the number of sexual pistils, the results obtained in this study were compared with the DE transcripts obtained in a previous study (sexual vs. apomictic [51]), where two transcriptomes of weeping lovegrass, a fully sexual and a fully apomictic one, were sequenced.

Ninety three out of the 501 DE transcripts showed homology in the unidirectional alignment with the 9750 DE transcripts reported by Garbus et al. [51]. Sixty nine out of the 93 DE transcripts belonged to the downregulated in the stress group and 24 to the upregulated one (Figure 3 and Table S2). From this analysis we can point out two interesting subgroups of transcripts that could be related to the increase in sexuality (Table 2 and Figure S3). The first one is composed of eight transcripts that were downregulated under stress conditions and upregulated in the apomictic genotype (in the apomixis/sexuality comparisons). The other group is composed of 20 transcripts that were upregulated under stress and downregulated in the apomictic genotype (Table 2).

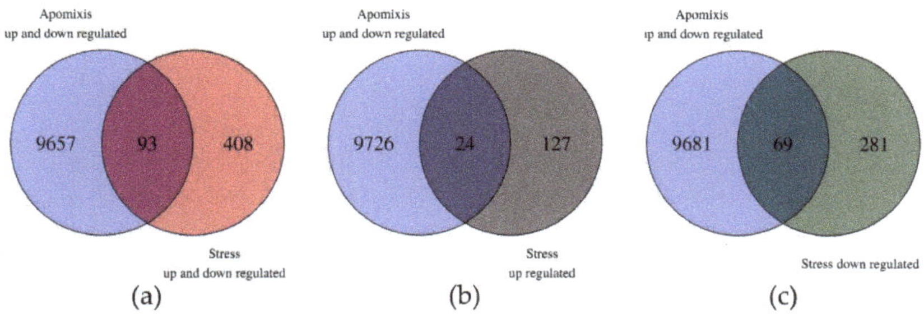

Figure 3. Venn diagrams comparing the weeping lovegrass differentially expressed (DE) transcripts between control vs. stressed plants with the DE transcripts between apomictic vs. sexual plants. Data for the comparison apo/sex come from Garbus et al. [51]. (**a**) In the intersection between both groups of transcripts, there are 93 in common; (**b**) in the intersection are the common transcripts that are DE in apomictic plants (up- and downregulated) and downregulated under stress conditions; (**c**) in the intersection are the common transcripts that are DE in apomictic plants (up- and downregulated) and upregulated under stress conditions.

Table 2. Differentially expressed transcripts between the control and stressed weeping lovegrass flowers showing homology with DE genes between apomixis/sexual flowers obtained previously in *E. curvula* by Garbus et al. [51]. In the description column the annotation provided by Blast2go was included.

Stress Downregulated/Apomixis Upregulated	
Transcript ID	**Description**
DN38408_c2_g4_i2	subtilisin-like protease SBT5.3
DN35576_c3_g4_i12	β-expansin
DN38000_c1_g1_i5	pollen allergen Cyn d 15
DN25763_c0_g1_i3	Non-annotated
DN40203_c3_g2_i3	fibroin heavy chain
DN25402_c0_g2_i1	cysteine protease
DN33014_c0_g2_i2	guanine nucleotide-binding protein α-1 subunit
DN36051_c2_g3_i1	pectinesterase inhibitor 10
Stress Upregulated/Apomixis Downregulated	
DN37495_c2_g2_i1	hypothetical protein EJB05_27044, partial
DN35806_c0_g1_i1	stem-specific protein TSJT1
DN17975_c0_g1_i2	predicted protein
DN39515_c0_g1_i11	NAC domain-containing protein 110
DN39515_c0_g1_i3	NAC domain-containing protein 110
DN39515_c0_g1_i4	NAC domain-containing protein 110
DN33023_c1_g1_i3	linoleate 9S-lipoxygenase 2
DN40610_c1_g1_i1	lipoxygenase 1.1
DN40610_c2_g3_i2	linoleate 9S-lipoxygenase 2
DN30943_c0_g1_i1	tryptamine hydroxycinnamoyltransferase 2
DN30943_c0_g3_i1	tryptamine hydroxycinnamoyltransferase 2
DN35816_c1_g1_i1	*grassy tillers1*
DN35816_c1_g1_i2	homeobox-leucine zipper protein HOX12
DN35242_c1_g4_i1	SNF1-type serine–threonine protein kinase
DN36533_c0_g1_i4	xylanase inhibitor protein 1-like
DN36533_c0_g1_i5	xylanase inhibitor protein 1-like
DN36533_c0_g3_i5	xylanase inhibitor protein 1-like
DN36484_c3_g4_i1	dormancy-associated protein 1
DN37042_c3_g3_i9	dormancy-associated protein 1
DN27261_c2_g1_i1	pathogenesis-related protein PR-4-like

Table 2 shows that there is more than one transcript for the same gene. It is important to remark that the β-expansin and *grassy tillers1* genes were also annotated under the GO terms associated with reproduction. Another result to be highlighted from this analysis is a subtilisin (SBT5.3) downregulated under stress and overexpressed in apomictic plants. A member of this gene family, SBT1.4, was successfully validated by qRT-PCR. Subtilisins are proteases that are associated with the early stages of seed development [84] and reproductive mode [85,86].

3.7. Long Noncoding RNAs

From 501 DE transcripts blasted against the Magnoliopsida database, 452 were homologous with a coding sequence whereas 49 did not find a hit (Figure S2). Over these 49 transcripts introduced in the CPC software, 33 were predicted as noncoding with a high confidence. In total, 24 out of the 33 putative lncRNA were downregulated under stress conditions and nine were upregulated, representing 6% of the corresponding categories. Interestingly, 3 out of 33 of these RNAs were also found DE between apomictic and sexual plants in our previous work [51].

3.8. Analysis of Differentially Expressed Transcripts

The annotation of the DE transcripts that were not highlighted by the GO terms or KEGG pathways was also analyzed. Table 3 summarizes the annotated DE transcripts up- or downregulated under the stress treatment that have previously been mentioned by different authors in apomictic species or mutants. Among them, we can mention the transcription factor ethylene-responsive AINTEGUMENTA-like 5 (AIL5), belonging to the AP2 family, the same family as BABY BOOM (BBM), one of the few candidate genes for apomixis that has been confirmed to play a role in parthenogenesis [15]. Three DE transcripts were found associated with the brassinosteroid pathway (EXORDIUM, guanine nucleotide-binding protein α-1 subunit and serine/threonine protein phosphatase 2A 55 kDa regulatory subunit B). This pathway has previously been mentioned by other authors as being associated with megagametogenesis in Poaceae [87] and apomixis induction [88].

Another DE transcript upregulated under the stress treatment showed homology with MO25, a protein present in the apomixis-determining region in *Hypericum perforatum* [13] and associated with signal transduction (GO terms: intracellular signal transduction and positive regulation of protein serine/threonine kinase activity).

A very interesting candidate also related to apomixis that was found to be DE (upregulated under stress conditions) in the present study is a transcript with homology to the repressor of silencing ROS1A, a DNA-glycosylase that removes methylated cytosines and replaces them with unmethylated cytosines. If we postulate that sexuality is silenced in facultative apomictic plants, this gene is a very attractive candidate to de-repress this function by a demethylation pathway under stress situations, allowing an increase in the number of sexual pistils.

Among other genes previously related to apomixis and downregulated under stress conditions are the SNF1-related protein kinase regulatory subunit β-1 and F-box proteins (At2g14290, At3g07870, and At5g07610). Regarding genes previously related to apomixis and upregulated under stress, the transcription factor NAC10 should be mentioned.

Table 3. Differentially expressed transcripts reported by different authors in different apomictic species compared with their sexual counterparts.

SeqName	Description	Condition	Probable Function	Reference
DN32086_c1_g2_i1	AP2-like ethylene-responsive transcription factor AIL5	Up	Transcription factor	[15]
DN30159_c1_g5_i1	protein EXORDIUM	Up	Brassinosteroid pathway	[89]
DN37585_c0_g1_i12	serine/threonine protein phosphatase 2A 55 kDa regulatory subunit B β isoform	Down	Brassinosteroid pathway	[89]
DN36281_c2_g1_i5	F-box protein	Down	F-box	[82]
DN36248_c0_g1_i2	F-box protein	Down	F-box	[82]
DN30985_c0_g1_i9	F-box protein At5g07610-like	Down	F-box	[82]
DN23587_c0_g1_i1	F-box domain containing protein	Up	F-box	[82]
DN36607_c0_g1_i5	protein ROS1	Up	Transcription factor	[90]
DN34979_c0_g1_i2	putative MO25-like protein At5g47540	Down	Signaling	[13]
DN30222_c0_g5_i1	transducin/WD40 repeat-like superfamily protein	Down	Histone binding	[38]
DN30614_c2_g1_i7	protein MARD1	Up	SnRK1 regulation pathway	[91]
DN29708_c0_g2_i2	putative vesicle-associated membrane protein 726	Down	Vesicle-mediated transport	[83]
DN27157_c0_g2_i1	LIM domain-containing protein PLIM2b	Down	Actin filament binding	[92]
DN38850_c0_g1_i10	Methylesterase 17	Down	Auxin pathway	[9]
DN36845_c1_g1_i17	kinesin-like protein KIN-14D isoform X1	Down	Microtubule, mitosis	[85]
DN35282_c0_g2_i2	zinc finger CCCH domain-containing protein 35	Up	RNA processing, cell cycle	[38]
DN38610_c0_g1_i2	Tyrosine-protein kinase BAZ1B	Up	Chromatin remodeling	[12]
DN32750_c0_g1_i2	protein TONSOKU	Up	Silencing	[12]
DN34106_c0_g1_i6	NLR family CARD domain-containing protein 3	Up	Ubiquitination, mTOR pathway	[9]
DN38412_c0_g4_i1	BTB/POZ and MATH domain-containing protein 2	Down	Ubiquitination	[40]

3.9. Validation by qRT-PCR

qRT-PCR assays were used to corroborate the in silico differential expression analysis of key genes (Figure 4a and Table S1), which were selected on the basis of their expression pattern and/or annotation. Heatmap (Figure 4b) was also used to show the significance of the in silico differential expression of the six selected genes.

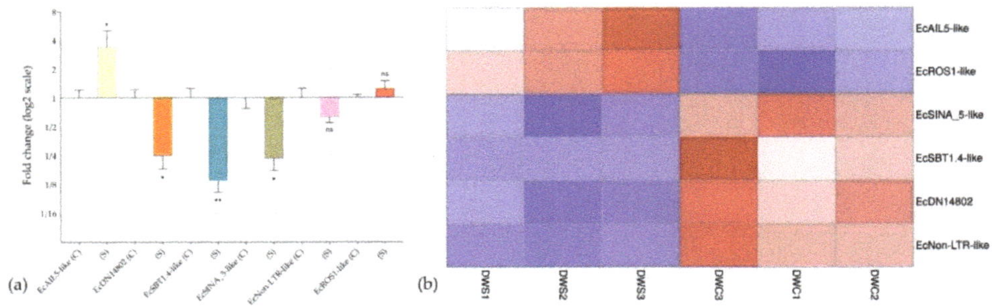

Figure 4. (**a**) qRT-PCR validation and (**b**) the Heatmap of weeping lovegrass selected for differentially expressed genes between the control (C) and water-stressed (S) plants. * Indicates significant or highly significant differences (**); ns indicates nonsignificant differences between treatments.

EcAIL5-like, EcROS1-like and EcSINA_5-like, a transcription factor that belong to the AP2 family, a DNA glycosylase, and a putative E3 ubiquitin-protein ligase, respectively, were chosen for validation because they have previously been mentioned to be involved in pathways recently identified as having regulatory roles in apomixis.

An interesting transcript among those selected for qRT-PCR validation was EcDN14802, a transcript that was predicted as lncRNA, downregulated under stress conditions and not found in *E. curvula* sexual genotypes (OTA-S transcriptome and Victoria genome). These observations point out this transcript as a possible candidate related to apomictic processes. With the same criteria, the transcript EcNon-LTR-like was also selected. Both transcripts were found only in the apomictic genotype Don Walter.

EcSBT1.4-like is a stress downregulated protease that was selected for validation because it was previously found in *E. curvula* transcriptomes (Table 2) and because it was included in the GO terms reproductive process and reproduction.

Although EcNon-LTR-like and EcROS1-like were not statistically validated by qRT-PCR, they were very close and the Heatmap analysis showed a very consistent expression across all the three samples in each treatment.

4. Discussion

Our experimental design was aimed at finding a relationship between stress, the increase in sexual processes and the candidate genes involved using a facultative apomictic cultivar of *E. curvula*. Prior to the analysis, it is important to consider four aspects of the experimental design. Firstly is the fact that the experiments were conducted with clonal plants, so both the control and stressed plants have the same background. Secondly, the treatment involved water stress, hence some of the responses could be exclusively due to stress itself. Thirdly, the biological samples used consisted of whole flowers (spikelets), which comprise raquis, glumes, lemma, palea, ovary and anthers, therefore, the whole set of transcripts characterized here are derived from a variety of cell types, including somatic cells and male and female reproductive cells from premeiosis to anthesis. Finally, both treatments were conducted with a facultative apomictic cultivar, hence we expected to find mostly apomictic/sexual regulators, not apomictic triggers because they must be present in plants from both treatments. On the other hand, we are working with the knowledge that apomixis in *E. curvula* is determined by a genomic region [54] with an epigenetic component modulating the traits [40,52], and apomixis and sexuality co-exist in facultative plants [54], allowing us to hypothesize that sexuality is being repressed at different levels in apomictic plants. This repression is deregulated by different endogenous [56] and exogenous stresses [40], probably mediated by transcriptional and post transcriptional regulatory mechanisms.

In the present study, as in previous ones using this grass, it was shown that stress significantly increased the percentage of sexual processes, as was observed in the facultative apomictic Tanganyika and Don Walter cultivars, where the increase in sexual pistils ranged from 1.8 to 14.4% and from 4 to 22%, respectively [40,56]. Similar results were obtained in this study (4 to 24% increase) using plants of the Don Walter cultivar. This situation has also been observed in other plant species, such as *Boechera*, where Mateo de Arias [38] demonstrated that drought and heat stress caused a shift from apomeiosis to meiosis in female meiocytes when plants are stressed and the frequency of sexual embryo sacs was increased from an average of 10 to 30% in the drought-stress-treated plants, with similar percentages to the ones observed here in *E. curvula*.

Previous studies conducted in sexual and apomictic *E. curvula* flowers using different strategies, like expressed sequence tags (ESTs) [48,49], differential displays [50] and 454 sequencing [51], allowed us to detect that many of the DE transcripts are homologous to genes related with different environmental stresses. This fact, plus the observation that stress increases the frequency of sexual processes, encouraged us to look for an association between both findings. The present study using the Illumina platform represents a deeper and complete transcriptomic sequencing of spikelets from *E. curvula* that provides a great amount of information that can be used to look for DE genes potentially involved in the modulation of apomixis/sexuality. Despite the large number of assembled transcripts (201,011), few were DE between treatments (501). An important proportion of these transcripts could be annotated and classified under the GO terms and KEGG pathways. The nonannotated transcripts (49) were evaluated with the CPC software and 33 were predicted as long noncoding RNAs. The identification of these genes and the lncRNAs are a benchmark for understanding the common pathways between stress and apomixis.

The Blast2GO analysis at level 2 showed that the same GO terms are present in both treatments, having more downregulated than upregulated transcripts in each and this could be due to the overall stress response. An interesting finding was the presence of seven transcripts in the GO terms reproduction and reproductive process that are related to stress responses [77] and were also present in apo/sex comparisons made earlier in *E. curvula* [49], such as GsSRK. Other transcripts included in these terms are involved in auxin pathways (YUCCA2, [9]), transcription regulation (*grassy tillers1*, [78]), embryogenesis (stromal processing peptidase, [79]), flowering time regulation (LFL1, [80]), cell wall biosynthesis and the pollen tube penetration of the stigma (β-expansin, [81]) and could also be associated with apomixis in weeping lovegrass.

Based on homology or annotations, some of the DE transcripts are involved in pathways recently identified as having regulatory roles during apomixis [13,82,83]. One of these pathways is ubiquitination, in which a putative E3 ubiquitin-protein ligase RING1a and a sulfite exporter TauE/SafE family protein 3 were present, among others. Ubiquitination regulates nearly every aspect of cellular events in eukaryotes. It modifies intracellular proteins with the 76-amino acid polypeptide ubiquitin and destines them for proteolysis or activity alteration. The proteasomal degradation-mediated control of key cell cycle regulators and other targets influence reproductive fate decisions and germline development [93,94]. In this study, nine transcripts, representing seven genes that belong to the ubiquitination pathway, were found DE. Five out of these seven genes (E3 ubiquitin-protein ligase SINA-like 5, E3 ubiquitin-protein ligase SINA-like 10, U-box domain-containing protein 35, ubiquitin-NEDD8-like protein RUB2 and U-box domain-containing protein 15) were downregulated under stress conditions and two (putative E3 ubiquitin-protein ligase RING1a and sulfite exporter TauE/SafE family protein 3) were upregulated. This finding provides further evidence that tightly regulated protein degradation affecting cell cycle progression might be of crucial importance for governing the distinct specification and differentiation of apomictic and sexual germlines.

Genes related to ubiquitin pathways were also observed in comparisons of apomictic and sexual weeping lovegrass [49–51] and *Paspalum notatum* flowers [95] and in ploidy related changes in the apomictic grass *P. notatum* [96].

Together with the ubiquitin pathway, F-box proteins were also mentioned by Zühl et al. [82] as differentially expressed and they might potentially play a role in the reproductive mode. Martelotto et al. [96] found transcripts homologous to F-box genes differentially expressed and related to ploidy changes in the apomictic species *P. notatum*. Here we found six downregulated F-box proteins and one upregulated. F-box proteins are part of Skp1–Cullin1–F-box protein (SCF) ligase complexes, acting in the polyubiquitination-mediated 26S proteasomal degradation [97]. Changes in this pathway due to the stress treatment could also be associated with the apomictic/sexual switch, as in ubiquitination.

Interestingly, when contrasted with previous data, four groups of DE transcripts were also found as DE in the comparison of apomixis vs. sexuality in weeping lovegrass [51]. In Table 2 we highlighted two groups. The first group is composed of eight transcripts that were downregulated under stress conditions and upregulated in the apomictic genotype (in the apo/sex comparisons). The second group is composed of 20 transcripts that were upregulated under stress and downregulated in the apomictic genotype. These 20 transcripts could be the ones that are associated to the increase in sexual embryo sacs. In the first group, transcripts presented homology with subtilisins, β-expansin, pollen allergen Cyn d 15, glycine-rich proteins, guanine nucleotide-binding proteins, pectine esterases inhibitors, cysteine proteases and a long noncoding RNA. These genes are mainly involved in cell wall modifications, proteolysis, signal transduction and a possible regulatory function (lncRNA). In the second group, we can mention transcription factors, such as the NAC domain containing protein, *grassy tillers1* and HOX12; lipoxygenases; tryptamine hydroxycinnamoyl transferases and SNF1-type serine–threonine protein kinases. The latter gene belongs to the SnRK complex and another component of this pathway was found as DE in the present study (MARD1). The SnRK1 kinases control metabolism, growth and development, and stress tolerance by the direct phosphorylation of metabolic enzymes and regulatory proteins and by extensive transcriptional regulation. SnRK1 is also part of a more elaborate metabolic and stress signaling network, which includes the TOR kinase and ABA-signaling. The plant SnRK1–TOR system is heavily intertwined also with hormone signaling pathways, mainly auxins and brassinosteriods. Many components of these complex pathways were found as DE in the present study, and although they are related to stress responses, they may also be involved in the apo–sex ratio change. Mateo de Arias [38] also finds the DE components of these pathways and Gao [89] proposes that the differences in gene expression between apomictic and sexual plants might be triggered by the presence or absence of stress signaling and showed strong evidence that the TOR–brassinosteroids molecular stress response pathway is involved in the apomeiosis/meiosis switch. Recently, Carman et al. [91] patented components of these pathways as apomixis related, which reinforced the involvement of these genes, not only in the stress response, but also in the reproductive mode regulation.

Several genes related to hormones, such as auxins and brassinosteriods, could also be acting, since a number of genes involved in hormone perception and homeostasis, including cytokinins, auxins, and brassinosteroids, were DE in apomictic pistils [9,85,88]. Auxins and brassinosteroids are pathways that respond to stress and are considered to be key components for cell dedifferentiation (totipotence phenomenon) related to somatic embryogenesis [98], a situation that is similar to apomixis.

Another transcript that is overexpressed under stress conditions in the present study is EcROS1-like, a DNA glycosylase with a high homology to ROS1 (repressor of silencing 1) a DNA demethylase that is indispensable in both male and female gametophyte development [90]. The mutation of this gene in *Arabidopsis* showed DNA hypermethylation (an increased level of methylated cytosine) at nearly 5000 loci involved in different pathways [99]. We suggest that EcROS1-like, as a specific or an unspecific stress response, could be demethylating a key target, thus de-repressing some gene or genes involved in sexuality pathways that are silenced in apomictic plants. Demethylation together with other pathways found here as DE could be part of the complex mechanism that regulates apomixis and sexuality in this grass, the ones in the intersection between control/stress and apo/sex being the strongest candidates.

In general, our results mainly point to the involvement of demethylation, protein degradation, transcriptional and post transcriptional regulatory mechanisms and regulation by plant hormones and signal transduction in the apomixis/sexuality switch regulation under stress situations.

5. Conclusions

These data, together with previous results [40], reinforce our previous hypothesis related to apomixis regulation in weeping lovegrass and its connections with stress. In this model, both pathways—the apomictic and the sexual one—co-exist in facultative apomictic plants, the sexuality mainly being repressed or expressed at very low levels under normal conditions. Stress situations can de-repress sexuality and the number of sexual embryo sacs increases in apomictic plants. We suggest that EcROS1-like can be demethylating, thus de-repressing some gene or genes involved in the sexuality pathways. Many of the other transcripts found as DE could be part of the complex mechanism that regulate apomixis and sexuality in this grass, the ones in the intersection between control/stress and apo/sex being the strongest candidates. Probably some of them are being upregulated under stress by this general demethylation response. Other related processes involved are ubiquitination, hormone and signal transduction pathways, transcription regulation and cell wall biosynthesis, some acting as a general response to stress and some that can be specific to the reproductive mode.

Finally, the availability of the sequence database reported here would make possible the characterization and validation of the numerous genes involved in apomixis and stress.

Supplementary Materials: The following are available online at http://www.mdpi.com/2073-4425/11/9/969/s1, Figure S1: Principal component analysis (PCA) of the first two principal components representing 90.6% of the total variance using differentially expressed transcripts from control (DWC1, DWC2 and DWC3) and water-stressed (DWS1, DWS2 and DWS3) samples of *E. curvula* plants; Figure S2: Classification of the weeping lovegrass differentially expressed transcripts according to gene ontology (GO): (a) biological process (b) molecular function and (c) cellular component. Each main category was classified at level 3. Blue bars show the transcripts that are upregulated and red bars represent the transcripts that are downregulated under stress conditions; Figure S3: Venn diagrams comparing weeping lovegrass differentially expressed (DE) transcripts between control vs. stressed plants with the DE transcripts between apomictic vs. sexual plants. Data for the comparison apo/sex come from Garbus et al. [51]. (a): In the intersection are the common transcripts that are downregulated under stress conditions and downregulated in apomictic plants, (b): in the intersection are the common transcripts that are upregulated under stress conditions and downregulated in apomictic plants, (c): in the intersection are the common transcripts that are downregulated under stress conditions and upregulated in apomictic plants, (d): in the intersection are the common transcripts that are upregulated under stress conditions and upregulated in apomictic plants; Table S1: Primer pairs used for the qRT-PCR validation of transcripts expressed in control and water-stressed apomictic plants of *E. curvula* Don Walter cultivar, Table S2: DE transcripts information: transcript ID, Blast, Blast2GO and KEGG analyses results.

Author Contributions: Conceptualization, J.P.S., J.M.R., D.Z. and V.E.; methodology, J.P.S., D.Z. and J.C.; software, J.C. and C.A.G.; cytoembryological analysis, J.P.S., J.M.R. and J.G.; validation, J.P.S. and A.B.; writing—original draft preparation, J.P.S., D.Z., J.C.; writing—review and editing, J.P.S., D.Z., J.C. and V.E.; project administration, V.E.; funding acquisition, V.E. All authors have read and agreed to the published version of the manuscript.

Funding: This research was funded by Agencia Nacional de Promoción Científica y Tecnológica (ANPCyT, PICT Raíces 2014–1243 and PICT Raíces 2017–0879), Universidad Nacional del Sur (PGI 24/A199), Consejo Nacional de Investigaciones Científicas y Técnologicas (CONICET) and the European Union's Horizon 2020 Research and Innovation Programme under the Marie Skłodowska-Curie Grant Agreement No. 645674 (PROCROP).

Conflicts of Interest: The authors declare no conflict of interest.

References

1. Carman, J.G. Asynchronous expression of duplicate genes in angiosperms may cause apomixis, bispory, tetraspory, and polyembryony. *Biol. J. Linn. Soc.* **1997**, *61*, 51–94. [CrossRef]
2. Hand, M.L.; Koltunow, A.M. The genetic control of apomixis: Asexual seed formation. *Genetics* **2014**, *197*, 441–450. [CrossRef] [PubMed]

3. Koltunow, A.M. Apomixis: Embryo sacs and embryos formed without meiosis or fertilization in ovules. *Plant Cell* **1993**, *5*, 1425–1437. [CrossRef] [PubMed]
4. Carman, J.G. Apomixis for crop production: Status of technology development and commercialization implications. *WJILDR* **2004**, *12*, 29–48.
5. Singh, M.; Burson, B.L.; Finlayson, S.A. Isolation of candidate genes for apomictic development in buffelgrass (*Pennisetum ciliare*). *Plant Mol. Biol.* **2007**, *64*, 673–682. [CrossRef]
6. Rodrigues, J.C.; Cabral, G.B.; Dusi, D.M.; de Mello, L.V.; Rigden, D.J.; Carneiro, V.T. Identification of differentially expressed cDNA sequences in ovaries of sexual and apomictic plants of *Brachiaria brizantha*. *Plant Mol. Biol.* **2003**, *53*, 745–757. [CrossRef]
7. Albertini, E.; Marconi, G.; Barcaccia, G.; Raggi, L.; Falcinelli, M. Isolation of candidate genes for apomixis in *Poa pratensis* L. *Plant Mol. Biol.* **2004**, *56*, 879–894. [CrossRef]
8. Yamada-Akiyama, H.; Akiyama, Y.; Ebina, M.; Xu, Q.; Tsuruta, S.I.; Yazaki, J.; Kishimoto, N.; Kikuchi, S.; Takhara, M.; Takamizo, T.; et al. Analysis of expressed sequence tags in apomictic guineagrass (*Panicum maximum*). *J. Plant Physiol.* **2009**, *166*, 750–761. [CrossRef]
9. Polegri, L.; Calderini, O.; Arcioni, S.; Pupilli, F. Specific expression of apomixis-linked alleles revealed by comparative transcriptomic analysis of sexual and apomictic *Paspalum simplex* Morong flowers. *J. Exp. Bot.* **2010**, *61*, 1869–1883. [CrossRef]
10. Okada, T.; Hu, Y.; Tucker, M.R.; Taylor, J.M.; Johnson, S.D.; Spriggs, A.; Tsuchiya, T.; Oelkers, K.; Rodrigues, J.C.M.; Koltunow, A.M. Enlarging cells initiating apomixis in *Hieracium praealtum* transition to an embryo sac program prior to entering mitosis. *Plant Physiol.* **2013**, *163*, 216–231. [CrossRef]
11. Pellino, M.; Hojsgaard, D.; Schmutzer, T.; Scholz, U.; Hörandl, E.; Vogel, H.; Sharbel, T.F. Asexual genome evolution in the apomictic *Ranunculus auricomus* complex: Examining the effects of hybridization and mutation accumulation. *Mol. Ecol.* **2013**, *22*, 5908–5921. [CrossRef] [PubMed]
12. Schmidt, A.; Schmid, M.W.; Klostermeier, U.C.; Qi, W.; Guthoerl, D.; Sailer, C.; Waller, M.; Resenstiel, P.; Grossniklaus, U. Apomictic and sexual germline development differ with respect to cell cycle, transcriptional, hormonal and epigenetic regulation. *PLoS Genet.* **2014**, *10*, e1004476. [CrossRef] [PubMed]
13. Schallau, A.; Arzenton, F.; Johnston, A.J.; Hähnel, U.; Koszegi, D.; Blattner, F.R.; Altschmied, L.; Haberer, G.; Barcaccia, G.; Bäumlein, H. Identification and genetic analysis of the APOSPORY locus in *Hypericum perforatum* L. *Plant J.* **2010**, *62*, 773–784. [CrossRef] [PubMed]
14. Galla, G.; Basso, A.; Grisan, S.; Bellucci, M.; Pupilli, F.; Barcaccia, G. Ovule gene expression analysis in sexual and aposporous apomictic *Hypericum perforatum* L. (Hypericaceae) accessions. *Front. Plant Sci.* **2019**, *10*, 654. [CrossRef] [PubMed]
15. Albertini, E.; Marconi, G.; Reale, L.; Barcaccia, G.; Porceddu, A.; Ferranti, F.; Falcinelli, M. SERK and APOSTART. Candidate genes for apomixis in *Poa pratensis*. *Plant Physiol.* **2005**, *138*, 2185–2199. [CrossRef] [PubMed]
16. Bhatt, A.M.; Lister, C.; Page, T.; Fransz, P.; Findlay, K.; Jones, G.H.; Dickinson, H.G.; Dean, C. The DIF1 gene of *Arabidopsis* is required for meiotic chromosome segregation and belongs to the REC8/RAD21 cohesin gene family. *Plant J.* **1999**, *19*, 463–472. [CrossRef]
17. Boutilier, K.; Offringa, R.; Sharma, V.K.; Kieft, H.; Ouellet, T.; Zhang, L.; Hattori, J.; Liu, C.M.; van Lammeren, A.A.; Miki, B.L.; et al. Ectopic expression of BABY BOOM triggers a conversion from vegetative to embryonic growth. *Plant Cell* **2002**, *14*, 1737–1749. [CrossRef]
18. Corral, J.M.; Vogel, H.; Aliyu, O.M.; Hensel, G.; Thiel, T.; Kumlehn, J.; Sharbel, T.F. A conserved apomixis-specific polymorphism is correlated with exclusive exonuclease expression in premeiotic ovules of apomictic *Boechera* species. *Plant Physiol.* **2013**, *163*, 1660–1672. [CrossRef]
19. Schoft, V.K.; Chumak, N.; Choi, Y.; Hannon, M.; Garcia-Aguilar, M.; Machlicova, A.; Slusarz, L.; Mosiolek, M.; Park, J.S.; Park, G.T.; et al. Function of the DEMETER DNA glycosylase in the *Arabidopsis thaliana* male gametophyte. *Proc. Natl. Acad. Sci. USA* **2011**, *108*, 8042–8047. [CrossRef]
20. Guitton, A.E.; Berger, F. Loss of function of multicopy suppressor of IRA1 produces nonviable parthenogenetic embryos in *Arabidopsis*. *Curr. Biol.* **2005**, *15*, 750–754. [CrossRef]
21. Peragine, A.; Yoshikawa, M.; Wu, G.; Albrecht, H.L.; Poethig, R.S. SGS3 and SGS2/SDE1/RDR6 are required for juvenile development and the production of trans-acting siRNAs in *Arabidopsis*. *Genes Dev.* **2004**, *18*, 2368–2379. [CrossRef] [PubMed]

22. Siddiqi, I.; Ganesh, G.; Grossniklaus, U.; Subbiah, V. The dyad gene is required for progression through female meiosis in *Arabidopsis*. *Development* **2000**, *127*, 197–207. [PubMed]
23. Ravi, M.; Marimuthu, M.P.A.; Siddiqi, I. Gamete formation without meiosis in *Arabidopsis*. *Nature* **2008**, *451*, 1121–1124. [CrossRef] [PubMed]
24. Olmedo-Monfil, V.; Durán-Figueroa, N.; Arteaga-Vázquez, M.; Demesa-Arévalo, E.; Autran, D.; Grimanelli, D.; Slotkin, R.K.; Martienssen, R.A.; Vielle-Calzada, J.P. Control of female gamete formation by a small RNA pathway in *Arabidopsis*. *Nature* **2010**, *464*, 628. [CrossRef]
25. Singh, M.; Goel, S.; Meeley, R.B.; Dantec, C.; Parrinello, H.; Michaud, C.; Leblanc, O.; Grimanelli, D. Production of viable gametes without meiosis in maize deficient for an ARGONAUTE protein. *Plant Cell* **2011**, *23*, 443–458. [CrossRef] [PubMed]
26. Siena, L.A.; Ortiz, J.P.A.; Calderini, O.; Paolocci, F.; Cáceres, M.E.; Kaushal, P.; Grisan, S.; Pessino, S.C.; Pupilli, F. An apomixis-linked ORC3-like pseudogene is associated with silencing of its functional homolog in apomictic *Paspalum simplex*. *J. Exp. Bot.* **2016**, *67*, 1965–1978. [CrossRef] [PubMed]
27. Ferreira, L.G.; de Alencar Dusi, D.M.; Irsigler, A.S.T.; Gomes, A.C.M.M.; Mendes, M.A.; Colombo, L.; de Campos Carneiro, V.T. GID1 expression is associated with ovule development of sexual and apomictic plants. *Plant Cell Rep.* **2018**, *37*, 293–306. [CrossRef]
28. Ohad, N.; Yadegari, R.; Margossian, L.; Hannon, M.; Michaeli, D.; Harada, J.J.; Goldberg, R.B.; Fischer, R.L. Mutations in FIE, a WD polycomb group gene, allow endosperm development without fertilization. *Plant Cell* **1999**, *11*, 407–416. [CrossRef]
29. Hehenberger, E.; Kradolfer, D.; Köhler, C. Endosperm cellularization defines an important developmental transition for embryo development. *Development* **2012**, *139*, 2031–2039. [CrossRef]
30. Siena, L.A.; Ortiz, J.P.A.; Leblanc, O.; Pessino, S. PnTgs1-like expression during reproductive development supports a role for RNA methyltransferases in the aposporous pathway. *BMC Plant Biol.* **2014**, *14*, 297. [CrossRef]
31. Zhao, L.; He, J.; Cai, H.; Lin, H.; Li, Y.; Liu, R.; Yang, Z.; Qin, Y. Comparative expression profiling reveals gene functions in female meiosis and gametophyte development in *Arabidopsis*. *Plant J.* **2014**, *80*, 615–628. [CrossRef] [PubMed]
32. Nedelcu, A.; Marcu, O.; Michod, R. Sex as a response to oxidative stress: A twofold increase in cellular reactive oxygen species activates sex genes. *Proc. R. Soc. B Biol. Sci.* **2004**, *271*, 1591–1596. [CrossRef] [PubMed]
33. Suomalainen, E.; Saura, A.; Lokki, J. *Cytology and Evolution in Parthenogenesis*; CRC Press, Inc.: Boca Raton, FL, USA, 1987.
34. Berman, J.; Hadany, L. Does stress induce (para)sex? Implications for *Candida albicans* evolution. *Trends Genet.* **2012**, *28*, 197–203. [CrossRef] [PubMed]
35. Nedelcu, A.; Michod, R. Sex as a response to oxidative stress: The effect of antioxidants on sexual induction in a facultatively sexual lineage. *Proc. R. Soc. B Biol. Sci.* **2003**, *270*, S136–S139. [CrossRef]
36. Hiruta, C.; Tochinai, S. How does the alteration of meiosis evolve to parthenogenesis?—Case study in a water flea, *Daphnia pulex*. In *Meiosis-Molecular Mechanisms and Cytogenetic Diversity*; Swan, A., Ed.; InTech: Rijeka, Croatia, 2012; pp. 109–122.
37. Carman, J.G.; Jamison, M.; Elliott, E.; Dwivedi, K.K.; Naumova, T.N. Apospory appears to accelerate onset of meiosis and sexual embryo sac formation in sorghum ovules. *BMC Plant Biol.* **2011**, *11*, 9. [CrossRef] [PubMed]
38. Mateo de Arias, M. Effects of Plant Stress on Facultative Apomixis in *Boechera* (Brassicaceae). Ph.D. Thesis, Utah State University, Logan, UT, USA, 2015.
39. Klatt, S.; Hadacek, F.; Hodac, L.; Brinkmann, G.; Eilerts, M.; Hojsgaard, D.; Hörandl, E. Photoperiod extension enhances sexual megaspore formation and triggers metabolic reprogramming in facultative apomictic *Ranunculus auricomus*. *Front. Plant Sci.* **2016**, *7*, 278. [CrossRef]
40. Rodrigo, J.M.; Zappacosta, D.; Selva, J.P.; Garbus, I.; Albertini, E.; Echenique, V. Apomixis frequency under stress conditions in weeping lovegrass (*Eragrostis curvula*). *PLoS ONE* **2017**, *12*, e0175852. [CrossRef]
41. Evans, L.; Knox, R. Environmental control of reproduction in *Themeda australis*. *Aust. J. Bot.* **1969**, *17*, 375. [CrossRef]
42. Quarin, C.L. Seasonal changes in the incidence of apomixis of diploid, triploid, and tetraploid plants of *Paspalum cromyorrhizon*. *Euphytica* **1986**, *35*, 515–522. [CrossRef]

43. Lutts, S.; Ndikumana, J.; Louant, B.P. Male and female sporogenesis and gametogenesis in apomictic *Brachiaria brizantha, Brachiaria decumbens* and F1 hybrids with sexual colchicine induced tetraploid *Brachiaria ruziziensis. Euphytica* **1994**, *78*, 19–25. [CrossRef]
44. Gounaris, E.; Sherwood, R.; Gounaris, I.; Hamilton, R.; Gustine, D. Inorganic salts modify embryo sac development in sexual and aposporous *Cenchrus ciliaris. Sex. Plant Reprod.* **1991**, *4*, 188–192. [CrossRef]
45. Schilling, M.P. Hybridization, Population Genetic Structure and Gene Expression in the Genus *Boechera*. Ph.D. Thesis, Utah State University, Logan, UT, USA, 2016.
46. Voigt, P.W.; Bashaw, E.C. Facultative apomixis in *Eragrostis curvula. Crop Sci.* **1976**, *16*, 803–806. [CrossRef]
47. Meier, M.; Zappacosta, D.; Selva, J.P.; Pessino, S.; Echenique, V. Evaluation of different methods for assessing the reproductive mode of weeping lovegrass plants, *Eragrostis curvula* (Schrad.) Nees. *Aust. J. Bot.* **2011**, *59*, 253–261. [CrossRef]
48. Cervigni, G.; Paniego, N.; Díaz, M.; Selva, J.P.; Zappacosta, D.; Zanazzi, D.; Landerreche, I.; Martelotto, L.; Felitti, S.; Pessino, S.; et al. Expressed sequence tag analysis and development of gene associated markers in a near-isogenic plant system of *Eragrostis curvula. Plant Mol. Biol.* **2008**, *67*, 1–10. [CrossRef]
49. Cervigni, G.D.; Paniego, N.; Pessino, S.; Selva, J.P.; Díaz, M.; Spangenberg, G.; Echenique, V. Gene expression in diplosporous and sexual *Eragrostis curvula* genotypes with differing ploidy levels. *Plant Mol. Biol.* **2008**, *67*, 11–23. [CrossRef]
50. Selva, J.P.; Pessino, S.; Meier, M.; Echenique, V. Identification of candidate genes related to polyploidy and/or apomixis in *Eragrostis curvula. Amer. J. Plant Sci.* **2012**, *3*, 403–416. [CrossRef]
51. Garbus, I.; Romero, J.; Selva, J.P.; Pasten, M.C.; Chinestra, C.; Carballo, J.; Zappacosta, D.; Echenique, V. *De novo* transcriptome sequencing and assembly from apomictic and sexual *Eragrostis curvula* genotypes. *PLoS ONE* **2017**, *12*, e0185595. [CrossRef]
52. Selva, J.P.; Siena, L.; Rodrigo, J.M.; Garbus, I.; Zappacosta, D.; Romero, J.; Ortiz, J.P.; Pessino, S.; Leblanc, O.; Echenique, V. Temporal and spatial expression of genes involved in DNA methylation during reproductive development of sexual and apomictic *Eragrostis curvula. Sci. Rep.* **2017**, *7*, 15092. [CrossRef]
53. Garbus, I.; Selva, J.P.; Pasten, M.C.; Bellido, A.M.; Carballo, J.; Albertini, E.; Echenique, V. Characterization and discovery of miRNA and miRNA targets from apomictic and sexual genotypes of *Eragrostis curvula. BMC Genom.* **2019**, *20*, 839. [CrossRef]
54. Zappacosta, D.; Gallardo, J.; Carballo, J.; Meier, M.; Rodrigo, J.M.; Gallo, C.; Selva, J.P.; Stein, J.; Ortiz, J.P.; Albertini, E.; et al. A high-density linkage map of the forage grass *Eragrostis curvula* and localization of the diplospory locus. *Front. Plant Sci.* **2019**, *10*, 918. [CrossRef]
55. Carballo, J.; Santos, B.; Zappacosta, D.; Garbus, I.; Selva, J.P.; Gallo, C.; Díaz, A.; Albertini, E.; Caccamo, M.; Echenique, V. A high-quality genome of *Eragrostis curvula* grass provides insights into Poaceae evolution and supports new strategies to enhance forage quality. *Sci. Rep.* **2019**, *9*, 10250. [CrossRef]
56. Zappacosta, D.; Ochogavía, A.; Rodrigo, J.M.; Romero, J.; Meier, M.; Garbus, I.; Pessino, S.C.; Echenique, V. Increased apomixis expression concurrent with genetic and epigenetic variation in a newly synthesized *Eragrostis curvula* polyploid. *Sci. Rep.* **2014**, *4*, 4423. [CrossRef]
57. Fritz, G. Measurement of water status by water content methods. In *An Introductory Plant Physiology*; Noggle, G., Fritz, G., Eds.; Prentice-Hall, Inc.: Englewood Cliffs, NJ, USA, 1976; pp. 445–447.
58. Johansen, D. *Plant Microtechnique*; Mc Graw-Hill Book Company Inc.: New York, NY, USA, 1940.
59. Bolger, A.M.; Lohse, M.; Usadel, B. Trimmomatic: A flexible trimmer for Illumina sequence data. *Bioinformatics* **2014**, *30*, 2114–2120. [CrossRef]
60. Grabherr, M.G.; Haas, B.J.; Yassour, M.; Levin, J.Z.; Thompson, D.A.; Amit, I.; Adiconis, X.; Fan, L.; Raychowdhury, R.; Zeng, Q.; et al. Full-length transcriptome assembly from RNA-Seq data without a reference genome. *Nat. Biotechnol.* **2011**, *29*, 644–652. [CrossRef]
61. Li, W.; Godzik, A. Cd-hit: A fast program for clustering and comparing large sets of protein or nucleotide sequences. *Bioinformatics* **2006**, *22*, 1658–1659. [CrossRef]
62. Fu, L.; Niu, B.; Zhu, Z.; Wu, S.; Li, W. CD-HIT: Accelerated for clustering the next-generation sequencing data. *Bioinformatics* **2012**, *28*, 3150–3152. [CrossRef]
63. Haas, B.J.; Papanicolaou, A.; Yassour, M.; Grabherr, M.; Blood, P.D.; Bowden, J.; Couger, M.B.; Eccles, D.; Li, B.; Lieber, M.; et al. *De novo* transcript sequence reconstruction from RNA-seq using the Trinity platform for reference generation and analysis. *Nat. Protoc.* **2013**, *8*, 1494–1512. [CrossRef]

64. Simão, F.A.; Waterhouse, R.M.; Ioannidis, P.; Kriventseva, E.V.; Zdobnov, E.M. BUSCO: Assessing genome assembly and annotation completeness with single-copy orthologs. *Bioinformatics* **2015**, *31*, 3210–3212. [CrossRef]
65. Waterhouse, R.M.; Seppey, M.; Simão, F.A.; Manni, M.; Ioannidis, P.; Klioutchnikov, G.; Kriventseva, E.V.; Zdobnov, E.M. BUSCO applications from quality assessments to gene prediction and phylogenomics. *Mol. Biol. Evol.* **2017**, *35*, 543–548. [CrossRef]
66. Li, B.; Dewey, C.N. RSEM: Accurate transcript quantification from RNA-Seq data with or without a reference genome. *BMC Bioinform.* **2011**, *12*, 323. [CrossRef]
67. Langmead, B.; Salzberg, S.L. Fast gapped-read alignment with Bowtie 2. *Nat. Methods* **2012**, *9*, 357–359. [CrossRef]
68. Robinson, M.D.; McCarthy, D.J.; Smyth, G.K. edgeR: A Bioconductor package for differential expression analysis of digital gene expression data. *Bioinformatics* **2010**, *26*, 139–140. [CrossRef]
69. McCarthy, J.D.; Chen, Y.; Smyth, K.G. Differential expression analysis of multifactor RNA-Seq experiments with respect to biological variation. *Nucleic Acids Res.* **2012**, *40*, 4288–4297. [CrossRef]
70. Conesa, A.; Götz, S.; García-Gómez, J.M.; Terol, J.; Talón, M.; Robles, M. Blast2GO: A universal tool for annotation, visualization and analysis in functional genomics research. *Bioinformatics* **2005**, *21*, 3674–3676. [CrossRef]
71. Kanehisa, M.; Goto, S. KEGG: Kyoto encyclopedia of genes and genomes. *Nucleic Acids Res.* **2000**, *28*, 27–30. [CrossRef]
72. Slater, G.S.; Birney, E. Automated generation of heuristics for biological sequence comparison. *BMC Bioinform.* **2005**, *6*, 31. [CrossRef]
73. Kong, L.; Zhang, Y.; Ye, Z.Q.; Liu, X.Q.; Zhao, S.Q.; Wei, L.; Gao, G. CPC: Assess the protein-coding potential of transcripts using sequence features and support vector machine. *Nucleic Acids Res.* **2007**, *35*, 345–349. [CrossRef]
74. Ruijter, J.M.; Ramakers, C.; Hoogaars, W.M.; Karlen, Y.; Bakker, O.; van den Hoff, J.B.; Moornman, A.F.M. Amplification efficiency: Linking baseline and bias in the analysis of quantitative PCR data. *Nucleic Acids Res.* **2009**, *37*, e45. [CrossRef]
75. Ramakers, C.; Ruijter, J.M.; Lekanne Deprez, R.H.; Moorman, A.F.M. Assumption-free analysis of quantitative real-time polymerase chain reaction (PCR) data. *Neurosci. Lett.* **2003**, *339*, 62–66. [CrossRef]
76. Andrews, S. FastQC: A Quality Control Tool for High Throughput Sequence Data. 2010. Available online: http://www.bioinformatics.babraham.ac.uk/projects/fastqc (accessed on 1 April 2019).
77. Sun, X.L.; Yu, Q.Y.; Tang, L.L.; Ji, W.; Bai, X.; Cai, H.; Liu, X.F.; Ding, X.D.; Zhu, Y.M. *GsSRK*, a G-type lectin S-receptor-like serine/threonine protein kinase, is a positive regulator of plant tolerance to salt stress. *J. Plant Physiol.* **2013**, *170*, 505–515. [CrossRef]
78. Wills, D.M.; Whipple, C.J.; Takuno, S.; Kursel, L.E.; Shannon, L.M.; Ross-Ibarra, J.; Doebley, J.F. From many, one: Genetic control of prolificacy during maize domestication. *PLoS Genet.* **2013**, *9*, e1003604. [CrossRef]
79. Trösch, R.; Jarvis, P. The stromal processing peptidase of chloroplasts is essential in *Arabidopsis*, with knockout mutations causing embryo arrest after the 16-cell stage. *PLoS ONE* **2011**, *6*, e23039. [CrossRef]
80. Peng, L.T.; Shi, Z.Y.; Li, L.; Shen, G.Z.; Zhang, J.L. Overexpression of transcription factor *OsLFL1* delays flowering time in *Oryza sativa*. *J. Plant Physiol.* **2008**, *165*, 876–885. [CrossRef]
81. Marowa, P.; Ding, A.; Kong, Y. Expansins: Roles in plant growth and potential applications in crop improvement. *Plant Cell Rep.* **2016**, *35*, 949–965. [CrossRef]
82. Zühl, L.; Volkert, C.; Ibberson, D.; Schmidt, A. Differential activity of F-box genes and E3 ligases distinguishes sexual versus apomictic germline specification in *Boechera*. *J. Exp. Bot.* **2019**, *70*, 5643–5657. [CrossRef]
83. Bocchini, M.; Galla, G.; Pupilli, F.; Bellucci, M.; Barcaccia, G.; Ortiz, J.P.A.; Pessino, S.C.; Albertini, E. The vesicle trafficking regulator PN_SCD1 is demethylated and overexpressed in florets of apomictic *Paspalum notatum* genotypes. *Sci. Rep.* **2018**, *8*, 1–11. [CrossRef]
84. D'erfurth, I.; Le Signor, C.; Aubert, G.; Sanchez, M.; Vernoud, V.; Darchy, B.; Lherminier, J.; Bourion, V.; Bouteiller, N.; Bendahmane, A.; et al. A role for an endosperm-localized subtilase in the control of seed size in legumes. *New Phytol.* **2012**, *196*, 738–751. [CrossRef]
85. Galla, G.; Zenoni, S.; Avesani, L.; Altschmied, L.; Rizzo, P.; Sharbel, T.F.; Barcaccia, G. Pistil transcriptome analysis to disclose genes and gene products related to aposporous apomixis in *Hypericum perforatum* L. *Front. Plant Sci.* **2017**, *8*, 79. [CrossRef]

86. Bräuning, S.; Catanach, A.; Lord, J.M.; Bicknell, R.; Macknight, R.C. Comparative transcriptome analysis of the wild-type model apomict *Hieracium praealtum* and its loss of parthenogenesis (lop) mutant. *BMC Plant Biol.* **2018**, *18*, 206. [CrossRef]
87. Gruszka, D. Exploring the brassinosteroid signaling in monocots reveals novel components of the pathway and implications for plant breeding. *Int. J. Mol. Sci.* **2020**, *21*, 354. [CrossRef]
88. Sherwood, D.A. A Simple Metabolic Switch May Activate Apomixis in *Arabidopsis thaliana*. Ph.D. Thesis, Utah State University, Logan, UT, USA, 2018.
89. Gao, L. Pharmacologically Induced Meiosis Apomeiosis Interconversions in *Boechera*, *Arabidopsis* and *Vigna*. Ph.D. Thesis, Utah State University, Logan, UT, USA, 2018.
90. Ono, A.; Yamaguchi, K.; Fukada-Tanaka, S.; Terada, R.; Mitsui, T.; Iida, S. A null mutation of *ROS1a* for DNA demethylation in rice is not transmittable to progeny. *Plant J.* **2012**, *71*, 564–574. [CrossRef]
91. Carman, J.G.; Sherwood, D.; Gao, L. Methods of Inducing Apomictic or Sexual Reproduction. U.S. Patent Application No. 16/273,132, 9 April 2020.
92. De Storme, N.; Geelen, D. The impact of environmental stress on male reproductive development in plants: Biological processes and molecular mechanisms. *Plant Cell Environ.* **2014**, *37*, 1–18. [CrossRef]
93. Dieterle, M.; Thomann, A.; Renou, J.P.; Parmentier, Y.; Cognat, V.; Lemonnier, G.; Muller, R.; Shen, W.H.; Kretsch, T.; Genschik, P. Molecular and functional characterization of *Arabidopsis* Cullin 3A. *Plant J.* **2005**, *41*, 386–399. [CrossRef]
94. Thomann, A.; Brukhin, V.; Dieterle, M.; Gheyeselinck, J.; Vantard, M.; Grossniklaus, U.; Genschik, P. *Arabidopsis CUL3A* and *CUL3B* genes are essential for normal embryogenesis. *Plant J.* **2005**, *43*, 437–448. [CrossRef]
95. Laspina, N.V.; Vega, T.; Seijo, J.G.; González, A.M.; Martelotto, L.G.; Stein, J.; Podio, M.; Ortiz, J.P.A.; Echenique, V.; Quarin, C.L.; et al. Gene expression analysis at the onset of aposporous apomixis in *Paspalum notatum*. *Plant Mol. Biol.* **2008**, *67*, 615–628. [CrossRef]
96. Martelotto, L.G.; Ortiz, J.P.A.; Stein, J.; Espinoza, F.; Quarin, C.L.; Pessino, S.C. A comprehensive analysis of gene expression alterations in a newly synthesized *Paspalum notatum* autotetraploid. *Plant Sci.* **2005**, *169*, 211–220. [CrossRef]
97. Skowyra, D.; Craig, K.L.; Tyers, M.; Elledge, S.J.; Harper, J.W. F-box proteins are receptors that recruit phosphorylated substrates to the SCF ubiquitin-ligase complex. *Cell* **1997**, *91*, 209–219. [CrossRef]
98. Elhiti, M.; Stasolla, C.; Wang, A. Molecular regulation of plant somatic embryogenesis. *In Vitro Cell. Dev. Biol. Plant* **2013**, *49*, 631–642. [CrossRef]
99. Qian, W.; Miki, D.; Zhang, H.; Liu, Y.; Zhang, X.; Tang, K.; Kan, Y.; Li, X.; Li, S.; Zhu, X.; et al. A histone acetyltransferase regulates active DNA demethylation in *Arabidopsis*. *Science* **2012**, *336*, 1445–1448. [CrossRef]

genes

MDPI

Article

Chasing the Apomictic Factors in the *Ranunculus auricomus* Complex: Exploring Gene Expression Patterns in Microdissected Sexual and Apomictic Ovules

Marco Pellino [1,2], Diego Hojsgaard [3], Elvira Hörandl [3,*] and Timothy F. Sharbel [1,2,†]

1. Leibniz Institute of Plant Genetics and Crop Plant Research (IPK), D-06466 Gatersleben, Germany; marco.pellino@gifs.ca (M.P.); tim.sharbel@usask.ca (T.F.S.)
2. Seed Developmental Biology Program, Global Institute for Food Security (GIFS), University of Saskatchewan 105 Administration Place, Saskatoon, SK S7N 5A2, Canada
3. Department of Systematics, Biodiversity and Evolution of Plants, Albrecht-von-Haller Institute for Plant Sciences, Georg-August-University of Göttingen, Untere Karspüle 2, D-37073-1 Göttingen, Germany; dhojsga@gwdg.de

* Correspondence: Elvira.hoerandl@biologie.uni-goettingen.de; Tel.: +49-551-39-7843

† Present address: Department of Plant Sciences, University of Saskatchewan, 51 Campus Drive, Saskatoon, SK S7N 5A8, Canada.

Received: 20 April 2020; Accepted: 27 June 2020; Published: 30 June 2020

Abstract: Apomixis, the asexual reproduction via seeds, is associated to polyploidy and hybridization. To identify possible signatures of apomixis, and possible candidate genes underlying the shift from sex to apomixis, microarray-based gene expression patterns of live microdissected ovules at four different developmental stages were compared between apomictic and sexual individuals of the *Ranunculus auricomus* complex. Following predictions from previous work on mechanisms underlying apomixis penetrance and expressivity in the genus, gene expression patterns were classified into three categories based on their relative expression in apomicts compared to their sexual parental ancestors. We found evidence of misregulation and differential gene expression between apomicts and sexuals, with the highest number of differences detected during meiosis progression and emergence of aposporous initial (AI) cells, a key developmental stage in the ovule of apomicts where a decision between divergent reproductive pathways takes place. While most of the differentially expressed genes (DEGs) could not be annotated, gene expression was classified into transgressive, parent of origin and ploidy effects. Genes related to gametogenesis and meiosis demonstrated patterns reflective of transgressive and genome dosage effects, which support the hypothesis of a dominant factor controlling apomixis in *Ranunculus* and modulated by secondary modifiers. Three genes with probable functions in sporogenesis and gametogenesis development are identified and characterized for future studies.

Keywords: apomixis; differentially expressed genes; hybridization; microarrays; polyploidy; *Ranunculus*; sexuality

1. Introduction

The evolution of asexuality from sexual progenitors has occurred repeatedly and independently in a wide range of plants and animals [1–3]. The evolutionary consequences of such a drastic functional trait change have been the subject of many studies and have shed light onto mechanisms of species dispersal and adaptation relevant to the evolutionary success of asexuality [4,5]. It has been suggested that asexual reproduction is advantageous because it reduces to a half the costs associated with sex [6] and because of a greater adaptability to stable environments [7]. In addition, asexuality would be beneficial for fixing advantageous mutations and genotypes highly adapted to the surrounding environment [8,9]. Despite these expected advantages, limitations arise from the lack of meiotic recombination and exploitation of recombinational DNA repair mechanisms [10]. The absence of meiosis has also drastic consequences on purging deleterious mutations [11]. Asexuals are expected to accumulate deleterious mutations in a non-reversible ratchet-like fashion [12], a process which accelerates in small populations and can drive a lineage to extinction due to mutational meltdown [13]. In contrast, sexual recombination allows the purging of deleterious mutations [12,14] and provides statistical chances for advantageous mutations from distinct individuals to recombine together in a single offspring [8]. Moreover, sexuality is advantageous in changing environments whereby the exploitation of genetic variability through meiotic and gametic recombination is central for adaptation in natural populations [15,16].

Asexuality in plants can happen by different forms of vegetative propagation (like runners, bulbils, tubers, etc.), or by apomixis (clonal seed formation [17]). Apomixis has been described in more than 400 species scattered throughout 293 angiosperms genera [18,19]. Apomixis is sporophytic when an embryo develops directly from a somatic cell of the ovary, or gametophytic when the embryo develops through the formation of an unreduced female gametophyte [17]. Two main types of gametophytic apomixis have been characterized following the formation of an unreduced embryo sac from a nucellar cell (apospory), or from the megaspore mother cell (MMC) via an altered meiosis (diplospory). In all cases of gametophytic apomixis, three developmental steps must sequentially occur to produce a functional clonal seed: (1) the development of a meiotically unreduced egg cell (apomeiosis), (2) the development of the seed's embryo without fertilization (parthenogenesis), and (3) the production of a functional endosperm with (pseudogamy) or without (autonomous) fertilization [20].

The molecular control of apomixis is still uncertain. Different hypotheses have been proposed to explain observations and experimental data from diverse apomictic plant species. Perhaps the two most accepted hypothesis are the idea of apomixis being a consequence of changes to developmental timing of reproduction, and the one considering apomixis as a mutation-based phenomenon [21–23].

In the first case, since polyploidy and hybridization are common traits of almost all apomictic plants [20], both have been hypothesized to be triggers for the reproductive switch from sexuality to apomixis. In this theoretical frame, it is proposed that global gene regulatory changes that follow polyploidization and hybridization in general, and specifically de-regulation of the standard sexual reproductive program lead to apomixis in plants [18], and can arise via chromosomal [24–26], genomic [27,28], and transcriptomic [26,29] changes in gene regulatory networks. As support for this hypothesis, studies demonstrated heterochronic gene expression (i.e., deregulation) between sexual and apomictic reproductive tissues in *Tripsacum* spp. [30] and *Boechera* spp. [31].

In the second case, apomixis in plants has been often identified with one or two dominant factors segregating according to Mendelian proportions, and which are altered via epistasis, genetic modifiers, polyploidy, etc. [32]. In several studies using different apomictic species of the *Ranunculus auricomus* complex throughout four generations of crossings and backcrossings, Nogler [33–35] established that apomixis was controlled by a dominant factor A^{-} with a recessive lethal behaviour. One of the main consequences of such conclusion was that A^{-} could not be transmitted though haploid gametes, and thus nicely provided a causality for the intriguing rarity of apomixis in natural diploid plants. In his studies, Nogler [35,36] further concluded that the frequency (i.e., expressivity) of apomixis in individual plant genotypes was dosage dependent and relative to the wild type A^{+}. More recently, part of Nogler's results were contradicted by crossing experiments on other, more distantly related species within the *R. auricomus* complex, in which experimental diploid hybrids displayed functional apomixis in low frequencies and inheritance of apospory in diploids [37]. In this study, the expression of apospory was dosage-dependent and higher in plants with two aposporous parents compared to those with just one aposporous parent [37]. The apospory-specific factor appears to be not necessarily lethal, but allelic dosage effects were confirmed.

The *Ranunculus auricomus* complex is becoming a model for studying genomic differences between sexual and apomictic taxa. Hundreds of apomictic species are derived from a handful of sexual progenitor species [38,39]. The closely-related diploid sexual species *R. carpaticola*, and diploid and autotetraploid *R. cassubicifolius* extend from Switzerland to west Slovakia [39], while hexaploid apomicts are located in central Slovakia. Analyses based upon AFLPs and SSRs, combined with DNA sequence [39,40] and RNAseq SNPs [41], support a hybrid origin for hexaploid apomicts resulting from a cross between a putative 4x *R. cassubicifolius* and a 2x *R. carpaticola* sometime during the last glacial period (60,000–80,000 years ago). Apomictic *Ranunculus* species are aposporous (*Hieracium*-type), pseudogamous and express apomixis facultatively [36,37,42–46]. Aposporous initial (AI) cells arise from a somatic cell of the nucellus during the time in which the megaspore mother cell (MMC) starts meiosis [35,43,44], and competes for the formation of the female gametophyte within the ovule. Once the mature female gametophyte is formed, fertilization of the egg and central cells is required in meiotic female gametophytes while in apomictic ones only the central cell needs to be fertilized to form the endosperm [36,43,44].

Taking advantage of a previous study using SNPs and comparing sexual and asexual *Ranunculus* genotypes using Illumina RNAseq data [41], we developed a custom expression microarray and used it to transcriptionally profile microdissected ovules at four developmental stages from sexual diploid *R. carpaticola* and tetraploid *R. cassubicifolius* genotypes, and hexaploid apomictic *R. carpaticola* × *cassubicifolius* genotypes. Using SNPs and RNAseq data might be of help to distinguish the molecular cause for apomixis in *Ranunculus*, i.e., whether is due to heterochronic gene expression or dominant factors. Based on levels of gene expression changes and dysregulated genes, we would expect global changes in gene expression in the first case or rather a few changes on master genes in the second case. This, however, is at least challenging since one or a few dominant factors regulating apomixis would likely trigger a wide range of downstream changes in gene regulatory networks, blurring possibilities to segregate expression patterns into one of both hypotheses. Either way, our experimental approach and in silico analyses were designed to overcome difficulties due to the occurrence of differing ploidies in our target *Ranunculus* plants, exploiting multiple genotypes as biological replicates. The goals of this study were to (1) identify transcripts showing differential expression between apomictic and sexual ovules, (2) quantify heterochronic gene expression patterns during apomictic and sexual ovule developments, (3) partition transcripts into groups whose expression signatures reflect hybridity, parent of origin effects, or polyploidy, and (4) frame our findings according to previous and ongoing studies to understand the inheritance of apomixis in *Ranunculus* and identify evidence of molecular evolution between reproductively relevant sequences within the complex.

2. Materials and Methods

2.1. Custom Microarray Development

The RNAseq approach used to generate the data for microarray design was explained in detail in Pellino et al. [41]. In short, flowers were collected from two sexual and two apomictic genotypes of the *Ranunculus auricomus* complex within a two-week time window (Table 1). For each individual, total RNA was isolated from pooled flowers of five different sizes (two to seven mm in length, from an early stage when sepal primordia have enclosed the floral meristem, to the fully developed unopened flower stage) using the Qiagen RNeasy Plant Mini Kit (Qiagen GmbH, Hilden, Germany). After isolation, any DNA contamination was removed using Qiagen RNase-Free DNase, while contamination from the DNase enzyme, polysaccharides, and proteins were removed with a second purification step using the Qiagen RNeasy Mini Kit. The manufacturer's instructions were followed in all purification steps.

Table 1. Diploid, tetraploid and hexaploid *Ranunculus* samples used in this study.

Taxon (Code)	Nind	Ploidy [1]	Reprod Mode	Apospory Expressivity [2]	Collection Locality, Collector, Collection Number, and Vouchers
R. carpaticola REV1	3	2x	Sex	0.0	Slovakia, Slovenské rudohorie, Revúca, hill Skalka (forest) Hörandl 8483 (WU)
R. carpaticola × *cassubicifolius* TRE	1	6x	Apo	0.226	Slovakia, Strázovské vrchy (near Trençín), between Kubra and Kubrica (margin of forest and meadow) Hörandl et al. C29 (SAV)
R. carpaticola × *cassubicifolius* VRU 2	2	6x	Apo	0.339	Slovakia, Turçianska kotlina, Vrútky-Piatrová (meadow) Hörandl et al. C35 (SAV)
R. cassubicifolius YBB 1	1	4x	Sex	0.0	Austria, Lower Austria, Wülfachgraben, SE Ybbsitz (forest) Hörandl 8472 (WU)

[1] For ploidy identification see [47,48]; [2] Data from [44]. WU: herbarium of the University of Vienna, SAV: herbarium of the Institute of Botany, Slovak Academy of science, Bratislava.

To avoid over-representation of highly transcribed genes during subsequent sequencing steps, full length normalized cDNA libraries were produced, whereby each step of the normalization was performed and optimized following the procedures described by Vogel et al. [49] and Vogel and Wheat [50].

15 µL of normalized cDNA (200 ng/µL) was sent to Fasteris SA (Geneva, Switzerland) for RNAseq, where a dual-sequencing approach (54-mer single read (SE) and 108-mer pair-end read (PE)) was chosen in order to balance cost and efficiency of a de novo assembly (see Pellino et al. [41]). Both sequencing strategies were conducted using the HiSeq^{TM} 2000 sequencing system (Illumina Inc., San Diego, CA, USA).

CLC Genomics Workbench (CLC bio version 4.9) was implemented for the sequencing assembly. At first, sequences were trimmed for vector contamination, length and quality score using CLC default values. As no reference genome for *Ranunculus* is available, de novo and iterated de novo approaches were used to assemble the data. For the de novo assembly, all libraries were pooled and assembled using CLC default parameters. In the iterated de novo assembly, each library was assembled individually, and the resultant contigs from each individual assembly re-assembled together, including all unassembled reads from each individual assembly. In both strategies the final assembly was trimmed for contigs shorter than 300 bp (see details in Pellino et al. [41]). The two approaches were evaluated based on the total number of matching reads (i.e reads that could be assembled into longer contigs) and N50 values, and the de novo assembly was selected for array design. Both contigs and singletons were forwarded to Roche NimbleGen Inc. (Madison, WI, USA) for design and manufacture. The NimbleGen selection strategy bioinformatically designed three different probes for each contig, and one to three probes for each singleton, followed by design of a 3 × 1.4 million-spot array. The final array therefore contained multiple technical replicates for each gene expressed during flower development in *Ranunculus*. All the microarrays data has been uploaded to Arrayexpress at EMBL-EBI ("Experiment E-MTAB-3316").

2.2. Transcriptomal Profiling of Sexual and Apomictic Ovules

2.2.1. Sample Selection, Ovule Microdissection, and RNA Extraction

Plants were grown from seedling to pre-flowering stages in outdoor plots at the Leibniz Institute of Plant Genetics and Crop Plant Research (IPK), and were then moved into a phytotron for flowering (day: 16 h, 21 °C; night: 8 h, 16 °C; humidity 70%; light intensity: 150 µmol/m^2). Based on the cytological observations of Hojsgaard et al. [44], ovules at four developmental stages were collected at the pre-meiotic, meiotic (tetrad stage/aposporous initial), 2–4 nuclei embryo sac and mature embryo sac stages (I, II, III, IV; Figure S1) at standardized times (between 7:00 and 9:00 AM) over multiple days from both sexual and apomictic *Ranunculus* under a sterile laminar-flow hood with a stereoscopic microscope (100 Stemi; Carl Zeiss AG, Oberkochen, Germany). Using sterile scalpel and forceps to open the flower, carpels were collected and immediately immersed in a sterile 0.55 M mannitol solution and placed on ice.

In the second step, microdissection of each single carpel was conducted in a in a sterile laminar air-flow hood under an inverted microscope (Axiovert 200M; Carl Zeiss AG, Oberkochen, Germany) using hand-crafted sterile glass needles (self-made with a Narishige PC-10 puller). Individual ovules were collected using an Eppendorf Cell Tram Vario connected to a 150 μm inner diameter glass capillary, immersed in 100 μL RNA-stabilizing buffer (RNA later; Sigma-Aldrich, St. Luis, MO, USA), and immediately frozen in liquid nitrogen. 20 to 40 ovules for each genotype/stage were dissected and stored at −80 °C. RNA extraction was carried out using a PicoPure isolation kit (Thermo Fisher Scientific, Carlsbad, CA, USA) and quantification and quality were assessed with RNA Pico chips on an Agilent 2100 Bioanalyzer (Agilent Technologies, Santa Clara, CA, USA).

2.2.2. cDNA Synthesis and Amplification

For each of the samples from the seven *Ranunculus* individuals (Table 1), cDNA synthesis and amplification were conducted using the Sigma TransPlex Complete WTA2 kit (Sigma-Aldrich) following the producer's instructions. Amplified cDNA was purified using the GenEluteTM PCR Cleanup kit (Sigma-Aldrich) following the manufacturer's protocol, and concentration and quality were measured with a NanoDropTM 1000 Spectrophotometer (Thermo Fisher Scientific, Wilmington, DE, USA).

2.2.3. Microarray Hybridization and Data Processing

Twenty-eight 1.4 M probe custom microarrays were outsourced to Roche NimbleGen for design and manufacture, each of which was used for an independent sample labeling and hybridization reaction for each of the 28 classes of microdissected ovule samples (four stages for seven genotypes; Table 1, Figure S1) following the NimbleGen microarray labeling protocol of their One-Color DNA Labeling Kit. The labeled samples where then individually hybridized in random order, using the NimbleGen Hybridization System 12 to the custom *Ranunculus* arrays according to the producer's instructions, and scanned using a NimbleGen MS 200 Microarray Scanner at 535 nm. Feature intensities were extrapolated using the DEVA software (version 1.1, Roche NimbleGen), and the raw expression data were normalized together using the DEVA implemented robust multiarray average (RMA) algorithm.

The normalized data were analyzed using the Qlucore Omics Explorer software (version 2.3). Principal Component Analysis (PCA) and Qlucore filtering by variance was implemented, whereby contigs with variation (σ/σ^{max}) < 0.7 between apomictic and sexual groups were removed from the dataset. This threshold was chosen, using the Qlucore software, such that the differentially expressed apomictic and sexual probe groups could be visually separated on a PCA, while at the same time retaining the maximum number of differentially expressed genes. Second, the resultant set of differentially expressed genes was ranked according to *p*-value (<0.01) using a paired *t*-test between the apomictic and sexual groups at each stage separately. In addition to the *p*-value filter, log2 fold change >2 and adjustment of the *p*-value for multiple tests using a false discovery rate (FDR) with q-value < 0.05 were applied.

2.2.4. Analyses of Gene Expression throughout Development

In order to calculate significant changes in gene expression patterns throughout ovule development between apomictic and sexual samples, the STEM software [51] was used to perform a similar analysis to that made by Sharbel et al. [31]. First, a data set of genes showing significant expression differences in at least one stage (but including the expression over all other stages) was constructed and used for STEM analysis. Except for the number of permutations, which was set to 1000 to increase accuracy [31], the profiling analysis was performed using default options with Bonferroni correction for multiple testing [51]. For the comparative analysis of different patterns of gene expression across the four developmental stages, the minimum number of intersected genes between sexual and apomictic samples was set to 1 [31] with maximum uncorrected intersection p-values < 0.05.

2.3. Analyses for Signatures of Ploidy, Parent of Origin Effects or Hybridization

Considering variable ploidy and evolutionary origins between sexual and hybrid apomictic *Ranunculus*, we sought to classify gene expression patterns into groups reflective of ploidy, parent of origin (sensu expression level dominance) or hybridization (sensu transgressive) effects. We first selected those genes showing statistically significant differential expression ($p < 0.05$, log2 fold change > 2, and FDR q-value < 0.05) between hexaploid apomicts and diploid sexuals, the two groups for which samples were available (Table 1) and hence for which statistical analysis was possible (due to sample size and lack of biological replicate, a similar statistical comparison using the tetraploid sample was not possible). Therefore, we used this subset of genes defined by the hexaploid apomict-diploid sexual comparison in all approaches. Then, in order to obtain information of partitioning of gene expression in the allopolyploid relative to each parent, we compared gene expression of each group, including data from the single tetraploid, estimating for each gene the log-fold expression difference of two contrasts: both parents to each other, and each parent to the allopolyploid, and controlling the distribution of p-values for each estimate using a false discovery rate of 0.05 [52].

2.3.1. Ploidy

On a gene-by-gene basis, the standard deviation of expression was calculated for the hexaploid apomicts and diploid sexuals. Then, genes in the tetraploid and hexaploids whose expression was (1) lower than two standard deviations in the diploid, and (2) higher than two standard deviations in the diploid, were classified as exhibiting a statistically significant pattern reflective of ploidy/additivity effects (sensu Yoo et al. [53]). Thus, genes in the hexaploid were classified as displaying additivity and expression level dominance compared to either parent (for transgressive expression see Section 2.3.3).

2.3.2. Parent of Origin Effects

Genes in sexual diploids and the tetraploid which (1) showed no expression differences with the hexaploid apomicts (i.e., <two standard deviations of mean gene expression), and (2) had expression levels which differed by >two standard deviations of the mean diploid sexual levels (the single tetraploid had no mean values per se), were classified as having patterns reflective of a parent of origin/expression dominance effect.

Lastly, we selected the group of all genes which did not show diploid-hexaploid parent of origin effects and were not initially selected (i.e., no significant differences in expression between hexaploid apomicts and diploid sexuals, see above). Next, we compared gene-by-gene the value of expression in the tetraploid against mean value of the hexaploid. Those genes in which the tetraploid sample had higher (i.e., >two standard deviations of the mean hexaploid apomict levels) or lower (i.e., <two standard deviations of the mean hexaploid apomict level) expression levels were classified as having patterns reflecting a tetraploid–hexaploid parent of origin/expression dominance effect.

In order to identify parental origin of alleles/factor(s) likely associated with apomixis (i.e., A^- putative genes), we compared high quality single nucleotide polymorphisms (SNPs) mined from a previous study [41] between the same individual plants analyzed here.

2.3.3. Hybridization

In the opposite case, genes were classified as having a hybrid/transgressive pattern when they (1) showed no expression differences between diploids and tetraploid sexuals (i.e., <two standard deviations of mean gene expression), and (2) had expression levels which differed by >two standard deviations compared to the mean of hexaploid apomicts.

2.4. Microarray Validation Using qRT-PCR

Ten genes showing differential expression between apomictic and sexual genotypes across the four developmental stages were randomly selected (see Section 3). After retrieving the sequences from the assembled cDNA database used in the array manufacture, PCR primers were designed using the PrimerSelect software (DNASTAR Inc., Madison, WI, USA) and selected, when possible, to overlap the microarray probes with the following parameters: product size <150 bp, GC content between 40 and 60%, annealing temperature ca. 60 °C.

Ranunculus-specific reference genes were selected by identifying sequences which were homologous to a selection of *Arabidopsis thaliana* reference genes (www.tair.com) using a blast analysis (blastX 2.2.30+ using the default NCBI parameters). Based on maximum similarity (similarity > 95%, *e*-value < 1 e^{-100}) homologous genes to UBQ (gb|ABH08754.1|) and ACTIN 11 (ref|NP_187818.1|) where chosen. Primers were designed following the procedures and parameters described above, and tested for amplification and expected product length in 10 µL PCR reactions including 25 ng of DNA, 1 µL of PCR Buffer II, 10 *p*mol for each primer, 0.025 U DNA Taq DNA Polymerase (Sigma-Aldrich, with 3.5 mM of $MgCl_2$ and 4.95 µL of H_2O. PCR reactions were performed in a Mastercycler ep384 (Eppendorf, Hamburg, Germany) using the following touchdown thermal cycling profile: 94 °C for 10 min; 9 cycles of 94 °C for 15 sec, 65 °C for 15 sec (1 degree decrease in temperature every cycle with a final temperature of 54 °C), 72 °C for 30 sec; 35 cycles of 94 °C for 30 sec, 57 °C for 15 sec, 68 °C for 2 min 30 sec; and a final 68 °C for 15 min.

qRT-PCR reactions, using UBQ and ACT11 (Dryad Id number: 82481 and 151955 Dryad entry doi:10.5061/dryad.nk151) as housekeeping genes, were run on a 7900HT FAST RT-PCR machine (Applied Biosystems, Foster City, CA, USA) using the SYBR Green Master Mix (Applied Biosystems) and the following program: initial denaturation at 90 °C for 10 min, followed by 40 cycles of 95 °C for 15 sec, and 60 °C for 1 min. Ct values (PCR cycle number where SYBR Green is detected) were extrapolated and used to infer initial copy number of the genes. Mean expression and standard deviation were calculated between two technical replicates and three biological replicates from apomictic and sexual genotypes using cDNA from the second stage of ovule micro-dissected tissues. Relative quantification was calculated using the ΔΔCt method and using the genes with higher CT values compared to the calibrator sample.

2.5. blastx, tblastx, Gene Ontology

Significantly differentially expressed genes obtained in all apomictic–sexual comparisons were selected for gene ontology (GO) analysis with Blast2Go (http://www.blast2go.com/b2ghome) using blastx (E-value cut off of $E \leq 1^{-5}$) and the default annotation parameters of the program. For genes where no blastx hits were obtained, an additional tblastx analysis was performed (E-value cut off of $E \leq 1^{-5}$). Overrepresentation analysis was not possible since only a small fraction of the *Ranunculus* transcriptome could be annotated (see Section 3).

3. Results

3.1. Gene Expression Differences

3.1.1. Principal Component Analysis (PCA)

In order to recognize graphically possible subsets of genes that may account for the maximum distinction between the apomictic and sexual samples, PCA was applied to the normalized expression values of probes from the original 62102 contigs (≥300bp; raw data first normalized using the RMA algorithm implemented in the DEVA software). The resulting PCA graph showed a clear separation between samples with respect to ploidy, distinguishing apomictic and sexual groups (Figure 1).

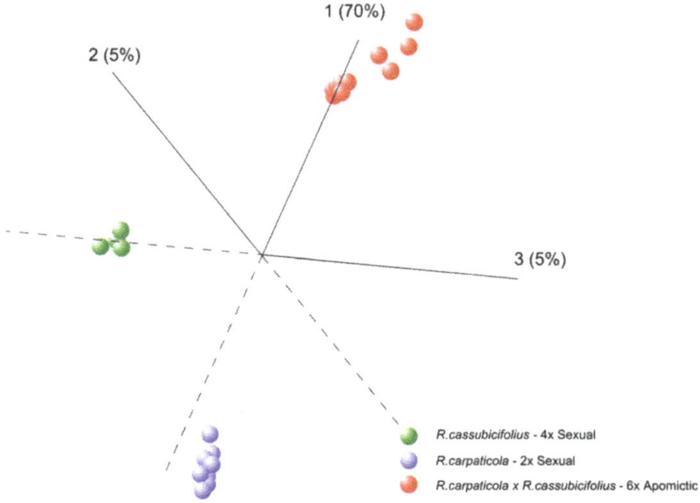

Figure 1. Principal component analysis (PCA) of gene expression on apomictic and sexual *Ranunculus*. PCA applied to normalized microarray data representing 4 ovule developmental stages from 3 apomictic (red) and 4 sexual (blue and green) *Ranunculus* genotypes showing ploidy and reproduction specific effects. Each dot represents one ovule stage, and frequencies in parentheses show the percentage of total variation explained by that principal component. The numbers on the axes represent the respective component (and the % of variation for each component).

3.1.2. Stage Specific Differential Expression Analysis

Transcriptome-wide gene expression variation through ovule development was compared between sexual diploids and apomictic hexaploids of *Ranunculus*, whereby four pre- to post-meiotic ovule stages were compared between six different genotypes. Overall, across all stages 439 and 339 transcripts were found to be significantly down- or up-regulated between apomictic versus sexual genotypes (Figure 2; Venn A and Venn B, respectively). The distribution of up- and downregulated transcripts in apomicts differed across the studied developmental stages I to IV (30, 48, 247, and 3 upregulated, 44, 98, 58, and 27 downregulated for each stage respectively) (Figure 2).

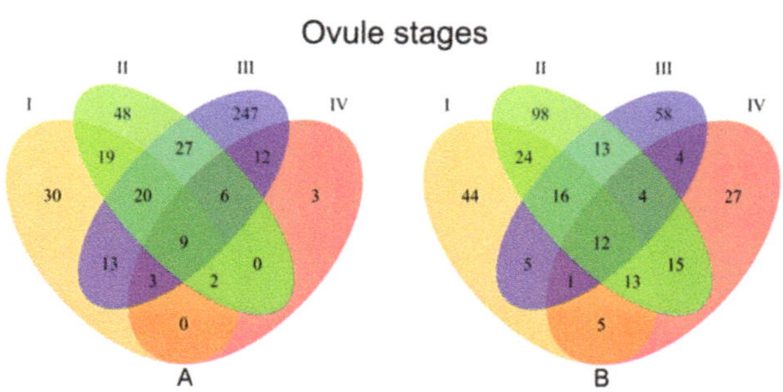

Figure 2. Venn diagrams showing number of differentially expressed transcripts between apomictic and sexual genotypes at each ovule developmental stage. A: Genes downregulated in the 6x apomicts compared to 2x sexuals B: Genes upregulated in the 6x apomicts compared to 2x sexuals. I = premeiotic, II = meiotic/aposporous, III = early embryo sac, IV = mature embryo sac.

Blast2Go analyses could be completed on only a small fraction of the differentially expressed transcripts, the limiting step being the inability to identify significant homologies to the NCBI nucleotide database(s). Of the number of genes which could be annotated (110 apomictically upregulated and 161 apomictically downregulated), only 59 (53.6%) upregulated and 78 (48.5%) downregulated transcripts from apomicts could be assigned a gene ontology (GO) term (Table S1; Table 2). Hence, representation analysis was not possible due to the significant bias introduced by insufficient GO categorization. A full list of genes can be accessed at GenBank (BankIt2351891: MT624108 - MT624295).

Table 2. Numbers of differentially expressed transcripts at each of four ovule developmental stages.

Number of Contigs	Number of Contigs, >300 bp	Differentially Expressed Transcripts			Annotated
462102	62102				
		Develop. stage	Upregulated [1]	Downregulated [1]	
		I	120 (44)	96 (30)	72
		II	195 (98)	131 (48)	107
		III	113 (58)	337 (247)	156
		IV	81 (3)	35 (3)	30

[1] data from 6x apomictic individuals compared to the normal expression in 2x sexuals; numbers in brackets represent number of stage specific transcripts.

A clustering-based STEM analysis [51] was performed to detect significant changes in patterns of transcript abundance across ovule development, by first grouping transcripts with similar expression profiles through development for the sexual and apomictic array datasets separately (Figure S2). In doing so, 46 and 62 transcripts could be assigned to three distinct patterns (STEM analysis, $p < 0.01$) in the sexual and apomictic groups, respectively. A comparison of patterns between these transcript sets identified eight transcripts as having significant differences (STEM analysis, $p < 0.01$) in the corresponding reproductive form, and showed a general trend of expression increase in developmental stage II followed by a sharp drop in stage III in apomicts (Figure S3). None of the eight genes could be assigned a GO term, and three had a significant homology to the transposon mutator sub-class protein (XP_006654086.1), salt overly sensitive 1b isoform 1 (EXC05020.1), and annexin-like protein (XP_007042996.1), respectively.

3.2. Transcriptome Wide Signatures of Hybridization, Ploidy Variation and Parent of Origin Effects

In order to understand and classify gene expression patterns of the apomictic hybrid (*R. carpaticola* × *R. cassubicifolius*) compared to the ancestral sexual parents (*R. carpaticola* and *R. cassubicifolius*), 304 differentially expressed transcripts were first identified in at least one developmental stage based on minimum fold change and statistical significance ($\log_2 > 2$, p-value < 0.01, FDR < 0.05) between the apomictic hybrids and the diploid sexuals (both groups having three genotypes each for statistical comparison). Secondly, a stage-by-stage comparison was performed for the expression of these 304 transcripts in the second tetraploid sexual parent (*R. cassubicifolius*; for which sample size precluded the first level statistical analysis—see Section 2) such that they could be classified into four different expression states (i.e., the hybrid relative to each ancestral parent). Genes showing expression bias with respect to the diploid or tetraploid parent were classified as showing a parent of origin effect, while those significantly over- or under-expressed in the apomict compared to both parents were classified as showing transgressive effects. Lastly, genes showing expression level changes in proportion to that of ploidy (e.g., increasing expression with increase of ploidy) were classified as showing a ploidy effect.

Examination of sequence similarity between genes showing a parent of origin effect and a high-quality SNP library (44) revealed only eight genes characterized by sufficient DNA sequence read coverage in all individuals, and of those only one (dyad contig id: 368515, Dryad entry doi:10.5061/dryad.nk151) showed 100% SNP similarity (for 16 SNPs) between the apomictic and the tetraploid genotypes only, implying a potential origin from the putative tetraploid parent (Table S2).

The same procedure was applied to 19,116 genes that were not significantly differentially expressed between the apomictic hexaploids and the sexual diploids in order to classify an additional state of gene expression (2x-parent of origin; see methods). This approach was necessary considering that only a single 4x sample was used, and allowed the comparison of gene expression of this one parent (4x) to the hybrid apomict. This classification was further subdivided with regards to whether expression of the hybrid and the diploid were higher or lower in comparison to the tetraploid. In total eight different groups of genes were classified, with most being downregulated in apomictic ovules at stage II (Figure 3; Table 3, Tables S3 and S4).

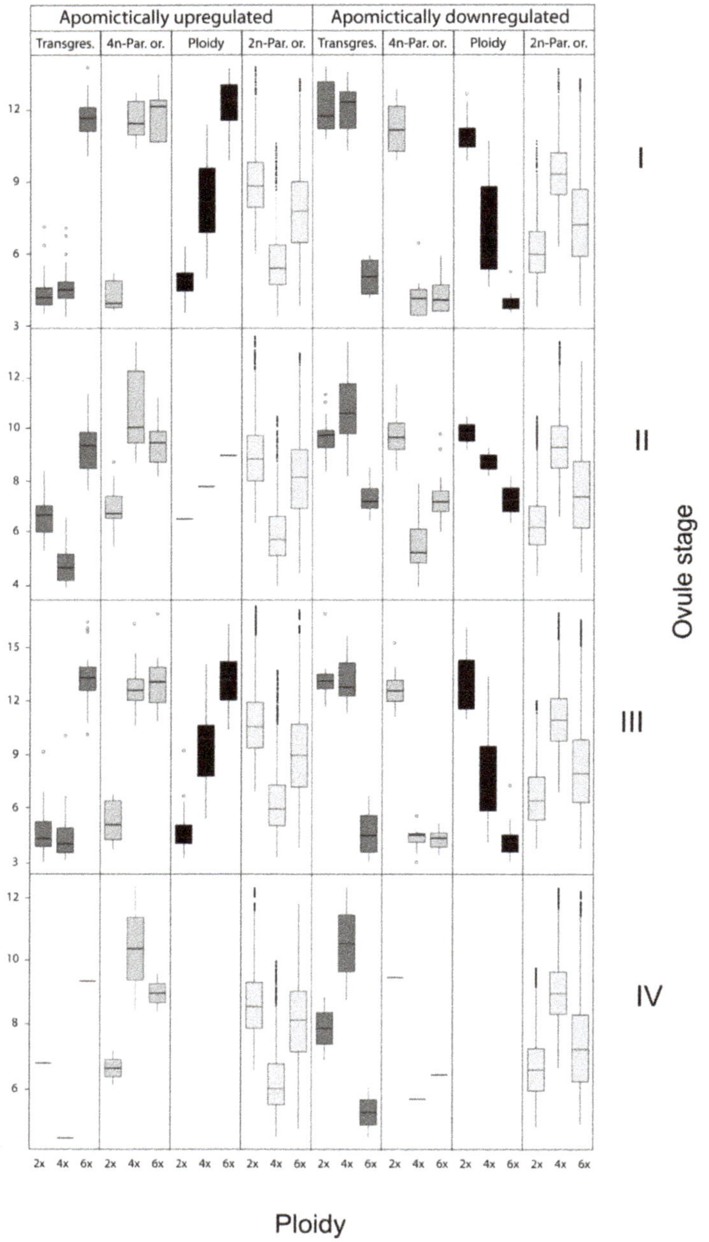

Figure 3. Box plots of expression distributions for genes which were significantly up- or downregulated in at least one apomictic ovule stage, as grouped into transgressive, ploidy-mediated and parent of origin patterns.

Table 3. Number of differentially expressed genes [1] associated with each ovule developmental stage and expression class in diploid–polyploid comparisons (brackets show relative percentage compared to the overall stage number of genes).

Expression Class [2]	Develop. Stage	Transgressive Effect	Parent of Origin Effect	Ploidy Effect	Total
Apomictic 6x upregulated [3]	I	5 (0.1)	27 (0.49)	23 (0.41)	55
	II	15 (0.58)	10 (0.38)	1 (0.04)	26
	III	16 (0.16)	44 (0.46)	36 (0.38)	96
	IV	2 (0.4)	1 (0.2)	2 (0.4)	5
	Total up	38 (0.21)	82 (0.45)	62 (0.34)	182
Apomictic 6x downregulated [3]	I	6 (0.26)	7 (0.3)	10 (0.44)	23
	II	18 (0.32)	37 (0.65)	2 (0.03)	57
	III	9 (0.24)	11 (0.3)	17 (0.46)	37
	IV	2 (0.4)	1 (0.2)	2 (0.4)	5
	Total down	35 (0.29)	56 (0.46)	31 (0.25)	122

[1] for seq. names, GO terms, enzyme codes and statistics refer to Tables S1 and S5; [2] compared to normal sexual diploid expression; [3] $p < 0.001$, FDR < 0.05, $\log_2$ change > 2.

Overall, most transcripts were expressed in a parent of origin pattern, followed by ploidy and then transgressive patterns (Table 3; Table S3). The strongest effect was apparently due to parent of origin, with a total of 82 upregulated and 138 downregulated apomictic-specific transcripts. Ploidy-mediated patterns of expression were revealed in 62 and 93 apomictic specific upregulated and downregulated transcripts, respectively. Transcripts showing transgressive patterns were the least abundant, with 38 and 73 upregulated and downregulated in apomicts, respectively (Table 3).

Of all transcripts showing such patterns, only 83 (49 and 34 up- and downregulated in apomicts respectively) could be annotated (Table S5). A simple screening of these genes revealed only three genes related to ovule development and reproduction. These three genes were coding for embryo sac developmental arrest protein (EDA), gamete expressed protein (GEX3), and a gene of the argonaute family (AGO) (e-value = 1.43 e-13, 3.28 e-32, 1.61 e-63, respectively and mean similarity of 47%, 70.5% and 84.4%, respectively; Table S5). Other annotated genes were related to expression of retrotransoposons, transposon proteins, transcription factors, ribonucleases, kinases, among others (Table S5).

3.3. *qPCR Validation*

Validation of the randomly selected genes showed concordance with the microarray analysis from the five upregulated and five downregulated genes. According to the REST software all 10 genes showed statistically significant up- and downregulation when measured with qRT-PCR (Table S6).

4. Discussion

Whether caused by a single "master gene", or through a polygenic complex, apomixis has long been a dilemma for scientists. Using different experimental approaches, the search for factor(s) underlying the switch in reproduction has led to the identification of a number of gene candidates in different species [22,54,55]. Such studies have also revealed a shift in gene expression through time for dozens or hundreds of genes normally involved in sexual reproduction [31,56,57]. Omic approaches have become useful for detecting expression shifts, although functional characterization and screening of the many candidates from such an experiment is a formidable task. This is especially true when dealing with wild species (as most apomicts are) in which annotated genomic information is not available. Additionally, the effects of polyploidy and hybridization [58] are often associated with the apomixis phenotype, adding an extra level of complexity to all analyses.

Despite hybridization and polyploidy introducing difficulties into data interpretations (see Mau et al. [59], their association with apomixis potentially reflects its mechanism of induction [18,60]. The "genomic shock", i.e., the genomic perturbation at both genetic and epigenetic levels produced by the union of two different genomes [61,62], could trigger the cascade of spatial and timing mis-expression of sex specific genes and lead to apomixis [20,63]. More recent studies suggest that polyploidy is not essential for expression of aposporous apomixis, because it appears in diploid hybrids [37] and non-hybrids [64,65], albeit in low frequencies. These and other studies suggest that origin of apomixis in wild populations probably starts in diploid populations and is just indirectly enhanced by side-effects of polyploidy [66,67].

Continuing our ongoing work [41,44] with wild allopolyploid apomictic *Ranunculus auricomus* and related sexuals, we wanted to shed light upon the molecular basis for apomixis in *Ranunculus* and possible factors underlying the transition from sexuality to apomixis. We have analyzed genome-wide gene expression through morphologically-defined ovule developmental stages in both apomictic and sexual genotypes, and using a specific sampling strategy to maximize biological diversity which enabled the identification of differences encompassing both reproductive mode (e.g., apomictic versus sexual) and genetic background (e.g., different sample clusters) (see Figure 1). Thus, our approach allowed us to disclose species' specific transcriptional variation connected to polyploidy and hybridity, and partition differentially expressed genes into patterns reflective of (a) reproduction specific expression, (b) heterochronic expression across ovule development, and (c) the expression of homologous genes in the apomicts and their ancestral (i.e., phylogenetic) parents.

4.1. Transcriptional Variation Reflect Contrasted Sexual vs. Apomictic Developments in the Ovule

In the *Ranunculus* cytotypes used here, Hojsgaard et al. [44] showed that the first stage of ovule primordium development is undifferentiated between apomictic and sexual *Ranunculus*. In contrast, diverse developmental irregularities accumulate during megasporogenesis, leading to perturbed and non-functional meiosis in apomicts (i.e., arrested development and/or altered megaspore selection; [44]), followed or accompanied by enlargement of a somatic cell that takes on the role of an aposporous initial (AI) cell. Deviations between the sexual and apomictic pathways continue during gametogenesis with subsequent development of the AI, parallel in timing with normal sexual gametogenesis in a few cases where meiosis progressed. The resultant unreduced megagametophyte (arising from the AI) then attains a similar multicellular structure to that of the sexuals at the final developmental stage [44]. Our tissue sampling design covered such developmental stages preceding and leading up to AI development till formation of female gametophytes (Table 1).

A reflection of the phenotypically-diverging development between pathways, gene expression differences between sexual and apomictic ovules show fewer differences in the total number of differentially expressed transcripts in the first and last stages of ovule sampling, whereas higher numbers of differentially expressed transcripts were found during the second and third stages (Table 2 and Figure 1). These findings agree with overall findings in other apomictic systems displaying global gene de-regulations in ovules at different developmental stages [68,69]. In our *Ranunculus* samples, stages II encompass AI appearance, usually in parallel with meiotic depletion, whereby the MMC did not enter meiosis or aborted during meiotic division. Stage III encompass the acquisition of a gametogenesis program whereby the AI enlarges to squeeze the forming meiotic megagametophyte (if any) and eventually take its place. Despite the relative weakness of any GO inference (see results), we note that no particular GO classes distinguished differentially expressed genes (Table S1) at any stage. Especially in stage II, the appearance of aposporous initials and parallel development/arrestment of meiotic products (with almost equal frequencies between hexaploids [44]) may blur the distinction of genes expressed between these two cell lineages. Later on, in stage III, the differences between sexual and apomictic developments become most distinct (Figure 3) as further developmental competition increase rates of abortion in the meiotic pathway (whether of meiotic products, functional megaspore or early embryo sacs), and only unreduced gametophytes develop further. Stage IV might be principally not so different in gene expression as there are no key developmental transitions as in stages II (AI appearance, sporogenesis) and III (gametogenesis), and mature sexual and apomictic embryo sacs do not differ phenotypically.

4.2. Apomixis Inheritance in Allopolyploid Ranunculus and Partitioned Gene-Expression Effects

Apomixis involves changes to a number of individual (meiotic recombination, purging of mutations) and population (gene flow, adaptation) attributes which contribute to shape the life history of species [17]. As complex as the genetic processes underlying apomixis can be, it is believed to be controlled by one or few master regulatory genes [54]. In the 1980s, after a series of experimental crossings and offspring analyses, Nogler published a seminal work on the genetic inheritance of apomixis in *Ranunculus*. The association of apomixis and ovule gene de-regulation has been described in many plant species [22], thus, here we focus on discussing the relevance of our present results in the frame of Nogler's ideas about apomixis inheritance, including more recent studies on different *Ranunculus* species.

4.2.1. Homologous Parent of Origin Effects in the Apomictic Hybrid

As mentioned above, the hexaploid natural hybrid studied here originated during the last glaciation from a cross between a *Ranunculus carpaticola* diploid and a *R. cassubicifolius* tetraploid [41,48]. Our comparative analyses of gene expression patterns between each putative parental cytotype and its derivative hybrid lineage reveal taxon-related differences.

In the putative diploid parent, we found a relatively low number of transcripts with similar expression levels as in hexaploids, and thus the bias in gene expression of hexaploids toward the tetraploid parent suggests a parent of origin effect. Contributing to this effect is gene dosage, whereby the hexaploid is composed of two haploid genomes from the diploid parent *R. carpaticola* [48]. In regard to Nogler's hypothesis, this observation suggests that A^- is absent (as expected) in diploids [35]. However, aposporous embryo sacs had been observed in very low frequencies in diploid *R. carpaticola* [44] which poses the alternative explanation that A^- is present in diploids but is not expressed, or that it is part of a multigenic network influencing the trait. Sequence similarity analysis using high quality single nucleotide polymorphisms (SNPs) mined from a previous study [41] revealed that only one of the genes showing a parent of origin effect and sufficient DNA sequence read coverage in all individuals showed 100% SNP similarity between alleles present in the apomictic and the tetraploid genotypes, but not to the diploid parent, implying a potential origin from the putative tetraploid parent. Considering the variability in read coverage between samples and the fact that the genotypes analyzed are, sensu stricto, not those involved in the original hybridization event circa 80,000 years before present [41], more cannot be inferred from the SNP dataset except that the diploid *R. carpaticola* probably lacks the factor A^-.

Regarding the putative tetraploid parent, over all developmental stages 220 genes were found to be similarly expressed between 6x apomictic *R. carpaticola* × *cassubicifolius* and 4x *R. cassubicifolius* plants, once again suggesting a parent of origin effect and genomic dosage (6x has expectedly four putative haploid genomes from *R. cassubicifolius* [48]). Even though the relatively higher number of genes displaying a tetraploid parent of origin pattern of expression in *Ranunculus* ovules is likely biased by the lack of biological replicates, the result is concordant with similar studies in natural populations and synthetic hybrids of cotton [70], maize [71], rice [72] and *Senecio* [73]. All such studies showed similarly higher levels of parent of origin expression relative to transgressive effects, which tended to decrease with the age of the hybrid.

Investigations into the causes of parental expression bias point to the effect of *cis-trans* regulation, epigenetic changes, and introgression in organisms with different life histories like *Cirsium arvense* (thistle) [74] and cotton [53]. In *Ranunculus* synthetic hybrids (between diploids *Ranunculus carpaticola* × *notabilis* and tetraploid and diploid *R. cassubicifolius* × *notabilis*), apospory appeared spontaneously in diploid and triploid F_1 hybrids of sexual species that did not previously show any signs of apomixis; however, functional apomictic seed were found in triploid F_1 hybrids [44], and in diploid F_2 hybrids [37], both in low frequencies.

Even though Nogler's [35,36] apospory-incurring A^- factor is causal in the genomes of hybrids, it is plausible that such a factor is present in one (or both) parental species but not expressed.

Previous morphological and genetic analyses of tetraploid *R. cassubicifolius* indicate that the cytotype is sexual [35,39,47,48], and embryological studies failed to show any sign of AI formation [44]. However, the high dosage of the wild-type A^+ from the tetraploid sexual parent might explain the relatively high frequencies of sexual ovule (69%) and seed formation (29%) in hexaploid apomicts [44] compared to other polyploid apomicts having divergent evolutionary histories in which sexuality is found at residual levels (±5%, e.g., [75,76]). Likewise, *Ranunculus variabilis*, a naturally-related tetraploid apomict of the *R. auricomus* complex, showed overall higher frequencies of apospory and apomictic seed formation than our hexaploid hybrids used here [46]. These results suggest that the dosage of apomixis-factors depends on the parentage and evolutionary origin of plants, and does not necessarily increase linearly with the level of ploidy. Considering this, the origin of apomixis in the hexaploids could be explained in two ways. First, apospory could have appeared for the first time spontaneously in natural, triploid hybrids between a diploid sexual *R. carpaticola* and a tetraploid sexual *R. cassubicifolius* (as observed in 2x × 4x crosses mentioned above [44]). Subsequent polyploidization of the triploid hybrid could result in an hexaploid genotype like that established in nature.

Second, A^- could have already been present but not phenotypically expressed in tetraploid *R. cassubicifolius* plants. This possibility was already discussed by Nogler [35] as a probable consequence of dosage effects (see discussion below) and supported by observations on progenies from several rounds of backcrossing to the sexual parent which show a progressive delay in the initiation of apomictic development [35]. In many natural hybrids and polyploids, the correct spatial and temporal expression of developmental programs is altered, and is hypothesized as a central point for the development of apomixis [18,30,60,63]. Thus, the expression of apomixis in the hexaploids *Ranunculus carpaticola* × *cassubicifolius* analyzed here is likely associated with the observed heterochronic expression of genes during sporogenesis and gametogenesis which could only arise after hybridization or polyploidization.

4.2.2. Ploidy Effect and Apomixis Expressivity in the Hybrid

Ploidy or additivity effects are due to the presence of multiple genome sets in pairs of diploid–polyploid species. Levels of genome and gene dosage can have relevant effects on different general plant attributes like heterosis [77], or reproductively related features like fitness [78]. However, disaggregating contributions of polyploidy versus hybridity to such attributes is usually not possible or require very specific experimental setups (e.g., using isogenic and hybrid genetic contexts [79]). In our analysis, the contribution of differentially expressed genes displaying apparent ploidy effects cannot be untangled from hybridity. Even though the relative contribution of ploidy effects to the total number of differentially expressed genes is intermediate (below those grouped under a parent of origin effect; Table 3), together with transgressive effects (hybridization) they contribute the most genes to the molecular changes driven the emergence of apomixis in *Ranunculus* ovules. At this point, the number of differentially expressed genes is not high enough to suggest global heterochrony rather than a few dominant factors is the molecular basis for apomixis in this group. Therefore, we will focus our discussion on possible genome dosage effects associated with ploidy and hybridity, and framed by Nogler's theoretical model of apomixis inheritance.

In his cytological and inheritance studies in *Ranunculus*, Nogler [36] further observed that in experimental crossings increased dosage of the controlling factor for apomixis (A^-) relative to the wild type sexual A^+ allele resulted in higher levels of apomixis penetrance, as evidenced by the occurrence of AI cells. Results of Barke et al. [37] on diploid hybrids also support an allelic dosage effect. Assuming that A^- is allelic (or epiallelic), it then becomes relevant to understand how dosage could vary (see [35]) between cytotypes. Tetraploids are fully sexual [44], implying either that they lack A^- altogether (and hence A^- must have appeared in hybrids only), or that A^- is present in tetraploids but because of wrong timing (see above), low dosage ($A^- A^+ A^+ A^+$ genotypes) and/or lack of a hybrid background, it is not expressed (i.e., low or lack of AI cells, as observed in natural diploid and tetraploid sexual *Ranunculus* spp. and experimental hybrids [37,44]).

In hexaploid apomicts, one or more copies of A^- in a hybrid background could result in increased expression levels with respect to the diploid and tetraploid parents of lower ploidy. In fact, both the frequencies of AI's (67%, ranging between 50–87%) in hexaploids and observed variability in expressivity of apomixis (71% ± 12%) [44] suggest variable dosage of factors directly or indirectly related to apomixis (i.e., A^-). The observed variability in levels of sexuality in these facultative hexaploids (mean value 69%, ranging between 56–96% of ovules with functional megaspores [44]) point to hexaploids being heterozygous for A^- and the wild type factor A^+. Since *Ranunculus* plants were grown under identical conditions, such variability may be due to a variable dosage for A^- and/or distinct genomic contexts. Multiple copies of A^- can be acquired by gamete recombination and formation of sexual offspring between facultative hexaploids. However, an increase in copies of A^- would be restricted by its dosage effects, as increasing expressivity (i.e., level of apomixis) will concomitantly reduce levels of residual sexuality, thus making it increasingly impossible to attain an obligate (100% expressivity) apomictic hexaploid individual via recombination. This explains observations in *Ranunculus*, *Paspalum* and many other apomictic genera in which no fully apomictic plants had been so far recovered ([46,65,80,81]; although, see [82] for cases of obligate apomixis in *Boechera*). The opposite situation is also possible, with occasional recombinant individuals having no copies of A^- that have lost their capacity to reproduce apomictically. Even though it has not been observed in natural apomictic hexaploid *Ranunculus auricomus* individuals [44–46,48], a reversal to sex in an apomictic lineage has been rarely observed [81,83] but is theoretically expected in apomictic populations [67,84,85].

From a phylogenetic viewpoint, in the species-rich *Ranunculus* genus most taxa have unique biogeographic and phylogenetic histories [86]. Apomixis appeared at least twice independently [39]. Hence the origin of the A^- allele in independent lineages can certainly be of diverse nature (i.e., via vertical or horizontal transfer, through polyploidy and/or hybridization). Considering our pedigree system of diploid and tetraploid parents and derivative hexaploid hybrid, we have partial evidence that the A^- factor originating apomixis in hexaploids was inherited from the putative tetraploid parent. Yet, one cannot assume with certainty whether A^- is absent in the diploid or polyploid plants studied here, whereby they display apomixis elements but lack functional apomixis (e.g., low levels of multiple embryo sacs where observed in diploid sexual *R. carpaticola*; [44]). In any case, expression patterns of A^- and its dosage would reflect ploidy effects, being absent/not expressed in diploids, lowly expressed in triploid carriers and highly expressed in the hexaploid apomict (as observed in Hojsgaard et al. [44]).

4.2.3. Transgressive Gene Expression in the Apomictic Hybrid

Transgressive segregation relates to the formation of extreme or transgressive phenotypes (falling beyond the parental range) often observed in segregating populations mainly for traits influenced by multiple quantitative loci (QTL) [87]. Studies on apomixis inheritance have demonstrated that apomixis factors deviate from Mendelian segregation due to modifiers and epistasis associated with secondary factors [32]. In our analysis, observed transgressive gene expression effects may be influencing the dominance of A^- (see discussion below), and the extent to which such transgressive expression attenuates apomixis expression would require another experimental design.

Expression patterns of an apomixis factor A^- and associated cofactors could, in addition to the parent of origin and ploidy effects discussed above, be associated with *cis* and *trans* regulatory dynamics that characterize hybrid genomes [88]. Aposporous initials have been observed in the tetraploid *R. megacarpus* carrying the postulated A^- factor [35], and in a low percentage in synthetic diploid and triploid hybrids between sexual parents [37,44]. Although A^- might control aposporous embryo sac formation as have been described in other aposporous systems [89], A^- cannot directly control the timing of AI induction as it is dosage dependent [35]. Thus, since the timing of AI induction is key for successful apomictic development [35], transgressive gene expression caused by genome merging during hybridization seems to have a relevant role in *Ranunculus*. Transgressive gene expression effects might well cause shifts in timing of floral development and reproductive programs underlying developmental heterochrony and the transition to apomixis in hybrids [18]. The identification of 35 and 38 homologs showing transgressive effects which were down- and up-regulated in the apomictic, respectively, support this view. Such a relatively low number of genes displaying transgressive expression could be a reflection of the old evolutionary age of the hexaploid lineage (about 80,000 yo [41]), as suggested by Hegarty et al. [73] whereby transgressive effects caused by genomic shock after hybridization are ameliorated by genetic mechanisms and evolutionary forces acting upon neopolyploids during first generations post-hybridization and polyploidization.

Despite differentially expressed genes falling in the transgressive expression effects group contributed the less to the overall gene dysregulation in hexaploid *Ranunculus*, the only three genes related to plant reproduction out of all genes that could be annotated were from this group. This also suggest transgressive rather than ploidy or parent of origin expression effects might play a more active role on apomixis emergence. These genes are involved in megagametogenesis and development, ones being the embryo sac developmental arrest protein (EDA), gamete expressed protein (GEX3), and a gene of the argonaute family (AGO). Interestingly, *Arabidopsis* EDA mutants show a series of defects during megagametogenesis, resulting in interrupted or abnormal meiotic division [90]. Furthermore, GEX3 is essential for pollen tube guidance during double fertilization, and its misregulation in *Arabidopsis* resulted in reduced seed set and undeveloped embryos [91]. If these genes had analogous phenotypic effects in *Ranunculus*, its transgressive expression in hexaploid apomicts is likely modulating sexual vs. apomixis expression and seed formation, which might explain observed developmental variability [44,46]. Finally, in *Arabidopsis* and Maize genes defective for AGO9 and AGO104 were shown to control female gamete formation via a small RNA pathway controlling methylation and transcription of many targets in the ovary [92,93]. Mutants displayed apomixis-like phenotypes including AI-like cells and unreduced gametes [92,93]. In *Ranunculus* in general and in the hexaploid apomictic *R. carpaticola* × *cassubicifolius* plants, the AGO gene might also be associated with a possible epigenetic control of apomixis and gamete formation. Likewise, observed variability of proportions of facultative apospory under different light stress conditions in hexaploid clone-mates of *R. carpaticola* × *cassubicifolius*, including residual levels in diploid and tetraploid cytotypes, suggests an influence of epigenetic regulatory mechanisms on reproductive phenotypes [45,46]. In *R. kuepferi*, differential cytosine-methylation patterns found in sexual and apomictic natural populations support this hypothesis [94]. In other apomictic plant groups different genes had been found linked to the apomixis phenotype or to individual components of apomixis (e.g., [95–101]). Even when these three genes might have important roles in the functional expression of apomixis in *Ranunculus*, it seems more likely they are being involved in the molecular cascade rather than being master genes associated to apomeiosis, parthenogenesis or endosperm development.

Global shifts in gene expression patterns associated with hybridization and polyploidy [102] have been hypothesized to underlie the switch from sex to apomixis [18]. Experimental support for this hypothesis has been found in apomictic *Boechera* [31,103] and between sexual species of *Tripsacum* [30]. The hexaploid apomictic *Ranunculus* lineage analyzed here had a hybrid origin around the last glacial maximum [41,47,48]. Hybridization as a trigger of apomixis in this lineage is supported by synthetically-derived hybrid *Ranunculus* mirroring the original hybridization event which show formation of apomeiotic embryo sacs and low rates of functional apomixis in the first two generations after hybridization [37,44]. This hypothesis is supported by our gene expression analyses in sexual and apomictic ovules pointing to at least three genes likely associated with the expression of the apomixis phenotype, with influences of hybridity and polyploidy reflected in dosage and transgressive effects. Similar reproductive phenotypes have been observed in sexual mutants [92] and other experimental hybrids in *Sorghum* and *Antennaria* [21]. Interestingly, despite having different forms of apomixis [104], diplosporous *Boechera* [31] and aposporous *Ranunculus* similarly show a negative spike in differential gene expression at similar stages of megaspore mother cell/aposporous initial cells progression during ovule development in comparisons between sexual–apomictic ovaries. Even though such parallelism on gene expression changes at key developmental stages may point to conserved processes leading to apomeiosis, specific studies will be required. Genetic and functional evaluation of EDA, GEX3, and AGO orthologues in *Ranunculus* will shed light on regulatory mechanisms through which apomixis and sexuality are modulated in polyploid plants.

5. Conclusions and Prospects

In this study we investigate gene expression variation in the apomictic *Ranunculus auricomus* complex and present a list of genes that show differential expression between aposporous and sexual ovules across four developmental stages. Considering the common difficulty in wild species for which genomic sequence information can be readily collected but little to no gene annotation can be made, we have considered the natural history of the *Ranunculus auricomus* complex in order to classify genes whose expression patterns reflect the evolutionary and genetic mechanisms hypothesized to underlie the switch from sex to apomixis. Despite this, our study cannot confirm a molecular basis for apomixis based either on global heterochronic gene expression or dominant genetic factors. The results suggest apomixis in *Ranunculus* might depend on changes of reproductive genes with downstream influence on different gene regulatory cascades. Thus, our observations based on the number of differentially expressed genes and their expression patterns are in concordance with the commonly accepted idea that apomixis is caused by gene de-regulation of the sexual reproductive program in connection with hybridization and polyploidization. Here we set the foundations for more specific studies on gene regulation with respect to apomixis expression and expressivity.

Supplementary Materials: The following are available online at http://www.mdpi.com/2073-4425/11/7/728/s1, Figure S1: Images of ovule staging (I, II, III, IV). The bar represents 100 μm, Figure S2: Change in expression patter according to the Short Time-series Expression Miner (STEM) program, Figure S3: Expression pattern changes in differentially expressed genes, Table S1: Blast2Go ontology for differentially expressed genes across stages, Table S2: SNPs Contig 368515, Table S3: Genes showing similar 2x-6x expression pattern and 4x parent of origin effects, Table S4: Normalized gene expression data of four ovule developmental stages used in Figure 3. A: stage I; B: stage II; C: stage III; D: stage IV. Rows correspond to gene IDs and columns to microarray chips, Table S5: Blast2go p1p2hyb effect, Table S6: List of Genes and primer used for microarray validation.

Author Contributions: Conceptualization, E.H. and T.F.S.; methodology and analysis, M.P. and D.H.; writing—original draft, M.P.; writing—review and editing, D.H., E.H. and T.F.S. All authors have read and agreed to the published version of the manuscript.

Funding: This research was funded by jointly by the Austrian Research Fund (FWF), grant number I 310 to EH, and the German Research Fund (Deutsche Forschungsgemeinschaft DFG) project 154994351 to TFS.

Conflicts of Interest: The authors declare no conflict of interest.

References

1. Barton, N.; Charlesworth, D. Why sex and recombination? *Science* **1998**, *281*, 1986–1990. [CrossRef]
2. Mittwoch, U. Parthenogenesis. *J. Med. Genet.* **1978**, *15*, 165–181. [CrossRef]
3. Suomalainen, E. Parthenogenesis in animals. In *Advances in Genetics*; Academic Press Inc.: New York, NY, USA, 1950; pp. 193–253.
4. Hörand, E.; Grossniklaus, U.; van Dijk, P.J.; Sharbel, T.S. *Apomixis: Evolution, Mechanisms and Perspectives*, 1st ed.; Regnum Vegetabile; Hörand, E., Grossniklaus, U., van Dijk, P.J., Sharbel, T.S., Eds.; A.R.G. Gantner Verlag: Rugell, Liechtenstein, 2007; Volume 147, ISBN 978-3-906166-60-5.
5. Schön, I.; Van Dijk, P.; Martens, K. *Lost Sex*; Schön, I., Martens, K., Dijk, P., Eds.; Springer Science & Business Media: Berlin, Germany, 2009; ISBN 978-90-481-2769-6.
6. Maynard Smith, J. *The Evolution of Sex*, 1st ed.; Cambridge University Press: Cambridge, NY, USA, 1978; ISBN 0521293022.
7. Bell, G. *The Masterpiece of Nature—The Evolution and Genetics of Sexuality*, 1st ed.; University of California Press: Berkeley, CA, USA, 1982; ISBN 10: 0520045831, ISBN 13: 9780520045835.
8. Crow, J.F.; Kimura, M. Evolution in sexual and asexual populations. *Am. Nat.* **1965**, *99*, 439–450. [CrossRef]

9. Vrijenhoek, R.C. Ecological differentiation among clones: The Frozen Niche Variation model. In *Population Biology and Evolution*; Woehrmann, K., Loeschcke, V., Eds.; Springer: Berlin/Heidelberg, Germany, 1984; pp. 217–231.
10. Hörandl, E. A combinational theory for maintenance of sex. *Heredity (Edinb.)* **2009**, *103*, 445–457. [CrossRef]
11. Lovell, J.T.; Williamson, R.J.; Wright, S.I.; McKay, J.K.; Sharbel, T.F. Mutation accumulation in an asexual relative of *Arabidopsis*. *PLoS Genet.* **2017**, *13*, 1006550. [CrossRef] [PubMed]
12. Muller, H.J. The relation of recombination to mutational advance. *Mutat. Res. Mol. Mech. Mutagen.* **1964**, *1*, 2–9. [CrossRef]
13. Lynch, M.; Butcher, D.; Bürger, R.; Gabriel, W. The mutational meltdown in asexual populations. *J. Hered.* **1993**, *84*, 339–344. [CrossRef] [PubMed]
14. Kondrashov, A.S. Selection against harmful mutations in large sexual and asexual populations. *Genet. Res.* **1982**, *40*, 325–332. [CrossRef] [PubMed]
15. Fisher, R.A. *The Genetical Theory of Natural Selection*; The Clarendon Press: Oxford, UK, 1930.
16. Lewontin, R.C. *The Genetic Basis of Evolutionary Change*; Columbia University Press: New York, NY, USA, 1974; ISBN 0-231-03392-3.
17. Asker, S.E.; Jerling, L. *Apomixis in Plants*, 1st ed.; CRC Press: Boca Raton, FL, USA, 1992; ISBN 0-8493-4545-6.
18. Carman, J.G. Asynchonous expression of duplicate genes in angiosperms may cause apomixis, bispory, tetraspory, and polyembryony. *Biol. J. Linn. Soc.* **1997**, *61*, 51–94. [CrossRef]
19. Hojsgaard, D.; Klatt, S.; Baier, R.; Carman, J.G.; Hörandl, E. Taxonomy and biogeography of apomixis in angiosperms and associated biodiversity characteristics. *CRC Crit. Rev. Plant Sci.* **2014**, *33*, 414–427. [CrossRef]
20. Koltunow, A.M.; Grossniklaus, U. Apomixis: A developmental perspective. *Annu. Rev. Plant Biol.* **2003**, *54*, 547–574. [CrossRef] [PubMed]
21. Carman, J.G. Do duplicate genes cause apomixis? In *Apomixis: Evolution, Mechanisms and Perspectives*; Regnum Vegetabile; Hörandl, E., Grossniklaus, U., van Dijk, P., Sharbel, T.S., Eds.; A.R.G. Gantner Verlag: Rugell, Liechtenstein, 2007; ISBN 978-3-906166-60-5.
22. Hand, M.L.; Koltunow, A.M.G.G. The genetic control of apomixis: Asexual seed formation. *Genetics* **2014**, *197*, 441–450. [CrossRef] [PubMed]
23. Schmidt, A. Controlling apomixis: Shared features and distinct characteristics of gene regulation. *Genes (Basel)* **2020**, *11*, 329. [CrossRef] [PubMed]
24. Kantama, L.; Sharbel, T.F.; Schranz, M.E.; Mitchell-Olds, T.; de Vries, S.; de Jong, H. Diploid apomicts of the *Boechera holboellii* complex display large-scale chromosome substitutions and aberrant chromosomes. *Proc. Natl. Acad. Sci. USA* **2007**, *104*, 14026–14031. [CrossRef] [PubMed]
25. Sánchez-Morán, E.; Benavente, E.; Orellana, J. Analysis of karyotypic stability of homoeologous-pairing (ph) mutants in allopolyploid wheats. *Chromosoma* **2001**, *110*, 371–377. [CrossRef]
26. Adams, K.L.; Percifield, R.; Wendel, J.F. Organ-specific silencing of duplicated genes in a newly synthesized cotton allotetraploid. *Genetics* **2004**, *168*, 2217–2226. [CrossRef]
27. Pikaard, C.S. Genomic change and gene silencing in polyploids. *Trends Genet.* **2001**, *17*, 675–677. [CrossRef]
28. Soltis, D.E.; Soltis, P.S. Polyploidy: Recurrent formation and genome evolution. *Trends Ecol. Evol.* **1999**, *14*, 348–352. [CrossRef]
29. Adams, K.L. Evolution of duplicate gene expression in polyploid and hybrid plants. *J. Hered.* **2007**, *98*, 136–141. [CrossRef]
30. Grimanelli, D.; García, M.; Kaszas, E.; Perotti, E.; Leblanc, O. Heterochronic expression of sexual reproductive programs during apomictic development in *Tripsacum*. *Genetics* **2003**, *165*, 1521–1531. [PubMed]
31. Sharbel, T.F.; Voigt, M.-L.; Corral, J.M.; Galla, G.; Kumlehn, J.; Klukas, C.; Schreiber, F.; Vogel, H.; Rotter, B. Apomictic and sexual ovules of *Boechera* display heterochronic global gene expression patterns. *Plant Cell* **2010**, *22*, 655–671. [CrossRef]
32. Ozias-Akins, P.; van Dijk, P.J. Mendelian genetics of apomixis in plants. *Annu. Rev. Genet.* **2007**, *41*, 509–537. [CrossRef] [PubMed]
33. Nogler, G.A. Genetik der Aposporie bei *Ranunculus auricomus* s.l. W. Koch. I. Embryologie. *Ber. Schweiz. Bot.* **1971**, *81*, 139–179.
34. Nogler, G.A. Genetik der aposporie bei *Ranunculus auricomus*. III. F2-Rückkreuzungsbastarde. *Ber. Schweiz. Bot.* **1973**, *83*, 295–305.

35. Nogler, G.A. Genetics of apospory in apomictic *Ranunculus auricomus*. V: Conclusion. *Bot. Helv.* **1984**, *94*, 411–422.
36. Nogler, G.A. Gametophytic apomixis. In *Embryology of Angiosperms*; Johri, B.M., Ed.; Springer: Berlin/Heidelberg, Germany, 1984; pp. 475–518. ISBN 3-540-12739-9.
37. Barke, B.H.; Daubert, M.; Hörandl, E. Establishment of apomixis in diploid F2 hybrids and inheritance of apospory from F1 to F2 hybrids of the *Ranunculus auricomus* complex. *Front. Plant Sci.* **2018**, *9*, 1111. [CrossRef]
38. Hodač, L.; Scheben, A.P.; Hojsgaard, D.; Paun, O.; Hörand, E. ITS Polymorphisms shed light on hybrid evolution in apomictic plants: A case study on the *Ranunculus auricomus* complex. *PLoS ONE* **2014**, *9*, 28–30. [CrossRef]
39. Hörandl, E.; Greilhuber, J.; Klímová, K.; Paun, O.; Temsch, E.; Emadzade, K.; Hodálová, I. Reticulate evolution and taxonomic concepts in the *Ranunculus auricomus* complex (Ranunculaceae): Insights from analysis of morphological, karyological and molecular data. *Taxon* **2009**, *58*, 1194–1215. [CrossRef]
40. Paun, O.; Ho, E. Evolution of Hypervariable Microsatellites in Apomictic Polyploid Lineages of Ranunculus carpaticola: Directional Bias at Dinucleotide Loci. *Genetics* **2006**, *174*, 387–398. [CrossRef]
41. Pellino, M.; Hojsgaard, D.; Schmutzer, T.; Scholz, U.; Hörandl, E.; Vogel, H.; Sharbel, T.F. Asexual genome evolution in the apomictic *Ranunculus auricomus* complex: Examining the effects of hybridization and mutation accumulation. *Mol. Ecol.* **2013**, *22*, 5908–5921. [CrossRef]
42. Hörandl, E.; Cosendai, A.C.; Temsch, E.M. Understanding the geographic distributions of apomictic plants: A case for a pluralistic approach. *Plant Ecol. Divers.* **2008**, *1*, 309–320. [CrossRef] [PubMed]
43. Izmailow, R. Megasporogenesis in the apomictic species *Ranunculus cassubicus*. *Acta Biol. Crac.* **1965**, *8*, 183–195.
44. Hojsgaard, D.; Greilhuber, J.; Pellino, M.; Paun, O.; Sharbel, T.F.; Elvira, H.; Hörandl, E. Emergence of apospory and bypass of meiosis via apomixis after sexual hybridisation and polyploidisation. *New Phytol.* **2014**, *204*, 1000–1012. [CrossRef] [PubMed]
45. Klatt, S.; Hadacek, F.; Hodač, L.; Brinkmann, G.; Eilerts, M.; Hojsgaard, D.; Hörandl, E. Photoperiod extension enhances sexual megaspore formation and triggers metabolic reprogramming in facultative apomictic *Ranunculus auricomus*. *Front. Plant Sci.* **2016**, *7*, 278. [CrossRef]
46. Ulum, F.B.; Costa Castro, C.; Hörandl, E. Ploidy-dependent effects of light stress on the mode of reproduction in the *Ranunculus auricomus* complex (Ranunculaceae). *Front. Plant Sci.* **2020**, *11*, 104. [CrossRef]
47. Hörandl, E.; Greilhuber, J. Diploid and autotetraploid sexuals and their relationships to apomicts in the Ranunculus cassubicus group: Insights from DNA content and isozyme variation. *Plant Syst. Evol.* **2002**, *234*, 85–100. [CrossRef]
48. Paun, O.; Stuessy, T.F.; Hörandl, E. The role of hybridization, polyploidization and glaciation in the origin and evolution of the apomictic *Ranunculus cassubicus* complex. *New Phytol.* **2006**, *171*, 223–236. [CrossRef]
49. Vogel, H.; Heidel, A.J.; Heckel, D.G.; Groot, A.T. Transcriptome analysis of the sex pheromone gland of the noctuid moth *Heliothis virescens*. *BMC Genom.* **2010**, *11*, 29. [CrossRef]
50. Vogel, H.; Wheat, C.W. Accessing the transcriptome: How to normalize mRNA pools. In *Molecular Methods for Evolutionary Genetics*; Humana Press: Totowa, NJ, USA, 2011; Volume 772, pp. 105–128. [CrossRef]
51. Ernst, J.; Bar-Joseph, Z. STEM: A tool for the analysis of short time series gene expression data. *BMC Bioinform.* **2006**, *7*, 191. [CrossRef]
52. Rapp, R.A.; Udall, J.A.; Wendel, J.F. Genomic expression dominance in allopolyploids. *BMC Biol.* **2009**, *7*, 18. [CrossRef] [PubMed]
53. Yoo, M.J.; Szadkowski, E.; Wendel, J.F. Homoeolog expression bias and expression level dominance in allopolyploid cotton. *Heredity (Edinb.)* **2013**, *110*, 171–180. [CrossRef] [PubMed]
54. Barcaccia, G.; Albertini, E. Apomixis in plant reproduction: A novel perspective on an old dilemma. *Plant Reprod.* **2013**, *26*, 159–179. [CrossRef] [PubMed]
55. Pupilli, F.; Barcaccia, G. Cloning plants by seeds: Inheritance models and candidate genes to increase fundamental knowledge for engineering apomixis in sexual crops. *J. Biotechnol.* **2012**, *159*, 291–311. [CrossRef]
56. Polegri, L.; Calderini, O.; Arcioni, S.; Pupilli, F. Specific expression of apomixis-linked alleles revealed by comparative transcriptomic analysis of sexual and apomictic *Paspalum simplex* Morong flowers. *J. Exp. Bot.* **2010**, *61*, 1869–1883. [CrossRef]

57. Bräuning, S.; Catanach, A.; Lord, J.M.; Bicknell, R.; Macknight, R.C. Comparative transcriptome analysis of the wild-type model apomict *Hieracium praealtum* and its *loss of parthenogenesis* (*lop*) mutant. *BMC Plant Biol.* **2018**, *18*, 206. [CrossRef]
58. Osborn, T.C.; Chris Pires, J.; Birchler, J.A.; Auger, D.L.; Jeffery Chen, Z.; Lee, H.-S.; Comai, L.; Madlung, A.; Doerge, R.W.; Colot, V.; et al. Understanding mechanisms of novel gene expression in polyploids. *Trends Genet.* **2003**, *19*, 141–147. [CrossRef]
59. Mau, M.; Lovell, J.T.; Corral, J.M.; Kiefer, C.; Koch, M.A.; Aliyu, O.M.; Sharbel, T.F. Hybrid apomicts trapped in the ecological niches of their sexual ancestors. *Proc. Natl. Acad. Sci. USA* **2015**, *112*, E2357–E2365. [CrossRef]
60. Grossniklaus, U. From sexuality to apomixis: Molecular and genetic approaches. In *The Flowering of Apomixis: From Mechanisms to Genetic Engineering*; Savidan, Y., Carman, J.G., Dresselhaus, T., Eds.; CIMMYT: El Batán, Mexico, 2001; pp. 168–211, ISBN 970-648-074-9.
61. McClintock, B. The significance of responses of the genome to challenge. *Science* **1984**, *226*, 792–801. [CrossRef]
62. Jackson, S.; Chen, Z.J. Genomic and expression plasticity of polyploidy. *Curr. Opin. Plant Biol.* **2010**, *13*, 153–159. [CrossRef]
63. Grimanelli, D.; Leblanc, O.; Perotti, E.; Grossniklaus, U. Developmental genetics of gametophytic apomixis. *Trends Genet.* **2001**, *17*, 597–604. [CrossRef]
64. Delgado, L.; Galdeano, F.; Sartor, M.E.; Quarin, C.L.; Espinoza, F.; Ortiz, J.P.A. Analysis of variation for apomictic reproduction in diploid *Paspalum rufum*. *Ann. Bot.* **2014**, *113*, 1211–1218. [CrossRef] [PubMed]
65. Schinkel, C.C.F.; Kirchheimer, B.; Dellinger, A.S.; Klatt, S.; Winkler, M.; Dullinger, S.; Hörandl, E. Correlations of polyploidy and apomixis with elevation and associated environmental gradients in an alpine plant. *AoB Plants* **2016**, *8*, plw064. [CrossRef] [PubMed]
66. Hojsgaard, D. Transient activation of apomixis in sexual neotriploids may retain genomically altered states and enhance polyploid establishment. *Front. Plant Sci.* **2018**, *9*, 230. [CrossRef] [PubMed]
67. Hojsgaard, D.; Hörandl, E. The rise of apomixis in natural plant populations. *Front. Plant Sci.* **2019**, *10*, 358. [CrossRef] [PubMed]
68. Sharbel, T.F.; Voigt, M.L.; Corral, J.M.; Thiel, T.; Varshney, A.; Kumlehn, J.; Vogel, H.; Rotter, B. Molecular signatures of apomictic and sexual ovules in the *Boechera holboellii* complex. *Plant J.* **2009**, *58*, 870–882. [CrossRef]
69. Galla, G.; Zenoni, S.; Avesani, L.; Altschmied, L.; Rizzo, P.; Sharbel, T.F.; Barcaccia, G. Pistil transcriptome analysis to disclose genes and gene products related to aposporous apomixis in *Hypericum perforatum* L. *Front. Plant Sci.* **2017**, *8*, 79. [CrossRef]
70. Flagel, L.E.; Wendel, J.F. Evolutionary rate variation, genomic dominance and duplicate gene expression evolution during allotetraploid cotton speciation. *New Phytol.* **2010**, *186*, 184–193. [CrossRef]
71. Swanson-Wagner, R.A.; Jia, Y.; DeCook, R.; Borsuk, L.A.; Nettleton, D.; Schnable, P.S. All possible modes of gene action are observed in a global comparison of gene expression in a maize F1 hybrid and its inbred parents. *Proc. Natl. Acad. Sci. USA* **2006**, *103*, 6805–6810. [CrossRef]
72. He, G.; Zhu, X.; Elling, A.A.; Chen, L.; Wang, X.; Guo, L.; Liang, M.; He, H.; Zhang, H.; Chen, F.; et al. Global epigenetic and transcriptional trends among two rice subspecies and their reciprocal hybrids. *Plant Cell* **2010**, *22*, 17–33. [CrossRef]
73. Hegarty, M.J.; Barker, G.L.; Wilson, I.D.; Abbott, R.J.; Edwards, K.J.; Hiscock, S.J. Transcriptome shock after interspecific hybridization in *Senecio* is ameliorated by genome duplication. *Curr. Biol.* **2006**, *16*, 1652–1659. [CrossRef] [PubMed]
74. Bell, G.D.M.; Kane, N.C.; Rieseberg, L.H.; Adams, K.L. RNA-Seq analysis of allele-specific expression, hybrid effects, and regulatory divergence in hybrids compared with their parents from natural populations. *Genome Biol. Evol.* **2013**, *5*, 1309–1323. [CrossRef] [PubMed]
75. Hojsgaard, D.H.; Martínez, E.J.; Quarin, C.L. Competition between meiotic and apomictic pathways during ovule and seed development results in clonality. *New Phytol.* **2013**, *197*, 336–347. [CrossRef] [PubMed]
76. Hand, M.L.; Vít, P.; Krahulcová, A.; Johnson, S.D.; Oelkers, K.; Siddons, H.; Chrtek, J.; Fehrer, J.; Koltunow, A.M.G. Evolution of apomixis loci in *Pilosella* and *Hieracium* (Asteraceae) inferred from the conservation of apomixis-linked markers in natural and experimental populations. *Heredity (Edinb.)* **2015**, *114*, 17–26. [CrossRef]

77. Washburn, J.D.; Birchler, J.A. Polyploids as a "model system" for the study of heterosis. *Plant Reprod.* **2014**, *27*, 1–5. [CrossRef]
78. Comai, L. The advantages and disadvantages of being polyploid. *Nat. Rev. Genet.* **2005**, *6*, 836–846. [CrossRef]
79. Fort, A.; Ryder, P.; Mckeown, P.C.; Wijnen, C.; Aarts, M.G.; Sulpice, R.; Spillane, C. Disaggregating polyploidy, parental genome dosage and hybridity contributions to heterosis in *Arabidopsis thaliana*. *New Phytol.* **2016**, *209*, 590–599. [CrossRef]
80. Hojsgaard, D.; Schegg, E.; Valls, J.F.M.; Martínez, E.J.; Quarin, C.L. Sexuality, apomixis, ploidy levels, and genomic relationships among four *Paspalum* species of the subgenus *Anachyris* (Poaceae). *Flora Morphol. Distrib. Funct. Ecol. Plants* **2008**, *203*, 535–547. [CrossRef]
81. Ortiz, J.P.A.; Quarin, C.L.; Pessino, S.C.; Acuña, C.; Martínez, E.J.; Espinoza, F.; Hojsgaard, D.H.; Sartor, M.E.; Cáceres, M.E.; Pupilli, F. Harnessing apomictic reproduction in grasses: What we have learned from *Paspalum*. *Ann. Bot.* **2013**, *112*, 767–787. [CrossRef]
82. Aliyu, O.M.; Schranz, M.E.; Sharbel, T.F. Quantitative variation for apomictic reproduction in the genus *Boechera* (Brassicaceae). *Am. J. Bot.* **2010**, *97*, 1719–1731. [CrossRef]
83. Chapman, H.; Houliston, G.J.; Robson, B.; Iline, I. A case of reversal: The evolution and maintenance of sexuals from parthenogenetic clones in *Hieracium pilosella*. *Int. J. Plant Sci.* **2003**, *164*, 719–728. [CrossRef]
84. Hörandl, E.; Hojsgaard, D. The evolution of apomixis in angiosperms: A reappraisal. *Plant Biosyst.* **2012**, *146*, 681–693. [CrossRef]
85. Hojsgaard, D.; Hörandl, E. A little bit of sex matters for genome evolution in asexual plants. *Front. Plant Sci.* **2015**, *6*, 82. [CrossRef] [PubMed]
86. Emadzade, K.; Hörandl, E. Northern Hemisphere origin, transoceanic dispersal, and diversification of Ranunculeae DC. (Ranunculaceae) in the Cenozoic. *J. Biogeogr.* **2011**, *38*, 517–530. [CrossRef]
87. Rieseberg, L.H.; Widmer, A.; Arntz, A.M.; Burke, J.M.; Carr, D.E.; Abbott, R.J.; Meagher, T.R. The genetic architecture necessary for transgressive segregation is common in both natural and domesticated populations. *Philos. Trans. R. Soc. B Biol. Sci.* **2003**, *358*, 1141–1147. [CrossRef]
88. Hegarty, M.J.; Hiscock, S.J. Genomic clues to the evolutionary success of polyploid plants. *Curr. Biol.* **2008**, *18*, 435–444. [CrossRef]
89. Mancini, M.; Permingeat, H.; Colono, C.; Siena, L.; Pupilli, F.; Azzaro, C.; de Alencar Dusi, D.M.; de Campos Carneiro, V.T.; Podio, M.; Seijo, J.G.; et al. The MAP3K-coding *QUI-GON JINN* (*QGJ*) gene is essential to the formation of unreduced embryo sacs in *Paspalum*. *Front. Plant Sci.* **2018**, *9*, 1547. [CrossRef]
90. Pagnussat, G.C.; Yu, H.J.; Ngo, Q.A.; Rajani, S.; Mayalagu, S.; Johnson, C.S.; Capron, A.; Xie, L.F.; Ye, D.; Sundaresan, V. Genetic and molecular identification of genes required for female gametophyte development and function in *Arabidopsis*. *Development* **2005**, *132*, 603–614. [CrossRef]
91. Ray, S.M.; Park, S.S.; Ray, A. Pollen tube guidance by the female gametophyte. *Development* **1997**, *124*, 2489–2498. [PubMed]
92. Olmedo-Monfil, V.; Durán-Figueroa, N.; Arteaga-Vázquez, M.; Demesa-Arévalo, E.; Autran, D.; Grimanelli, D.; Slotkin, R.K.; Martienssen, R.A.; Vielle-Calzada, J.-P. Control of female gamete formation by a small RNA pathway in *Arabidopsis*. *Nature* **2010**, *464*, 628–632. [CrossRef]
93. Singh, M.; Goel, S.; Meeley, R.B.; Dantec, C.; Parrinello, H.; Michaud, C.; Leblanc, O.; Grimanelli, D. Production of viable gametes without meiosis in maize deficient for an ARGONAUTE protein. *Plant Cell* **2011**, *23*, 443–458. [CrossRef] [PubMed]
94. Schinkel, C.C.F.; Syngelaki, E.; Kirchheimer, B.; Dullinger, S.; Klatt, S.; Hörandl, E. Epigenetic patterns and geographical parthenogenesis in the alpine plant species *Ranunculus kuepferi* (Ranunculaceae). *Int. J. Mol. Sci.* **2020**, *21*, 3318. [CrossRef] [PubMed]
95. Corral, J.M.; Vogel, H.; Aliyu, O.M.; Hensel, G.; Thiel, T.; Kumlehn, J.; Sharbel, T.F. A conserved apomixis-specific polymorphism is correlated with exclusive exonuclease expression in premeiotic ovules of apomictic *Boechera* species. *Plant Physiol.* **2013**, *163*, 1660–1672. [CrossRef] [PubMed]
96. Mau, M.; Corral, J.M.; Vogel, H.; Melzer, M.; Fuchs, J.; Kuhlmann, M.; de Storme, N.; Geelen, D.; Sharbel, T.F. The conserved chimeric transcript *UPGRADE2* is associated with unreduced pollen formation and is exclusively found in apomictic *Boechera* species. *Plant Physiol.* **2013**, *163*, 1640–1659. [CrossRef] [PubMed]

97. Schallau, A.; Arzenton, F.; Johnston, A.J.; Hähnel, U.; Koszegi, D.; Blattner, F.R.; Altschmied, L.; Haberer, G.; Barcaccia, G.; Bäumlein, H. Identification and genetic analysis of the *APOSPORY* locus in *Hypericum perforatum* L. *Plant J.* **2010**, *62*, 773–784. [CrossRef] [PubMed]
98. Conner, J.A.; Mookkan, M.; Huo, H.; Chae, K.; Ozias-Akins, P. A parthenogenesis gene of apomict origin elicits embryo formation from unfertilized eggs in a sexual plant. *Proc. Natl. Acad. Sci. USA* **2015**, *112*, 11205–11210. [CrossRef] [PubMed]
99. Albertini, E.; Marconi, G.; Reale, L.; Barcaccia, G.; Porceddu, A.; Ferranti, F.; Falcinelli, M. *SERK* and *APOSTART*. Candidate genes for apomixis in *Poa pratensis*. *Plant Physiol.* **2005**, *138*, 2185–2199. [CrossRef]
100. Bocchini, M.; Galla, G.; Pupilli, F.; Bellucci, M.; Barcaccia, G.; Ortiz, J.P.A.; Pessino, S.C.; Albertini, E. The vesicle trafficking regulator *PN-SCD1* is demethylated and overexpressed in florets of apomictic *Paspalum notatum* genotypes. *Sci. Rep.* **2018**, *8*, 3030. [CrossRef]
101. Rodrigues, J.C.M.; Luo, M.; Berger, F.; Koltunow, A.M.G. Polycomb group gene function in sexual and asexual seed development in angiosperms. *Sex. Plant Reprod.* **2010**, *23*, 123–133. [CrossRef]
102. Hegarty, M.J.; Barker, G.L.; Brennan, A.C.; Edwards, K.J.; Abbott, R.J.; Hiscock, S.J. Changes to gene expression associated with hybrid speciation in plants: Further insights from transcriptomic studies in *Senecio*. *Philos. Trans. R. Soc. B Biol. Sci.* **2008**, *363*, 3055–3069. [CrossRef]
103. Schmidt, A.; Schmid, M.W.; Klostermeier, U.C.; Qi, W.; Guthörl, D.; Sailer, C.; Waller, M.; Rosenstiel, P.; Grossniklaus, U. Apomictic and sexual germline development differ with respect to cell cycle, transcriptional, hormonal and epigenetic regulation. *PLoS Genet.* **2014**, *10*, e1004476. [CrossRef] [PubMed]
104. Carman, J.G.; Mateo de Arias, M.; Gao, L.; Zhao, X.; Kowallis, B.M.; Sherwood, D.A.; Srivastava, M.K.; Dwivedi, K.K.; Price, B.J.; Watts, L.; et al. Apospory and diplospory in diploid *Boechera* (Brassicaceae) may facilitate speciation by recombination-driven apomixis-to-sex reversals. *Front. Plant Sci.* **2019**, *10*, 724. [CrossRef] [PubMed]

genes

MDPI

Article

The Role of APOSTART in Switching between Sexuality and Apomixis in *Poa pratensis*

Gianpiero Marconi [1], Domenico Aiello [1], Bryan Kindiger [2], Loriano Storchi [3,4], Alessandro Marrone [3], Lara Reale [1], Niccolò Terzaroli [1] and Emidio Albertini [1,*]

1 Dipartimento di Scienze Agrarie, Alimentari e Ambientali, Università degli Studi di Perugia, Borgo XX Giugno 74, 06121 Perugia, Italy; gianpiero.marconi@unipg.it (G.M.); aiellodomenico@yahoo.it (D.A.); lara.reale@unipg.it (L.R.); niccolo.terzaroli@studenti.unipg.it (N.T.)

2 USDA-ARS, Grazinglands Research Laboratory, 7207 West Cheyenne St., El Reno, OK 73036, USA; bryan.kindiger@ars.usda.gov

3 Dipartimento di Farmacia, Università G. d'Annunzio, via dei Vestini 31, 66100 Chieti, Italy; loriano@storchi.org (L.S.); alessandro.marrone@unich.it (A.M.)

4 Molecular Discovery Limited, Elstree WD6 3FG, UK

* Correspondence: emidio.albertini@unipg.it

Received: 25 June 2020; Accepted: 11 August 2020; Published: 14 August 2020

Abstract: The production of seeds without sex is considered the holy grail of plant biology. The transfer of apomixis to various crop species has the potential to transform plant breeding, since it will allow new varieties to retain valuable traits thorough asexual reproduction. Therefore, a greater molecular understanding of apomixis is fundamental. In a previous work we identified a gene, namely APOSTART, that seemed to be involved in this asexual mode of reproduction, which is very common in *Poa pratensis* L., and here we present a detailed work aimed at clarifying its role in apomixis. In Situ hybridization showed that PpAPOSTART is expressed in reproductive tissues from pre-meiosis to embryo development. Interestingly, it is expressed early in few nucellar cells of apomictic individuals possibly switching from a somatic to a reproductive cell as in aposporic apomixis. Moreover, out of 13 APOSTART members, we identified one, APOSTART_6, as specifically expressed in flower tissue. APOSTART_6 also exhibited delayed expression in apomictic genotypes when compared with sexual types. Most importantly, the SCAR (Sequence Characterized Amplified Region) derived from the APOSTART_6 sequence completely co-segregated with apomixis.

Keywords: apomixis; APOSTART; plant reproduction; *Poa pratensis*

1. Introduction

Angiosperms reproduce primarily through sexual reproduction, however in plants, this requires a series of developmental steps that culminate in the formation of the seed. Meiosis and fertilization as the main signs of sexuality, secure the formation of genetically variable diploid progeny [1]. However, this is not the only way to produce seed. In an asexual mode of reproduction, namely various forms of apomixis [2], successful apomictic events typically silence sexual development within the same ovule, allowing for the generation of unreduced gametes [2]. Apomixis generally causes germline cells to completely avoid meiosis or, by modification of meiosis, the full restoration of the sporophyte chromosome number.

Apomixis is a challenging trait that offers unique opportunities for developing superior, true breeding cultivars irrespective of their level of heterozygosity [3–5]. The conversion of a sexual reproducing genotype to one that reproduces by apomixis would make it possible to fix the genotype of a superior plant variety selected for a particular environment or market niche, whereby the clonally reproduced seeds could be continuously and inexpensively produced, independent of pollination or pollinator [6].

The main advantages in having an apomictic genotype are that it reduces cost and breeding time for a new cultivar, avoiding various complications associated with sexual reproduction, such as incompatibility barriers and viral transfer in plants typically propagated vegetatively [7]. Moreover, farmers within the developing world could exploit the advantages of maintaining the genetic integrity of local, high-yielding varieties from generation to generation, removing the tendency of open-pollinated local varieties to lose their genetic advantages through sexual reproduction and outcrossing over time. Utilizing local varieties that are adapted to local mechanized agricultural systems could also lower production costs.

Today, the general consensus regarding apomixis and sexual reproduction indicates these as evolutionarily related processes, sharing many regulatory components. Once apomictic genes initiate embryo development and the initial cell forms and divides, the genes controlling embryo cell formation and patterning are most likely the same as those required for sexual embryo development. These similarities recently have been debated [2,8,9], and a new hypothesis has been proposed that considers apomicts anciently polyphenic with sex. This polyphenic viewpoint suggests that apomixis fails to occur in obligate sexual eukaryotes because genetic or epigenetic modifications have silenced the primitive sexual-apomictic switch and/or disrupted molecular capacities for apomixis [2]. According to the polyphyletic hypothesis, apomixis occurs in plants due to specific mutations [10,11]. If apomixis were well understood and harnessed, it could be utilized to indefinitely propagate superior hybrids or specific genotypes bearing complex gene sets. Until the gene(s) that promote and control apomixis are understood at the molecular level, this trait can only be introgressed into agricultural crops through traditional breeding methods, most of which are slow and laborious. Progeny testing for the selection of apomictic genotypes following each round of backcrossing also slows the introgression of apomixis into a cultivar.

Apomixis occurs in thousands of species across all kingdoms of the Eukaria [9], and this includes several important forage grasses, such as *Poa*, *Brachiaria*, *Cenchrus/Pennisetum*, *Panicum*, and *Paspalum*. Among these, we elected to investigate *Poa pratensis* L. Kentucky bluegrass (*P. pratensis*) is a highly polyploid, hardy, persistent, and attractive forage and turf grass adapted to a wide range of soils and climates. Its mode of reproduction is extremely versatile ranging from sexuality to natural obligate apomixis (apospory and parthenogenesis). In *P. pratensis*, apospory involves the development of embryo sacs from somatic cells that differentiate into the nucellus. If unreduced polar nuclei positioned centrally within the embryo sac fuse with a sperm cell released from the pollen tube (pseudogamy), the unreduced egg can develop autonomously through parthenogenesis to form viable apomictic seeds [12]. Moreover, overall results indicate that in this species, apospory and parthenogenesis are controlled by two distinct genetic factors that are genetically uncoupled [12]. Therefore, the possibility to find or obtain recombinant genotypes (aposporic but non-parthenogenetic or vice-versa) could be very useful for investigating the impact of different candidate genes. In a previous work we [13] applied a cDNA-AFLP transcriptional profiling assay to developmental staged inflorescences of *P. pratensis* and isolated 2.248 ESTs, most of which (60%) were specific of floral organs and/or were involved in megasporogenesis and seed development. In particular, one expressed sequence tag (EST) clone showed a high similarity with an EST isolated from a pistil-specific cDNA library of apomictic *Pennisetum ciliare* and with the human *StAR* gene that can cause pseudohermaphroditism. Due to its START domain and putative involvement in apomixis, we named this gene *APOSTART* [14]. In this previous work we described the cloning of two APOSTART members (*APOSTART_1* and *APOSTART_2*). Here, we report the isolation of additional 13 cDNA and six genomic clones for which we determined genomic organization, temporal, and spatial expression and, computationally, infer on the binding of phytosterols at the START domain of their coded proteins. Moreover, we used the sequence of our best candidate (*APOSTART_6*) to search for similar proteins and found 207 proteins belonging to 158 species. Their relationships based on distances in the protein amino acid sequences composition were used to build a phylogenetic tree. The putative involvement of APOSTART in apomixis is also reported and discussed.

2. Materials and Methods

2.1. Plant Material

Cloning of cDNA and genomic full-length clones: a completely sexual clone (S1/1-7) and two highly apomictic genotypes (RS7-3 and L4) of *Poa pratensis* were used to amplify, clone, and sequence the full-lengths of APOSTART members (Table S1).

SCAR marker test: 48 genotypes from a *P. pratensis* segregating F_1 population (PG-F1) derived from a cross between S1/1-7 and RS7-3 [15] and their parents; forty-three F_1 interspecific hybrids from a cross between a sexual *P. arachnifera* (Pa1FM) and a highly apomictic *P. pratensis* genotype, evaluated for their mode of reproduction by progeny tests (Table S1); thirteen genotypes of *P. pratensis* from USA with known reproductive behavior, and one commercially available *P. arachnifera* × *P. pratensis* hybrid (Reveille) (Table S1). These Poa materials were selected for the APO-SCAR marker to trait segregation analysis because the *P. pratensis* germplasm was characterized as a monoecious, obligate apomictic, where no sexuality is expressed. *P. arachnifera*, representing an indigenous species to the Central USA, is dioecious and has a purely sexual form of reproduction. Individuals representing the *P. arachnifera* × *P. pratensis* F_1 hybrid population utilized in the APO-SCAR analysis had been previously characterized for their expression of either a sexual or apomictic form of reproduction. The F_1 population provided a segregating sexual/apomictic sample with which to evaluate the utility and validity of the APOSTART_6 marker.

qPCR (quantitative PCR): S1/1-7 (sexual), RS7-3 and L4 (apomictic), together with three progeny plants from the S1/1-7 × RS7 cross (2 apomictic and 1 sexual) and four progeny plants from the S1/1-7 × L4 cross (2 sexual, one apomictic, and one parthenogenetic, but not aposporic) of *P. pratensis* (Table S1). Reproduction mode and chromosome numbers of these plants were described in Raggi et al. [16].

2.2. Rapid Amplification of cDNA Ends Analysis

Clone-specific primers (Table S2) were used for performing both 5′-and 3′-rapid amplification of cDNA ends (RACE) to obtain the full-length genes. The SMART RACE cDNA amplification kit (BD Biosciences) was applied to the mRNA poly(A)+ of the stage where the cDNA was scored according to the manufacturer's instructions. Twelve colonies for each RACE experiment were sequenced and full-length cDNA sequences were reconstructed from RACE fragments using the VECTOR NTI® Suite 8 Contig Express (Invitrogen TM, Carlsbad, CA, USA). Sequences were aligned using VECTOR NTI® Suite 8 AlignX software (Invitrogen TM, Carlsbad, CA, USA). Based on this alignment, specific primers were designed to perform end-to-end PCR on cDNA and genomic DNAs (Table S2).

2.3. Cloning of Full-Length Genomic and cDNAs

Specific primers, as reported above, were used for performing end-to-end amplifications of cDNAs and genomic DNAs to obtain the entire transcriptional units. A 0.7 μL aliquot of PCR-derived products was sticky-end ligated into a pCR4-TOPO vector using the TOPO TA cloning kit for sequencing (Invitrogen TM, Carlsbad, CA, USA). The plasmid DNA was purified from 5 mL of an overnight culture on LB medium of *E. coli* using the GenElute plasmid miniprep kit (Sigma-Aldrich, St. Louis, MO, USA). After a first confirmation sequence was performed using M13 forward and reverse primers, a primer walking approach was adopted to sequence the complete clones. Alignment between the full-length cDNAs and genomic clones (Table S3) disclosed the intron/exon structures of the genes (Figure S1).

2.4. RNA Isolation and cDNA Synthesis

Florets, harvested at five developmental stages (pre-meiosis, meiosis, post-meiosis, anthesis, and post-anthesis) according to cytohistological investigations [12,14], leaves, and roots of five apomictic, four sexual, and one parthenogenetic genotype of *P. pratensis*, were collected. Nucleic acids were isolated from about 0.1 g of fresh tissue using the GenElute total RNA purification kit (Sigma-Aldrich, St. Louis, MO, USA), according to the manufacturer's instructions, with some modifications to adapt it to plants. Total RNA was purified from residual genomic DNA by using the DNA-free (Ambion, Austin, TX, USA). Reverse transcription and second-strand synthesis were carried out with 1 μg of total RNA, and the standard procedure was followed [17].

2.5. qPCR

All qPCR analyses were performed using an Mx3000P qPCR (Stratagene, La Jolla, CA, USA) system with the SYBR green JumpStart Taq ReadyMix for quantitative PCR (Sigma-Aldrich, St. Louis, MO, USA). Specific primers were designed within the sequences of each allele (Table S2). The PCR fragments were analysed using a dissociation protocol to ensure that each amplicon was a single product. Amplicons were also sequenced to verify the specificities of the targets. The amplification efficiency was calculated from raw data using the LingRegPCR software. All RT-qPCRs were performed using three biological replicates in a final volume of 25 μL containing 5 μL of cDNA template (previously diluted 1:10), 0.2 μM of each primer, and 12.5 μL of 2 × SYBR Green PCR Master Mix (Sigma-Aldrich, St. Louis, MO, USA), according to the manufacturer's instructions. The following thermal cycling profile was used: 95 °C for 10 min, followed by 50 cycles of 95 °C for 10 s, 57 °C for 15 s, and 72 °C for 15 s. Following cycling, the melting curve was determined in the range 57–95 °C, with a temperature increment of 0.01 °C/sec. Each reaction was run in triplicate (technical replicates). Negative controls included in each run were a reaction conducted in the absence of reverse transcriptase and a reaction with no template (2 μL of nuclease-free water instead of 2 μL of cDNA). For negative controls, no signals were observed (data not shown). Raw C_t data from the MX3000P instrument were exported to a data file and analysed using GeneEx Pro software (bioMCC, Freising, Germany). During the pre-processing phase, data were corrected for PCR efficiencies and the three technical repeats were averaged. The selected reference genes, *P. pratensis* β-tubulin [13] subsequently used to normalize C_t values [18,19], and quantities were calculated relative to the maximum C_t value. Because our interest was in fold changes in gene expression between groups, we ultimately converted quantities to a logarithmic scale using a log base 2 conversion, which also allowed us to test the normal distribution of values.

2.6. SCAR Marker Development and Testing

Genomic sequences of APOSTART alleles (Table S3) were aligned using the VECTOR NTI® Suite 8 AlignX software (Invitrogen TM, Carlsbad, CA, USA) to disclose differences in nucleotide composition. Several primer pairs were designed on the basis of genomic differences and tested on the F_1 population segregating for the mode of reproduction (Table S1). Amplifications of genomic DNAs with primer pairs were done in a final volume of 25 μL containing 1× PCR buffer (Invitrogen TM, Carlsbad, CA, USA), 5 mM dNTPs, 20 pmol of each primer, 25 ng of genomic DNA, and 1 U recombinant Taq DNA polymerase (Invitrogen). PCR was carried out in an initial denaturation step of 94 °C for 5 min, followed by 30 cycles of 94 °C for 30 s, 60 °C for 30 s, 72 °C for 1 min, and a final extension step of 72 °C for 10 min. Amplification products were separated by electrophoresis in 2% agarose gels. The primer pair that co-segregate with apomixis (hereafter named APO-SCAR, Table S2) was then tested also on other genotypes. The APO-SCAR PCR was carried out in an initial denaturation step of 94 °C for 5 min, followed by 40 cycles of 94 °C for 30 s, 59 °C for 30 s, 72 °C for 30 s, and a final extension step of 72 °C for 10 min. In-PCR positive controls were obtained by amplifying a chloroplast sequence using a chloroplast-specific pair of primers (Table S2).

2.7. In Situ Hybridization

During inflorescence development, in sexual and apomictic *P. pratensis* genotypes, we distinguished four different stages (pre-meiosis, meiosis, post-meiosis, anthesis, and post-anthesis). At each stage, single spikelets were collected, fixed in ethanol-formaldehyde-acetic acid, embedded in paraffin, and used for ISH experiments. Tissue preparation and hybridization conditions were the same as described by Angenent et al. [20].

Sense and antisense probes were obtained by in vitro transcription using cloned PCR-derived fragments APOSTART as templates. In particular, APOSTART riboprobes were obtained from a single cDNA fragment of 1060 bp, which comprised only a small part of the START and DUF1336 domains and the sequence between the two domains. APOSTART riboprobes did not discriminate between alleles due to the very small differences in sequences.

DIG-UTP sense and antisense riboprobes were synthesized by the T3 and T7 RNA polymerase. Transcripts were partially hydrolyzed by incubation at 60 °C in 0.2 M Na_2CO_3/$NaHCO_3$ buffer, pH 10.2, for about 35 min. Immunological detection was performed as described by Cañas et al. [21].

2.8. Sequence Data Analysis and Phylogenetic Trees

Using *APOSTART_6* cDNA and amino acid sequences as a query, similarities were searched in the National Center for Biotechnology Information (NCBI; www.ncbi.nlm.nih.gov) [22]. In both cases, multiple sequence alignment and phylogenetic tree were obtained using the function ClustalW with default parameters in the Molecular Evolutionary Genetics Analysis Version 7.0 (MEGA7). Protein sequences were first analyzed with MEGA function "Find Best DNA/Protein Models (ML)", which indicated Jones-Taylor-Thornton as the best method to construct the maximum-likelihood phylogenetic tree [23]. A discrete γ distribution was used to model evolutionary rate differences among sites (5 categories (+Gamma parameter)) with 1000 of Bootstrap as the resampling method [24].

2.9. Phosphorylation Site, Secondary Structure Predictions, and Subcellular Localization Prediction

Phosphorylation sites analyses of the 15 APOSTART amino acids were performed with NetPhos 3.1 Server [25]. Secondary structures, solvent accessibility, disordered regions, and trans-membrane helices were predicted using RaptorX servers (http://raptorx.uchicago.edu/) and PredictProtein servers (http://www.predictprotein.org) [26–29]. Subcellular localization prediction of APOSTART proteins was carried out using in-silico prediction tools, such as: CELLO2GO [30], DeepLoc [31], MultiLoc2 [32], Plant-mPloc [33], and SherLoc2 [34], Yloc [35].

2.10. Molecular Modeling

The protein sequences coded by APOSTART_1, APOSTART_6, and APOSTART_8 genes were subjected to the computational workflow depicted in Figure S1. Preliminary, reliable template structures for the initiation of the homology modeling investigations were searched for by using the HHpred web-server toolkits [36]. In particular, we carried out template searches on all available databases. The protein structures with higher scores and reported in the pdb archive were identified as plausible entries to be used as homology modeling templates. The graphical inspection of these structures was then performed by the use of the Maestro software [37] and allowed to identify the pdb coded 1LN1 entry as the most reliable for our purposes (vide infra). Homology modeling calculations were, then, carried out on the protein sequences of APOSTART_1, APOSTART_6, and APOSTART_8 with the 1LN1 protein structure as a template by using the Modeller software [38]. Modeller models 3D structures of proteins and their assemblies by the use of spatial restraints is most frequently used for homology or comparative protein structure modeling. Once provided with an alignment of a sequence to be modeled with a proper template, Modeller calculates a 3D model with all non-hydrogen atoms. The APOSTART_1-1LN1 and APOSTART_6-1LN1 sequence alignments used in the Modeller runs were reported in S2. The resulting 3D models were obtained by optimizing the molecular probability density function (pdf) subjected to only spatial restraints in Cartesian space and by employing the conjugate gradients and molecular dynamics with simulated annealing procedures implemented in the Modeller software.

The 3D structures of APOSTART_1, APOSTART_6, and APOSTART_8 provided at the homology modeling stage were refined by using the Protein Preparation Wizard implemented in Maestro [39]. This step allowed us to gain an all-atom and OPLS force field [40] with consistent protein structures to be processed in the further stages of the workflow. To gain a proper orientation of the binding site protein side chains, the structure of the co-crystallized ligand retrieved from the 1ln1 pdb entry was included in the model during the refinement steps. Optimized structures of APOSTART_1 and APOSTART_6 corresponding to local minima of the potential energy surface (PES) were calculated with the OPLS force field by using the GB/SA method [41] to simulate the water environment and the LBFGS minimization algorithm at a maximum iteration limit of 10,000 steps and a gradient-based convergence threshold of 0.05 kJ mol^{-1} $Å^{-1}$. A further refinement of the obtained protein structures of APOSTART_1 and APOSTART_6 was carried out by the use of the large-scale low-mode (LLMOD) algorithm implemented in MacroModel [42,43]. This procedure allowed sampling the PES by moving along the direction of the low-energy eigenvectors of the hessian matrix to search for lower energy structures, thus corresponding to a "local" conformational search. A maximum of 1000 LLMOD steps along the directions (eigenvectors) associated to the 30 lowest eigenvalues of the hessian matrix is used in the conformational sampling; each sampled structure is filtered according to the Monte Carlo criterium and eventually minimized in the OPLS force field by the use of loose convergence criteria (gradient-based convergence threshold of 0.5 kJ mol^{-1} $Å^{-1}$). Each LLMOD run provided 20 conformations that were further optimized with MacroModel, and the resulting lowest energy structure was eventually selected as the final 3D model of the considered protein (upon removing the 1ln1 co-crystallized ligand).

The 3D structures of stigmasterol, brassicasterol and campesterol were sketched in the Maestro workspace and the corresponding local minimum structures were calculated using the OPLS force field with the TNCG minimization algorithm and a convergence threshold of 0.05 kJ mol^{-1} $Å^{-1}$.

Subsequently, docking calculations were performed by the use of Glide software [44] in two stages. Initially, the grid generation module was employed to delimit the binding site region on APOSTART_1 and APOSTART_6 protein structure and to calculate the molecular interaction fields to be used for the following docking evaluation. The binding site was represented by a box of 28 Å per dimension and a smaller sub-box of 10 Å per dimension to delimit the space in which the center of mass of each ligand can be positioned. The docking SP procedure was then employed to dock flexibly the considered phytosterols in the binding site of APOSTART_1 and APOSTART_6 to obtain 10 poses per ligand.

2.11. Consensus Scoring and Molecular Dynamics

The bound complexes of APOSTART_1, APOSTART_6, and APOSTART_8 obtained from the docking calculations underwent a further refinement step consisting of a multiple minimization performed with MacroModel using the same settings as protein structure minimization (see Section 2.10). The 90 structures of bound complexes obtained after the post-docking refinement were processed to estimate the corresponding target-ligand binding energy by using three different methods. First, we performed "Rigid" docking calculations with Glide at the XP level of precision using the GlideScore fitness function [45]. Then, an estimation of the energy for the "snapping" of bound complexes obtained through the single-point energy calculation on complex, free protein, and free ligands with MacroModel were generated. Finally, we estimated the target-ligand affinity by using the FLAP (fingerprints for ligands and proteins) procedure [46] that provides a common reference framework for comparing molecules, using GRID molecular interaction fields (MIFs). The GRID MIFs (i.e., GRID molecular interaction fields) [47], originally developed for structure-based drug design [48], have been applied to a variety of drug discovery areas over the years, such as *pKa* [49] and tautomers modeling [50], scaffold-hopping [51], 3D-QSAR [52], and metabolism prediction [53]. Using the GRID MIFs one can easily obtain information related to non-covalent bonding between the selected probe and the target. The target may either be a small molecule or a protein. Probes, on the other hand, represent different chemical moieties (e.g., OH2 a water molecule, DRY the hydrophobic probe, N1 a neutral lat N-H, O a carbonyl oxygen, etc.) that are located in a 3D grid surrounding the target. At each point of the 3D grid the interaction energy is computed by determining and summing up the Lennard-Jones (ELJ), electrostatic (EEL), hydrogen-bonding (EHB), and entropic (ES) terms. We estimated the target-ligand affinity using the GP value returned by FLAP that is basically a product of the major GRID probes (H, N1, O, and DRY) similarity (i.e., FLAP similarity is a kind of Tanimoto similarity, weighted by energy).

The consensus scoring was obtained by averaging the estimations of Glide, MacroModel, and FLAP. For this purpose, each set of estimations was transformed to assign the same metrics, based on standard deviation units (i.e., they have been normalized), to the three-scoring set. The deviation of each value from the minimum score (penalty) was calculated with:

$$P_{ij} = \frac{S_{ij} - \min\{S_{1j}, S_{2j} \ldots S_{Nj}\}}{\sigma_j}$$

where S_{ij} is the i-th scoring value of j-th scoring, and σ_j is the standard deviation on the j-th set of N scoring values. The j-th normalized score is then obtained as the deviation of penalty from its maximum:

$$\overline{S}_{ij} = \max\{p_{1j}, p_{2j} \ldots p_{Nj}\} - p_{ij}$$

The consensus scoring for the i-th bound system is eventually expressed by:

$$S_i = \frac{\sum_{j-i}^{3} \overline{S}_{ij}}{3}$$

The bound complexes of APOSTART_1, APOSTART_6, and APOSTART_8 with stigmasterol, brassicasterol, and campesterol gaining the highest consensus scoring (two top-scored per bound complex) were further investigated by molecular dynamics (MD) simulations with the Gromacs package [54]. Each bond complex was placed in a cubic box whose dimensions prevent self-interaction and, then, solvated with up to 16,300 water molecules, depending on both ligand and binding mode, at the typical density of water at 298 K and 1.0 atm and by employing the single point charge (SPC) model [55]. A proper number of counterions (sodium and chloride ions) were added to ensure the electrical neutrality of the whole system and induce salt concentrations of 0.15 M. All the simulations were performed adopting the same computational scheme: (i) after an energy minimization, the whole system was slowly heated up to 300 K using short (100.0 ps) MD runs, (ii) the simulation was extended up to 25 ns for all simulated systems, at 300 K in an isothermal/isobaric ensemble, using the velocity rescaling scheme (temperature) and the isotropic Parinello-Rahman coupling scheme (pressure) [56,57], (iii) the investigated bound complexes were simulated in the GROMOS, a united-atom force field [58], the LINCS algorithm was adopted to constrain all bond lengths [59], and the long-range electrostatics were computed by the Particle Mesh Ewald method [60]. Trajectory analyses were carried out by using the available Gromacs utilities and by the support of VMD graphical interface [61].

The system configuration obtained at 500 ps of each production MD run (vide infra) was selected for a final consensus scoring of each ligand pose. For this purpose, only the water molecules within 5 Å from the center of mass of the bound ligand were included in the model, which then underwent a local minimization with MacroModel (see above). The minimized structures were thus processed using the same approach employed to estimate the ligand–protein affinities in the bound complexes obtained by docking, and, thus, providing a value of consensus scoring for each bound complex.

3. Results

3.1. Cloning of APOSTART Alleles/Members

Starting from the sequences reported in [13], forward and reverse primers were designed for both 5′ and 3′ rapid amplification of cDNA ends (RACE). Several RACE experiments were required to obtain the entire 5′-end of APOSTART. RACE identified 13 other members of APOSTART in addition to APOSTART1 and APOSTART2 already published in [13]. Following the isolation of the full lengths, we performed end-to-end PCRs with primers specific for each cDNA reconstructed sequence, and obtained complete cDNA and DNA clones of each allele (Table S3). Allele specificity was confirmed in replicated experiments by directly sequencing the amplified products. Therefore, we were able to obtain 13 full-length cDNAs (APOSTART_3 to APOSTART_15) and the relative full-length genomic sequences for six of them (APOSTART_5, APOSTART_6, APOSTART_7, APOSTART_8, APOSTART_10, APOSTART_12) as reported in Table S3.

The final length of the cDNA clones ranged from 2158 nt for APOSTART_14 to 2256 nt for APOSTART_15, while the total length of the eight genomic clones ranged from 5030 nt (APOSTART_6) to 5531 nt (APOSTART_8). The alignment between genomic and full-length cDNA clones confirmed what was already known for APOSTART_1 and APOSTART_2 [13], that all APOSTART members are structured into 21 introns and 22 exons (Figure S2).

Moreover, the average identity of the 15 cDNA clones was 96.87% ranging from 91.48% to 100% (Table S4). The most diverse clone was APOSTART_5 that shared a 93.47% mean identity with other clones (min 91.48% and max 96.18%). The average identity of the 8 genomic clones was 91.70% ranging from 85.33% to 99.58%. The two most diverse clones were APOSTART_6 and APOSTART_5 that shared a mean identity with other clones of 87.11% and 87.49%, respectively (Table S4).

3.2. APOSTART Members Are also or Exclusively Expressed in Flower Tissues

Expression of six APOSTART members was assayed in genotypes with different modes of reproduction by using allele specific primers in the qPCR analyses (Table S2). Allele specificity was verified by directly sequencing an aliquot of the amplified products of each experiment. Reactions were performed in triplicate on independently isolated and retrotranscribed mRNAs from three apomictic and three sexual genotypes. APOSTARTs expression levels were investigated in inflorescences collected during five developmental stages (pre-meiosis, meiosis, post-meiosis, anthesis, and post-anthesis) from sexual and apomictic genotypes as assessed by cytohistological investigations [12,14]. Moreover, two other tissues (leaves and roots) were used in the analysis.

Overall, three of the six members investigated (APOSTART_7, APOSTART_10, and APOSTART_12) were expressed in floral tissues as well as in roots and leaves. In all three cases the higher expression was recorded in flower tissues at anthesis, while in leaves and roots it was, on average, lower than in flowers (Figure 1). The expression pattern of the other two members (APOSTART_5 and APOSTART_8) was almost identical as it was expressed in flowers and roots but not in leaves (Figure 1). The most interesting member was APOSTART_6, not only because it showed to be flower specific as its expression was absent both in leaves and roots but also for the marked differences in its expression between apomictic and sexual genotypes (Figure 1).

While in sexual genotypes APOSTART_6 shows an increasing expression level from pre-meiosis to anthesis after which its expression drops to an undetectable level; in apomictic genotypes it is possible to note a shift in expression. In fact, the apomictic genotypes APOSTART_6 starts to be expressed at post-meiosis, and it remains expressed throughout meiosis until the post-anthesis stage.

Since our population included one recombinant genotype (non-aposporic and parthenogenetic), we performed a new qPCR analysis for APOSTART_6 including this genotype. As shown in Figure 2, the parthenogenetic genotypes exhibited the same behavior as the apomictic genotypes.

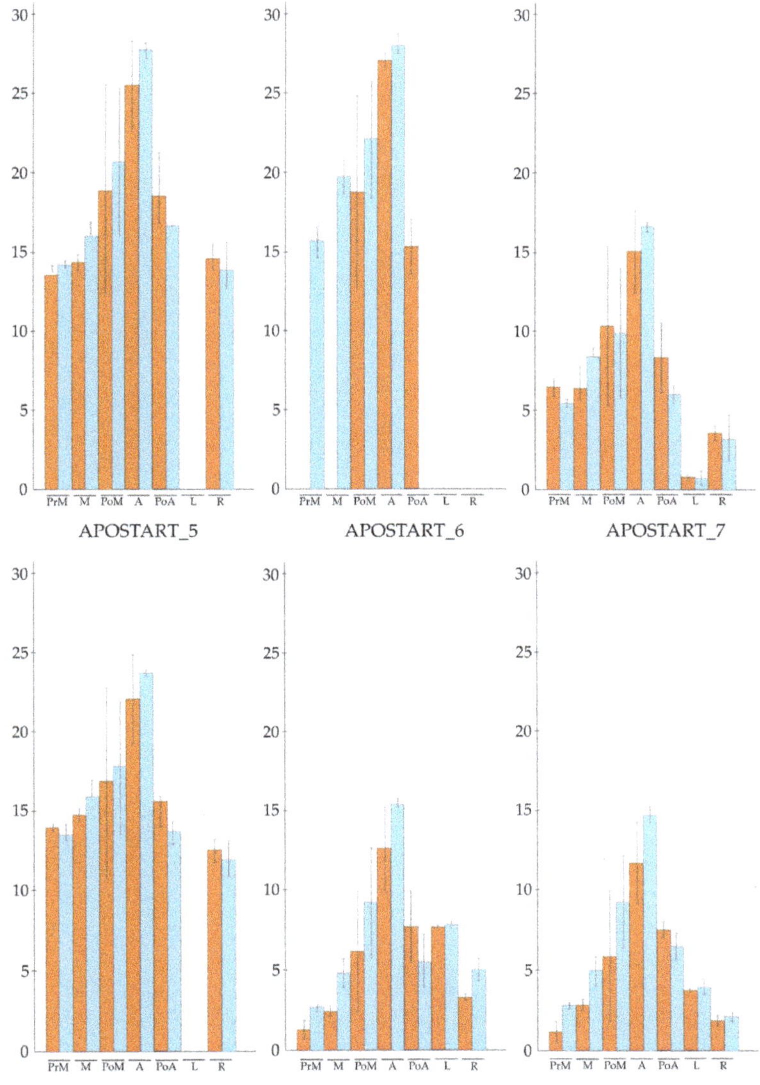

Figure 1. qPCR (quantitative PCR) profile of six APOSTARTs in two different sets of genotypes of *P. pratensis* (orange, apomictic genotypes; blue, sexual genotypes). The expression level was evaluated in five different flowering stages (pre-meiosis; meiosis; post-meiosis; anthesis; post-anthesis) and two tissues (leaves and roots).

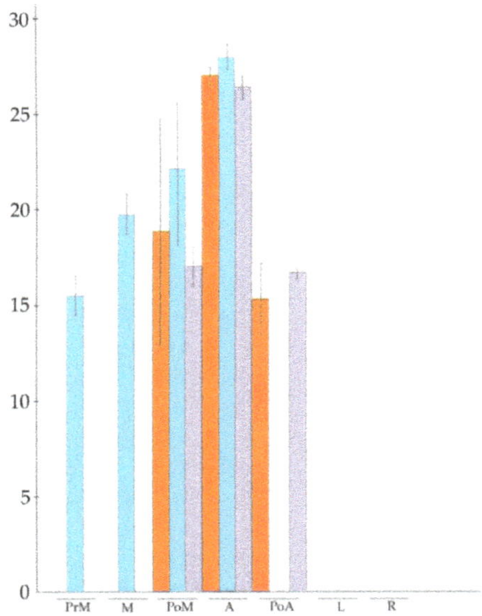

Figure 2. qPCR profile of APOSTART_6 member in three different sets of genotypes (orange, apomictic genotypes; blue, sexual genotypes; purple, parthenogenetic recombinant genotype).

3.3. *APO-SCAR Cosegregate with Apomixis*

Based on the results of the qPCRs, we were interested in determining if APOSTART_6 co-segregated with apomixis. To do so, we aligned all APOSTART genomic clones (Figure S2) and designed several APOSTART_6-specific primer pairs to test the segregation of the derived marker and its putative association with apomixis. Of all primer pairs evaluated, one, hereafter named APO-SCAR (Table S2, Figure S2 underlined in red), showed to be the best candidate based on preliminary testing. The APO-SCAR primer pair was then tested on an F_1 population of 68 individuals segregating for the mode of reproduction [12,15,62] and on its parental genotypes (Figure 3A). The APO-SCAR pair of primers produced a single amplification product of 225 bp present in the apomictic paternal genotype but absent in the maternal sexual plant. Figure 3A also shows the complete co-segregation of the APO-SCAR compared with reproductive behavior as assessed by progeny testing [12,15,62]. To further test the efficiency of APO-SCAR in detecting the mode of reproduction, APO-SCAR primers were tested on apomictic and sexual genotypes of *P. pratensis* or *Poa aracnifera* × *P. pratensis* provided by the USDA, Grazinglands Research Laboratory, El Reno, OK, USA all having adaptation to the contrasting environments of Southern and Eastern Europe and the Southern Plains Region of the United States. The reproductive behavior, and the provenance of the analyzed genotypes, are reported in Table S1. The amplification pattern generated by the APO-SCAR primer pair strongly resembled that observed in the mapping progeny and unequivocally distinguished the sexual from the highly apomictic individuals (Figure 3B).

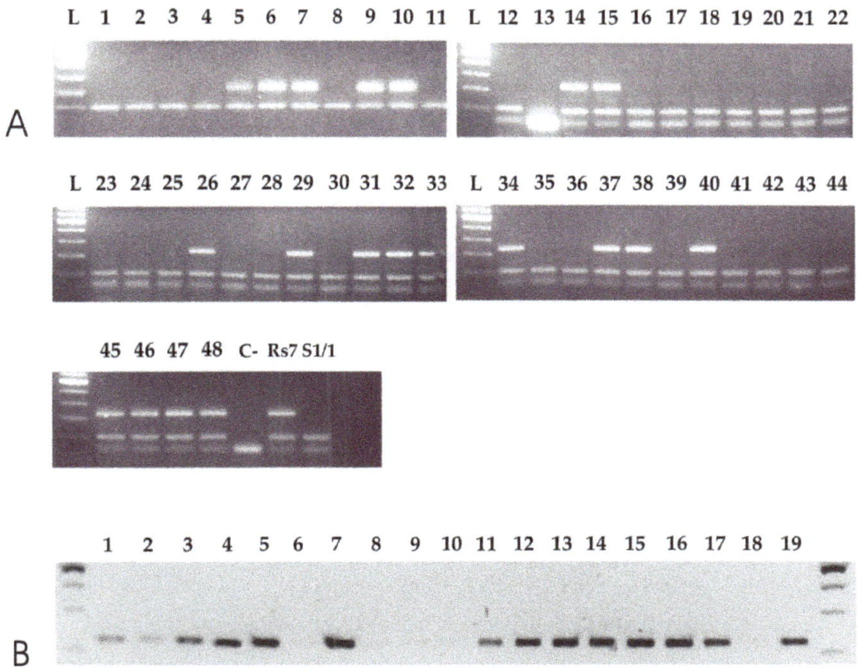

Figure 3. (**A**) APO-SCAR primer pair tested on 48 F1 individuals (lanes 1–48) from a segregating population for the mode of reproduction and obtained by crossing an apomictic (RS7, lane 50) and a sexual (S1/1, lane 51) genotype, as reported in Albertini et al. 2001 [12]. APO-SCAR completely co-segregate with apomixis; (**B**) on exotic germplasm sources of known reproductive behavior (Table S1): lanes 1–5 and 11, apomictic PaPp F_1 hybrids; lane 6, sexual *P. arachnifera* maternal parent (Pa1FM); lane 7, apomictic KB3 *P. pratensis* pollen parent; lanes 8–10, non-apomictic PaPp F_1 hybrids; lanes 12–13, apomictic 9601, 1915 *P. pratensis*.

3.4. Phylogenetic Cluster Analysis

Since APOSTART_6 completely co-segregated with apomixis, we searched in the BLAST database (https://blast.ncbi.nlm.nih.gov/Blast.cgi) [22] for protein sequences similar to APOSTART_6 and found 208 proteins belonging to 158 species. Protein sequences were aligned with MEGA 7 and resulted in an alignment of 1208 positions. By estimating the pairwise distances obtained using a JTT model of the aligned protein sequences, a maximum-likelihood phylogenetic tree was constructed. The unrooted condensed original tree (Figure 4, Figure S3) is drawn to scale, with branch lengths measured in the number of substitutions per site and a bootstrap value cut-off of 70.

From a first look at the tree, *Physcomitrella patens* and *Marchantia polymorpha*, two non-vascular plants belonging to Bryophyta and Marchantiophyta, respectively, clustered together in an outgroup clade. Genera with more than one sequence (*Oryza*, *Triticum*, *Gossypium*, *Coffea*, etc.) exhibited consistent clustering. *Poa* sequences clustered together as well. The first main cluster in fact includes monocots, in particular the Poaceae family (in red). Continuing counterclockwise, monocots are followed by a few independent taxa (*Amborella*, *Nymphaea*, *Cinnamomum*) belonging to the Magnoliopsida class, intermediate between mono- and dicotyledons. The last large clade contains all the *Eudicots* class plants, divided into very homogeneous groups corresponding to the orders and then to the botanical families. As a matter of fact, it is possible to Solanales, divided in Solanaceae (in light blue) and Convolvulaceae, Lamiales with several families, Asteraceae (in green), Malvaceae (in yellow), Malpighiales with

Salicaceae and Euphorbiaceae, Brassicales, divided in Brassicaceae to which *Arabidopsis* belongs (in purple), Cleomaceae and Caricaceae, Myrtales, Fagales, Cucurbitaceae, and Fabaceae (in dark blue).

Overall, only one sequence of *Dichanthelium oligosanthes* is not in the Poaceae cluster and Rosales (in black), with Rhamnaceae on one side and Rosaceae, Cannabaceae, and Moraceae on the other, clustered in independent clades, suggesting that APOSTART_6 phylogeny is strictly dependent on systematic botany.

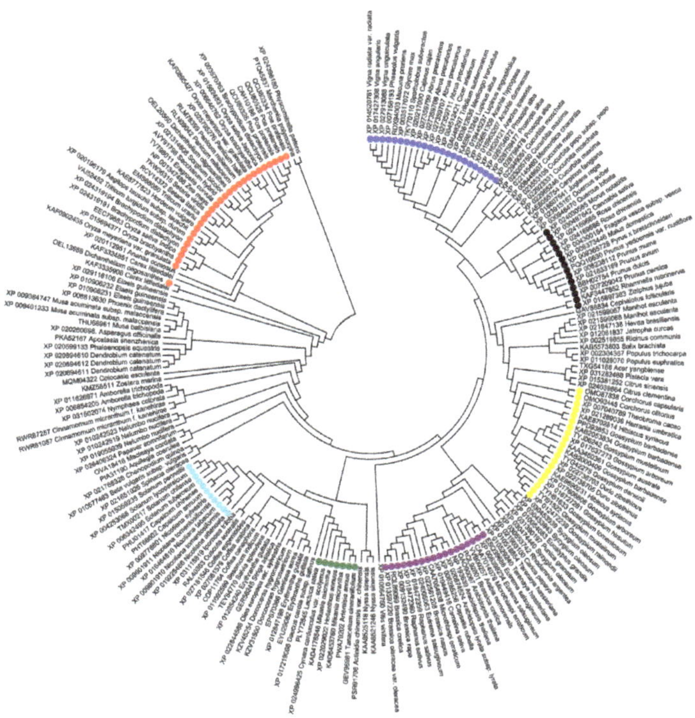

Figure 4. Maximum-likelihood phylogenetic unrooted tree based on APOSTART_6 similar amino acid sequences. The tree has the highest log likelihood (-45391.1656) and is condensed with bootstrap value >70. Phylogenetic analysis was performed using Neighbor-Join and BioNJ algorithms to a matrix of pairwise distances estimated using a JTT model and a discrete γ distribution (5 categories (+G, parameter = 0.6937)). The analysis involved 208 amino acid sequences and was conducted in MEGA7 [23]. Bootstrap values are reported in Figure S3 for more accuracy. Poaceae are indicated in red, Solanaceae in light blue, Asteraceae in green, Malvaceae in yellow, Brassicaceae in purple, Rosaceae in black, and Fagaceae in dark blue.

3.5. *In Situ Hybridization (ISH)*

APOSTART expression was investigated by in situ hybridization (ISH) in longitudinal sections of sexual and apomictic flowers. In a previous analysis [14], APOSTART transcripts were detected both in male and female meiosis in micro- and megaspores. APOSTART was expressed both during megasporogenesis and megagametogenesis, from megaspore mother cells (MMCs) to the mature embryo sac developmental stages. Moreover, it was also expressed during embryo sac development. The new analysis, herein reported, studied the expression of APOSTART considering a wider range in reproduction processes, from pre-meiosis to embryo stage. During pre-meiosis in both reproductive behaviors (apomictic and sexual) a poor hybridization signal was observed in all tissues of the ovule and in the MMCs (Figure 5A–C). During meiosis the hybridization signal detected in the MMC was the same as that of the neighboring nucellar cells (Figure 5E,F), but a strong signal was observed in one or more nucellar cells, detected in the apomictic genotypes near the MMC, that can be the cells that putatively change their fate and become aposporous initials (Figure 5F, black arrow). At post-meiosis stage, before the degeneration of non-functional megaspores, both genotypes' dyads and tetrads exhibited the identical signal in the background; instead, in the degenerating cells of dyads (Figure 5H, black arrow) or tetrads the hybridization signal was much weaker than that of the nucellar cells (Figure 5H,I). At anthesis (Figure 5K,L) APOSTART expression was strong in all tissues of the ovule but particularly in the embryo sac. Moreover, the expression of the APOSTART appeared stronger in the multiple apomictic embryo sac (Figure 5L). After anthesis, a strong hybridization signal was observed in the embryos of sexual and apomictic genotypes (Figure 5N,O), while a weaker nonspecific signal was detected in the aleuronic layer due to endogenous phosphatase activity.

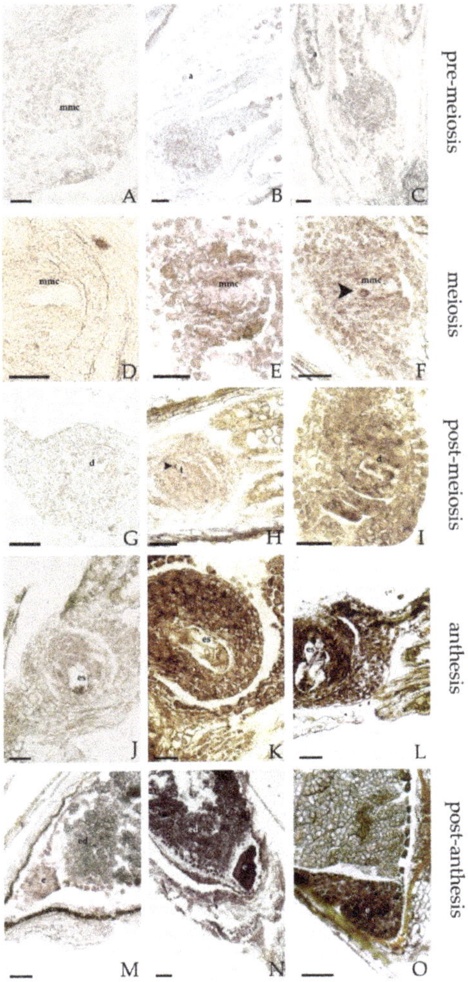

Figure 5. APOSTART expression in longitudinal sections of flowers studied by ISH. Sections of sexual and apomictic genotypes were probed with DIG-labeled antisense (**B**,**C**,**E**,**F**,**H**,**I**,**K**,**L**,**N**,**O**) or sense (**A**,**D**,**G**,**J**,**M**) RNAs and viewed under a microscope bright field that gives a purple label. No signals

were detected in the sections hybridized with the sense probes ((**A**,**J**,**M**) sexual genotypes; (**D**,**G**) apomictic genotypes). (**B**,**C**), longitudinal sections of ovules of sexual (**B**) and apomictic (**C**) genotypes containing the MMC, in both genotypes a poor hybridization signal was observed. In sexual (**E**) and apomictic (**F**) genotype ovules during megasporogenesis, the hybridization signal detected in the MMC was the same background in both genotypes, while in apomictic ovules a strong signal was observed in the aposporic cells (indicated by a black arrow) near the MMC (**F**). (**H**,**I**), dyads and tetrads observed in ovules of sexual (**H**) and apomictic (**I**) genotypes have the same signal the background unless they are destined to degenerate. (**K**,**L**), at the anthesis the hybridization signal was strong in the sexual embryo sac (**K**) and stronger in the apomictic embryo sacs (**L**). (**N**,**O**), a strong hybridization signal was observed also in the embryo of sexual (**N**) and apomictic (**O**) genotypes. **a** = anther; **d** = dyad; **e** = embryo; **ed** = endosperm; **es** = embryo sac; **mmc** = megaspore mother cell; **t** = tetrad. Bars = 40 μm.

3.6. Localization, Phosphorylation Sites, and Secondary Structure Predictions of PpAPOSTART Proteins

We predicted the subcellular localizations of the 15 APOSTART proteins based on amino acid sequence motifs. We used six different software, and their outputs were not coincident. Despite this, it is evident that the most predicted localization for PpAPOSTART proteins is the mitochondrion and the nucleus (Table S5).

Moreover, using NetPhos 3.1 software [25] we significantly predicted (with a threshold >0.5) serine, threonine, and tyrosine phosphorylation on the 15 APOSTARTs isolated from *P. pratensis* (Table S6). For serine, we observed an average of 46.86 loci (ST.DEV ± 2.09) ranging from 41 to 50 for APOSTART_14 and APOSTART_15, respectively. A lower number of significant loci (24 ± 1.85 sites) were recorded for threonine, and this ranged from 22 (APOSTART_11 and APOSTART_8) to 28 (APOSTART_15). The lowest number of phosphorylation sites (8.2 ± 1.52) that ranged from 7 (APOSTART_4, APOSTART_7, APOSTART_12, APOSTART_13, and APOSTART_14) to 13 (APOSTART_10) was recorded for tyrosines (Table S6).

The datasets were analyzed using the PCoA ordination method and by employing the Bray–Curtis distance matrix [63]. In the PCoA for serine phosphorylation eleven out of 15 APOSTARTs were clearly separated by coordinate 1 (26.44%) in two groups. The other four proteins (APOSTART_6, APOSTART_1, APOSTART_10 and APOSTART_8) were separated (Figure S4A).

Coordinate 2 (15.83%) distinguished APOSTART_13, APOSTART_14, and APOSTART_11 from the others. In the PCoA of threonines phosphorylation coordinate 1 (29.63%) separated 13 out of 15 APOSTART proteins into three groups (Figure S4B) from which APOSTART_6 and APOSTART_2 were excluded. Coordinate 2 (18.29%) was able to differentiate APOSTART_10 and APOSTART_15 from the others. In the PCoA for tyrosine, coordinate1 (36.88%) separated all APOSTART proteins with the exception of APOSTART_10 into three main groups (Figure S4C), while the second coordinate (22.31%) separated APOSTART_2, APOSTART_6, and APOSTART_15 from the others.

Sequences of all the proteins were also analyzed for secondary structure similarities. The secondary structure is defined by the pattern of hydrogen bonds of the protein, such as α-helices, β-sheets, and coils that are observed in an atomic-resolution structure. All the *P. pratensis* APOSTART proteins and all APOSTART-like proteins (in monocots and eudicots) scored in the phylogenetic analysis were tested for protein structure prediction analyses.

PpAPOSTART proteins consist on average of 12.8% α-helix, 26.6% β-sheet, and 57.9% coil (Table S7). In monocots, APOSTART-like proteins are composed on average of 13.2% α-helix, 28.4% β-sheet, and 57% coil, while in eudicots are on average composed of: 13.2% α-helix, 28.7% β-sheet, and 56.7% coil (Table S8). These data do not show statistical differences of secondary structure between monocots and eudicots within the *P. pratensis* APOSTART protein structures. In addition, solvent accessibility was considered as another parameter to characterize the surface area of the protein. The relevance of this measure may explain how different protein shapes have different solvent accessibilities (distinct in exposed and buried sites). This may reveal that the examined genes do not share the same 3D protein structure, even with high sequence homology. PpAPOSTART proteins showed an average percentage of exposed sites of 48.13%, 23.5% of buried sites, and 26.6% of intermediate sites (Table S7). Monocots, solvent exposed sites indicated an average of 48.5%, 24% of buried sites, and 25.9% of intermediate zones. Eudicots show an average of 48.9% of exposed sites and a percentage of 25.7% of buried sites, with 24.7% of undetermined zones. When compared, no statistical differences were found between monocots and eudicots and between the PpAPOSTART and APOSTART-like proteins (Table S8).

A third parameter considered was the intrinsic protein disorder (IDP). IDP regions fail to form a stable structure and are characterized by a low content of bulky hydrophobic amino acids and a high proportion of polar and charged amino acids. This is usually referred to as low hydrophobicity [43]. Furthermore, high net charges promote disorder because of electrostatic repulsion resulting from equally charged residues. A lack of this structure provides a larger interaction surface area that allows bindings with several other proteins. Because of this, IDPs are enriched in signaling and regulatory functions. Up to 33% of eukaryotic proteins are thought to have disordered segments to some degree [44]. In this context, disorder content prediction (DISO) comparison was employed as a tool to speculatively delineate proteins that share similar gene sequences. This analysis (see Tables S7 and S8) shows that *P. pratensis* have a disordered segment percentage of 26.8% higher with respect to the monocots and eudicots paralogs genes with a DISO average of 25.5% and 25%, respectively.

3.7. Molecular Modeling and Dynamics

Several lines of evidence have indicated the role of START domain in the shuttling of steroids and other biological lipids between the mitochondria compartments and may suggest a possible relationship between the expression of APOSTART proteins and the trafficking of phytosterols.

To corroborate this hypothesis, the 3D structures of APOSTARTs START domains and their interaction with phytosterols, namely stigmasterol, brassicasterol, and campesterol, were modeled by using a computational workflow (Figure S1) based on the combination of comparative homology modeling and molecular mechanics approaches. To do so, the protein sequences coded by APOSTART_1, APOSTART_6, and APOSTART_8 genes were chosen as representative of the three main clusters extracted by the multiple sequence alignment restricted to the START domain (see Figure S5). Initially 3D structures of the START domain were generated through the comparative modeling of APOSTART proteins to a template structure reported in the RCSB-PDB (Research Collaboratory for Structural Bioinformatics Protein Data Bank) archive (https://www.rcsb.org/) [64], coded 1LN1, whose START domain was discovered to be co-crystallized with a membrane lipid [65]. Thus, 203 residues were correctly modeled, thus corresponding to approximately 73% of the sub-sequences labelled as START domains. The 3D structures of APOSTART_1, APOSTART_6, and APOSTART_8 obtained by the homology modeling investigations are almost superimposable and are very close to the template model (Figure S6). The large binding site region on both proteins is composed of a β-sheet pavement comprising residues from 20 to 46 (residue numbering referred to the protein sequence processed by homology modeling) (β1-3) and from 136 to 167 (β8-9), which is buried on the top by two small helices (52 to 61 and 64 to 70 assigned to α 2 and 3, respectively) and surrounded by a loop from 168 to 171 (ω11) and by the extended C-terminal helix from 172 to 198 (α4). On these protein models, the most plausible binding modes, i.e., poses, of the three considered phytosterols, were identified by the use of docking approaches and a consensus scoring procedure for the evaluation of the protein-ligand affinity on each calculated bound system. Two top scored poses per ligand were eventually detected on either modeled protein thus yielding to 18 bound complexes. Overall, two different orientations of the steroid ligand were appreciated in the top scored poses characterized by the distal or vicinal position of hydroxyl group with respect to the C-terminus; these orientations resembled the IN or OUT mode, respectively, reported by Murcia at al. [66]. MD (Molecular Dynamics) calculations were then carried out by soaking the bimolecular complexes obtained at the docking stage in a box simulating the physiological aqueous medium, which yielded a more realistic description of the solvation effect on both the protein structure and the ligand binding. MD simulations corresponding to 250 ns trajectories were performed for the

18 bound complexes; however, we found a substantial loss of secondary structure, mainly affecting the N-terminal and the C-terminal domains in most of the considered systems after only 500 ps.

This structural detriment may be explained by assuming that the START domain of these proteins could be stabilized by the interaction with other domains of the APOSTART coded protein, such as the PH domain. Although this result substantially reduced the "informative potential" of the MD analyses, we employed the molecular systems obtained at 500 ps of each trajectory (treated as reported in the method section) in which the secondary structure was conserved in all-molecular systems to assay the effect of explicit solvation on the physterol binding properties. At this point, consensus scoring calculations were performed on the 18 sampled bound complexes yielding to the top scored poses of stigmasterol, brassicasterol, and campesterol at APOSTART_1, APOSTART_6, and APOSTART_8. The results of consensus scoring (performed as described in the previous section) are reported in Figure 6.

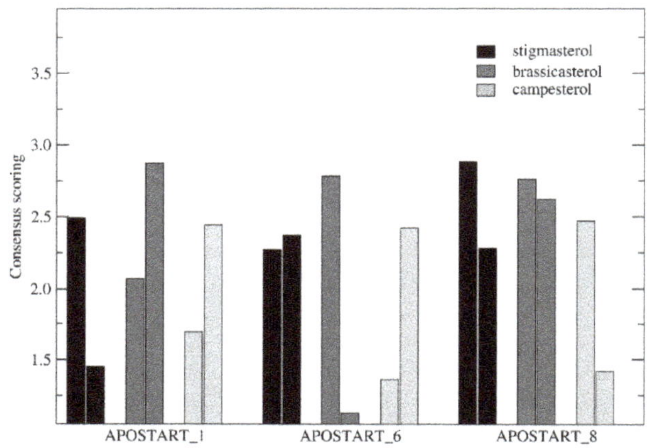

Figure 6. Consensus scoring of stigmasterol (black), brassicasterol (dark gray), and campesterol (light gray) poses binding at APOSTART_1 (left), APOSTART_6 (middle), and APOSTART_8 (right) obtained after 500 ps of MD (Molecular Dynamics) simulation (vide infra).

As shown, ten of the 18 bound systems were characterized by score values higher than 2.3 (corresponding to approximately the difference between the maximum score, 2.88, and the standard deviation on the whole set, 0.56) and were identified as top-scoring and analyzed at major details. The poses of the investigated phytosterols binding at APOSTART_1 and APOSTART_6 is reported in Figure 7.

It is worth noting that two binding site regions were eventually detected in the calculated molecular systems of either APOSTART_1 and APOSTART_6 (Figure 7). Four of the six top scored complexes, including the bound complexes of APOSTART_1 and stigmasterol binding at APOSTART_6, presented a hydrophobic binding region, which hosted the polycyclic steroid moiety of ligands independently on its IN or OUT orientation, with a majority formed in the inner wall of the C-terminal helix domain. In this binding region, residues L179, C184, L188, and Y191 on $\alpha4$ were identified as mainly involved in the hydrophobic interaction with these bound steroids, whereas the hydrophilic interactions of the ligand hydroxyl group may involve H156 or water molecules depending on the IN or OUT, respectively, binding mode (Figure S7). On the other hand, the bound complexes of brassicasterol and campesterol at APOSTART_6 were found to also be positioned in another region comprising the hydrophobic residues M89, V91, and P93 on $\beta5$ and residues L167, Y175, and L179 on $\omega11$. Hydrophilic contacts of the ligand OH group with residue E70 and/or with water molecules were also detected. It is worth noting that the two bound complexes of stigmasterol at APOSTART_6 were both characterized by a moderately high score. By this consideration, it may be concluded that the APOSTART_1 binding site is mainly localized in proximity of the inner wall of the C-terminal helix domain, whereas in the bound complexes of APOSTART_6 steroids are more likely to be positioned in the region between the β5 and ω11 domains.

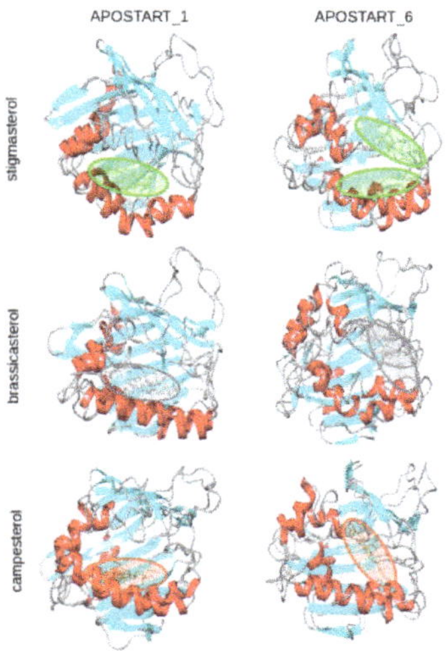

Figure 7. Representation of the binding complexes of APOSTART_1 and APOSTART_6 (cartoon) with the considered phytosterols (sticks). Binding sites of stigmasterol (green), brassicasterol (gray), and campesterol (orange) at the two proteins are also displayed.

4. Discussion

A successful transfer of apomixis to non-apomictics will allow clonal seed production in a myriad of crop species. Hopefully, this event will reduce input costs of fruit and vegetable production, while also raising yield [67]. A better understanding of the inheritance patterns for apomixis is fundamental for facilitating the identification of candidate genes, which in turn, is essential for engineering apomixis into sexual crops [68].

In a previous work [13] we have isolated APOSTART, a gene containing three domains, one of which is START whose name was given after the discovery of the *StAR* gene that is involved in human congenital lipoid adrenal hyperplasia. The clinical phenotype of this disease includes male pseudo hermaphroditism resulting from deficient fetal testicular testosterone synthesis. In apospory, a cell of the nucellus becomes an aposporous initial and then develops into a nonreduced embryo sac, which through parthenogenesis, gives rise to a viable embryo. How and why somatic cells of the ovule change their developmental fate and gain embryogenic potency is unknown. This question, together with the features of the *StAR* gene, made APOSTART our primary candidate for the control of apomixis. As a consequence, we decided to further characterize its expression, structure, and inheritance. Our data seem to confirm our vision. If we take all PpAPOSTARTs into consideration, we can see that while some of them are expressed in all tissues even if sometimes differentially between apomictic and sexual genotypes (i.e., APOSTART_10 and APOSTART_12), APOSTART_6 is the only PpAPOSTART specifically expressed in tissue flowers. Moreover APOSTART_6 showed a delayed expression in apomictic genotypes when compared with sexual ones, and this tendency was confirmed in a parthenogenetic individual. Both sexual reproduction and apomixis involve life-cycle renewal from gamete or gamete-like cells following ploidy restitution. The difference in expression timing between apomictic and sexual genotypes may imply that APOSTART_6 is the key regulator that switches a "normal" nucellar cell into an aposporous initial.

In addition, as shown by ISH, in apomictic genotypes a strong signal was detected only in some nucellar cells neighboring the MMC. This confirmed the hypothesis that *APOSTART* is involved in the change of fate of these cells and, therefore, in embryo sac development from nucellar cells.

Based on specific polymorphisms between PpAPOSTARTs, we developed some member-specific primer pairs, and the one designed for APOSTART_6 co-segregated with apomixis and was renamed APO-SCAR. This, together with the qPCR data, made APOSTART_6 the candidate to control apomixis in *P. pratensis*.

Moreover, an important quality of SCAR markers is their suitability on different genetic backgrounds. We then tested APO-SCAR on material belonging to different areas and continents and confirmed its capability to discriminate apomictic versus sexual genotypes. Therefore, APO-SCAR can be valuable tool for use in breeding and selection programs, considering that among the practical choices for minimizing time and costs of breeding programs, an easy method for early selection based on molecular markers would be of major advantage.

The subcellular localization predicts that PpAPOSTART might be localized in the mitochondrion or nucleus, although the six different software programs we used provided contrasting results. Nevertheless, a comparison with two proteins belonging to *Arabidopsis thaliana*, known as AtAPOSTART_1 and AtEDR2 (or AtAPOSTART_2), that share a high amino acid similarity and identity score with PpAPOSTART proteins could be made. In fact, AtAPO1 protein has been predicted to be localized in mitochondria and plastids [69], while Vorwerk et al. [70] could not co-localize the chimeric EDR2:HA:eGFP with the mitochondrial dye MitoTracker.

Conversions from diplospory to normal tetrad formation was observed in *Boechera* when pistils were subjected to variety of stress treatments, including carbohydrate starvation, osmotic stress, exposure to H_2O_2, and inhibition of brassinosteroid synthesis [2,71]. In addition, Antennaria-type diplospory occurred when DNA methylation was suppressed prior to MMC formation [72]. The conversions, from apomictic to sexual, and from one apomeiotic type to another, are evidence that the metabolic status of ovules (and the genes that affect this status) regulate the developmental sex/apomixis decision, and that the type of apomixis expressed is a function of the temporal and spatial expression of perhaps a single apomixis-conferring signal [2]. It has been proposed [73] that *EDR2*, one of the Arabidopsis orthologs of *PpAPOSTART*, may play a role in lipid signaling, mitochondria, and the activation of PCD (Programmed Cell Death) in plants. The prediction of PpAPOSTART localizes in mitochondria matches with this hypothesis. We also suggested [13] that APOSTART expression in *P. pratensis* may be related to the PCD that is involved in the non-functional megaspores and nucellar cell degeneration events that permit enlargement of maturing embryo sacs. The strong signal detected by ISH in degenerating cells of tetrads and in developing embryo sacs and nucellar cells in close proximity would appear to confirm this hypothesis.

Moreover, even if our tests to identify phosphorylation sites are predictions, they provided interesting hints about the potential post-translational different functions of the 15 PpAPOSTART identified. In serine, threonine, and tyrosine PCoA some PpAPOSTART members showed a clear differentiation from all others and, in particular, APOSTART_6 was separated from other members in all tests. This suggests a potential unique post-translational molecular function, in line with its flower-specific expression, and could represent a sign of its specific, important role in the cell changing fate and/or the apomictic embryo sac development.

Due to cell proliferation during evolution, the START domain acquired two distinct functions: the first appears to be generally related to the stress response [74–76] and the second to signaling mediated by lipid binding. In the last, the START domain underwent the lineage-specific fusions to other effector domains that are typical of multidomain eukaryotic signaling proteins. At least 60 START-containing multidomain proteins have been found in *Arabidopsis* and rice.

Sterols are essential membrane components and are critical for many physiological processes in all eukaryotes. In plants, it has been demonstrated that sexual reproduction is highly affected when the ratio of campesterol to sitosterol differs by a factor of 10, and yet vegetative growth such as stem elongation is only moderately affected when the ratio of campesterol to sitosterol differs by a factor of 30 to the wild-type [77]. Moreover, when stigmasterol is above a threshold level in the diets of grasshoppers [78] and aphids [79], these insects show high mortality and low reproduction. Therefore, determining the lipid or sterol molecule bound by the START domain would be an important step in unraveling the role of PpAPOSTART. Sterols are essential molecules for embryogenesis, and the high signal observed in an embryo during ISH could be related to this role of APOSTART. Therefore, the possible binding of stigmasterol, campesterol, and brassicasterol at the START domain of APOSTART_1, APOSTART_6, and APOSTART_8 coded proteins were investigated by means of a multilayered computational procedure. Calculations indicated that these three phytosterols effectively bind at the START domain of the modeled proteins, although differences were unveiled in the binding site structures. In APOSTART_1, all tested sterols bind at the same hydrophobic region, while a different binding pocket was found in APOSTART_12, thus probably reflecting a different responsiveness of the modeled proteins to phytosterols.

Recently, using CRISPR-Cas9, some researchers attempted to set up synthetic apomixis in rice but a maximum of 29% of offspring were maternal clones [80–82]. Here we have demonstrated that PpAPOSTART is expressed in different cells (MMC vs. Nucellar), and for one of its members (APOSTART_6), the expression is delayed confirming the hypothesis that the shift in timing and place of expression may cause apomixis or, at least, one of the two components of apomictic development. Since our results showed the closeness of three *Oryza* spp. sequences to PpAPOSTARTs, this makes APOSTART a gold candidate for future editing approaches for improving the maternal-like rate in rice progenies.

Supplementary Materials: The following are available online at http://www.mdpi.com/2073-4425/11/8/941/s1, Figure S1: Computational workflow used to model phytosterol-APOSTART_# binding complexes, Figure S2: Nucleotidic alignment of 15 APOSTART cDNA and 8 APOSTART genomic clones, Figure S3: Maximum-likelihood phylogenetic unrooted tree based on APOSTART_6 similar amino acid sequences, Figure S4: Bidimensional PCoA ordination method on predicted phosphorylation sites, Figure S5: Evolutionary relationships of the START domain aminoacidic sequences belonging to the 15 APOSTART members, Figure S6: Comparison between the calculated structure of the START domain of APOSTART_1, Figure S7: Maps of the ligand–protein interactions detected in the binding complexes of APOSTART_1 and APOSTART_6, Table S1: Information on germplasm source mode of reproduction, number of genotypes analyses and results of the APO-SCAR, Table S2: Sequence of primers used, Table S3: List of PpAPOSTART cDNA and genomic clones, Table S4: Identity values between either APOSTART genomic or APOSTART cDNA clones, Table S5: APOSTART proteins subcellular localization, Table S6: Predicted phosphorylation sites, Table S7: PpAPOSTART protein secondary structures predictions, Table S8: APOSTART and APOSTART-like proteins' secondary structures predictions.

Author Contributions: Conceptualization, G.M. and E.A.; methodology, G.M., L.S., and E.A.; validation, G.M., A.M., and E.A.; data analysis, G.M., D.A., L.S., A.M., and N.T.; investigation, G.M., B.K., L.R., and E.A.; resources, G.M., B.K., and E.A.; writing—original draft preparation, G.M., D.A., B.K., L.S., A.M., and E.A.; writing—review and editing, G.M., B.K., L.S., A.M., L.R., N.T., and E.A.; supervision, G.M. and E.A.; funding acquisition, E.A. All authors have read and agreed to the published version of the manuscript.

Funding: This project has received funding from the European Union's Horizon 2020 Research and Innovation Programme under the Marie Skłodowska-Curie Grant Agreement No 645674 and by the Ministry of University and Research (project "Genetic aspects of seed production: an integrated approach towards the understanding of apomixis").

Conflicts of Interest: The authors declare no conflict of interest. The funders had no role in the design of the study, in the collection, analyses, or interpretation of data, in the writing of manuscript or in the decision to publish the results.

References

1. Barcaccia, G.; Palumbo, F.; Sgorbati, S.; Albertini, E.; Pupilli, F. A Reappraisal of the Evolutionary and Developmental Pathway of Apomixis and Its Genetic Control in Angiosperms. *Genes* **2020**, *11*, 859. [CrossRef] [PubMed]
2. Albertini, E.; Barcaccia, G.; Carman, J.G.; Pupilli, F. Did apomixis evolve from sex or was it the other way around? *J. Exp. Bot.* **2019**, *70*, 2951–2964. [CrossRef] [PubMed]

3. Bashaw, E.C. Problems and possibilities of apomixis in the improvement of tropical forage grasses. *Trop. Forages Livest. Prod. Syst.* **1975**, *24*, 23–30.
4. Hanna, W.; Bashaw, E.C. Apomixis: Its identification and use in plant breeding 1. *Crop Sci.* **1987**, *27*, 1136–1139. [CrossRef]
5. Savidan, Y.; Dujardin, M. Apomixie: La prochaine révolution verte? *Rech. Paris 1970* **1992**, *23*, 326–334.
6. Koltunow, A.M.; Bicknell, R.A.; Chaudhury, A.M. Apomixis: Molecular strategies for the generation of genetically identical seeds without fertilization. *Plant Physiol.* **1995**, *108*, 1345. [CrossRef]
7. Bicknell, R.A.; Koltunow, A.M. Understanding apomixis: Recent advances and remaining conundrums. *Plant Cell* **2004**, *16*, S228–S245. [CrossRef]
8. Carman, J.G.; Jamison, M.; Elliott, E.; Dwivedi, K.K.; Naumova, T.N. Apospory appears to accelerate onset of meiosis and sexual embryo sac formation in sorghum ovules. *BMC Plant Biol.* **2011**, *11*, 9. [CrossRef]
9. Hojsgaard, D.; Greilhuber, J.; Pellino, M.; Paun, O.; Sharbel, T.F.; Hörandl, E. Emergence of apospory and bypass of meiosis via apomixis after sexual hybridisation and polyploidisation. *New Phytol.* **2014**, *204*, 1000–1012. [CrossRef]
10. Hand, M.L.; Koltunow, A.M. The genetic control of apomixis: Asexual seed formation. *Genetics* **2014**, *197*, 441–450. [CrossRef]
11. Brukhin, V. Molecular and genetic regulation of apomixis. *Russ. J. Genet.* **2017**, *53*, 943–964. [CrossRef]
12. Albertini, E.; Porceddu, A.; Ferranti, F.; Reale, L.; Barcaccia, G.; Romano, B.; Falcinelli, M. Apospory and parthenogenesis may be uncoupled in *Poa pratensis*: A cytological investigation. *Sex. Plant Reprod.* **2001**, *14*, 213–217. [CrossRef] [PubMed]
13. Albertini, E.; Marconi, G.; Reale, L.; Barcaccia, G.; Porceddu, A.; Ferranti, F.; Falcinelli, M. SERK and APOSTART. Candidate genes for apomixis in *Poa pratensis*. *Plant Physiol.* **2005**, *138*, 2185–2199. [CrossRef] [PubMed]
14. Albertini, E.; Marconi, G.; Barcaccia, G.; Raggi, L.; Falcinelli, M. Isolation of candidate genes for apomixis in *Poa pratensis* L. *Plant Mol. Biol.* **2004**, *56*, 879–894. [CrossRef]
15. Porceddu, A.; Albertini, E.; Barcaccia, G.; Falistocco, E.; Falcinelli, M. Linkage mapping in apomictic and sexual Kentucky bluegrass (*Poa pratensis* L.) genotypes using a two way pseudo-testcross strategy based on AFLP and SAMPL markers. *Theor. Appl. Genet.* **2002**, *104*, 273–280. [CrossRef]
16. Raggi, L.; Bitocchi, E.; Russi, L.; Marconi, G.; Sharbel, T.F.; Veronesi, F.; Albertini, E. Understanding genetic diversity and population structure of a *Poa pratensis* worldwide collection through morphological, nuclear and chloroplast diversity analysis. *PLoS ONE* **2015**, *10*. [CrossRef]
17. Sambrook, J.; Russell, D.W. *Molecular Cloning: A Laboratory Manual*, 3rd ed.; Cold Spring Harbor Laboratory Press: Cold Spring Harbor, NY, USA; New York, NY, USA, 2001.
18. Øvergaard, A.-C.; Nerland, A.H.; Patel, S. Evaluation of potential reference genes for real time RT-PCR studies in Atlantic halibut (*Hippoglossus Hippoglossus* L.); during development, in tissues of healthy and NNV-injected fish, and in anterior kidney leucocytes. *BMC Mol. Biol.* **2010**, *11*, 36. [CrossRef]
19. Ruan, W.; Lai, M. Actin, a reliable marker of internal control? *Clin. Chim. Acta* **2007**, *385*, 1–5. [CrossRef]
20. Angenent, G.C.; Franken, J.; Busscher, M.; van Dijken, A.; van Went, J.L.; Dons, H.J.; van Tunen, A.J. A novel class of MADS box genes is involved in ovule development in petunia. *Plant Cell* **1995**, *7*, 1569–1582.
21. Cañas, L.A.; Busscher, M.; Angenent, G.C.; Beltrán, J.-P.; Van Tunen, A.J. Nuclear localization of the petunia MADS box protein FBP1. *Plant J.* **1994**, *6*, 597–604. [CrossRef]
22. Altschul, S.F.; Gish, W.; Miller, W.; Myers, E.W.; Lipman, D.J. Basic local alignment search tool. *J. Mol. Biol.* **1990**, *215*, 403–410. [CrossRef]
23. Jones, D.T.; Taylor, W.R.; Thornton, J.M. The rapid generation of mutation data matrices from protein sequences. *Bioinformatics* **1992**, *8*, 275–282. [CrossRef] [PubMed]
24. Kumar, S.; Stecher, G.; Tamura, K. MEGA7: Molecular evolutionary genetics analysis version 7.0 for bigger datasets. *Mol. Biol. Evol.* **2016**, *33*, 1870–1874. [CrossRef] [PubMed]
25. Blom, N.; Gammeltoft, S.; Brunak, S. Sequence and structure-based prediction of eukaryotic protein phosphorylation sites. *J. Mol. Biol.* **1999**, *294*, 1351–1362. [CrossRef] [PubMed]
26. Wang, S.; Li, W.; Liu, S.; Xu, J. RaptorX-Property: A web server for protein structure property prediction. *Nucleic Acids Res.* **2016**, *44*, W430–W435. [CrossRef]
27. Wang, S.; Ma, J.; Xu, J. AUCpreD: Proteome-level protein disorder prediction by AUC-maximized deep convolutional neural fields. *Bioinformatics* **2016**, *32*, i672–i679. [CrossRef]

28. Wang, S.; Peng, J.; Ma, J.; Xu, J. Protein secondary structure prediction using deep convolutional neural fields. *Sci. Rep.* **2016**, *6*, 18962. [CrossRef]
29. Wang, S.; Sun, S.; Xu, J. AUC-maximized deep convolutional neural fields for protein sequence labeling. In *Machine Learning and Knowledge Discovery in Databases*; Springer: Cham, Switzerland, 2016; Volume 9852, pp. 1–16.
30. Yu, C.-S.; Cheng, C.-W.; Su, W.-C.; Chang, K.-C.; Huang, S.-W.; Hwang, J.-K.; Lu, C.-H. CELLO2GO: A web server for protein subCELlular LOcalization prediction with functional gene ontology annotation. *PLoS ONE* **2014**, *9*, e99368. [CrossRef]
31. Armenteros, J.J.A.; Sønderby, C.K.; Sønderby, S.K.; Nielsen, H.; Winther, O. DeepLoc: Prediction of protein subcellular localization using deep learning. *Bioinformatics* **2017**, *33*, 3387–3395. [CrossRef]
32. Blum, T.; Briesemeister, S.; Kohlbacher, O. MultiLoc2: Integrating phylogeny and Gene Ontology terms improves subcellular protein localization prediction. *BMC Bioinform.* **2009**, *10*, 274. [CrossRef]
33. Chou, K.-C.; Shen, H.-B. Plant-mPLoc: A top-down strategy to augment the power for predicting plant protein subcellular localization. *PLoS ONE* **2010**, *5*. [CrossRef] [PubMed]
34. Briesemeister, S.; Blum, T.; Brady, S.; Lam, Y.; Kohlbacher, O.; Shatkay, H. SherLoc2: A high-accuracy hybrid method for predicting subcellular localization of proteins. *J. Proteome Res.* **2009**, *8*, 5363–5366. [CrossRef] [PubMed]
35. Briesemeister, S.; Rahnenführer, J.; Kohlbacher, O. Going from where to why—Interpretable prediction of protein subcellular localization. *Bioinformatics* **2010**, *26*, 1232–1238. [CrossRef] [PubMed]
36. Söding, J.; Biegert, A.; Lupas, A.N. The HHpred interactive server for protein homology detection and structure prediction. *Nucleic Acids Res.* **2005**, *33*, W244–W248. [CrossRef] [PubMed]
37. *Maestro*, version 9.2; Schrödinger, LLC: New York, NY, USA, 2011.
38. Šali, A.; Blundell, T.L. Comparative protein modelling by satisfaction of spatial restraints. *J. Mol. Biol.* **1993**, *234*, 779–815. [CrossRef] [PubMed]
39. Sastry, G.M.; Adzhigirey, M.; Day, T.; Annabhimoju, R.; Sherman, W. Protein and ligand preparation: Parameters, protocols, and influence on virtual screening enrichments. *J. Comput. Aided Mol. Des.* **2013**, *27*, 221–234. [CrossRef]
40. Jorgensen, W.L.; Maxwell, D.S.; Tirado-Rives, J. Development and testing of the OPLS all-atom force field on conformational energetics and properties of organic liquids. *J. Am. Chem. Soc.* **1996**, *118*, 11225–11236. [CrossRef]
41. Qiu, D.; Shenkin, P.S.; Hollinger, F.P.; Still, W.C. The GB/SA continuum model for solvation. A fast analytical method for the calculation of approximate Born radii. *J. Phys. Chem. A* **1997**, *101*, 3005–3014. [CrossRef]
42. Kolossváry, I.; Keserü, G.M. Hessian-free low-mode conformational search for large-scale protein loop optimization: Application to c-jun N-terminal kinase JNK3. *J. Comput. Chem.* **2001**, *22*, 21–30. [CrossRef]
43. Watts, K.S.; Dalal, P.; Tebben, A.J.; Cheney, D.L.; Shelley, J.C. Macrocycle conformational sampling with MacroModel. *J. Chem. Inf. Model.* **2014**, *54*, 2680–2696. [CrossRef]
44. Halgren, T.A.; Murphy, R.B.; Friesner, R.A.; Beard, H.S.; Frye, L.L.; Pollard, W.T.; Banks, J.L. Glide: A new approach for rapid, accurate docking and scoring. 2. Enrichment factors in database screening. *J. Med. Chem.* **2004**, *47*, 1750–1759. [PubMed]
45. Friesner, R.A.; Murphy, R.B.; Repasky, M.P.; Frye, L.L.; Greenwood, J.R.; Halgren, T.A.; Sanschagrin, P.C.; Mainz, D.T. Extra precision glide: Docking and scoring incorporating a model of hydrophobic enclosure for protein- ligand complexes. *J. Med. Chem.* **2006**, *49*, 6177–6196. [PubMed]
46. Cross, S.; Ortuso, F.; Baroni, M.; Costa, G.; Distinto, S.; Moraca, F.; Alcaro, S.; Cruciani, G. GRID-based three-dimensional pharmacophores II: PharmBench, a benchmark data set for evaluating pharmacophore elucidation methods. *J. Chem. Inf. Model.* **2012**, *52*, 2599–2608. [CrossRef] [PubMed]
47. Goodford, P. Multivariate characterization of molecules for QSAR analysis. *J. Chemom.* **1996**, *10*, 107–117. [CrossRef]
48. von Itzstein, M.; Wu, W.-Y.; Kok, G.B.; Pegg, M.S.; Dyason, J.C.; Jin, B.; Van Phan, T.; Smythe, M.L.; White, H.F.; Oliver, S.W. Rational design of potent sialidase-based inhibitors of influenza virus replication. *Nature* **1993**, *363*, 418–423. [CrossRef] [PubMed]
49. Milletti, F.; Storchi, L.; Sforna, G.; Cruciani, G. New and original pKa prediction method using grid molecular interaction fields. *J. Chem. Inf. Model.* **2007**, *47*, 2172–2181. [CrossRef]

50. Milletti, F.; Storchi, L.; Sforna, G.; Cross, S.; Cruciani, G. Tautomer enumeration and stability prediction for virtual screening on large chemical databases. *J. Chem. Inf. Model.* **2009**, *49*, 68–75. [CrossRef]
51. Ahlström, M.M.; Ridderström, M.; Luthman, K.; Zamora, I. Virtual screening and scaffold hopping based on grid molecular interaction fields. *J. Chem. Inf. Model.* **2005**, *45*, 1313–1323. [CrossRef]
52. Pastor, M.; Cruciani, G.; McLay, I.; Pickett, S.; Clementi, S. GRid-INdependent descriptors (GRIND): A novel class of alignment-independent three-dimensional molecular descriptors. *J. Med. Chem.* **2000**, *43*, 3233–3243.
53. Cruciani, G.; Carosati, E.; De Boeck, B.; Ethirajulu, K.; Mackie, C.; Howe, T.; Vianello, R. MetaSite: Understanding metabolism in human cytochromes from the perspective of the chemist. *J. Med. Chem.* **2005**, *48*, 6970–6979.
54. Van Der Spoel, D.; Lindahl, E.; Hess, B.; Groenhof, G.; Mark, A.E.; Berendsen, H.J.C. GROMACS: Fast, flexible, and free. *J. Comput. Chem.* **2005**, *26*, 1701–1718. [CrossRef] [PubMed]
55. Berendsen, H.J.C.; Postma, P.M.; van Gunsteren, W.F.; Hermans, J. Interaction models for water in relation to protein hydration. In *Intermolecular Forces*; Pullman, B., Ed.; D. Reidel Publishing Company: Dordrecht, The Netherlands, 1981; pp. 331–342.
56. Bussi, G.; Donadio, D.; Parrinello, M. Canonical sampling through velocity rescaling. *J. Chem. Phys.* **2007**, *126*, 014101. [CrossRef] [PubMed]
57. Parrinello, M.; Rahman, A. Polymorphic transitions in single crystals: A new molecular dynamics method. *J. Appl. Phys.* **1981**, *52*, 7182–7190. [CrossRef]
58. Van Gunsteren, W.F.; Eising, A.A.; Hunenberger, P.H.; Kruger, P.; Mark, A.E.; Scott, W.R.P.; Tironi, I.G. *Biomolecular Simulation: The GROMOS96 Manual and User Guide*; vdf Hochschulverlag AG an der ETH Zürich and BIOMOS b.v.: Zürich, Switzerland; Groningen, The Netherlands, 1996.
59. Hess, B.; Bekker, H.; Berendsen, H.J.; Fraaije, J.G. LINCS: A linear constraint solver for molecular simulations. *J. Comput. Chem.* **1997**, *18*, 1463–1472. [CrossRef]
60. Darden, T.; York, D.; Pedersen, L. Particle mesh Ewald: An N·log (N) method for Ewald sums in large systems. *J. Chem. Phys.* **1993**, *98*, 10089–10092. [CrossRef]
61. Humphrey, W.; Dalke, A.; Schulten, K. VMD: Visual molecular dynamics. *J. Mol. Graph.* **1996**, *14*, 33–38. [CrossRef]
62. Albertini, E.; Barcaccia, G.; Porceddu, A.; Rosellini, D.; Falcinelli, M. Genetic control of parthenogenesis in Kentucky bluegrass: Results from a sexual x apomictic cross. In Proceedings of the International Grassland Congress, Sao Pedro, Sao Paulo, Brazil, 11–21 February 2001; Volume 19, p. 138.
63. Ramette, A. Multivariate analyses in microbial ecology. *FEMS Microbiol. Ecol.* **2007**, *62*, 142–160. [CrossRef]
64. Berman, H.M.; Westbrook, J.; Feng, Z.; Gilliland, G.; Bhat, T.N.; Weissig, H.; Shindyalov, I.N.; Bourne, P.E. The Protein Data Bank. *Nucleic Acids Res.* **2000**, *28*, 235–242. [CrossRef]
65. Roderick, S.L.; Chan, W.W.; Agate, D.S.; Olsen, L.R.; Vetting, M.W.; Rajashankar, K.R.; Cohen, D.E. Structure of human phosphatidylcholine transfer protein in complex with its ligand. *Nat. Struct. Biol.* **2002**, *9*, 507–511. [CrossRef]
66. Murcia, M.; Faráldo-Gómez, J.D.; Maxfield, F.R.; Roux, B. Modeling the structure of the StART domains of MLN64 and StAR proteins in complex with cholesterol. *J. Lipid Res.* **2006**, *47*, 2614–2630. [CrossRef]
67. Barcaccia, G.; Albertini, E. Apomixis in plant reproduction: A novel perspective on an old dilemma. *Plant Reprod.* **2013**, *26*, 159–179. [CrossRef] [PubMed]
68. Pupilli, F.; Barcaccia, G. Cloning plants by seeds: Inheritance models and candidate genes to increase fundamental knowledge for engineering apomixis in sexual crops. *J. Biotechnol.* **2012**, *159*, 291–311. [CrossRef] [PubMed]
69. Resentini, F.; Vanzulli, S.; Marconi, G.; Colombo, L.; Albertini, E.; Masiero, S. AtAPOSTART1, an *Arabidopsis thaliana* PH-START domain protein involved in seed germination. *Plant Biosyst. Int. J. Deal. Asp. Plant Biol.* **2014**, *148*, 1178–1186.
70. Vorwerk, S.; Schiff, C.; Santamaria, M.; Koh, S.; Nishimura, M.; Vogel, J.; Somerville, C.; Somerville, S. EDR2 negatively regulates salicylic acid-based defenses and cell death during powdery mildew infections of *Arabidopsis thaliana*. *BMC Plant Biol.* **2007**, *7*, 35. [CrossRef]
71. Gao, L. Pharmacologically induced meiosis apomeiosis interconversions in *Boechera*, *Arabidopsis* and *Vigna*. *Grad. Theses Diss.* **2018**, *7222*, 169.
72. Hojsgaard, D.; Klatt, S.; Baier, R.; Carman, J.G.; Hörandl, E. Taxonomy and biogeography of apomixis in angiosperms and associated biodiversity characteristics. *Crit. Rev. Plant Sci.* **2014**, *33*, 414–427. [CrossRef]

73. Tang, D.; Ade, J.; Frye, C.A.; Innes, R.W. Regulation of plant defense responses in Arabidopsis by EDR2, a PH and START domain-containing protein: Regulation of plant defenses by EDR2. *Plant J.* **2005**, *44*, 245–257. [CrossRef]
74. Osmark, P.; Boyle, B.; Brisson, N. Sequential and structural homology between intracellular pathogenesis-related proteins and a group of latex proteins. *Plant Mol. Biol.* **1998**, *38*, 1243–1246. [CrossRef]
75. Gamas, P.; de Billy, F.; Truchet, G. Symbiosis-specific expression of two *Medicago truncatula* nodulin genes, *MtN1* and *MtN13*, encoding products homologous to plant defense proteins. *Mol. Plant. Microbe Interact.* **1998**, *11*, 393–403. [CrossRef]
76. Ponting, C.P.; Aravind, L. START: A lipid-binding domain in StAR, HD-ZIP and signalling proteins. *Trends Biochem. Sci.* **1999**, *24*, 130–132. [CrossRef]
77. Schaller, H. The role of sterols in plant growth and development. *Prog. Lipid Res.* **2003**, *42*, 163–175. [CrossRef]
78. Behmer, S.T.; Elias, D.O. The nutritional significance of sterol metabolic constraints in the generalist grasshopper *Schistocerca americana*. *J. Insect Physiol.* **1999**, *45*, 339–348. [CrossRef]
79. Bouvaine, S.; Faure, M.-L.; Grebenok, R.J.; Behmer, S.T.; Douglas, A.E. A dietary test of putative deleterious sterols for the aphid *Myzus persicae*. *PLoS ONE* **2014**, *9*, e86256. [CrossRef] [PubMed]
80. Khanday, I.; Skinner, D.; Yang, B.; Mercier, R.; Sundaresan, V. A male-expressed rice embryogenic trigger redirected for asexual propagation through seeds. *Nature* **2019**, *565*, 91–95. [CrossRef] [PubMed]
81. Wang, C.; Liu, Q.; Shen, Y.; Hua, Y.; Wang, J.; Lin, J.; Wu, M.; Sun, T.; Cheng, Z.; Mercier, R.; et al. Clonal seeds from hybrid rice by simultaneous genome engineering of meiosis and fertilization genes. *Nat. Biotechnol.* **2019**, *37*, 283–286. [CrossRef]
82. Xie, E.; Li, Y.; Tang, D.; Lv, Y.; Shen, Y.; Cheng, Z. A strategy for generating rice apomixis by gene editing. *J. Integr. Plant Biol.* **2019**, *61*, 911–916. [CrossRef]

genes

Article

Genetic Dissection of Apomixis in Dandelions Identifies a Dominant Parthenogenesis Locus and Highlights the Complexity of Autonomous Endosperm Formation

Peter J. Van Dijk [1,*], Rik Op den Camp [1] and Stephen E. Schauer [2]

[1] Keygene N.V., Agro Business Park 90, 6708 PW Wageningen, The Netherlands; rik.op-den-camp@keygene.com
[2] Keygene Inc., Rockville, MD 20850, USA; stephen.schauer@keygene.com
* Correspondence: peter.van-dijk@keygene.com; Tel.: +31-317-466-866

Received: 20 July 2020; Accepted: 18 August 2020; Published: 20 August 2020

Abstract: Apomixis in the common dandelion (*Taraxacum officinale*) consists of three developmental components: diplospory (apomeiosis), parthenogenesis, and autonomous endosperm development. The genetic basis of diplospory, which is inherited as a single dominant factor, has been previously elucidated. To uncover the genetic basis of the remaining components, a cross between a diploid sexual seed parent and a triploid apomictic pollen donor was made. The resulting 95 triploid progeny plants were genotyped with co-dominant simple-sequence repeat (SSR) markers and phenotyped for apomixis as a whole and for the individual apomixis components using Nomarski Differential Interference Contrast (DIC) microscopy of cleared ovules and seed flow cytometry. From this, a new SSR marker allele was discovered that was closely linked to parthenogenesis and unlinked to diplospory. The segregation of apomixis as a whole does not differ significantly from a three-locus model, with diplospory and parthenogenesis segregating as unlinked dominant loci. Autonomous endosperm is regularly present without parthenogenesis, suggesting that the parthenogenesis locus does not also control endosperm formation. However, the high recovery of autonomous endosperm is inconsistent with this phenotype segregating as the third dominant locus. These results highlight the genetic complexity underlying apomixis in the dandelion and underline the challenge of introducing autonomous apomixis into sexual crops.

Keywords: apomixis; diplospory; parthenogenesis; autonomous endosperm; genetics; *Taraxacum*; dandelion

1. Introduction

Apomixis is a form of reproduction in the flowering plants in which the seeds are clones of the mother plant [1,2]. Apomixis, if introduced into the hybrids of otherwise sexually reproducing crops, will revolutionize plant breeding and agriculture because apomixis allows the one-step fixation of any valuable trait (e.g., heterosis or hybrid vigor), irrespective of the genetic complexity of the trait, for all subsequent generations [3–6]. This will reduce the time and cost of varietal development and is necessary for finding solutions to the immense and acute problems of population growth, changing climates, and the biodiversity crisis. Apomixis, however, does not occur in major crops and is rare in wild plant species. One promising way to introduce apomixis into crops is to reverse-engineer wild apomictic species: to genetically dissect natural apomixis, to clone natural apomixis genes, to identify the sexual orthologs of apomixis genes, and to modify these orthologs to apomictic versions in the crops of interest.

In the case of gametophytic apomixis, which may be the most straightforward approach to introducing apomixis into crops, a diploid egg cell is formed that is genetically identical to the mother plant and then transitions into an embryo without fertilization. Gametophytic apomixis is a rare reproductive system but widely distributed phylogenetically. One of the best-known apomicts is the common dandelion, *T. officinale* in the *Asteraceae* family, in which sexual diploid ($2n = 2x = 16$) and apomictic polyploid (mainly triploid, $2n = 3x = 24$) cytotypes occur. In developmental terms, three main components of apomixis in *Taraxacum* can be distinguished: 1. diplospory—a modified form of meiosis in which chromosomal recombination and reduction are absent, resulting in non-haploid female gametophyte cells; 2. parthenogenesis—the direct development of an embryo from an egg cell without fertilization; and 3. the autonomous development of endosperm—endosperm arising from the dividing, hexaploid central cell of the female gametophyte. The need for an endosperm may at first appear non-obvious; however, the endosperm is a tissue that nourishes the developing embryo; without it, the embryo will abort [7]. In diploid sexual flowering plants, the triploid endosperm develops from the fertilization of the diploid central cell by a haploid sperm cell deposited by the pollen tube. Many apomicts also require the fertilization of the central cell to trigger the development of the endosperm (pseudogamy), making autonomous endosperm formation dispensable in these species. However, most, if not all, apomictic *Asteraceae* species produce autonomous endosperm [8].

With an understanding of the three developmental components of apomixis, the genes responsible need to be cloned in order to enable the introduction of apomixis into sexual crops, for which genetic analysis is necessary so that the number and complexity of the genetic loci underlying the developmental components are understood [2,9,10]. For example, are the developmental components controlled by separate, unlinked loci; is there a single chromosomal cluster of genes encoding different components; or is there a primary gene with different pleiotropic effects? Because apomicts undergo pollen meiosis, they can be used to investigate the genetics of apomixis in crosses with sexual seed plants, which has been used in the investigation of apomixis from two other genera of the *Asteraceae*. Based on various crosses between sexuals and apomicts of *Erigeron annuus* (fleabane), Noyes and colleagues proposed that that apomixis was controlled by a diplospory locus (*D*), whereas parthenogenesis and autonomous endosperm were controlled by a single fertilization factor locus (termed *F*, [11,12]). Similarly, deletion mutagenesis in *Hieracium* subgenus *Pilosella* (hawkweed) identified loci containing two of the components of apomixis: the locus for apospory, termed *Loss-of-Apospory* (*LOA*), and a locus containing the genes for both parthenogenesis and autonomous endosperm, termed *Loss-of-Parthenogenesis* (*LOP*); [13,14]. However, further investigation of the *LOP* locus uncovered rare recombination events between the parthenogenesis phenotype and the autonomous endosperm phenotype, consistent with two tightly linked genes (*LOP* and Autonomous Endosperm, abbreviated as *AutE*). Unexpectedly, the genetic fine mapping of *AutE* in more crosses showed that this locus was on a different linkage group from the *LOP* locus, with a low penetrance (18%) and no additive effect [15], suggesting that the *AutE* phenotype in *Hieracium* is controlled by a more complex genetic mechanism than *LOA* or *LOP*.

Parthenogenesis has had extensive research focus in an attempt to understand the genes that drive the phenotype, both in natural apomicts as well as in model sexual systems (extensively reviewed in [16]), culminating in the cloning of the parthenogenesis-inducing transcription factor *BABY BOOM-Like* (*PsBBL*) from the monocot apomict *Pennisetum squamulatum* [17]. No other functionally validated parthenogenesis-controlling genes from apomicts have been described to date, so it is unclear if *PsBBL* is a universal mechanism for parthenogenesis in apomicts. By analysis of mutations in *Arabidopsis thaliana*, parthenogenesis was reported in mutants in the Polycomb Group 2 (PcG2) complex member *MULTICOPY SUPPRESSOR OF IRA1* (*MSI1*) [18] and in lines recovered from genetic screens performed by Fenby and colleagues [19].

Unlike parthenogenesis, the genetic control of the autonomous endosperm component of apomixis has not been thoroughly investigated outside of the apomicts in *Erigeron* and *Hieracium*. This is in part because many well-studied apomicts, such as *Pennisetum*, use pseudogamy and thus lack this component. Additionally, if present, the penetrance of the phenotype can be low, with *Boechera holboelli* showing a maximum penetrance of just 15% [20]. In contrast, the penetrance of autonomous endosperm development in *Taraxacum* is complete, making this system ideal for uncovering the basis of this component of apomixis, in combination with the wealth of information known about endosperm formation in general, as well as specific mutations that can induce its autonomous formation.

In *Arabidopsis*, mutations that give a Fertilization Independent Seed (FIS)-formation phenotype are found in genes encoding the PcG2 complex and act by initiating autonomous endosperm development, although the endosperm fails to cellularize and, eventually, the seeds abort development [21–23]. Therefore, *FIS* orthologs would appear to be clear candidate genes for autonomous endosperm development in autonomous apomicts and potentially more, as *MSI1* affects parthenogenesis as well. However, while the silencing of *FIS*-genes by RNAi in *Hieracium* affected endosperm development in sexuals, the transgene by itself did not induce autonomous endosperm [24], suggesting that alterations in the *FIS* genes may not be the causative lesions behind the autonomous endosperm formation in apomicts of the *Asteraceae*. In addition, the *Hieracium MSI1* gene did not map to the *LOP* locus, eliminating this gene as a candidate for parthenogenesis and autonomous endosperm [25]. Thus, one way to get insights into autonomous endosperm is to find the causative genes from apomicts such as *Taraxacum*.

Earlier studies in *Taraxacum* have focused on the inheritance of diplospory by using a tetraploid diplospory pollen donor [26]. The tetraploid diplospory pollen donor lacked parthenogenesis and was derived from a cross between a sexual seed plant and an apomictic pollen donor [27]. Co-dominant simple-sequence repeat (SSR) markers genetically linked to the diplospory-encoding *DIP* locus showed that this dominant component of apomixis was tetrasomically inherited, and the genotype had the *Dddd* simplex constitution [26]. Additionally, the *DIP* locus was weakly linked to the *18S–25S rDNA* locus, which was later confirmed by the fluorescent in situ hybridization of Bacterial Artificial Chromosome (BAC) probes on one of the Nucleolus Organizer Region (NOR) chromosomes [28]. Vijverberg and colleagues [29] used the tetraploid diplospory pollen donor described above for the genetic fine-mapping of the *DIP* locus and estimated the distance between the SSR *MSTA78-a* allele and the *DIP* locus to be 3.5 centimorgans (cM). Positional information of the *DIP* locus was used to identify the *DIP* gene as a *Vacuolar Protein Sorting-associated 13* (*VPS13*)-like gene [30]. The molecular function of this gene in apomixis is currently under investigation.

The fact that the tetraploid pollen donor plant above was diplosporous, but lacked parthenogenesis, suggested that diplospory and the parthenogenesis-encoding locus (*PAR*) in *Taraxacum* were controlled by at least two separate genetic loci. Van Dijk and colleagues [31,32] further investigated the breakdown of apomixis into its developmental components in non-apomictic offspring from crosses between diploid sexuals and triploid apomicts. Using Nomarski Differential Interference Contrast (DIC) microscopy of cleared ovules and SSR-marker analysis of progeny after pollination with diploid sexuals, they distinguished three non-apomictic phenotypes: type A, lacking diplospory and lacking parthenogenesis; type B, with diplospory and incomplete penetrance of parthenogenesis; and type C, with diplospory and autonomous endosperm but lacking parthenogenesis. Type B did not set seed in isolation, and because no autonomous endosperm development was seen with DIC microscopy, it was speculated that this type would need fertilization for endosperm development (pseudogamy). Mártonfiová et al. [33] found in crosses between diploid sexuals and tetraploid apomicts both type A and type C among the non-apomictic offspring but no type B. These studies suggest that apomixis in *Taraxacum* could be controlled by two or three major loci. However, the numbers of offspring in these crosses were too small for detailed segregation analysis.

To find the genes controlling parthenogenesis and autonomous endosperm in *Taraxacum*, the first question of if autonomous endosperm and parthenogenesis always co-occur and are controlled by a common dominant genetic factor needs to be addressed. As outlined in Figure 1, the inheritance (following a cross between a diploid sexual seed plant and an apomictic triploid pollen donor plant) of functional apomixis as a phenotype, along with the inheritance of the three components of apomixis, can be conceived in two alternative hypotheses: as either two or three unlinked genetic loci. In the two-locus model, parthenogenesis and autonomous endosperm formation are determined by a single common locus (such as the *F* locus of *Erigeron*); in the three-locus-model, each apomictic component is determined by a separate unlinked locus. Assuming that the loci are uncoupled and have the same genotypic constitution as the diplospory locus, 44 percent of the offspring from the two-locus model are expected to be apomictic, while in the three-locus model, 30 percent of the offspring are expected to be apomictic.

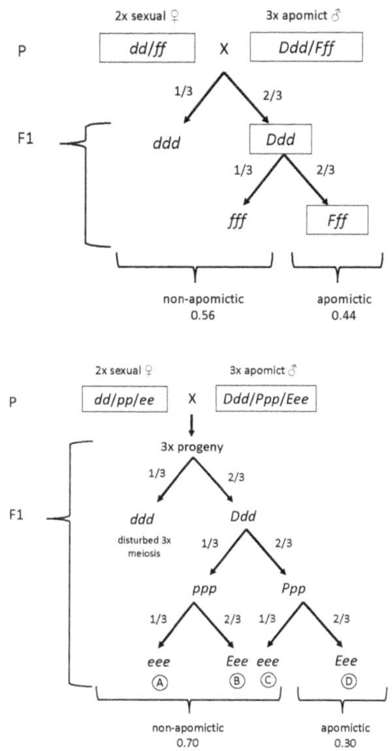

Figure 1. Two alternative genetic models for the polysomic inheritance of apomixis in *Taraxacum*. Both start with a cross between a diploid seed plant and a triploid apomictic pollen donor, and only triploid offspring are shown (produced by diploid pollen grains). The top panel shows a two-locus model with one factor *F* controlling both parthenogenesis and autonomous endosperm development. The bottom panel shows a three-locus model with separate factors for parthenogenesis and autonomous endosperm development. *D* is diplospory, *d* is meiotic, *F* is the fertilization factor, *f* is fertilization-dependent, *P* is parthenogenesis, *p* is fertilization-dependent, *E* is autonomous endosperm, and *e* is sexual endosperm.

In this article, the inheritance of the individual components of apomixis in a cross between a diploid sexual plant and an apomictic triploid plant is described. Although pollen fertility due to triploidy was very low, by making a large number of crosses, it was possible to generate sufficient triploid offspring, allowing the quantification of the segregation of different elements of apomixis. As previously found, the SSR *MSTA78-a* allele was tightly linked to diplospory. This allele was used as a molecular marker for diplospory so that the phenotyping could focus on elucidating parthenogenesis and autonomous endosperm formation in the diplosporous progeny. It is critical to focus on the diplosporous progeny because both parthenogenesis and autonomous endosperm are gametophytic traits, which will segregate on a plant with normal meiosis. Unfortunately, meiosis will be highly disturbed due to the chromosomal imbalance inherent in the triploid F_1 plants, causing the seed phenotypes to show segregation distortion, which impedes clear analysis. When focusing the analysis on the diplosporous progeny, the problem of segregation distortion is avoided because these F_1 plants do not undergo meiosis. Analysis of the F_1 showed that the SSR *MSTA44B-d* allele is tightly linked to parthenogenesis: the first molecular marker for parthenogenesis in *Taraxacum*. Additionally, while 29% of the diplosporous offspring contained all three of the components of apomixis, which is consistent with the three-locus model, nearly 95% of the diplosporous offspring showed some degree of autonomous endosperm formation, independent of whether or not the egg cell underwent parthenogenesis. The absence of the clear segregation of autonomous endosperm development in *Taraxacum* is different from the coupling between autonomous endosperm and parthenogenesis previously reported in other apomictic *Asteraceae*.

2. Materials and Methods

2.1. The Cross between a Sexual Diploid and an Apomictic Triploid

Pollen from a pollen-fertile apomictic *T. officinale* line triploid A68 was crossed with stigmas from the pollen-sterile sexual diploid *T. officinale* line TJX3-20 [34]. TJX3-20 originated from Langres in France, while A68 originated from Heteren in the Netherlands. TJX3-20 produces small, infertile pollen due to a cytoplasmic male sterility system [35]. The seed parent's pollen infertility avoids the breakdown of the sporophytic Self Incompatibility System, which occurs in crosses when triploids are used as pollen donors and would generate many diploid selfed offspring ("mentor pollen effect" [36]). From a total of 62 crossed capitula (inflorescences), only 192 viable seeds were produced in total. The F_1 seeds were sown, and the resulting F_1 plants were screened by flow cytometry to determine ploidy.

2.2. Apomixis Phenotyping

To induce flowering, eight-week-old F_1 plants were vernalized for nine weeks in a cold room at 4 °C. Like the seed parent TJX3-20, all the F_1 plants were male sterile and physically isolated from other wild-type *Taraxacum*, eliminating the possibility of sexual seed setting. Under these conditions, the development of a large seed head is a sign of apomictic seed setting. To determine the degree of apomictic seed setting, for each F_1 plant, 50 randomly chosen brown seeds were germinated, and the numbers of seedlings germinating were counted. Germinating seeds implied that the mother was apomictic and had all components of apomixis.

2.3. Microsatellite Genotyping

Twelve codominant SSR loci were screened (Micro Satellite TAraxacum (MSTA) [37,38]: *MSTA31*, *MSTA44B*, *MSTA53*, *MSTA64*, *MSTA67*, *MSTA73*, *MSTA74*, *MSTA78*, *MSTA85*, *MSTA101*, *MSTA105*, and *MSTA131* (Table 1). DNA extraction and SSR assays were performed as described in [26]. The PCR products were analyzed on an ALF express II automatic sequencer (Amersham, Pharmacia Biotech, Charfonte, UK).

Table 1. Primer sequences of twelve codominant simple-sequence repeat (SSR) markers used in this study. For each primer pair, the repeat motif, primer sequences, optimal annealing temperature (TA, in °C) and expected size (in bp) are indicated, along with the reference where the SSR marker was first described.

Marker	Repeat Motif	Forward Primer Sequence	Reverse Primer Sequence	T_A	Size	Reference
MSTA31	$(CT)_{17}$	CCTCAAAGCCCGAACTT	ACGACCCCAACTGATTTTTAC	51.0	240	[37]
MSTA44B	$(CT)_{19}$	AGTTTCTCTAAAATGGGAAGAT	TGTCAGGTATATTCAAAAGATTC	51.0	191	[37]
MSTA53	$(TC)_{12}(GT)_8$	CAATTATTATGGTCTCGTCCTT	CCAGTTGAAGCAAAAACAGT	55.0	203	[37]
MSTA64	$(TC)_4TT(TC)_2TT(TC)_2TT(TC)_5A(TC)_4$-$(A)_{16}$	TGCTTTTTGAACGACAGTG	TTTGCTTGGTTATTAGTGAACAT	55.0	191	[37]
MSTA67	$(TC)_{22}T(CA)_{12}$	TTCGGATATGACCCTTCACT	GACATCTTGCACCTAAAAACAAT	56.0	219	[37]
MSTA73	$(TC)_{21}CTG(TC)_8$	CCGCATGAGGTTGTCT	TGGGCTGTTTAATAGAACTTA	53.0	216	[37]
MSTA74	$(CT)_{10}$	GAGGTCTTTTTATTCGGTTTT	GGATGCCTTTACAGTTACAAT	49.0	223	[37]
MSTA78	$(CT)_9$	TGATTGATTCTGCCCTAAACC	TGCCAAGACATCCGAAAAG	52.0	151	[37]
MSTA85	$(CT)_{20}$	TGCATGTTCGTTCTACTGGT	ACGTAATAAAATTGGAAGTCAGG	55.0	196	[37]
MSTA101	$(CCT)_2TCT(TC)_{16}$	GCATGGGGGTCGAGGGGTAT	CCGCGATGGACTTATTCTTGGTTG	57.8	198	[38]
MSTA105	$(TC)_{23}$	CACCGTTCAAAAATAAAGATAAAA	AGAATAGCTCCGTCAAGTAGG	54.3	203	[38]
MSTA131	$(AT)_7$	TACCCTGCAAACATTACTCTTCTG	GTTGGCCTGTTAATACTTGATACG	55.0	181	[38]

2.4. Nomarski DIC Microscopy Phenotyping

Methyl-salicylate cleared ovules from plants that were *MSTA78-a* positive but did not show seed setting (i.e., were apomixis-negative) were investigated by Nomarski DIC microscopy, as described in [39]. At least ten different ovules were analyzed per plant at anthesis or one day after anthesis.

2.5. Seed Flow Cytometry

Matzk and colleagues [40] have shown that autonomous endosperm production in ripe 3 × *Taraxacum* apomictic seeds can be detected with a flow cytometer as a 6 × peak, derived from the unfertilized central cell, which contains two fused 3 × polar nuclei (flow cytometric seed screen, FCSS). Matzk and colleagues also reported that endopolyploidization peaks were absent in the autonomous apomictic seeds of *T. officinale* and *Hieracium pilosella*, suggesting that the presence of 6 × peaks in isolated seeds would be due to autonomous endosperm formation. To assess the ploidy of the endosperm, developing seeds from the apomictic F_1 plants were collected five days after anthesis and directly analyzed by flow cytometry. Ten developing seeds were homogenized in Otto I buffer [41] by chopping the seeds with a sharp razor blade. Ploidy levels were determined with a flow cytometer (Ploidy analyzer, Partec, Münster, Germany) using 4′, 6-diamidino-2-phenylindole (DAPI) as a fluorescent stain as described in [36].

3. Results

3.1. Segregation of Apomixis as a Whole

As previously described by Van Dijk and colleagues [34], stigmas from a pollen-sterile sexual diploid dandelion (TJX3-20) were crossed with pollen from a pollen-fertile apomictic triploid (A68). From a total of sixty-two crossed capitula (inflorescences, see Figure 2A), only 192 viable F_1 seeds were produced in total.

Figure 2. (**A**) A longitudinal section of a capitulum (inflorescence) of *T. officinale*, showing the florets attached to the receptacle. The inferior ovaries have a single ovule and produce a single-seeded fruit (achene), often referred to as "seed". (**B**) An opened seed head of an apomict that produces both light and dark brown seeds (about ten days after flowering). Light brown seeds are empty ((**C**), left), while dark brown seeds ((**C**), right) generally germinate.

The average seed set per seed head from this cross was only 2.1 seeds, while, in the diploid × diploid sexual crosses, an average of ~100 seeds per seed head were obtained. Flow cytometry showed that 96 of the 192 F_1 plants were diploid (50%), 95 were triploid (49.5%), and one (0.5%) was tetraploid. These plants were the products of the fertilization of a haploid egg cell by a haploid, diploid, or triploid pollen grain, respectively. A68 produced many collapsed and small pollen grains, reflecting the high frequency of inviable aneuploid pollen grains produced because of unbalanced triploid pollen meiosis. However, no viable aneuploid offspring was recovered, which was fortunate, as the lack of aneuploidy in the F_1 simplifies the interpretation of the genetic analysis of apomixis. None of the diploid F_1 plants were apomictic, while the triploid F_1 plants segregated for apomixis. The only tetraploid plant recovered was also capable of apomictic reproduction.

To gain further insights into the genetics of apomixis, Van Dijk and colleagues [34] (2009) also reported on the transmission of the diplospory-linked marker *MSTA78-a* to diploid and triploid hybrids of this cross. While the segregation ratio of the three alleles in the triploid progeny was not significantly different from the Mendelian 1:1:1 ratio, the ratio was highly distorted in the diploid hybrids (0.01:0.47:0.52). Assuming that the single case of a *MSTA78-a* allele transmitted to a diploid hybrid was a recombination event between the *a*-allele and the *DIP* allele, it was postulated that haploid pollen grains carrying the diplospory allele were lethal [34]. To expand this analysis, the further investigation of the inheritance of the developmental components of apomixis in triploid progeny described here was undertaken.

As one triploid F_1 plant died before its reproductive system could be assessed, the remaining 94 3 × plants were used for further analysis. Apomictic seed setting was variable, as can be seen in the examples shown in Figure 3. As shown in Figure 4, 69 percent of the triploid F_1 plants produced no germinating seeds (64/93; for one apomictic F_1 plant (H69), seed setting was not quantified). By contrast, 24 percent of the triploid F_1 plants produced more than 80% germinating seeds (22/93). The remaining six plants produced seeds with a germination rate between 6 and 78 percent, which may be due to incomplete penetration of apomixis factors. The two-locus model outlined in Figure 1 predicts 41 of the 94 triploid offspring plants to be apomictic, while the three-locus model predicts a significantly fewer 28 apomicts. A total of twenty-nine triploid apomicts were recovered, fitting the three-locus model well ($\chi^2 = 0.05$, degrees of freedom (d.f.) = 1; the p-value is 0.82, not significant at $p < 0.05$) and contradicting the two-locus model ($\chi^2 = 6.23$, d.f. = 1; the p-value is 0.01, significant at $p < 0.05$).

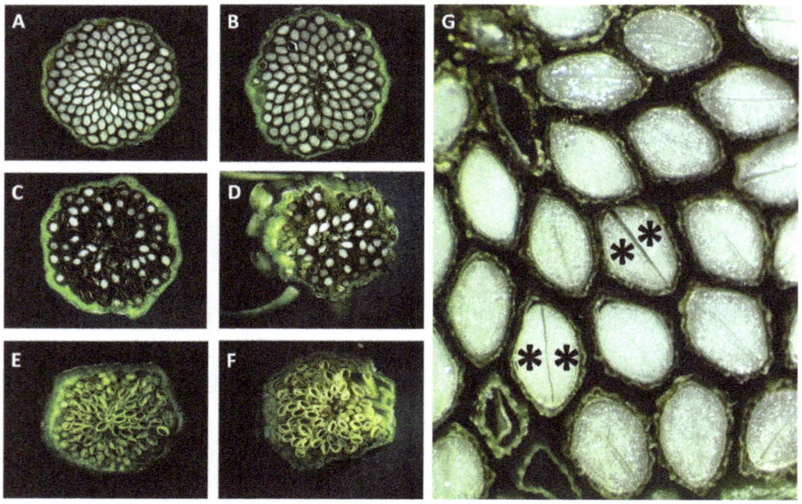

Figure 3. Variation in apomictic seed setting of F_1 triploids. Cross sections of seeds in (closed) seed heads of some 3 × hybrid progeny plants, shortly before opening and seed shatter (about nine days after anthesis). (**A**,**B**) are apomicts with high penetrance of apomixis (H2 and H18); (**C**,**D**) are apomicts with low penetrance (H37 and H99); (**E**,**F**) are two non-apomicts (H70 and H86). At this stage, the viable seeds are mainly filled with the two cotyledons (asterisks, panel (**G**)). The apomicts with low penetrance produce a seed coat, even in empty seeds; the non-apomicts produce no seed coat.

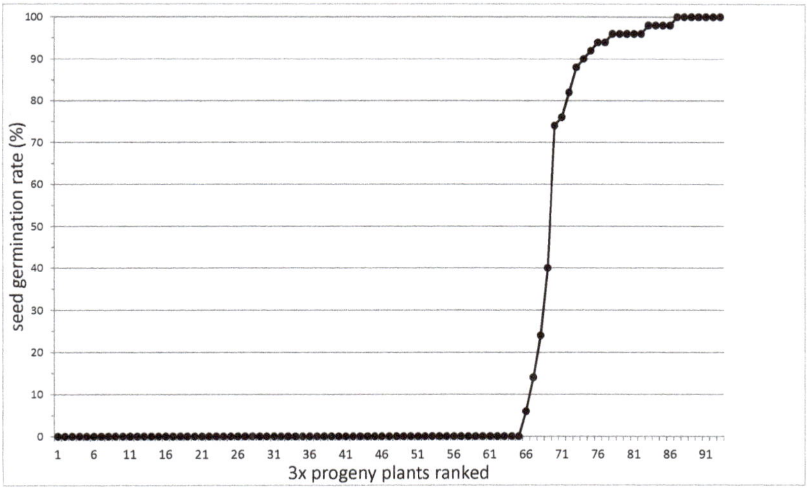

Figure 4. The germination rate of 93 3 × progeny plants.

3.2. Association between SSR Markers and Apomixis

To further understand the segregation of the components of apomixis, DNA from the progeny was screened with additional molecular markers (as described in [37,38]). In Table 2, the allele distributions of eleven SSR loci in the 29 apomictic F_1 triploids and the 66 non-apomictic F_1 triploids are listed. *MSTA31* could not be scored due to the presence of null alleles in the parental genotypes. As expected, all 29 F_1 triploids that reproduced via apomixis carried the paternal *MSTA78-a* (random association: $\chi^2 = 12.65$; d.f. = 1; the p-value is 0.0004, significant at $p < 0.05$), supporting the previously reported tight linkage between the *MSTA78-a* allele and the *DIP* locus [26,29]. Due to this, the *MSTA78-a* SSR was used as a molecular marker for the presence of diplospory induced by the *DIP* locus (independent of the phenotypic assessment of apomixis) and will be referred to as the "Dip-marker" below. Molecular analysis showed that 63 of the 94 3 × F_1 plants carried the *MSTA78-a* allele and were considered to be diplosporous. The observed frequency of 65 percent of F_1 plants carrying the Dip-marker is also consistent with the expected two-thirds of the diploid A68 pollen grains carrying the *D* allele ($\chi^2 = 0.19$; d.f. = 1; the p-value is 0.66, not significant at $p < 0.05$). Due to the function of the *DIP* locus blocking meiosis, no gametophytic segregation of the parthenogenesis or autonomous endosperm loci is expected in these plants. To see if molecular markers for either parthenogenesis or autonomous endosperm formation could be uncovered, these F_1 plants were genotyped for additional markers for linkage to the apomictic phenotype (as shown in Table 2). Of the remaining 10 SSR markers, one, *MSTA44B*, showed a significant association with apomixis. The genotype of the triploid pollen donor A68 for *MSTA44B* was *b* (167 bp)/*d* (171 bp)/*e* (183 bp).

Of the 29 apomictic F_1 plants, 27 had the *d* allele and two did not (H23 and H133, see Table 3). Furthermore, the *MSTA44B-d* allele is unlinked to the Dip-marker, consistent with it being genetically unlinked to the *DIP* locus (random association: $\chi^2 = 1.36$; d.f. = 1; the p-value is 0.85, not significant at $p < 0.05$). To determine if the *MSTA44B-d* allele was linked to either parthenogenesis or autonomous endosperm formation, Nomarski DIC microscopy phenotyping of developing seeds was performed.

Table 2. Allele frequencies of 11 paternal A68 SSR markers in apomictically reproducing progeny and non-apomictically reproducing progeny. A68 is a triploid and carries two or three different alleles per locus, which are indicated by a different letter (a–e). The *MSTA44B-d* (Par-marker) and the *MSTA78-a* (Dip-marker) alleles are significantly overrepresented in the apomicts.

SSR Locus	Allele 1	Allele 2	Allele 3	Chi-Square	d.f.	p-Value	Significance
MSTA44B	*b*	*d*	*e*				
apomicts	20	27	11				
non-apomicts	47	31	50	11.36	2	0.00	$p < 0.05$
MSTA53	*a*	*b*	*c*				
apomicts	14	20	16				
non-apomicts	44	44	38	0.81	2	0.67	n.s.
MSTA64	*a*	*b*	*c*				
apomicts	17	14	17				
non-apomicts	37	36	41	0.55	2	0.93	n.s.
MSTA67	*a*	*d*	*e*				
apomicts	24	17	18				
non-apomicts	38	46	46	2.43	2	0.30	n.s.
MSTA73	*a*	*a*	*c*				
apomicts	19		9				
non-apomicts	38		19	0.01	1	0.91	n.s.
MSTA74	*b*	*c*	*e*				
apomicts	22	17	15				
non-apomicts	44	34	46	1.45	2	0.48	n.s.
MSTA78	*a*	*c*	*d*				
apomicts	29	18	11				
non-apomicts	33	46	43	9.98	2	0.01	$p < 0.05$
MSTA85	*a*	*b*	*b*				
apomicts	20	36					
non-apomicts	32	90		1.67	1	0.20	n.s.
MSTA101	*a*	*c*	*d*				
apomicts	27	12	15				
non-apomicts	51	36	25	1.84	2	0.40	n.s.
MSTA105	*b*	*c*	*d*				
apomicts	17	18	21				
non-apomicts	36	44	42	0.28	2	0.87	n.s.
MSTA131	*a*	*c*	*d*				
apomicts	13	21	24				
non-apomicts	44	41	45	2.49	2	0.29	n.s.

Table 3. Analysis of apomixis and its developmental components in 62 Dip-marker positive 3 × F_1 plants.

	Ploidy	Germin. %	Apomixis	Dip Marker (*MST78-a*)	Par Marker (*MST44B-d*)	DIC Microscopy			
						PAR	AUT	FCSS AUT	AUT Combined
TJX 320	2 ×		no	ab	ac	non	non	non	–
68	3 ×	98	yes	acd	bde	yes	yes	yes	+
2	3 ×	96	yes	ac	bd	n.d.	n.d.	n.d.	+
16	3 ×	100	yes	ac	bd	n.d.	n.d.	n.d.	+
18	3 ×	96	yes	ad	bd	n.d.	n.d.	n.d.	+
23	3 ×	14	yes	ac	be	n.d.	?	n.d.	+
24	3 ×	100	yes	ad	bd	n.d.	?	n.d.	+
31	3 ×	100	yes	ad	bd	n.d.	n.d.	n.d.	+
34	3 ×	98	yes	ac	bd	n.d.	n.d.	n.d.	+
37	3 ×	40	yes	ac	bd	+	+	n.d.	+
48	3 ×	78	yes	ac	bd	+	+	n.d.	+
50	3 ×	98	yes	ac	bd	n.d.	n.d.	n.d.	+
69	3 ×	n.q.	yes	ad	de	+	+	n.d.	+
73	3 ×	100	yes	ac	de	+	+	n.d.	+
76	3 ×	94	yes	ad	bd	n.d.	n.d.	n.d.	+
98	3 ×	100	yes	ac	bd	n.d.	n.d.	n.d.	+
99	3 ×	74	yes	ad	bd	+	+	n.d.	+
113	3 ×	90	yes	ad	bd	n.d.	n.d.	n.d.	+
115	3 ×	96	yes	ad	de	n.d.	n.d.	n.d.	+
127	3 ×	82	yes	ad	bd	+	+	n.d.	+
132	3 ×	94	yes	ac	bd	n.d.	n.d.	n.d.	+
133	3 ×	98	yes	ac	be	n.d.	n.d.	n.d.	+
136	3 ×	96	yes	ac	de	n.d.	n.d.	n.d.	+
144	3 ×	88	yes	ac	de	n.d.	n.d.	n.d.	+
154	3 ×	100	yes	ac	bd	n.d.	n.d.	n.d.	+
158	3 ×	96	yes	ac	de	n.d.	n.d.	n.d.	+
159	3 ×	24	yes	ac	de	n.d.	n.d.	n.d.	+
165	3 ×	98	yes	ac	bd	n.d.	n.d.	n.d.	+
183	3 ×	92	yes	ac	bd	n.d.	n.d.	n.d.	+
194	3 ×	6	yes	ad	de	n.d.	n.d.	n.d.	+
201	3 ×	100	yes	ad	de	n.d.	n.d.	n.d.	+
163	3 ×	0	no	ad	de	+	+	+	+
70	3 ×	0	no	ac	bd	+	?	+	+
22	3 ×	0	no	ac	de	+	?	n.d.	?
56	3 ×	0	no	ac	de	+	?	n.d.	?
148	3 ×	0	no	ac	de	?	+	+	+
95	3 ×	0	no	ac	bd	?	?	+	+
114	3 ×	0	no	ac	de	n.d.	n.d.	+	+
123	3 ×	0	no	ad	de	?	+	n.d.	+
195	3 ×	0	no	ac	de	?	?	+	+
175	3 ×	0	no	ac	bd	n.d.	n.d.	–	–
193	3 ×	0	no	ac	de	n.d.	n.d.	–	–
30	3 ×	0	no	ad	be	?	+	+	+
45	3 ×	0	no	ad	be	?	+	+	+
139	3 ×	0	no	ad	be	?	?	+	+
170	3 ×	0	no	ad	be	?	+	+	+
181	3 ×	0	no	ac	be	?	+	+	+
185	3 ×	0	no	ac	be	?	+	+	+

Table 3. *Cont.*

	Ploidy	Germin. %	Apomixis	Dip Marker (*MST78-a*)	Par Marker (*MST44B-d*)	DIC Microscopy			
						PAR	AUT	FCSS AUT	AUT Combined
190	3 ×	0	no	ac	be	?	+	+	+
177	3 ×	0	no	ac	be	?	+	n.d.	+
68	3 ×	0	no	ad	be	n.d.	n.d.	+	+
78	3 ×	0	no	ad	be	n.d.	n.d.	+	+
79	3 ×	0	no	ac	be	?	?	+	+
86	3 ×	0	no	ad	be	n.d.	n.d.	+	+
117	3 ×	0	no	ad	be	n.d.	n.d.	+	+
140	3 ×	0	no	ad	be	?	?	+	+
149	3 ×	0	no	ac	be	?	?	+	+
176	3 ×	0	no	ad	be	?	?	+	+
207	3 ×	0	no	ad	be	?	?	+	+
65	3 ×	0	no	ac	be	?	?	–	–
112	3 ×	0	no	ac	be	?	?	n.d.	?
157	3 ×	0	no	ad	bc	?	?	n.d.	?
33	3 ×	0	no	ad	be	?	?	n.d.	?
43	3 ×	0	no	ac	be	?	?	n.d.	?
208	3 ×	0	no	ac	be	?	?	n.d.	?

The first two rows show the 2 × seed parent and the 3 × pollen parent, respectively. Only the paternal alleles of the SSR genotypes are shown, since apomixis is inherited from the pollen donor. PAR "+" means that parthenogenesis is confirmed by Differential Interference Contrast (DIC) microscopy. AUT "+" means that autonomous endosperm development is confirmed by DIC microscopy. A question mark (?) means inconclusive; n.q. means not quantified; n.d. means non-determined; FCSS means flow cytometric seed screen. The gray cell fill color shows a positive apomixis component.

3.3. *Nomarski DIC Microscopy Phenotyping of Parthenogenesis and Autonomous Endosperm*

For the triploid F_1 plants that produced viable seeds, it is reasonable to assume that all apomixis components are present (see Figure 5A, autonomous apomict H99). To further elucidate the impacts on seed development, developing seeds from several complete apomicts were cleared and imaged with Nomarski DIC microscopy for the presence of embryos and autonomous endosperm. The developing seeds of the F_1 were highly asynchronous in parthenogenetic embryo development and autonomous endosperm formation. Figure 5B shows an embryo sac with an advanced embryo but a non-divided central cell nucleus of the non-apomictic plant H70. The hexaploid peak in the FCSS indicates that some level of autonomous endosperm developed later, but ultimately, no viable seeds were produced from this F_1 plant. By contrast, Figure 5C shows an embryo sac with an undivided egg cell and an advanced cellularized endosperm; again, no viable seeds were produced from this F_1 plant. As a result, it is possible that, at the time of fixation, the development of either has not yet started, which would provide an erroneous negative score. To account for the asynchrony, the presence of a multicellular embryo was scored as positive, while the presence of an undeveloped egg was scored as "inconclusive". Similarly, the formation of endosperm was also scored as positive, while the presence of an undeveloped central cell was scored as "inconclusive".

There were four non-apomictic F_1 plants (H22, H56, H70, and H163) carrying the *MSTA44B-d* allele, in which parthenogenetic embryos were observed with DIC microscopy but that had no observed autonomous endosperm formation and no viable seed set (Figure 5D; H22). Interestingly, the *MSTA44B-d* allele was previously shown to be transmitted by haploid pollen grains to the diploid progeny (in contrast to the Dip-marker, [34]), although at a ratio significantly different from the expected Mendelian ratio (allele frequencies: b = 0.37; d = 0.17; e = 0.45 (N = 94); χ^2 = 12.20; d.f. = 2; the p-value is 0.002, significant at $p < 0.05$). Additionally, in two diploid male-sterile progeny-plants carrying the *MSTA44B-d* allele, embryo-like structures, and autonomous endosperm-like tissues were visible with Nomarski DIC microscopy (see Figure 6). These diploid plants, however, did not produce viable seeds. Taken together, these lines of evidence suggest that this SSR allele is genetically linked to the parthenogenesis locus, and this allele will be referred to as the "Par-marker" below.

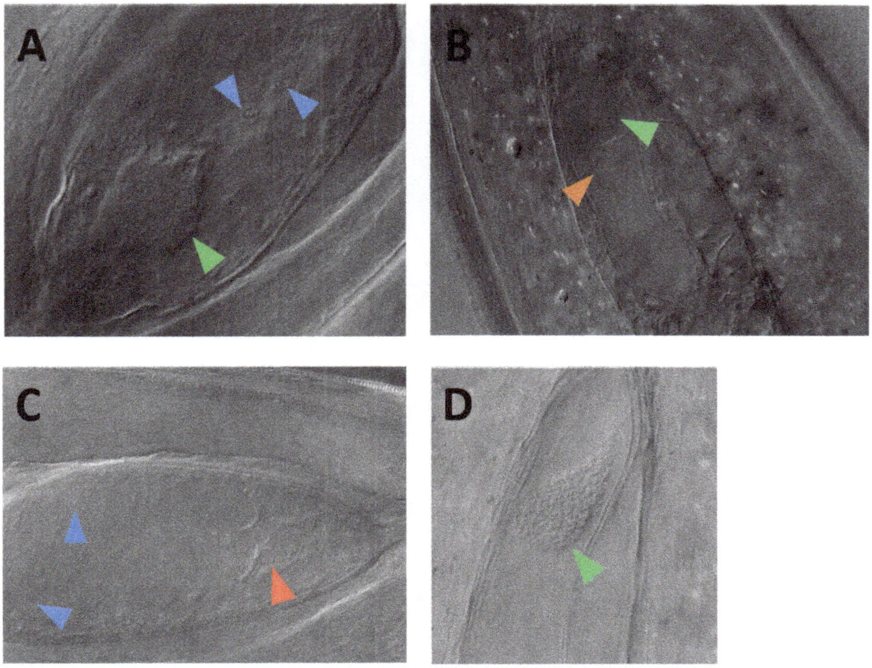

Figure 5. Examples of embryo sac development in 3 × plants. (**A**) A developing seed with a parthenogenetic embryo and autonomous endosperm from the apomictic plant H99. Globular embryo (green arrow) and endosperm nuclei (blue arrows) are visible. (**B**) A developing seed with a multicellular embryo (green arrow) with an undivided central cell of the gametophyte (orange arrow) from the non-apomictic plant H70. The formation of endosperm was seen in this line via FCSS as a hexaploid peak, showing autonomous endosperm was initiated at a later stage. (**C**) A developing seed without parthenogenesis but with cellularized autonomous endosperm development from the non-apomictic plant H123. The red arrow points to the egg cell nucleus; the blue arrows, to endosperm nuclei. (**D**) A developing seed with a parthenogenetic embryo (green arrow) with no detectable autonomous endosperm formation from the non-apomictic plant H22, which carries the Par-marker. Since no FCSS data were available for this line, the information about autonomous endosperm development is non-conclusive.

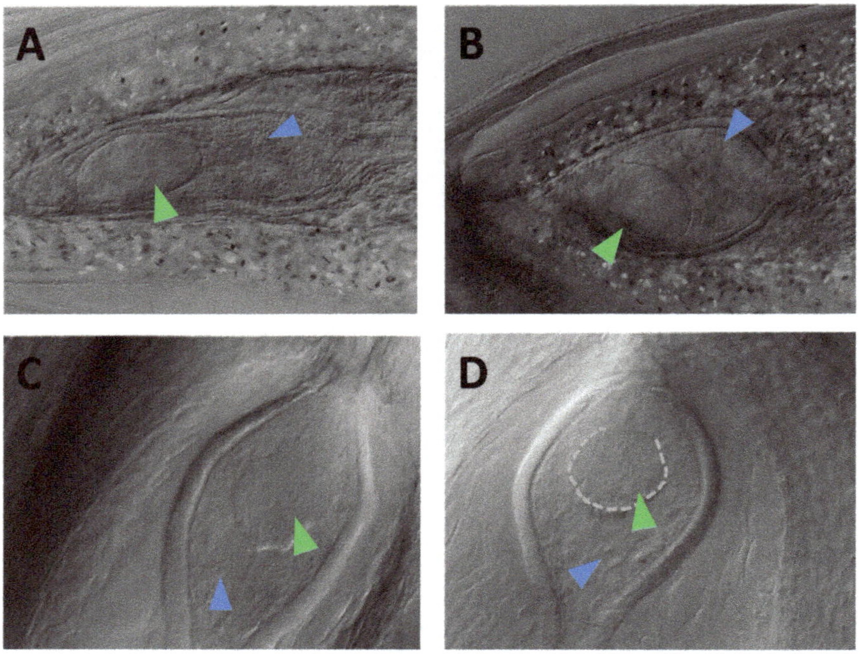

Figure 6. Examples of parthenogenetic embryo-like structures in 2 × non-apomictic plants that carry the Par-marker. (**A**) and (**B**): plant H47; panels (**C**) and (**D**): plant H17. The green arrows point to the embryo-like structures; the blue arrows, to the endosperm nuclei/cells.

3.4. Seed Flow Cytometry to Assess Autonomous Endosperm Formation

Due to the previously mentioned asynchronous initiation of autonomous endosperm, in order to perform a more robust assessment of the autonomous endosperm phenotype, the F_2 seeds of 24 $3 \times F_1$ plants were screened via flow cytometry (FCSS) to determine their ploidy (Table 3; Figure 7). By combining the number of apomictic $3 \times F_1$ (29, which, by definition, have autonomous endosperm formation) along with the non-apomictic $3 \times F_1$ that had the DIP-marker allele and in which autonomous endosperm was directly observed (24, by either Nomarski DIC microscopy or FCSS), a total of 53 diplosporous $3 \times F_1$ plants demonstrated autonomous endosperm formation. Only three did not show any sign of autonomous endosperm formation, and the remaining seven could not be scored (due to either failure to induce flowering or premature death). According to the three-locus model with a single dominant autonomous endosperm formation locus (Figure 1), only 38 plants with autonomous endosperm are expected (67 percent of 56; goodness-of-fit χ^2 = 18.42; d.f.= 1; the p-value is 0.00, significant at $p < 0.05$). Additionally, for the 20 diplosporous (Dip-marker present) $3 \times F_1$ plants that lacked the parthenogenesis-linked Par-marker allele, 19 $3 \times F_1$ plants showed autonomous endosperm formation (95 percent). Thus, significantly more plants with autonomous endosperm were found than were expected, which suggests that the developmental function of autonomous endosperm formation in apomixis is more complex in *Taraxacum* than can be determined by a single dominant genetic locus.

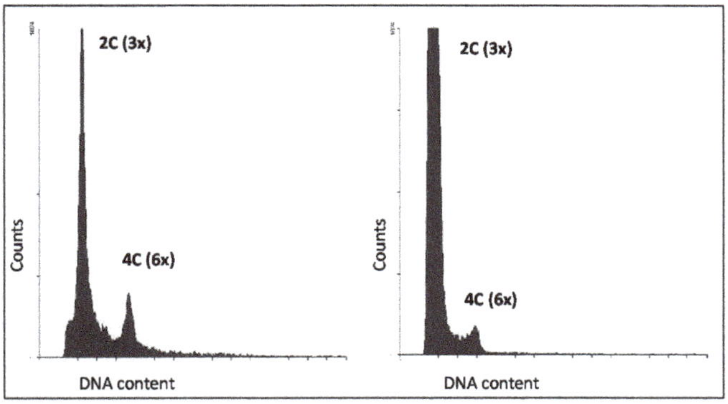

Figure 7. Examples of the Flow Cytometric Seed Screen. (**Left**): An apomictic triploid F_1 plant, with a triploid somatic (maternal and embryo tissue) and a hexaploid endosperm peak showing autonomous endosperm. (**Right**): A non-apomictic triploid F_1 plant, with a triploid somatic (maternal tissue) and a lower hexaploid endosperm peak, also showing the presence of autonomous endosperm.

4. Discussion

Apomixis has since long ago been a holy grail of agriculture [3,4]; now, over a quarter of a century later, the molecular nature of the genes driving apomixis is starting to come into view, largely due to the advances in apomictic model systems such as *Hieracium, Pennisetum*, and *Taraxacum*, starting with a fulsome understanding of the underlying genetic system [42,43]. The progeny of the sexual x apomict *Taraxacum* cross examined here segregated for apomixis as a whole trait as well as for the individual components of apomixis. When considered as a single trait, apomixis was consistent with a three-locus model controlling apomixis. In addition to the already-known SSR marker for diplospory (*MSTA78-a*; Dip-marker), a co-dominant marker tightly linked to the parthenogenesis allele was found (*MSTA44B-d*; Par-marker) that was not genetically linked to the Dip-marker. Thus, the genotype for parthenogenesis is dominant simplex *Ppp*, and the parthenogenesis and diplospory loci are not genetically linked. A dominant simplex parthenogenesis genotype has also been reported in other members of the *Asteraceae*, such as *Erigeron* [11,12] and *Hieracium* [13,14], perhaps with broader developmental roles then shown here for *Taraxacum*. The presence of a single, major parthenogenesis locus significantly enables the cloning of the parthenogenesis gene in *Taraxacum*, as was done for the *VPS13-like* gene in the *DIPLOSPOROUS* locus [30]. Reducing the number of potential apomixis genes in a forward genetic mutation screen or a comparative transcriptomic study to a small set of positional candidates significantly increases the chances of finding causal apomixis genes.

While having the Par-marker will aid in finding the causal gene, the introduction of apomixis into crop plants also requires endosperm formation: either pseudogamy or autonomous formation. In this study, the segregation of autonomous endosperm was investigated in the segregating 3 × F_1 through a combination of cytological observations, seed flow cytometry, and molecular marker analysis. That autonomous endosperm did not appear to segregate was unexpected, because, in an earlier cytological study, phenotypically complementary recombinants between parthenogenesis and autonomous endosperm had been found [32]. An explanation could be that parthenogenesis and/or autonomous endosperm developed late and that the cytological observations were performed too early for parthenogenesis or autonomous endosperm development to become expressed. Fagerlind [44] found a wide variation in developmental stages between florets within a capitulum (inflorescence) of a completely apomictic *T. officinale*. In addition, Cooper and Brink [45] described a high degree of independent development of the embryo and endosperm within florets in a completely apomictic *T. officinale*, while development was very synchronous in the sexual diploid related species *T. koksaghyz*. As extreme examples for *T. officinale*, Cooper and Brink [45] describe a unicellular endosperm (central cell) with a 112-celled embryo and, conversely, an egg cell with a 128-celled endosperm. An advantage of FCSS in this regard is that it pools several developing seeds (hence leveling out developmental variation between florets that is genotype independent) and that the ploidy level indicates whether there is autonomous endosperm (6 × peak). In the present study, microscopic observations in combination with seed flow cytometry of developing seeds five days after flowering suggests that the autonomous endosperm does not segregate in Dip-marker positive triploid progeny plants. The three plants in which no autonomous endosperm could be demonstrated may be explained by recombination between the Dip-marker and the *Diplosporous* locus/gene, because at a distance of 3.5 cM [29], 2–3 meiotic plants are expected among 62 Dip-marker positive plants. An alternative explanation for the failure to find the expected segregation of autonomous endosperm in these experiments could be that the locus for autonomous endosperm is genetically closely linked to the diplospory locus, as functional diplospory was assessed via a linked Dip-marker. In this case, the autonomous endosperm phenotype would be "fixed" in the 61 3 × F_1 lines used for analysis or, at most, ~5 cM away. However, there is no sign of the suppression of recombination in the diplospory locus [27,29], making this explanation unlikely. Lastly, it is possible that the apomictic pollen donor A68 used in these experiments was homozygous for genetic autonomous endosperm factors or that autonomous endosperm is not genetically determined but is, for example, a direct consequence of polyploidy. If autonomous endosperm formation is fixed in this population, then a developmental function needs to be assigned to the third segregating dominant apomixis locus, yet to be identified.

Our results strongly suggest an independent control of autonomous endosperm and parthenogenesis in *T. officinale*. This is supported by the study of Mártonfiová and colleagues [33]. They investigated the components of apomixis in 3 × progeny of a sexual x apomictic cross, using test crosses with a diploid sexual pollinator and FCSS of the developing seeds. Three plants were found that had lost parthenogenesis but made hexaploid autonomous endosperm, showing that parthenogenesis and autonomous endosperm formation do not have a common genetic control. These observations in *Taraxacum* are clearly different from what was seen in *Erigeron* and *Hieracium*. In these other apomictic *Asteraceae* species, autonomous endosperm co-segregated with parthenogenesis, either due to genetically linked genes in a single locus or due to the pleiotropic effects of a single gene. Since the activation of the egg to develop into an embryo and activation of the central cell to develop into an autonomous endosperm are in some ways similar processes, pleiotropy is a possibility, consistent with the observations of Guitton and Berger [18] in *Arabidopsis*. It is also possible that in some cases, at least, the parthenogenetic embryo triggers autonomous endosperm development, as has been reported in *Arabidopsis* for a mutant with a single sperm cell [46].

To aid the isolation of apomixis components, Van Dijk and colleagues [30] generated a γ radiation deletion population of A68, the same apomictic clone used in the crossing described here. In addition to mutants for the diplospory locus, mutants for the parthenogenesis locus were found (Van Dijk et al., in preparation). That these mutants had lost the dominant parthenogenesis allele but still had the dominant diplospory allele was shown by the fact that test crosses with a diploid pollen donor only produced tetraploid progeny. However, these PAR deletion mutants still made autonomous endosperm, as shown by FCSS and Nomarski DIC microscopy. This is in line with the crossing results that parthenogenesis and autonomous endosperm are independently genetically regulated and again underscores the difference in mechanism between *Taraxacum* and other members of the *Asteraceae*.

5. Conclusions

Diplospory and parthenogenesis are clearly major loci of apomixis in *Taraxacum*, each of which is controlled by a single dominant locus. However, apomixis as a whole does not fit a two-factor model, suggesting the presence of a third, critical dominant locus. Based on the observations presented here, this locus does not appear to be the single dominant autonomous endosperm formation locus that might have been expected. Apomixis may be co-controlled by a third unknown major locus of unknown developmental function or by several smaller background or modifier genes that jointly appear as a third main factor in this cross. In order to distinguish between these various hypotheses, further research is needed, which is critical if there is to be any progress in cloning the genes involved in the autonomous endosperm formation component of apomixis. While autonomous endosperm is a major component of apomixis in the *Asteraceae*, it is still the least understood.

Author Contributions: Conceptualization, methodology, investigation, and analysis, P.J.V.D.; writing—original draft preparation and review and editing, P.J.V.D., R.O.d.C., and S.E.S. All authors have read and agreed to the published version of the manuscript.

Funding: This research received no external funding.

Acknowledgments: We thank Marcel van Culemborg, Erik Wijnker, Tanja Bakx-Schotman, Slavica Milanovic-Ivanovic, Radim Vašut, and Sabine van Liere for technical assistance and Hans de Jong and Radim Vašut for discussions.

Conflicts of Interest: The authors declare no conflict of interest. P.J.V.D. and R.O.d.C. are employees of Keygene N.V., while S.E.S. is an employee of Keygene Inc.

References

1. Nogler, G.A. Gametophytic apomixis. In *Embryology of Angiosperms*; Johri, B.A., ed.; Springer: Berlin, Germany, 1984; pp. 475–518.
2. Grossniklaus, U.; Nogler, G.A.; Van Dijk, P.J. How to avoid sex: The genetic control of gametophytic apomixis. *Plant Cell* **2001**, *13*, 15–21. [CrossRef] [PubMed]
3. Jefferson, R.A. Apomixis: A social revolution for agriculture? *Biotechnol. Dev. Monit.* **1994**, *19*, 14–16.
4. Vielle-Calzada, J.-P.; Crane, C.F.; Stelly, D.M. Apomixis: The asexual revolution. *Science* **1996**, *274*, 1322–1323. [CrossRef]
5. Spillane, C.; Curtis, M.D.; Grossniklaus, U. Apomixis technology development—Virgin births in farmers' fields? *Nat. Biotechnol.* **2004**, *22*, 687–691. [CrossRef] [PubMed]
6. Van Dijk, P.J.; Rigola, D.; Schauer, S.E. Plant Breeding: Surprisingly less sex is better. *Curr. Biol.* **2016**, *26*, 122–124. [CrossRef] [PubMed]
7. Hands, P.; Rabiger, D.S.; Koltunow, A. Mechanisms of endosperm initiation. *Plant Reprod.* **2016**, *29*, 215–225. [CrossRef]
8. Noyes, R.D. Apomixis in the *Asteraceae*: Diamonds in the rough. *Funct. Plant Sci. Biotechnol.* **2007**, *1*, 207–222.
9. Ozias-Akins, P.; Van Dijk, P.J. Mendelian genetics of apomixis in plants. *Ann. Rev. Genet.* **2007**, *41*, 509–537. [CrossRef]
10. Hand, M.L.; Koltunow, A.M.G. The genetic control of apomixis: Asexual seed formation. *Genetics* **2014**, *197*, 441–450. [CrossRef]
11. Noyes, R.D.; Rieseberg, L.H. Two independent loci control agamospermy (apomixis) in the triploid flowering plant *Erigeron annuus*. *Genetics* **2000**, *155*, 379–390.
12. Noyes, R.D.; Baker, R.; Mai, B. Mendelian segregation for two-factor apomixis in *Erigeron annuus* (*Asteraceae*). *Heredity* **2007**, *98*, 92–98. [CrossRef] [PubMed]
13. Catanach, A.S.; Erasmuson, S.K.; Podivinsky, E.; Jordan, B.R.; Bicknell, R. Deletion mapping of genetic regions associated with apomixis in *Hieracium*. *Proc. Natl. Acad. Sci. USA* **2006**, *103*, 18650–18655. [CrossRef] [PubMed]
14. Koltunow, A.M.G.; Johnson, S.D.; Rodrigues, J.C.M.; Okada, T.; Hu, Y.; Tsuchiya, T.; Wilson, S.; Fletcher, P.; Ito, K.; Suzuki, G.; et al. Sexual reproduction is the default mode in apomictic *Hieracium* subgenus *Pilosella*, in which two dominant loci function to enable apomixis. *Plant J.* **2011**, *66*, 890–902. [CrossRef] [PubMed]
15. Henderson, S.T.; Johnson, S.D.; Eichmann, J.; Koltunow, A.M.G. Genetic analyses of the inheritance and expressivity of autonomous endosperm formation in *Hieracium* with different modes of embryo sac and seed formation. *Ann. Bot.* **2017**, *119*, 1001–1010. [CrossRef]
16. Vijverberg, K.; Ozias-Akins, P.; Schranz, M.E. Identifying and Engineering Genes for Parthenogenesis in Plants. *Front Plant Sci.* **2019**, *10*, 128. [CrossRef]
17. Conner, J.A.; Mookkan, M.; Huo, H.; Chae, K.; Ozias-Akins, P. Induction of parthenogenesis by a BBM-like gene. *Proc. Natl. Acad. Sci. USA* **2015**, *112*, 11205–11210. [CrossRef]
18. Guitton, A.-E.; Berger, F. Loss of Function of *MULTICOPY SUPPRESSOR OF IRA 1* Produces Nonviable Parthenogenetic Embryos in *Arabidopsis*. *Curr. Biol.* **2005**, *15*, 750–754. [CrossRef]

19. Fenby, N.; Pu, H.; Pennell, R.; Praekelt, U.; Day, R.; Scott, R. An uncoupling screen for autonomous embryo mutants in *Arabidopsis thaliana*. *Sex Plant Reprod.* **2010**, *4*, 255–264. [CrossRef]
20. Aliyu, O.M.; Schranz, M.E.; Sharbel, T.F. Quantitative variation for apomictic reproduction in the genus *Boechera* (*Brassicaceae*). *Am. J. Bot.* **2010**, *97*, 1719–1731. [CrossRef]
21. Chaudhury, A.M.; Ming, L.; Miller, C.; Craig, S.; Dennis, E.S.; Peacock, W.J. Fertilization-independent seed development in *Arabidopsis thaliana*. *Proc. Natl. Acad. Sci. USA* **1997**, *94*, 4223–4228. [CrossRef]
22. Grossniklaus, U.; Vielle-Calzada, J.P.; Hoeppner, M.A.; Gagliano, W.B. Maternal control of embryogenesis by *MEDEA*, a polycomb group gene in *Arabidopsis*. *Science* **1998**, *280*, 446–450. [CrossRef] [PubMed]
23. Luo, M.; Bilodeau, P.; Koltunow, A.; Dennis, E.S.; Peacock, W.J.; Chaudhury, A.M. Genes controlling fertilization independent seed development in *Arabidopsis thaliana*. *Proc. Natl. Acad. Sci. USA* **1999**, *96*, 296–301. [CrossRef] [PubMed]
24. Rodrigues, J.C.; Tucker, M.R.; Johnson, S.D.; Hrmova, M.; Koltunow, A.M. Sexual and apomictic seed formation in *Hieracium* requires the plant polycomb-group gene *FERTILIZATION INDEPENDENT ENDOSPERM*. *Plant Cell* **2008**, *20*, 2372–2386. [CrossRef] [PubMed]
25. Rodrigues, J.C.M.; Okada, T.; Johnson, S.D.; Koltunow, A.M. A *MULTICOPY SUPPRESSOR OF IRA1* (*MSI1*) homologue is not associated with the switch to autonomous seed development in apomictic (asexual) *Hieracium* plants. *Plant Sci.* **2010**, *179*, 590–597. [CrossRef]
26. Van Dijk, P.J.; Bakx-Schotman, J.M.T. Formation of unreduced megaspores (diplospory) in apomictic dandelions (*Taraxacum*) is controlled by a sex-specific dominant locus. *Genetics* **2004**, *166*, 483–492. [CrossRef] [PubMed]
27. Vijverberg, K.; Van der Hulst, R.; Lindhout, P.; Van Dijk, P.J. A genetic linkage map of the diplosporous chromosomal region in *Taraxacum* (common dandelion; Asteraceae). *Theor. Appl. Genet.* **2004**, *108*, 725–732. [CrossRef]
28. Vašut, R.; Vijverberg, K.; Van Dijk, P.J.; de Jong, H. Fluorescent In Situ Hybridization shows *DIPLOSPOROUS* located on one of the NOR-chromosomes in apomictic dandelions in the absence of a large hemizygous chromosomal region. *Genome* **2015**, *57*, 609–620. [CrossRef]
29. Vijverberg, K.; Milanovic-Ivanovic, S.; Bakx-Schotman, J.M.T.; Van Dijk, P.J. Genetic fine-mapping of *DIPLOSPOROUS* in *Taraxacum* (dandelion, Asteraceae) indicates a duplicated DIP-gene. *BMC Plant Biol.* **2010**, *10*, 154. [CrossRef]
30. Van Dijk, P.J.; Rigola, D.; Prins, M.W.; van Tunen, A.J. Diplospory Gene. WO2017/039452 A1, 9 March 2017.
31. Van Dijk, P.J.; Tas, I.C.Q.; Falque, M.; Bakx-Schotman, J.M.T. Crosses between sexual and apomictic dandelions (*Taraxacum*). II. The breakdown of apomixis. *Heredity* **1999**, *83*, 715–721. [CrossRef]
32. Van Dijk, P.J.; Van Baarlen, P.; De Jong, J.H. The occurrence of phenotypically complementary apomixis-recombinants in crosses between sexual and apomictic dandelions (*Taraxacum officinale*). *Sex. Plant Reprod.* **2003**, *16*, 71–76. [CrossRef]
33. Mártonfiová, L.; Majeský, L.; Mártonfi, P. Polyploid progeny from crosses between diploid sexuals and tetraploid apomictic pollen donors in *Taraxacum* sect. *Ruderalia*. *Acta Biol. Crac. Ser. Bot.* **2007**, *49*, 47–54.
34. Van Dijk, P.J.; De Jong, J.H.; Vijverberg, K.; Biere, A. An apomixis-gene's view on dandelions. In *Lost Sex, the Evolutionary Biology of Asexual Reproduction*; Schön, I., Van Dijk, P.J., Martens, K., Eds.; Springer: Berlin, Germany, 2009; pp. 475–493.
35. Van der Hulst, R.G.M.; Meirmans, P.; Van Tienderen, P.H.; Van Damme, J.M.M. Nuclear-cytoplasmic male-sterility in diploid dandelions. *Heredity* **2004**, *93*, 43–50. [CrossRef] [PubMed]
36. Tas, I.C.Q.; Van Dijk, P.J. Crosses between sexual and apomictic dandelions (*Taraxacum*). I. The inheritance of apomixis. *Heredity* **1999**, *83*, 707–714. [CrossRef] [PubMed]
37. Falque, M.; Keurentjes, J.; Bakx-Schotman, J.M.T.; Van Dijk, P.J. Development and characterization of microsatellite markers in the sexual-apomictic complex *Taraxacum officinale* (dandelion). *Theor. Appl. Genet.* **1998**, *97*, 283–292. [CrossRef]
38. Vašut, R.J.; Van Dijk, P.J.; Falque, M.; Travnicek, B.; De Jong, J.H. Development and characterisation of nine new microsatellite markers in *Taraxacum* (*Asteraceae*). *Mol. Ecol. Notes* **2004**, *4*, 645–648. [CrossRef]
39. Van Baarlen, P.; De Jong, J.H.; Van Dijk, P.J. Comparative cyto-embryological investigations of sexual and apomictic dandelions (*Taraxacum*) and their apomictic hybrids. *Sex. Plant Reprod.* **2002**, *15*, 31–38. [CrossRef]
40. Matzk, F.; Meister, A.; Schubert, I. An efficient screen for reproductive pathways using mature seeds of monocots and dicots. *Plant J.* **2000**, *21*, 97–108. [CrossRef]

41. Ulrich, I.; Ulrich, W. High-resolution flow cytometry of nuclear DNA in higher plants. *Protoplasma* **1991**, *165*, 212–215. [CrossRef]
42. Hojsgaard, D. Apomixis Technology: Separating the Wheat from the Chaff. *Genes* **2020**, *11*, 411. [CrossRef]
43. Ozias-Akins, P.; Conner, J.A. Clonal reproduction through seeds in sight for crops. *Trends Genet.* **2020**, *36*, 215–226. [CrossRef]
44. Fagerlind, F. Makrosporogenese und Embryosackbildung bei agamospermischen *Taraxacum*-Biotypen. *Svensk. Bot. Tidskr.* **1947**, *41*, 365–390.
45. Cooper, D.C.; Brink, R.A. The endosperm-embryo relationship in the autonomous apomict, *Taraxacum officinale*. *Bot. Gaz.* **1949**, *111*, 139–152. [CrossRef]
46. Nowack, M.K.; Grini, P.E.; Jakoby, M.J.; Lafos, M.; Koncz, C.; Schnittger, A. A positive signal from the fertilization of the egg cell sets off endosperm proliferation in angiosperm embryogenesis. *Nat. Gen.* **2006**, *38*, 63–67. [CrossRef] [PubMed]

genes

MDPI

Article

Haploidy in Tobacco Induced by *PsASGR-BBML* Transgenes via Parthenogenesis

Zhifen Zhang, Joann Conner, Yinping Guo and Peggy Ozias-Akins *

Department of Horticulture and Institute of Plant Breeding, Genetics & Genomics, University of Georgia, Tifton, GA 31793, USA; zhifen@uga.edu (Z.Z.); jconner@uga.edu (J.C.); yguo@uga.edu (Y.G.)
* Correspondence: pozias@uga.edu; Tel.: +1-229-386-3902

Received: 7 August 2020; Accepted: 10 September 2020; Published: 12 September 2020

Abstract: Background: Engineering apomixis in sexually reproducing plants has been long desired because of the potential to fix hybrid vigor. Validating the functionality of genes originated from apomictic species that contribute to apomixis upon transfer to sexually reproducing species is an important step. The *PsASGR-BABYBOOM-like* (*PsASGR-BBML*) gene from *Pennisetum squamulatum* confers parthenogenesis in this apomict, and its functionality was demonstrated in several sexually reproducing monocots but not in any dicots. Methods: We introduced the *PsASGR-BBML* gene regulated by egg cell-specific promoters, either *AtDD45* or *AtRKD2*, into tobacco, and analyzed progeny of the transgenic lines resulting from self-pollination and crossing by flow cytometry. Results: We identified haploid progeny at a frequency lower than 1% in the $AtDD45_{pro}$ lines, while at a frequency of 9.3% for an octoploid ($2n = 8x$) $AtRKD2_{pro}$ line. Haploid production in the T_2 generation, derived from the tetraploid T_1 offspring of this original octoploid $AtRKD2_{pro}$ line, was also observed. Pollinated by homozygous transgenic tobacco carrying a *DsRed* marker gene, $4x$ progeny of the $AtRKD2_{pro}$ line yielded parthenogenetic embryos identified as DsRed negative. We verified that the DsRed negative seedlings recovered were haploid ($2x$). Conclusion: The *PsASGR-BBML* gene regulated by egg cell-specific promoters could enable parthenogenesis in tobacco, a dicotyledon species.

Keywords: apomixis; parthenogenesis; haploid progeny; dicotyledon; *PsASGR-BBML*; pseudogamy

1. Introduction

Flowering plants (angiosperms) can reproduce both sexually and asexually. In female sexual reproduction, a nucellar cell goes through meiosis to produce four megaspores, one of which divides by mitosis to develop into the female gametophyte (embryo sac) containing an egg cell with two flanking synergid cells at the micropylar end, a central cell with two polar nuclei, and antipodal cells at the chalazal end. Microspore mother cells in anthers go through meiotic division and develop into the male gametophytes (pollen grains). The haploid gametophytes complete the life cycle by developing gametes which fuse to initiate the diploid sporophyte. Angiosperms require the process of double fertilization; the egg cell ($1n$) merges with a sperm cell ($1n$) to form a zygote ($2n$) that develops into an embryo encapsulated within a seed, while the central cell ($2n$) fuses with the second sperm cell ($1n$) from the same pollen grain to initiate endosperm ($3n$) development. In dicots, the endosperm typically is absorbed during seed maturation, while in monocots it provides nutrition for germinating seeds [1]. In contrast, as an asexual reproduction pathway, apomixis produces seeds without double fertilization. The mechanisms of apomixis can be complex and diverse across species and have been reviewed intensively [2–5]. In general, various forms of gametophytic apomixis share three common components: (1) apomeiosis by which unreduced female gametophytes form; (2) parthenogenesis by which female gametes develop into embryos without fertilization; (3) formation of endosperm autonomously or following fertilization of the central cell by a sperm cell (pseudogamy). If parthenogenesis occurs independently from apomeiosis, haploid progeny can be obtained when viable endosperm develops autonomously or after the central cell is fertilized.

Parthenogenesis is genetically controlled as reviewed previously [6]. As a member of the *AP2*-domain transcription factor superfamily, a *PsASGR-BABYBOOM*-like (*PsASGR-BBML*) gene isolated from an apomictic grass, *Pennisetum squamulatum*, within the Apospory-Specific Genomic Region (ASGR) [7], was the first gene shown to confer parthenogenesis in plants with its functionality validated in sexual pearl millet (*P. glaucum*) [8]. The promoter of the *PsASGR-BBML* gene was found to control gene expression restricted to the egg apparatus (egg/synergid complex) of the unfertilized embryo sac [8], indicating that egg-specific expression of the *PsASGR-BBML* gene could be critical for the resulting parthenogenesis phenotype. Later, the functionality of the *PsASGR-BBML* gene regulated by either its native promoter or an egg cell-specific promoter *DD45* (*DOWN REGULATED IN DETERMINANT INFERTILE* (*DD*) 45) from *Arabidopsis* [9] was validated in two additional monocot species, rice (*Oryza sativa*) and maize (*Zea mays*) [10]. However, when the *PsASGR-BBML* gene regulated by the *At*DD45 promoter was introduced into *Arabidopsis*, no significant haploid production was observed, even though the gene was expressed within ovules and no splicing changes were identified [10]. The exact reason why the *PsASGR-BBML* gene failed to enable parthenogenesis in *Arabidopsis*, a dicot species, remains unknown.

Given its amphiploid origin [11], tobacco (*Nicotiana tabacum*; $2n = 4x = 48$) can provide an alternative scenario to test the functionality of the *PsASGR-BBML* gene in a dicot species, where the gene will express in a di-haploid egg cell. The recovery/survival of haploid progeny was reported at a very low frequency in tobacco [12–14]. When seeds resulted from self-pollination, only one individual of a surviving twin was identified as haploid out of 104,000 germinated seeds representing 12 varieties, indicating a frequency of approximately one in 10,000 or lower [12]. When seeds resulted from intraspecific crosses, haploid progeny were identified with the aid of genetic markers at a frequency ranging from 0.02% to 0.2% [12–14], which varied depending on the combinations of selected parents. Our objective was to test whether introducing the *PsASGR-BBML* gene could significantly increase the tendency of haploid production in amphiploid tobacco. In order to test the functionality of the *PsASGR-BBML* gene in tobacco, eudicot egg cell-specific promoters are preferred to regulate the transgene, as the specific expression of the *PsASGR-BBML* gene in the egg cell was considered essential for the parthenogenesis phenotype and its endogenous promoter was not functional in dicots [8,10]. Over-expression of the *PsASGR-BBML* by *CaMV35S* (*Cauliflower Mosaic Virus 35S*) promoter in *Arabidopsis* led to the formation of somatic embryo-like structures on the adaxial leaf surface and ectopic shoot/leaf development [15] rather than parthenogenesis, comparable to the observation in transgenic *Arabidopsis* and *Brassica* where the *BnBBM1* and *BnBBM2* genes were over-expressed [16]. The promoters of *AtRKD1* and *AtRKD2*, members of a class of transcription factors containing a *RWP-RK domain* (*RKD*) that are important to maintain the egg identity in *Arabidopsis*, drove gene expression specifically in egg cells [17] so that they provide alternatives to the *AtDD45* promoter to regulate the *PsASGR-BBML* gene in dicot plants.

The study herein reported was an effort to test the functionality of the *PsASGR-BBML* transgene regulated by two egg cell-specific promoters in tobacco and validate whether this gene isolated from a monocot apomict could lead to haploid production via parthenogenesis in a dicot species.

2. Materials and Methods

2.1. Vector Construction

Two cassettes of the genomic *PsASGR-BBML* sequence (EU559280) fused with egg cell-specific promoters were generated. The cassette *AtDD45*$_{pro}$*:gPsASGR-BBML* contained a 1008 bp promoter of *Arabidopsis DD45/EC1.2* (At2g21740) [9], the 3540 bp of the *PsASGR-BBML* gene from BAC p208 [18,19], including 8 exons and 7 introns, and 609 bp 3′ of the stop codon plus the predicted poly(A) signal. The cassette was subsequently ligated into pCambia1300 (CAMBIA, Canberra, Australia) for hygromycin selection in transgenic plants [10]. The recombinant vector pCambia1300-*AtDD45$_{pro}$:gPsASGR-BBML* (ddBR1) was introduced into *Agrobacterium* strains AGL1 and LBA4404 for plant transformation.

The 521 bp promoter region of *AtRKD2* (At1g74480), previously verified as egg-cell expressing in Arabidopsis, [17] was amplified from genomic DNA of *Arabidopsis* Columbia using the primer combination p3905/3906 (Table S1). The sequence of the *AtRDK2* promoter was verified. The *AtDD45* promoter in the cassette *AtDD45$_{pro}$:gPsASGR-BBML* was replaced with the *AtRKD2* promoter and the new cassette *AtRKD2$_{pro}$:gPsASGR-BBML* was ligated into pCambia2300 for kanamycin selection in transgenic plants. Kanamycin selection offered a simpler scheme to separate transformant T1 progeny from non-transformant T1, compared with the use of hygromycin selection (see Section 2.5). The new recombinant vector pCambia2300-*AtRKD2$_{pro}$:gPsASGR-BBML* (*RKD2gBB*) was verified by additional restriction mapping, and introduced into *Agrobacterium* strain EHA105, as AGL1 led to relatively severe tissue browning during the transformation of ddBR1.

2.2. Plant Transformation

Two vectors were introduced into tobacco by *Agrobacterium*-mediated transformation adapted from Clemente [20]. Leaves from 1-month-old tobacco plants (PI 552484, *N. tabacum* cv. Xanthi NN, seeds purchased from Lehle Seeds, Round Rock, TX, USA) grown in the greenhouse were collected, immersed in reverse osmosis deionized (RODI) water, and then surface sterilized with 10% commercial bleach (Clorox, Oakland, CA, USA, (6% sodium hypochlorite)) for 10 min, followed by 5× rinse with sterile RODI water. After removal of the midribs, the leaves were cut into 1–2 cm^2 explants and cultured on a shoot induction medium (SIM) consisting of Murashige and Skoog salts and vitamins (M519, PhytoTechnology Laboratories, Lenexa, KS, USA), 3% (*w*/*v*) sucrose (Research Products International, Mt Prospect, IL, USA), 1 mg/L 6-benzylaminopurine (BA), 0.1 mg/L 1-naphthaleneacetic acid (NAA), and 8 g/l agar with pH adjusted to 5.7. After 2 days of culture, the leaf disc explants were inoculated with *Agrobacterium* by immersing the explants in *Agrobacterium* suspension (OD_{600} = 0.5~0.6, in liquid SIM containing 200 μM acetosyringone and 0.02% (*v*/*v*) Silwet-77 (Lehle Seeds, Round Rock, TX, USA)) for 15~20 min. After blotting dry with sterile filter paper, explants were placed on fresh SIM for co-culture. After 2 days of co-culture, explants were transferred to SIM containing 30 mg/L meropenem (ABBLIS Chemicals, Houston, TX, USA), plus 30 mg/L hygromycin or 200 mg/L kanamycin for ddBR1 and RKD2gBB, respectively. Each plate contained 5–7 explants which were transferred to fresh SIM with antibiotic selection every 2 weeks. When shoots reached 2–3 cm, they were excised and transferred to a root induction medium (RIM) containing MS salts and vitamins, 3% (*w*/*v*) sucrose, 0.1 mg/L NAA, and 8 g/L agar with pH adjusted to 5.7, plus 30 mg/L meropenem and the same selective agent used in SIM. Rooted plantlets were transferred to the Magenta™ GA7 vessels (Sigma-Aldrich, St. Louis, MO, USA), containing 0MS medium (SIM without plant growth regulators) with 30 mg/L meropenem and the same selective agent used in SIM, for further development. Plantlets derived from different leaf disc explants were considered independent transformation events while those from the same explants were considered potentially the same lines. Plantlets of 5–8 cm height were transferred to soil for acclimation. After genotyping, at least two plants from each independent line with a full-length *PsASGR-BBML* transgene were grown in the greenhouse to set seeds. All tissue cultures were maintained at 26 °C with a 16/8 h light cycle. Unless otherwise noted, all chemicals were from Sigma-Aldrich (St. Louis, MO, USA).

2.3. Genotyping

Genomic DNA was isolated from T0 transgenic plants and their progeny by using the CTAB method [21]. The presence of the full-length *gPsASGR-BBML* transgene in the T0 transgenic lines and the haploid T1 progeny was verified by polymerase chain reaction (PCR) using the primer combination p1792/1801 (Table S1) to amplify the region from the start of the open reading frame (ORF) through 159 bp of the 3′ UTR. PCR reactions consisted of 1× PrimeSTAR GXL Buffer, 200 μM dNTP, 0.2 μM primers, 0.625 U PrimeSTAR GXL DNA Polymerase (Takara Bio USA, Inc. Mountain View, CA, USA), and 50–100 ng genomic DNA in a 25 μL reaction followed by 35 cycles of amplification with a cycling condition of 15 s at 98 °C, 15 s at 60 °C, and 4 min at 68 °C. In order to confirm the presence of the transgenes in the haploid T2 progeny of line RKD2gBB, progeny were genotyped by PCR using primer combinations p3767/4127 and p4303/4304 (Table S1). PCR reactions consisted of 1× GoTaq® Master Mix (M7123, Promega, Madison, WI) and 50–100 ng genomic DNA in a 20 μL reaction followed by 32 cycles of amplification with a cycling condition of 15 s at 95 °C, 15 s at 60 °C, and 30 s at 72 °C. PCR products were visualized under UV light after electrophoresis in 1% agarose gels and staining with ethidium bromide.

2.4. Estimation of Transgene Copy Numbers by Quantitative PCR

In order to estimate the copy number of the transgenes in T0 and T1 plants, genomic DNA from transgenic tobacco was digested overnight with *Eco*RI-HF® (NEB, Ipswich, MA, USA), followed by purification using the DNA Clean and Concentrator Kit (D4033, Zymo Research, Irvine, CA, USA). The purified DNA, after dilution 20-fold, was used as a template for quantitative PCR (qPCR) using a LightCycler® 480 system (Roche, Basel, Switzerland). PCR reactions were set up following the manufacturer's instruction of the LightCycler® 480 SYBR Green I Master mix V13 (Roche) and conducted using the SYBR green I/HRM dye program with the primer combination p4303/4304 for the *PsASGR-BBML* transgene and the primer combination p4133/4134 for the *tubulin* gene (NCBI Reference Sequence: XM_016623993) as a reference gene [22]. The amplification efficiency of each assay was estimated based on the qPCR data of a 5-log serial dilution (0.0016, 0.008, 0.04, 0.2, 1×) of a DNA mixture that contained an equal amount of DNA from each tested sample by using the absolute quantification method in the software of the Lightcycler® 480 system. Each reaction had two technical replicates. The qPCR data were analyzed by using the advanced relative quantification method in the software of the Lightcycler® 480 system to estimate the copy numbers of the *PsASGR-BBML* transgene in transgenic tobacco plants.

2.5. Seed Germination Assay for Transgenic Progeny

Seeds were surfaced sterilized using a 16-h chlorine gas treatment (2 mL 12 M HCl into 200 mL commercial bleach in a 10 L desiccator) or immersing in 10% commercial bleach (Clorox) for 10 min followed by 5× rinse with sterile RODI water. Since cotyledons could turn green despite the lack of transgene when seeds were germinated on hygromycin-containing medium and unreduced transgenic progeny potentially could outcompete the haploid plants with prolonged culture on hygromycin-containing medium, a selection scheme was developed to quickly separate unreduced transgenic progeny from their haploid siblings and non-transgenic with a short culture period on hygromycin-containing medium, based on a method developed for *Arabidopsis* [23]. Seeds of the ddBR1 lines were sown on 0MS medium containing 60 mg/L hygromycin and incubated in the dark for 7–10 days. Seedlings displaying elongated hypocotyls (>0.5 cm, mostly 0.8~1 cm) were considered to be transgenic while seedlings that did not elongate (<0.5 cm, mostly 0.1~0.3 cm) were considered to be either non-transgenic or potentially haploid transgenic. After transfer to 0MS for further growth with a 16/8 h light cycle, short seedlings that turned green were considered potential transgenic haploids while seedlings that failed to turn green and grow further were considered to be non-transgenic. The ploidy levels of those potentially haploid transgenic plants were examined by flow cytometry within a month. Adapted from [24], seeds of the RKD2gBB lines were sown on 0MS containing 600 mg/L kanamycin which was intended to prevent non-transgenic escape. After a 2-week culture period, green seedlings were considered to be transgenic while seedlings showing chlorosis (i.e., fully or partially whitening) at cotyledons or the shoot apex were considered to be non-transgenic. Those resistant transgenic seedlings were transferred to 0MS for further growth and their ploidy levels were examined by flow cytometry within a month. In order to estimate the baseline of autonomous haploid production of the PI 552484, non-transgenic seeds were sown on 0MS and the seedlings that did not elongate after 7 days in the dark were considered to be potentially haploid. Those small seedlings were transferred to fresh 0MS for further growth with a 16/8 h light cycle, and their ploidy levels were examined by flow cytometry within a month.

2.6. Flow Cytometry

Possible parthenogenesis in T0 transgenic plants was first determined by flow cytometric seed screen (FCSS) [25]. Preliminary microscopy study found most of the endosperm was absorbed by the embryo when capsules started to turn brown but were not completely desiccated (21–28 days after pollination). Capsules at this developmental stage were selected for FCSS as seeds were not too hard to chop while the flow cytometric signal would be mostly from the developing embryos rather than the endosperm. Bulked seed samples from two individual capsules were processed for each transgenic plant adapted from Conner [8]. In brief, 50–100 developing seeds together with young leaf tissue of sorghum (*Sorghum bicolor*) or cowpea (*Vigna unguiculata*) as a genome-size standard were chopped in 200 μL LB01 lysis buffer consisting of 15 mM Tris, 2 mM Na_2EDTA, 0.5 mM spermine tetrahydrochloride, 80 mM KCl, 20 mM NaCl, 0.1% (*v*/*v*) Triton X−100, pH 7.5 and 16 mM 2-mercaptoethanol, and filtered with a 30 μm CellTrics disposable filter (Sysmex Partec, Görlitz, Germany). After propidium iodide solution containing RNase (Cat# 550825, BD Biosciences, San Jose, CA, USA) was added at half volume, the filtered samples were incubated on ice for at least 15 min, followed by analysis using a BD Accuri C6 flow cytometer (BD Biosciences) with gating set by the selection of objects with a strong correlation between FL2 and FL3 signals using a flow rate of 35 μL sample per min. Events were collected within the gated region for each sample.

In order to identify haploid seedlings, selected tobacco seedlings (see Section 2.5) were analyzed by flow cytometry when they reached at least the 4-leaf stage. Leaf tissues (1 mm^2 or less) from five seedlings were pooled together and processed as described previously by flow cytometry. If a bulk sample showed a haploid signal, leaf tissue from each individual comprising the bulk was collected and processed by flow cytometry to identify the haploid individuals. The efficiency of haploid production was determined based on the percentage of haploid seedlings among the total seedlings germinated.

2.7. Transcription Analysis of the PsASGR-BBML Transgene in Tobacco Ovules

Ovules were isolated from 6–7 flowers at stage 11 of tobacco flower development [26] and stored in the RNA*later*™ Storage Solution (Invitrogen, Carlsbad, CA, USA). Total RNA was extracted using the RNeasy® Plant Mini Kit (QIAGEN, Hilden, Germany), followed by a DNase (Cat# 18068015, ThermoFisher Scientific, Waltham, MA, USA) treatment according to the manufacturer's recommendation to remove genomic DNA. Single strand cDNA was synthesized with the SuperScript III First-strand Synthesis System (Invitrogen) by reverse transcription PCR (RT-PCR). The presence of the *PsASGR-BBML* transcript including 8 exons was verified by PCR using the primer combination p1792/1793 (Table S1). PCR reactions consisted of 2 μL of the first-strand cDNA synthesis reaction, 1× PrimeSTAR GXL Buffer, 200 μM dNTP, 0.2 μM primers, 0.625 U PrimeSTAR GXL DNA Polymerase in a 25 μL reaction (Takara Bio USA, Inc.) followed by 35 cycles of amplification with a cycling condition of 15 s at 98 °C, 15 s at 60 °C, and 2 min at 68 °C. PCR products were visualized under UV light after electrophoresis in 1% agarose gels and staining with ethidium bromide.

2.8. Sequence Analysis of the PsASGR-BBML Transgene from Line RKD2gBB_6.1

Total RNA from ovules of flowers at stage 11 from two haploid-producing 4*x* T1 progeny (K6.1_20 and K6.1_23) of line RKD2gBB_6.1 was extracted, DNase treated and converted into single strand cDNA as previously stated. PCR reactions consisted of 2 μL of the first-strand cDNA synthesis reaction, 1× PrimeSTAR GXL Buffer, 200 μM dNTP, 0.2 μM primers, and 0.625 U PrimeSTAR GXL DNA Polymerase in a 25 μL reaction (Takara Bio USA, Inc) followed by 35 cycles of amplification with a cycling condition of 15 s at 98 °C, 15 s at 60 °C, and 2 min at 68 °C. Primers p1792/1801 (Table S1) were used to amplify the transcript from the start of the ORF through to the 3′ UTR. Amplified PCR products from each sample were directly cloned into PCR4-TOPO (Invitrogen) vector with cloned inserts sequenced at Psomagen (Rockville, MD, USA). Sequencing data were analyzed using Geneious prime software (Biomatters Limited, Auckland, New Zealand).

2.9. Tobacco Crossing

Flowers were emasculated at stage 10 [26] in the evening before anthesis. The emasculated flowers were pollinated the following morning, tagged and capped with tailored pollination envelopes. The $4x$ T1 plants of line RKD2gBB_6.1 (an octoploid, $2n = 8x$) were pollinated by non-transgenic tobacco. The $4x$ T2 plants of line RKD2gBB_6.1 were pollinated by homozygous T3 plants of a transgenic tobacco line ($2n = 4x$) carrying a cassette of a *DsRed* gene regulated by the promoter of soybean (*Glycine max*) *elongation factor 1a* (*GmEF1a*, Glyma.17G186600, Phytozome v12) and the *nopaline synthase* (*NOS*) terminator that segregated as a single locus (Zhang unpublished). The embryos and seedlings without DsRed fluorescence were considered to be parthenogenetic, since these progenies did not inherit the *DsRed* transgene from the male parent.

2.10. Embryo Isolation and Observation

Capsules were harvested 9 through 17 days after pollination. Ovules were isolated and macerated in an enzymatic solution [27] composed of 12% (*w*/*v*) mannitol, 3 mM MES, 1% (*w*/*v*) cellulose R-10 (Research Products International), and 0.8% (*w*/*v*) *Rhizopus sp.* pectinase, with pH adjusted to 5.7, for 30 min with agitation at 60 rpm. The ovules were rinsed in a washing buffer composed of 12% (*w*/*v*) mannitol and 3 mM MES (pH 5.7) at least 3 times and then gently ground with a small pestle to release the developing embryos. Without removing integument debris, isolated embryos were mounted on a glass slide in the washing buffer and observed under a microscope (Zeiss, Thornwood, NY, USA) equipped with a PhotoFluor LM-75 illuminator (89 North, Williston, VT, USA) and a Ds-RED filter (excitation: 545/25 nm, emission: 605/70 nm, Chroma Technology, Bellows Falls, VT, USA). Images were taken using an AxioCam camera (Carl Zeiss, Oberkochen, Germany) and the AxioVision LE64 software. Embryos without visible damage were counted. Unless otherwise noted, all chemicals were obtained from Sigma-Aldrich.

3. Results

3.1. Recovery of Transgenic Tobacco

Thirty and seventeen T0 lines that contained all eight exons, seven introns and 3′UTR were recovered, respectively, for the cassettes *AtDD45*$_{pro}$*:gPsASGR-BBML* (ddBR1) and *AtRKD2*$_{pro}$*:gPsASGR-BBML* (RKD2gBB). In general, all transgenic lines displayed wild type vegetative growth, except for one ddBR1 line that was severely stunted. Upon entering the reproductive stage, most T0 lines showed normal flower development and produced seeds by self-pollination. In some T0 lines, short stamens and poor pollen production were observed. By hand-pollination with their own pollen when available, all lines with abnormal phenotypes produced viable seeds except for the severely stunted ddBR1 line. The *PsASGR-BBML* transcript including eight exons was detected in ovules of ten transgenic lines that were randomly selected for each construct at stage 11 of tobacco flower development (Table S2).

3.2. Haploid Production of T0 Transgenic Tobacco

As tobacco is an allotetraploid ($2n = 4x$), progeny were considered "haploid" when their genome size was calculated at half of their maternal parent. Five independent ddBR1 lines (5/29, 17.2%) and three independent RKD2gBB lines (3/17, 17.6%) displayed a haploid ($2x$ since tobacco is tetraploid) signal when immature seeds were analyzed by FCSS (Figure S1). All T0 transgenic lines remained tetraploid ($2n = 4x$) except for three RKD2gBB lines (RKD2gBB_4.1, RKD2gBB_6.1 and RKD2gBB_7.1) which became octoploid ($2n = 8x$), probably due to genome duplication during tissue culture. According to FCSS, two of the haploid-producing RKD2gBB T0 lines (RKD2gBB_6.1 and RKD2gBB_7.1) were octoploid ($2n = 8x$, Figure S1). The ploidy levels of the haploid-producing lines were confirmed by flow cytometry analysis using leaf tissue of the T0 plants, and these haploid-producing lines were analyzed in further detail.

In a preliminary study, T1 seedlings of a haploid-producing ddBR1 line that germinated on 0MS containing 60 mg/L hygromycin and elongated after seven days of culture in the dark were all tetraploid ($4x$) as their T0 parent. Haploid ($2x$) seedlings were only identified among seedlings that failed to elongate during the first seven days on 0MS containing hygromycin. Thus, we decided to focus on the T1 seedlings of the ddBR1 lines that failed to elongate on 0MS containing 60 mg/L hygromycin within 10 days where haploid plants would be more likely to be identified. Haploid ($2x$) seedlings were identified among T1 seedlings of three haploid-producing ddBR1 lines ($2n = 4x$), ddBR1_204.2, ddBR1_205.1 and ddBR1_240.2, at a frequency lower than 1%, while no haploid T1 seedlings were recovered from the other two haploid-producing ddBR1 lines, ddBR1_13.1 and ddBR1_204.1, showing FCSS haploid peaks (Table 1, Figure 1). This could be due to the frequency of haploid seedling production against the number of seedlings screened. In contrast, among the three haploid-producing RKD2gBB lines, haploid ($4x$) T1 seedlings were identified from one octoploid ($2n = 8x$) line (RKD2gBB_6.1) at a frequency of 9.3% while no haploid plant was identified in the other two lines, RKD2gBB_3.2 and RKD2gBB_7.1, showing FCSS haploid peaks (Table 1, Figure 1). In comparison, 397 seedlings that remained stunted within the first two weeks of growth on 0MS from 3700 non-transgenic seedlings were analyzed by flow cytometry, and no haploid seedlings were identified (Figure 1).

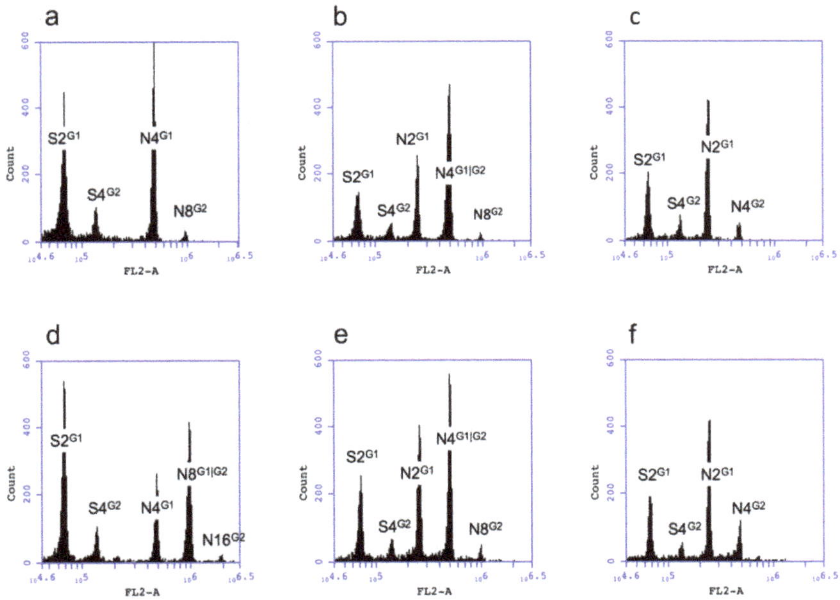

Figure 1. Representative flow cytometry analysis. (**a**) A bulked sample of five non-transgenic tobacco plants; (**b**) a bulked sample of T1 progeny of a haploid-producing $4x$ ddBR1 line (ddBR1_240.2) with $2x$ signal; (**c**) a $2x$ T1 progeny of line ddBR1_240.2; (**d**) a bulked sample of T1 progeny of haploid-producing $8x$ line RKD2gBB_6.1 with $4x$ signal; (**e**) a bulked sample of T2 progeny of a $4x$ T1 plant (K6.1_20) of line RKD2gBB_6.1 with $2x$ signal; (**f**) a $2x$ T2 progeny of a $4x$ T1 plant K6.1_20. $S2^{G1}$ and $S4^{G2}$ designate $2n/2x/2c$ and $2n/2x/4c$ peaks of sorghum or cowpea, $N2^{G1}$, $N4^{G1}$, $N4^{G2}$, $N4^{G1|G2}$, $N8^{G2}$, $N8^{G1|G2}$, and $N16^{G2}$ designate $2n/2x/2c$, $2n/4x/4c$, $2n/2x/4c$, $2n/4x/4c|2n/2x/4c$, $2n/4x/8c$, $2n/8x/8c|2n/4x/8c$, and $2n/8x/16c$ peaks of tobacco, respectively.

Table 1. Efficiency of haploid production in the haploid-producing T0 after self-pollination.

T0 line ID	Ploidy	# Total Seedlings	# Seedlings Screened [1]	# Haploid T1 (%)
ddBR1_13.1	4x	1065	186	0 (0)
ddBR1_204.1	4x	785	101	0 (0)
ddBR1_204.2	4x	585	153	4 (0.7%)
ddBR1_205.1	4x	582	68	1 (0.2%)
ddBR1_240.2	4x	575	123	5 (0.9%)
RKD2gBB_3.2	4x	127	80	0 (0)
RKD2gBB_6.1	8x	129	117	12 (9.3%)
RKD2gBB_7.1	8x	223	65	0 (0)

[1] For the ddBR1 lines, the seedlings screened were from a sub-population of seedlings which did not elongate when germinated on 0MS medium containing 60 mg/L hygromycin and remained shorter than 0.5 cm within 10 days.

All haploid-producing lines had a relatively simple integration of the transgene. Based on the elongation of T1 seedlings germinated in the dark, indicating resistance to hygromycin, the transgene likely segregated as a single locus in all five haploid-producing T0 ddBR1 lines except for line ddBR1_204.1 showing segregation distortion of the transgene and having an inflated number of T1 seedlings characterized as "susceptible" (Table S3). Among the haploid-producing RKD2gBB lines, the transgene segregated as a single locus in lines RKD2gBB_3.2 and RKD2gBB_7.1, and as two loci in the line RKD2gBB_6.1 which showed the highest efficiency in haploid production (Table S3). When genotyped by PCR, all haploid T1 plants identified carried the full-length sequence of *PsASGR-BBML* transgene. Quantitative PCR further suggested that the haploid-producing T0 lines should have one or two copies of the transgene integrated except for one line (Table S3).

3.3. Haploid Production of the Progeny Derived from Line RKD2gBB_6.1

As tobacco is an allotetraploid, haploid T1 plants derived from the ddBR1 lines could not form normal embryo sacs due to the lack of homologous chromosomes for pairing at meiosis. As a result, they were sterile and produced no seeds. Since line RKD2gBB_6.1 was 8*x*, its haploid (4*x*) T1 were fertile and capable of producing seeds, providing an opportunity to verify the functionality of the *PsASGR-BBML* transgene among progeny lines even when the ploidy level reduced by half in the second generation. Similar to the T0 generation, T1 plants of line RKD2gBB_6.1 tended to have short stamens and many empty anthers. Nevertheless, seeds could still be produced by hand-pollination with their own pollen as long as some pollen grains were available. Among nine 4*x* T1 plants derived from line RKD2gBB_6.1 ($2n = 8x$), three were able to produce T2 seeds after hand-pollination. The other six plants produced no T2 seeds after attempted hand-pollination as functional pollen grains were not produced from these plants. Haploid (2*x* and 4*x*) T2 seedlings were identified from the three 4*x* T1 plants (K6.1_11, K6.1_20 and K6.1_23) and four 8*x* T1 plants (K6.1_1, K6.1_12, K6.1_17, and K6.1_22) analyzed, respectively (Table 2, Figure 1). The efficiency of haploid production ranged from 2.7% to 27.3%, showing no clear correlation to the numbers of loci inherited or the copy numbers of the transgene (Table 2 and Table S4). The T1 plants that inherited one locus or both were able to produce haploids (Table 2 and Table S4). In general, the efficiency of haploid production tended to be higher in the 8*x* T1 than the 4*x* T1. The 8*x* T1 plants tended to set seeds more poorly with higher frequencies of haploid progeny associated with poorer seed setting and where fewer seeds were available for analysis (Table 2).

Haploid (2*x*) progeny were obtained at a frequency of 5.6% when a 4*x* T1 plant of line RKD2gBB_6.1 was pollinated by non-transgenic tobacco (Table 3). The number of seedlings susceptible to kanamycin inflated in the F1 to about three times as many as the resistant seedlings (Table S5). Segregation distortion was also observed in progeny from the 4*x* T1 plants of line RKD2gBB_6.1 that inherited one or both transgene loci. The number of resistant seedlings was much lower than expected when these T1 plants were self-pollinated and a relatively large proportion of seeds failed to germinate (Table S4). When genotyped, all haploid T2 progeny identified carried the *PsASGR-BBML* transgene and the *nptII* marker gene. Attempts to recover homozygous T1 plants of any haploid-producing ddBR1 lines were unsuccessful. A homozygous T1 plant of line RKD2gBB_6.1 was likely recovered (K6.1_7, Table S4) according to the copy number estimated by qPCR, but it only produced distorted flowers and no seeds.

Table 2. Efficiency of haploid production in the RKD2gBB_6.1 T1 progeny after self-pollination.

T1 Plant ID	Ploidy	# Total Seedlings	# Green Resistant Seedlings	# Seedlings Screened	# Haploid T2 (%)
K6.1_11	4*x*	103	57	56	13 (12.6%)
K6.1_14 [1]	4*x*	-	-	-	-
K6.1_20	4*x*	387	111	101	14 (3.6%)
K6.1_23	4*x*	211	80	52	12 (5.7%)
K6.1_1	8*x*	51	40	40	8 (15.7%)
K6.1_7 [1]	8*x*	-	-	-	-
K6.1_12	8*x*	224	173	150	50 (22.3%)
K6.1_17	8*x*	22	21	20	6 (27.3%)
K6.1_22	8*x*	75	73	60	2 (2.7%)

[1] No flowers with pollen produced during the growing season, resulting in no progeny.

Table 3. Haploid production from a 4*x* RKD2gBB_6.1 T1 plant pollinated with wildtype pollen.

♀	♂	# Total Seedlings	# Green Resistant Seedlings	# Seedlings Screened	# Haploid Progeny (%)
K6.1_20	wildtype	285	71	56	16 (5.6%)

3.4. *Transcription of the PsASGR-BBML Transgene in the 4x T1 of Line RKD2gBB_6.1*

The presence of the *PsASGR-BBML* transcript including eight exons was verified in all four 4*x* T1 plants of line RKD2gBB_6.1 (Figure S2) along with four T1 plants not displaying parthenogenesis. Sequencing data showed that the *PsASGR-BBML* transcript from two 4*x* T1 plants (K6.1_20 and K6.1_23) of line RKD2gBB_6.1 was identical to the original transcript, showing no alternative splicing or sequence alteration to explain the higher rate of parthenogenesis in this line.

3.5. *Observation of Parthenogenetic Embryos*

Attempts to observe parthenogenetic embryos via ovule clearing from emasculated flowers of the haploid-producing lines were unsuccessful. This method had been used to successfully identify parthenogenesis in pearl millet transgenic lines [8]. Tobacco flowers abscised approximately six days after emasculation with no defined embryos observed in cleared ovules. Pollination was needed to prevent flower abscission, and to ensure that haploid embryos could further develop and be identified. In preliminary studies to enzymatically isolate embryos from the several hundred ovules of a capsule, only a small number of four-celled embryos were observed seven days after pollination. This suggested that either tobacco embryos at the four-cell stage were too fragile to isolate or that the majority of embryos might be at an earlier developmental stage that could be easier to damage during isolation or harder to identify after isolation. We decided to examine capsules at least nine days after pollination when the majority of embryos entered the eight-cell stage or later. After crossing with pollen from tobacco plants homozygous for a DsRed fluorescence marker gene driven by a strong constitutive promoter GmEF1apro, 4*x* T2 progeny derived from three 8*x* T1 plants of line RKD2gBB_6.1 that inherited one or both loci of the transgene were able to produce embryos or seeds without DsRed fluorescence (Figure 2, Table 4). The DsRed negative embryos were considered to be parthenogenetic due to the lack of the DsRed transgene transmitted by the paternal parent. Developing parthenogenetic embryos without DsRed fluorescence were observed at a frequency from 3.2% to 8.0% among different 4*x* T2 plants of line RKD2gBB_6.1 (Table 4) which was comparable to the frequency of haploid production observed when a 4*x* T1 plant was pollinated by wildtype pollen (Table 3). Those DsRed negative parthenogenetic embryos tended to develop as normally as the DsRed positive embryos from the same capsule, though abnormality (e.g., slow-growth, deformity, etc.) was occasionally observed in some of the DsRed negative embryos. Seedlings without DsRed fluorescence were also identified when the mature seeds from capsules obtained by crossing were germinated, and they were verified as haploid (2*x*) by flow cytometry (Table 5). According to PCR genotyping, these 2*x* haploid seedlings carried the *PsASGR-BBML* transgene but not the DsRed gene.

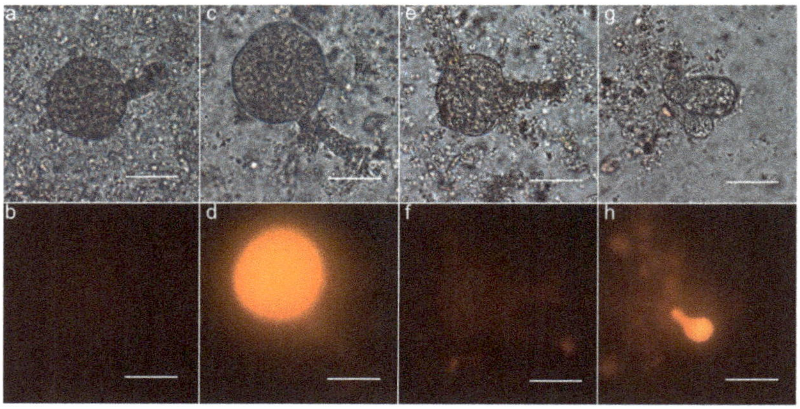

Figure 2. Expression of DsRed in developing embryos of the progeny from crosses with a homozygous transgenic tobacco line carrying a DsRed marker gene. (**a**,**c**,**e**,**g**) Bright field; (**b**,**d**,**f**,**h**) Ds-RED filter. (**a**,**b**) Progeny of the non-transgenic obtained by self-pollination. (**c**,**d**) Progeny of the non-transgenic crossed with the homozygous DsRed transgenic plant; (**e**,**f**) DsRed negative progeny of K6.1_12_26 crossed with the homozygous DsRed transgenic plant; (**g**,**h**) DsRed positive and negative progeny of K6.1_17_7 crossed with the homozygous DsRed transgenic plant. Bar = 50 μm.

Table 4. Expression of DsRed in developing embryos of the crosses between 4*x* T2 plants of line RKD2gBB_6.1 and homozygous transgenic tobacco carrying a DsRed cassette.

Maternal Parent	# Crosses Investigated	# Total Embryos Observed	# DsRed + Embryos	# DsRed − Embryos	% Parthenogenetic (DsRed − Embryos)
wildtype	9	1497	1497	0	0
K6.1_1_35	4	446	428	18	4.0%
K6.1_12_26	8	537	504	33	6.1%
K6.1_12_75	1	250	242	8	3.2%
K6.1_12_104	1	155	150	5	3.2%
K6.1_17_7	6	550	506	44	8.0%

Table 5. Germination of progeny from the crosses between 4*x* T2 plants of line RKD2gBB_6.1 and homozygous transgenic tobacco carrying a DsRed cassette.

Maternal Parent	# Total Seedlings	# DsRed + Seedlings	# DsRed − Seedlings	# DsRed − Seedlings with Reduced Ploidy (Parthenogenesis %)
wildtype	220	220	0	0
K6.1_1_35	216	212	4	4 (1.8%)
K6.1_12_35	23	22	1	1 (4.3%)
K6.1_12_26	152	148	4	4 (2.6%)

4. Discussion

This report demonstrates that the *PsASGR-BBML* transgene, regulated by egg cell-specific promoters, enabled tobacco to produce haploid progeny via parthenogenesis as previously observed in sexual pearl millet [8], rice and maize [10]. While haploid production by self-pollination has been reported in some tobacco varieties at a frequency of approximately 0.01% [12], the haploid frequency observed in the current study with transgenic tobacco was 20 to 900 times higher. In addition, haploid progeny was not identified among seeds resulting from self-pollination of the non-transgenic PI line used to produce the transgenic lines. This indicates that haploid production was a gain of function attributable to the transgene rather than a propensity caused by the genetic composition of the PI. Expression of *PsASGR-BBML* in egg cells presumably transforms the egg cells into a zygote-like state, inducing embryogenesis from unfertilized eggs. Use of an egg cell-specific promoter *AtDD45* to regulate the gene *PsASGR-BBML* or *OsBBM1* whose protein sequence clustered with the PsASGR-BBML protein in a phylogenetic subclade [8] was able to induce parthenogenetic embryos in rice [10,28]. Compared with the monocot species tested [10], the efficiency of haploid production in transgenic tobacco where the *PsASGR-BBML* transgene was regulated by the *AtDD45* promoter was low (Table 1). Given that embryogenesis in monocots significantly diverges from dicots [29], the *PsASGR-BBML* transgene by itself may be inefficient, if sufficient, to transform the egg cell into a zygote-like cell in tobacco, which was also implied in the study of *Arabidopsis* where haploid progeny were not recovered from any transgenic lines expressing the *PsASGR-BBML* transgenes [10]. The reasons why the *PsASGR-BBML* transgene functioned more efficiently in tobacco than *Arabidopsis* remains unknown.

The higher efficiency of haploid production in the RKD2gBB line than the ddBR1 lines was likely attributable to temporal or spatial differences in the promoter activity. When haploid progeny were identified, the RKD2gBB line (RKD2gBB_6.1) yielded haploid progeny at a frequency at least 10 times higher than the ddBR1 lines (Table 1). While $AtDD45_{pro}$ and $AtRKD2_{pro}$ were both considered to be egg cell-specific, gene expression regulated by the $AtRKD2_{pro}$ was tightly restricted to the egg cell [17] while the expression regulated by the $AtDD45_{pro}$ was observed in the egg cell and synergids (the egg apparatus), as well as the early developing embryos from the zygote stage to the eight-cell stage [30]. If the activity of these two promoters in tobacco was consistent with what was observed in *Arabidopsis*, such temporal and spatial differences in the *PsASGR-BBML* expression likely resulted in more haploid progeny recovered in the RKD2gBB line than the ddBR1 lines, which was comparable to the observation where egg ablation by using the *barstar* gene happened more frequently when the gene was regulated by the $AtRKD2_{pro}$ than the $AtDD45_{pro}$ [30]. While RKD transcription factors play an essential role in maintaining the quiescent state of egg cells [31], expression of the *PsASGR-BBML* transgene in a temporal and spatial manner similar to the endogenous *RKD* genes appeared to relax the quiescent state of egg cells and promote parthenogenesis (inducing embryogenesis from unfertilized egg cells).

In addition to the differences of the promoter activity, it could be argued that the genome duplication in line RKD2gBB_6.1 might contribute to the higher efficiency of haploid production. Since the 4*x* T1 progeny of line RKD2gBB_6.1 still produced more haploid progeny than the ddBR1 lines (Tables 1 and 2) and two other octoploid (8*x*) RKD2gBB lines, line RKD2gBB_4.1 and line RKD2gBB_7.1, recovered in this study did not produce any haploid progeny or at a very low frequency, respectively, the increase in ploidy would not be the only cause of more haploid production. We are unable to exclude the possibility that the particular integrations of the transgene in the genome could magnify the transgene effect in line RKD2gBB_6.1 which would require further investigation. Regardless of a similar frequency of haploid-producing lines identified, none of the ddBR1 lines reached the same level of haploid production as line RKD2gBB_6.1.

The efficiency of haploid production varied among lines, though a full-length of the *PsASGR-BBML* transgene was present in all these lines. Such variation was also observed in the previous study of rice and maize [10], probably due to positional differences in integration of the transgene. The transgenes seemed to segregate as a single locus in all haploid-producing lines except for line RKD2gBB_6.1 that had the transgenes integrated at two independent loci (Table S3). Based on the haploid production in the limited number of RKD2gBB_6.1 T1 progeny tested, the transgene at one locus was sufficient to induce haploid production (Table 2 and Table S4). A segregating population would be needed to determine whether both loci were functional and whether an additive effect of the two loci might lead to a higher frequency of haploid production. The segregating population could also help to unravel potentially unknown changes in the genome if any occurred that contributed to haploid production. Characterizing the integration sites of the transgene would be helpful to examine whether the transgene integration interrupted any genes associated with reproduction which might have enhanced the transgene effect on parthenogenesis.

Homozygous progeny have not been recovered for any haploid-producing lines, except for one T1 plant (K6.1_7) of line RKD2gBB_6.1 that appeared to be a homozygote based on the qPCR data but was unable to set seeds due to the severe malformation of flowers. Previous study in rice and maize also had difficulty in recovering plants homozygous for the *PsASGR-BBML* transgenes [10]. Homozygosity potentially brought about embryo lethality given the segregation ratio of the T1 progeny derived from some haploid-producing lines (i.e., a 2:1 ratio in Table S3). The lack of fertile homozygous plants made it difficult to address the question of why not all egg cells that inherited the transgene showed the parthenogenesis phenotype (penetrance). The inflated numbers of non-transgenic progeny and a large proportion of seeds that failed to germinate in some T0 and T1 plants (Tables S3 and S4) suggested that the transgene or its integration site might affect embryo development and survival, which was observed in rice carrying the *PsASGR-BBML* transgenes [10].

Unlike sexual pearl millet carrying the *PsASGR-BBML* transgene where parthenogenetic embryos were observed in ovules of emasculated flowers without pollination [8], developing embryos were not observed in ovules of the haploid-producing transgenic tobacco lines unless emasculated flowers were hand-pollinated. Pseudogamous parthenogenesis observed in the *PsASGR-BBML* transgenic tobacco might be similar to what was observed in *Boechera* apomicts where self-pollination is still required to initiate the development of apomictic embryos as in sexual reproduction but the egg cell does not fuse with either sperm cell [32]. Since parthenogenetic progeny were obtained as long as the 4*x* progeny of line RKD2gBB_6.1 were pollinated regardless of whether the pollen carried the *PsASGR-BBML* transgene or not (Tables 2 and 3), pollination could simply lead to fertilization of the central cell and endosperm formation to support development of parthenogenetic embryos into viable seeds, which was also observed and required in monocot species [8,10]. Fertilization of the central cells could generate certain critical signals to regulate early embryo development as observed in *Arabidopsis* [33] that would affect embryo survival in tobacco. Pollination also could deliver some paternal signals [34,35] that potentially influence parthenogenetic egg or zygote development. If parthenogenetic embryos began to develop without pollination, the slowness of embryo development in tobacco made it difficult to observe them, as most embryos only completed the first division in ovules seven days after pollination whereas flowers without pollination usually abscised six days after emasculation, as previously reported [27,36]. Taking advantage of a homozygous tobacco line carrying a *GmEF1a:DsRed:NOS* cassette as a pollen donor enabled us to identify parthenogenetic embryos that could be distinguished from the sexually reproducing embryos due to their lack of the paternal trait DsRed fluorescence. The haploidy characteristic of the DsRed negative seedlings was verified, indicating that the DsRed negative embryos did result from parthenogenesis.

5. Conclusions

Despite its origin from a monocot species, the *PsASGR-BBML* gene was shown to be functional in a dicot species, inducing haploid production in tobacco. Tuning the *PsASGR-BBML* transgene by altering the promoter seemed to improve the efficiency of haploid production. Compared with the monocot species previously studied, the lower penetrance of the gene in the dicot species tobacco probably stems from the divergence in the reproductive process and embryogenesis between monocots and dicots. Further investigation on the genetic network and protein interactions with the PsASGR-BBML transcription factor in egg cells may reveal the absence of monocot-specific co-factors or the presence of suppressors in dicots that hinder the full potential of the *PsASGR-BBML* gene to promote parthenogenesis.

Supplementary Materials: The following are available online at http://www.mdpi.com/2073-4425/11/9/1072/s1, Figure S1: Flow cytometric seed screen (FCSS) analysis identified T0 plants that produced haploid progeny; Figure S2: Expression of the PsASGR-BBML transgene in the ovules from the haploid-producing 4*x* T1 of RKD2gBB_6.1 compared with non-haploid-producing RKD2gBB T0 lines; Table S1: Sequences of primers used for PCR; Table S2: T0 lines randomly selected to verify expression of the PsASGR-BBML transgene in ovules of flowers at developmental stage 11 by RT-PCR; Table S3. Genetic segregation and copy number estimate of the transgenes in the haploid-producing T0 lines; Table S4. Genetic segregation and copy number estimate of the transgenes in T1 progeny of line RKD2gBB_6.1; Table S5. Genetic segregation in progeny of a cross between a 4*x* RKD2gBB_6.1 T1 and non-transgenic tobacco.

Author Contributions: Conceptualization, Z.Z., J.C., and P.O.-A.; methodology, Z.Z., J.C., and P.O.-A.; formal analysis, Z.Z., J.C., and Y.G.; investigation, Z.Z., J.C., and Y.G.; writing—original draft preparation, Z.Z.; writing—review and editing, J.C. and P.O.-A.; supervision, P.O.-A.; project administration, P.O.-A.; funding acquisition, P.O.-A. All authors have read and agreed to the published version of the manuscript.

Funding: This research was funded through a sub-award from the CSIRO under the Capturing Heterosis for Smallholder Farmers grant from the Bill and Melinda Gates Foundation.

Acknowledgments: We thank Rebecca Grantham and Kathleen Monfero Marasigan for providing technical assistance in flow cytometry and genotyping.

Conflicts of Interest: The authors declare no conflict of interest. The funders had no role in the design of the study; in the collection, analyses, or interpretation of data; in the writing of the manuscript, or in the decision to publish the results.

References

1. Johri, B.M. (Ed.) *Embryology of Angiosperms*, 1st ed.; Springer: Berlin/Heidelberg, Germany; New York, NY, USA; Tokyo, Japan, 1984.
2. Nogler, G.A. Gametophytic apomixis. In *Embryology of Angiosperms*, 1st ed.; Johri, B.M., Ed.; Springer: Berlin/Heidelberg, Germany, 1984; pp. 475–518.
3. Koltunow, A.M.; Grossniklaus, U. Apomixis: A developmental perspective. *Annu. Rev. Plant Biol.* **2003**, *54*, 547–574. [CrossRef] [PubMed]
4. Ozias-Akins, P.; van Dijk, P.J. Mendelian genetics of apomixis in plants. *Annu. Rev. Genet.* **2007**, *41*, 509–537. [CrossRef] [PubMed]
5. Hand, M.L.; Koltunow, A.M. The genetic control of apomixis: Asexual seed formation. *Genetics* **2014**, *197*, 441–450. [CrossRef] [PubMed]
6. Vijverberg, K.; Ozias-Akins, P.; Schranz, M.E. Identifying and engineering genes for parthenogenesis in plants. *Front. Plant Sci.* **2019**, *10*, 128–144. [CrossRef]
7. Ozias-Akins, P.; Roche, D.; Hanna, W.W. Tight clustering and hemizygosity of apomixis-linked molecular markers in *Pennisetum squamulatum* implies genetic control of apospory by a divergent locus that may have no allelic form in sexual genotypes. *Proc. Natl. Acad. Sci. USA* **1998**, *95*, 5127–5132. [CrossRef]
8. Conner, J.A.; Mookkan, M.; Huo, H.; Chae, K.; Ozias-Akins, P. A parthenogenesis gene of apomict origin elicits embryo formation from unfertilized eggs in a sexual plant. *Proc. Natl. Acad. Sci. USA* **2015**, *112*, 11205–11210. [CrossRef]
9. Steffen, J.G.; Kang, I.H.; Macfarlane, J.; Drews, G.N. Identification of genes expressed in the *Arabidopsis* female gametophyte. *Plant J.* **2007**, *51*, 281–292. [CrossRef]
10. Conner, J.A.; Podio, M.; Ozias-Akins, P. Haploid embryo production in rice and maize induced by *PsASGR-BBML* transgenes. *Plant Reprod.* **2017**, *30*, 41–52. [CrossRef]
11. Leitch, I.J.; Hanson, L.; Lim, K.Y.; Kovarik, A.; Chase, M.W.; Clarkson, J.J.; Leitch, A.R. The ups and downs of genome size evolution in polyploid species of *Nicotiana* (Solanaceae). *Ann. Bot.* **2008**, *101*, 805–814. [CrossRef]
12. de Nettancourt, D.D.; Stokes, G.W. Haploidy in tobacco. *J. Hered.* **1960**, *51*, 102–104. [CrossRef]
13. Burk, L.G. Haploids in genetically marked progenies of tobacco. *J. Hered.* **1962**, *53*, 222–226. [CrossRef]
14. Lewis, R.S.; Rose, C. Identification of tobacco haploids on the basis of transgenic overexpression of *PAP1* from *Arabidopsis thaliana*. *Crop. Sci.* **2011**, *51*, 1491–1497. [CrossRef]

15. Huo, H. Genetic Analysis of the Apospory-Specific Genomic Region. (ASGR) in *Pennisetum squamulatum*: From Mapping to Candidate Gene. Ph.D. Thesis, University of Georgia, Athens, GA, USA, August 2008; pp. 123–189.
16. Boutilier, K.; Offringa, R.; Sharma, V.K.; Kieft, H.; Ouellet, T.; Zhang, L.; Hattori, J.; Liu, C.M.; van Lammeren, A.A.; Miki, B.L.; et al. Ectopic expression of BABY BOOM triggers a conversion from vegetative to embryonic growth. *Plant Cell* **2002**, *14*, 1737–1749. [CrossRef]
17. Kőszegi, D.; Johnston, A.J.; Rutten, T.; Czihal, A.; Altschmied, L.; Kumlehn, J.; Wüst, S.E.; Kirioukhova, O.; Gheyselinck, J.; Grossniklaus, U.; et al. Members of the *RKD* transcription factor family induce an egg cell-like gene expression program. *Plant J.* **2011**, *67*, 280–291. [CrossRef] [PubMed]
18. Gualtieri, G.; Conner, J.A.; Morishige, D.T.; Moore, L.D.; Mullet, J.E.; Ozias-Akins, P. A segment of the apospory-specific genomic region is highly microsyntenic not only between the apomicts *Pennisetum squamulatum* and Buffelgrass, but also with a rice chromosome 11 centromeric-proximal genomic region. *Plant Physiol.* **2006**, *140*, 963–971. [CrossRef] [PubMed]
19. Conner, J.A.; Goel, S.; Gunawan, G.; Cordonnier-Pratt, M.M.; Johnson, V.E.; Liang, C.; Wang, H.; Pratt, L.H.; Mullet, J.E.; Debarry, J.; et al. Sequence analysis of bacterial artificial chromosome clones from the apospory-specific genomic region of Pennisetum and Cenchrus. *Plant Physiol.* **2008**, *147*, 1396–1411. [CrossRef]
20. Clemente, T. Nicotiana (*Nicotiana tobaccum*, *Nicotiana benthamiana*). In *Agrobacterium Protocols*, 2nd ed.; Wang, K., Ed.; Humana Press: Totowa, NJ, USA, 2006; Volume 1, pp. 143–154.
21. Doyle, J.J.; Doyle, J.L. Isolation of plant DNA from fresh tissue. *Focus* **1990**, *12*, 39–40.
22. Głowacka, K.; Kromdijk, J.; Leonelli, L.; Niyogi, K.K.; Clemente, T.E.; Long, S.P. An evaluation of new and established methods to determine T-DNA copy number and homozygosity in transgenic plants. *Plant Cell Environ.* **2016**, *39*, 908–917. [CrossRef]
23. Harrison, S.J.; Mott, E.K.; Parsley, K.; Aspinall, S.; Gray, J.C.; Cottage, A. A rapid and robust method of identifying transformed *Arabidopsis thaliana* seedlings following floral dip transformation. *Plant Methods* **2006**, *2*, 1–7. [CrossRef]
24. Odell, J.T.; Hoopes, J.L.; Vermerris, W. Seed-Specific gene activation mediated by the Cre/lox site-specific recombination system. *Plant Physiol.* **1994**, *106*, 447–458. [CrossRef]
25. Matzk, F.; Meister, A.; Schubert, I. An efficient screen for reproductive pathways using mature seeds of monocots and dicots. *Plant J.* **2000**, *21*, 97–108. [CrossRef] [PubMed]
26. Koltunow, A.M.; Truettner, J.; Cox, K.H.; Wallroth, M.; Goldberg, R.B. Different temporal and spatial gene expression patterns occur during anther development. *Plant Cell* **1990**, *2*, 1201–1224. [CrossRef]
27. Fu, C.; Sun, M.; Zhou, C.; Yang, H. Isolation of fertilized embryo sacs and zygotes and triggering of zygote division in *Nicotiana tabacum*. *Acta Bot. Sin.* **1996**, *38*, 262–267.
28. Khanday, I.; Skinner, D.; Yang, B.; Mercier, R.; Sundaresan, V. A male-expressed rice embryogenic trigger redirected for asexual propagation through seeds. *Nature* **2019**, *565*, 91–95. [CrossRef] [PubMed]
29. Zhao, P.; Begcy, K.; Dresselhaus, T.; Sun, M.X. Does early embryogenesis in eudicots and monocots involve the same mechanism and molecular players? *Plant Physiol.* **2017**, *173*, 130–142. [CrossRef]
30. Lawit, S.J.; Chamberlin, M.A.; Agee, A.; Caswell, E.S.; Albertsen, M.C. Transgenic manipulation of plant embryo sacs tracked through cell-type-specific fluorescent markers: Cell labeling, cell ablation, and adventitious embryos. *Plant. Reprod.* **2013**, *26*, 125–137. [CrossRef] [PubMed]
31. Rövekamp, M.; Bowman, J.L.; Grossniklaus, U. Marchantia MpRKD regulates the gametophyte-sporophyte transition by keeping egg cells quiescent in the absence of fertilization. *Curr. Biol.* **2016**, *26*, 1–8. [CrossRef]
32. Kirioukhova, O.; Shah, J.N.; Larsen, D.S.; Tayyab, M.; Mueller, N.E.; Govind, G.; Baroux, C.; Federer, M.; Gheyselinck, J.; Barrell, P.J.; et al. Aberrant imprinting may underlie evolution of parthenogenesis. *Sci. Rep.* **2018**, *8*, 1–7. [CrossRef]
33. Costa, L.M.; Marshall, E.; Tesfaye, M.; Silverstein, K.A.; Mori, M.; Umetsu, Y.; Otterbach, S.L.; Papareddy, R.; Dickinson, H.G.; Boutiller, K.; et al. Central cell-derived peptides regulate early embryo patterning in flowering plants. *Science* **2014**, *344*, 168–172. [CrossRef]
34. Bayer, M.; Nawy, T.; Giglione, C.; Galli, M.; Meinnel, T.; Lukowitz, W. Paternal control of embryonic patterning in *Arabidopsis thaliana*. *Science* **2009**, *323*, 1485–1488. [CrossRef]

35. Ueda, M.; Aichinger, E.; Gong, W.; Groot, E.; Verstraeten, I.; Dai Vu, L.; De Smet, I.; Higashiyama, T.; Umeda, M.; Laux, T. Transcriptional integration of paternal and maternal factors in the *Arabidopsis* zygote. *Genes Dev.* **2017**, *31*, 617–627. [CrossRef] [PubMed]
36. Tian, H.Q.; Russell, S.D. Calcium distribution in fertilized and unfertilized ovules and embryo sacs of *Nicotiana tabacum* L. *Planta* **1997**, *202*, 93–105. [CrossRef]

genes

MDPI

Article

Efficient CRISPR/Cas9-Mediated Knockout of an Endogenous *PHYTOENE DESATURASE* Gene in T1 Progeny of Apomictic *Hieracium* Enables New Strategies for Apomixis Gene Identification

Sam W. Henderson *,†, **Steven T. Henderson**, **Marc Goetz** and **Anna M. G. Koltunow** *,‡

Commonwealth Scientific and Industrial Research Organisation (CSIRO) Agriculture and Food, Glen Osmond, SA 5064, Australia; steven.henderson@csiro.au (S.T.H.); marc.goetz@csiro.au (M.G.)

* Correspondence: sam.henderson@adelaide.edu.au (S.W.H.); a.koltunow@uq.edu.au (A.M.G.K.); Tel.: +61-407-323-260 (A.M.G.K.)

† Present address: Adelaide Medical School, University of Adelaide, Adelaide, SA 5005, Australia.

‡ Present address: Queensland Alliance for Agriculture and Food Innovation (QAAFI), The University of Queensland, St Lucia, QLD 4069, Australia.

Received: 7 August 2020; Accepted: 8 September 2020; Published: 10 September 2020

Abstract: Most *Hieracium* subgenus *Pilosella* species are self-incompatible. Some undergo facultative apomixis where most seeds form asexually with a maternal genotype. Most embryo sacs develop by mitosis, without meiosis and seeds form without fertilization. Apomixis is controlled by dominant loci where recombination is suppressed. Loci deletion by γ-irradiation results in reversion to sexual reproduction. Targeted mutagenesis of genes at identified loci would facilitate causal gene identification. In this study, the efficacy of CRISPR/Cas9 editing was examined in apomictic *Hieracium* by targeting mutations in the endogenous *PHYTOENE DESATURASE* (*PDS*) gene using *Agrobacterium*-mediated leaf disk transformation. In three experiments, the expected albino dwarf-lethal phenotype, characteristic of *PDS* knockout, was evident in 11% of T0 plants, 31.4% were sectorial albino chimeras, and the remainder were green. The chimeric plants flowered. Germinated T1 seeds derived from apomictic reproduction in two chimeric plants were phenotyped and sequenced to identify *PDS* gene edits. Up to 86% of seeds produced albino seedlings with complete *PDS* knockout. This was attributed to continuing Cas9-mediated editing in chimeric plants during apomictic seed formation preventing *Cas9* segregation from the *PDS* target. This successful demonstration of efficient CRISPR/Cas9 gene editing in apomictic *Hieracium*, enabled development of the discussed strategies for future identification of causal apomixis genes.

Keywords: *Hieracium piloselloides*; apomixis; CRISPR/Cas9; *PHYTOENE DESATURASE* (*PDS*); amplicon sequencing; genome editing; tissue culture

1. Introduction

Increasing seed yields in major crops requires new insights into the function of genes that regulate plant reproduction. Plants typically form seeds via a sexual pathway requiring meiosis to form male and female gametes and fertilization to initiate seed formation (Figure 1A). A number of non-agronomic plants can also form seed asexually via apomixis. Apomixis is typically a dominant trait. However, most apomicts are facultative, meaning the sexual pathway remains intact in some ovules. During apomixis, genetically identical seeds of a maternal genotype form because meiosis is avoided during female gametophyte or embryo sac formation, and fertilization is not required for embryo formation [1–3]. Harnessing apomixis in plant breeding would have significant benefits for seed production and accelerate the delivery of hybrid crops with improved traits [4]. Attempts to introgress apomixis into sexual seed crops from apomictic relatives, which are typically polyploid, have been unsuccessful [5]. Furthermore, the suppression of recombination around apomixis loci in most molecularly studied apomicts has consistently hindered positional cloning to identify apomixis genes [2]. To date, the only known gene identified from a gametophytic apomict is that conferring fertilization-independent embryogenesis (parthenogenesis) from *Pennisetum squamulatum* [6].

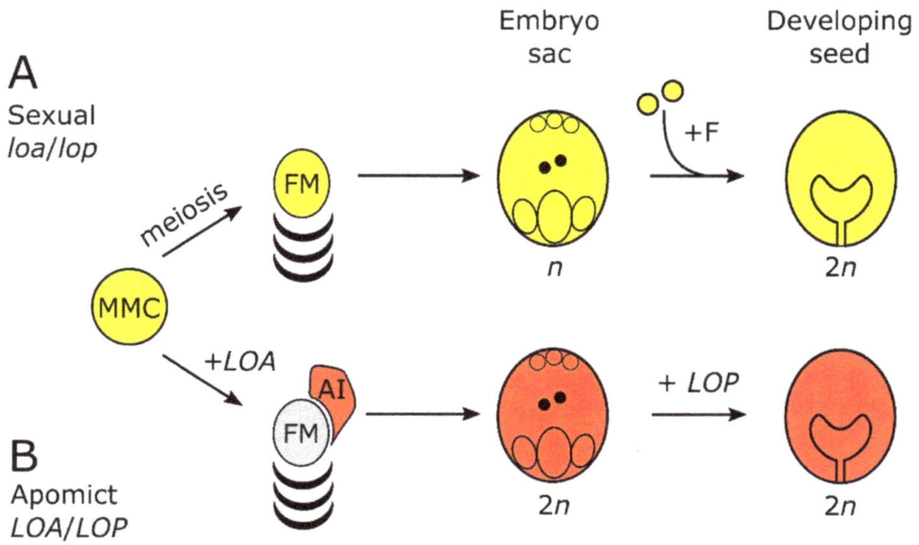

Figure 1. Simplified schematic of sexual and apomictic reproduction in *Hieracium* subgenus *Pilosella* (**A**) The sequence of events in sexual *Hieracium pilosella* indicated in yellow. MMC indicates megaspore mother cell, +F = fertilization, FM = functional megaspore; (**B**) events of apomixis (red) in *Hieracium praealtum* (R35) and events controlled by activity of the *LOSS OF APOSPORY* (*LOA*) and *LOSS OF PARTHENOGENESIS* (*LOP*) loci. AI = aposporous initial cell.

Targeted genomic deletions, together with mutations in genes at apomixis loci would facilitate identification of additional apomixis genes. However, CRISPR/Cas9-mediated genome editing, which takes advantage of the cellular DNA double-strand break repair pathways to generate small indels, targeted substitutions and multiplex genome modifications [7], has not yet been reported in a gametophytic apomict. The application of CRISPR has been demonstrated to alter gene function in polyembryonic *Citrus* [8,9], wherein nucellar embryos form via fertilization-independent sporophytic apomixis [2]. However, those studies did not focus on identification of causal genes for apomixis utilizing this technology.

Hieracium subgenus *Pilosella* species, members of the Asteraceae, have been developed as a eudicot model for the molecular analysis of gametophytic apomixis. Sexual and apomictic species are self-incompatible, therefore seeds formed in the absence of fertilization are easily identified. In characterized apomicts, more than 95% of seed is formed via the facultative apomictic pathway [10], which is dependent on the initiation of the sexual pathway (Figure 1B). During apomixis, somatic aposporous initial (AI) cells differentiate near cells undergoing meiosis, the sexual pathway terminates, and the AI cells undergo mitosis forming the (aposporous) embryo sac within which embryo and endosperm formation are fertilization-independent [11]. In apomictic *Hieracium praealtum* (R35) and *Hieracium piloselloides* (D36) the dominant, *LOSS OF APOMEIOSIS* (*LOA*) locus regulates AI cell formation, suppression of the sexual pathway and aposporous embryo sac formation. In R35 the dominant *LOSS OF PARTHENOGENESIS* (*LOP*; Figure 1B) locus controls fertilization-independent embryogenesis and endosperm formation [11,12]. An additional locus has been identified in D36 termed *AutE* which enables fertilization-independent endosperm formation [13,14].

Recombination is suppressed at identified apomixis loci. For example, the *LOA* locus is located near the distal end of a single chromosome and is surrounded by extensive repeats and transposons [15]. This is also a feature of the apomixis carrying chromosome in monocots *P. squamulatum* and *Cenchrus ciliaris* [16,17]. Despite these genetic features of apomicts, extensive genome and transcriptome resources have been developed for apomictic *Hieracium* [18,19], and the interrogation of *Hieracium* cell-type-specific transcriptomes has identified candidate genes and pathways for apomixis that require functional testing [19–21]. Chromosome walking has enabled the identification of additional genes at both the *LOA* and *LOP* loci [15,22]. Thus, it would be timely to develop rapid new methods to test candidate apomixis gene function.

Interestingly, γ irradiation-induced deletions of the *LOA* and *LOP* loci in apomictic *H. praealtum* (R35) have shown phenotypic reversion to the sexual mode of reproduction. This indicates that sexual reproduction is the default reproductive mode [11,12]. Targeted deletions in genomic regions and genes linked to and within at *LOA* and *LOP* loci from chromosome walking would conceivably enable the functional characterization of apomixis genes. Importantly, unlike many apomict model species, the characterized sexual and apomictic subgenus *Pilosella* species can be efficiently transformed using *Agrobacterium* mediated leaf disc transformation [23].

CRISPR/Cas9 gene editing has been assessed in a small number of sexual Asteraceae species. The thermoinhibition gene *NCED4* was successfully edited using CRISPR/Cas9 in diploid, self-compatible lettuce (*Lactuca sativa*) with editing stability and *Cas9* segregation demonstrated in the next seed generation [24]. In both sexual diploid *Tragopogon porrifolius* and tetraploid *Tragopogon mirus,* a low-efficiency *Agrobacterium*-mediated transformation system was developed to examine CRISPR/Cas9 gene editing efficiency in T0 calli and regenerated shoots, however plants were not taken to the next generation [25]. In hexaploid self-incompatible *Chrysanthemum morifolium,* low-efficiency editing of an introduced fluorescent marker gene was observed [26]. CRISPR/Cas9 has also been evaluated in T0 plants of the rubber producing dandelion (*Taraxacum kok-saghyz*), a sexual relative of *Hieracium*, however due to self-incompatibility the analyses did not progress to the T1 generation [27]. The *Cas9* gene should be retained in progeny derived via the apomictic pathway and not segregate as typically occurs among sexually derived progeny. A knowledge of the efficiency of CRISPR/Cas9 gene editing is therefore required in apomictic *Hieracium* before embarking on large scale editing approaches to identify and characterize apomixis genes.

In this study, we targeted knockout of the endogenous phytoene desaturase (PDS) enzyme in tetraploid self-incompatible apomictic *Hieracium piloselloides*, D36 to assess CRISPR/Cas9 gene editing efficiency. PDS catalyzes the desaturation of phytoene to ζ-carotene during carotenoid biosynthesis and loss of *PDS* function in *Arabidopsis* results in a visible albino dwarf seedling-lethal phenotype [28]. *PDS* has been used to develop and evaluate CRISPR/Cas9 gene editing in crops including rice [29], cassava [30], wheat [31] and banana [32]. Due to the obvious albino phenotype, plants exhibiting complete *PDS* knockout in all tissues can be phenotypically identified, in addition to chimeric plants exhibiting albino sectors. *Agrobacterium*-mediated leaf disk transformation was used to introduce constructs targeting *Hieracium PDS* (*HPDS*). T0 regenerants and subsequent T1 progeny derived via apomictic reproduction in flowering chimeric T0 plants were analyzed. Editing events were assessed using amplicon deep sequencing. Albino seedlings increased in frequency in T1 progeny suggesting that editing continues during apomictic seed formation due to non-segregation of *Cas9*.

2. Materials and Methods

2.1. Plant Growth

Hieracium piloselloides (4x = 2n = 36) was maintained by vegetative micropropagation in vitro and grown in a glasshouse, as described previously [11].

2.2. Identification of an Hieracium Phytoene Desaturase (HPDS) Gene Ortholog and Confirmation of Expressed Leaf cDNA Sequence

The tetraploid *H. piloselloides*, D36 is closely related to diploid lettuce and shares the same base chromosome number (n = 9). Comparative genomic analysis between *Hieracium* species and lettuce has revealed partial macrosynteny for six linkage groups [33]. To identify lettuce PDS, a reciprocal best hit BLASTP search was performed with the *Arabidopsis* PDS amino acid sequence (NCBI accession AF360196) [28] against the lettuce genome. This revealed a 581 amino acid (aa) hypothetical protein (accession PLY83262.1) that showed 79% identity to *Arabidopsis* PDS. The amino acid sequence of lettuce PDS was then used in TBLASTN analysis against the *Hieracium* diploid plant D18 genomic DNA scaffold and tetraploid apomictic *Hieracium* D36 transcriptomes at http://hieracium.csiro.au [19]. This revealed a genomic DNA fragment (D18-gDNA-s59305) encoding a putative protein with 74.87% identity to lettuce PDS and additional transcript fragments (Figure S1A). Phylogenetic analysis confirmed the close relationship between lettuce and *Hieracium* PDS (Figure S2). The identified *HPDS* gene sequence was confirmed by isolating and sequencing the predicted gene from D36 leaf cDNA using the primers listed in Table S1. *Hieracium* leaf total RNA was isolated using the RNeasy Plant Mini Kit (Qiagen, Hilden, Germany). The RNA was treated with Turbo DNAse I (Ambion, Austin, TX, USA) to remove contaminating genomic DNA, ethanol precipitated and resuspended in nuclease-free water. First-strand cDNA was synthesized from 5 µL of RNA using SuperScript® III First-Strand Synthesis SuperMix (Thermo Fisher Scientific, Waltham, MA, USA). The predicted full-length coding sequence of the *HPDS* gene was amplified by polymerase chain reaction (PCR) using 0.5-µM forward (SWH12) and reverse (SWH13) primers (Table S1), 1 µL of diluted (1:5) cDNA and 0.2 U of Phusion® High-Fidelity DNA Polymerase (New England Biolabs, Ipswich, MA, USA) in a 20-µL reaction. The PCR product was purified using the Illustra GFX PCR DNA and Gel Band Purification Kit (GE Healthcare, Chicago, IL, USA) and Sanger sequenced.

2.3. Development of a CRISPR/Cas9 Gene Editing Construct Targeting Hieracium PDS

Three 20-nucleotide guide RNAs, beginning with a guanidine and binding directly upstream of an NGG protospacer adjacent motif (PAM), were designed using Geneious v. 11.0.4 (Biomatters, Auckland, New Zealand). Guides were analyzed for offsite targeting against the *Hieracium* diploid D18 genome sequence [19]. Complementary oligonucleotides encoding the three different sgRNAs (Table S1) were diluted to 4 µM in sterile water and self-annealed in a PCR cycler using 70 cycles of 95 °C for 35 s (−1 °C/cycle). Annealed oligos were ligated into the BsaI-HF site of the gateway entry vectors pEN-Comaira.1, pEN-Comaira.2 and pEN-Comaira.3. Multisite gateway was performed, using LR Clonase II plus (Thermo Fisher Scientific), to recombine the three U6: sgRNA constructs into pDE-Cas9 [34] and to produce a single expression construct containing three guides targeting *HPDS*.

2.4. In Vitro Cleavage Assay

The functional efficacy of the three sgRNAs to target *HPDS* was screened in vitro using the Guide-it Complete sgRNA Screening System (Takara Bio, Kusatsu, Shiga, Japan). The forward primers used for amplifying the templates for in vitro transcription of the sgRNAs are shown in Table S1. The substrate of the in vitro cleavage assays was a 2716 base pair (bp) PCR fragment of the *HPDS* gene generated using the primers SWH44 and SWH46 (Table S1). Briefly, the cleavage template (2.5 µL) was incubated with 50 ng of recombinant Cas9 protein and 250 ng of in vitro transcribed sgRNA in

Cas9 Reaction Buffer at 37 °C for 1 h. Reactions were stopped by incubating at 80 °C for 5 minutes and resolved on a 1.5% agarose gel (Figure S3).

2.5. Plant Transformation

Constructs were electroporated into *Agrobacterium tumefaciens* strain AGL1. Plants were transformed using the leaf disk protocol of Bicknell and Borst, [23] with the exception that phosphinothricin (5 mg L^{-1}) was used as the plant selectable marker instead of kanamycin for selecting positive transformants. To confirm the presence and integrity of the T-DNA in the phosphinothricin-resistant primary T0 transformants, the Extract-N-Amp Kit (Sigma-Aldrich, St. Louis, MO, USA) was used to amplify a 906 bp fragment of the *Cas9* coding sequence by PCR using primers SWH190 and SWH191 (Table S1), following manufacturer's procedures.

2.6. Identification of Edits in HPDS Genes of Transformed Hieracium Plants by Sequencing

Genomic DNA was extracted from emerging leaves of young T0 plantlets using the Extract-N-Amp Kit (Sigma-Aldrich, MO, USA) for sequencing. Primers (SWH186 and SHW187), surrounding the predicted editing sites (Table S1), were used to amplify a 540 bp genomic DNA fragment of *HPDS* from a selection of transformed plants exhibiting green and white leaf phenotypes, with KAPA-HiFi hotstart high-fidelity polymerase (Roche, MA, USA). PCR fragments were Sanger sequenced at the Australian Genome Research Facility (AGRF). Sequence chromatograms with mixed spectra were analyzed using the Inference of CRISPR Edits (ICE) web tool at https://ice.synthego.com/.

For next generation sequencing (NGS), 18 independent T0 transformants and an untransformed D36 control, were analyzed. Genomic DNA was isolated from *Hieracium* leaf tissue using the DNeasy Plant Mini Kit (Qiagen), following the manufacturer's protocol. Libraries were prepared using a two-step PCR protocol. For the first round PCR, primers were designed to generate two amplicons of ~200 bp each. The primers SWH200/201/202/203 spanned sgRNA1 and sgRNA2, while the primers SWH204/205/206/207 spanned sgRNA3 (Table S1). First-round PCR reactions contained 5 μM of each primer, 1 × KAPA HiFi Hotstart ReadyMix (Roche, Boston, MA, USA) and 10 ng of genomic DNA, in 20 μL. The PCR consisted of 35 cycles of 98 °C for 20 s, 60 °C for 15 s and 72 °C for 15 s. Amplicons were purified using Agencourt AMPure XP (Beckman Coulter, Brea, CA, USA). The second round PCR was carried out as the first round, except that Illumina Nextera i5 and i7 indexing primers were used, and 15 cycles were performed. Purified amplicons were pooled together, and the library was sequenced using a MiSeq (Illumina, Foster City, CA, USA) by the AGRF using the nano flow cell and 300 cycles. All next generation sequencing data were uploaded to the sequence read archive (SRA) at NCBI (Accession Number PRJNA636229).

Illumina paired-end reads were imported into Geneious (Biomatters) as fastq files. The sequences were paired, trimmed and then aligned to the amplicon reference sequence using Geneious Read Mapper with fast sensitivity. All sequence variants occurring at a frequency ≥1% from the reference amplicon were selected for analyses. These variants were further analyzed using CRISPResso [35] and CRISPResso2 [36].

2.7. Analyses of T1 Progeny Derived from Chimeric Apomictic T0 Transformed Plants

Seeds were harvested from mature plants and surface-sterilized with 12.5 g/L sodium hypochlorite and 0.05% (*v*/*v*) Triton X-100 for 5 min. Seeds were washed 4 times in sterile water and then plated onto media consisting of 0.5 × Murashige and Skoog Basal Salts, 3% (*w*/*v*) sucrose, pH 5.7 (KOH), 0.8% (*w*/*v*) plant agar and 1 × Gamborg's vitamins (Sigma-Aldrich) without selection. Whole seedlings were harvested 10 days after germination. DNA was extracted from whole seedlings using the method described by Edwards et al. [37]. Edits in the T1 progeny were determined using amplicon deep sequencing on a MiSeq (Illumina) as described above.

3. Results

3.1. PDS Edited Apomictic Hieracium T0 Plants Show Dwarfism, Albinism and Chimeric Phenotypes in Vegetative and Floral Tissues

The 3.8 kb *HPDS* gene used as a phenotypic marker for gene editing in tetraploid apomictic *Hieracium* was predicted to contain 11 exons and encode a protein of 589 amino acids (Figure S1A). The CRISPR/Cas9 editing constructs designed to disrupt endogenous *HPDS* gene function, contained three sgRNAs targeting a 252 bp region within exon 10 (Figure S1A,B), which is similar to the region that was used to target *PDS* in cassava [30]. Expression of each guide was independently regulated by the *Arabidopsis* U6-26 promoters (Figure S1C). The completed *HPDS*-targeting CRISPR/Cas9 construct was introduced to cells of D36 leaf explants by *Agrobacterium*-mediated transformation.

A total of 35 primary (T0) transgenic plants containing the construct targeting *HPDS* were generated from three independent experiments (Table 1; Figure 2). All transgenic lines tested positive by PCR for a 906 bp fragment corresponding to the *Cas9* gene (Figure 3A, upper panel). Four of the transgenic plants (11.4%) were completely albino showing the expected loss-of-function *pds* phenotype comprising loss of chlorophyll pigments (Table 1). Their rosette leaves were white and narrower relative to untransformed control seedlings (Figure 2A,B). Eleven T0 transgenics (31.4%) exhibited chimeric phenotypes with leaves that showed sectorial patterning (Figure 2C–E) and patchy loss of pigmentation (Figure 2F). In some cases, during early in vitro growth stages, albinism coincided with the accumulation of anthocyanin pigments (Figure 2B–D). The remaining twenty transgenic plants (57.1%) showed no obvious albino phenotype (Table 1).

Table 1. Phenotypic frequency of recovered transformants after CRISPR/Cas9-mediated editing of the *Hieracium PDS* gene.

	Experiment 1		Experiment 2		Experiment 3		Combined Total	
	No. Plants	% Total	No. Plants	% Total	No. Plants	% Total	No. Plants	% Total
Green	9	64.3	6	50.0	5	55. 6	20	57.1
Albino	1	7.1	2	16.7	1	11.1	4	11.4
Chimera	4	28.6	4	33.3	3	33.3	11	31.4

Figure 2. *Cont.*

Figure 2. Vegetative and floral phenotypes of CRISPR/Cas9-induced *PDS* mutations in T0 *Hieracium* plants. (**A**) Non-edited transgenic plant with green shoots; (**B**) full albino plant; (**C**) sectorial chimera of a single emerging leaf with accumulation of pink anthocyanin pigments; (**D**) patch of albino cells on the edge of an expanding leaf; (**E**) chimeric leaf visible within a whole rosette; (**F**) leaf with variegated patchy albino appearance; (**G**) flower from a nontransgenic plant at anthesis; (**H**) flower from a chimeric *pds* transgenic at anthesis displaying pale-yellow to white petals within each floret, but normal yellow-colored bilobed stigmas; (**I**) full albino flower showing bleached florets, including white stigmas; (**J**) chimeric plant with both normal, sectorial and full albino capitula, visualized as white sepal coloration. Scale bar = 5 mm (**A–C**,**E–J**) or 1 mm (**D**).

Completely albino plants did not survive to develop flowers. Therefore, the chimeric transgenics were grown to the flowering stage, and the phenotypes within the capitula were assessed. Compared to untransformed D36 control capitula (Figure 2G), florets in capitula of chimeric transgenics displayed weak carotenoid pigmentation in the petals, possessing yellow pigmented stigmas (Figure 2H). In addition, fully albino composite flowers were observed (albeit less frequently) that showed a total loss of pigmentation in the sepals, petals, stamens and carpels (Figure 2I). Panicles of flowers in chimeric *HPDS*-edited plants contained a mixture of green and white capitula, as well as sectorially mutated capitula that displayed a mixture of white and green tissues (Figure 2J). These results indicate that CRISPR/Cas9-induced mutations in chimeric *Hieracium* transformants occurs in reproductive organs, which may include sporophytic and gametophytic cells.

3.2. *CRIPSR/Cas9-Induced Indels and Deletions Can Be Rapidly Detected in Transgenic Hieracium Using PCR and Direct Sequencing*

A rapid screening approach combining PCR amplification of the target region followed by Sanger sequencing and analysis was used to initially detect edits. The power of this approach is demonstrated here in a representative analysis of a putative tetra-allelic albino knockout (line #3). A 540 bp region of the *HPDS* gene targeted for mutagenesis by all three guide RNAs was amplified from the T0 transgenic lines (Figure 3A, lower panel). In albino line #3, two distinct bands were seen on the agarose gel (540 bp and 322 bp), suggesting that a deletion event had occurred (Figure 3A, lane 3). Sanger sequencing of the 322 bp amplicon revealed a clean chromatogram with single peaks, suggesting that the deletion had occurred in a single allele (Figure 3B). Alignment with the reference sequence showed that a 218 bp deletion had occurred precisely between the region 3 bp upstream of the PAMs in sgRNA1 and sgRNA3 (Figure 3B). Conversely, Sanger sequencing of the larger 540-bp band revealed regions with polymorphic sequence peaks commencing from within the sgRNA2-binding site, which indicated that this band consisted of multiple alleles (Figure 3B). Similar polymorphic sequence peaks were observed for other 540-bp bands (Figure S4). Analysis of the 540-bp chromatogram from Line #3 was performed using the Inference of CRISPR Edits (ICE) tool [38] (Figure 3C,D). Three alleles highly likely to contain indels (R^2 = 0.96) were found, including single base deletions and insertions, plus a 49 base deletion between guides sgRNA1 and sgRNA2 (Figure 3B,D). These data suggest that the albino Line #3 contained tetra-allelic disruptions, one of which was a 218 bp deletion between the two most distant sgRNAs. These analyses demonstrated that PCR in conjunction with Sanger sequencing can rapidly detect indels and infer editing outcomes in *Hieracium*. However, as all three guides were complementary to the target amplicon in this study, not all possible editing outcomes could be inferred using this method. A comprehensive assessment was, therefore, performed.

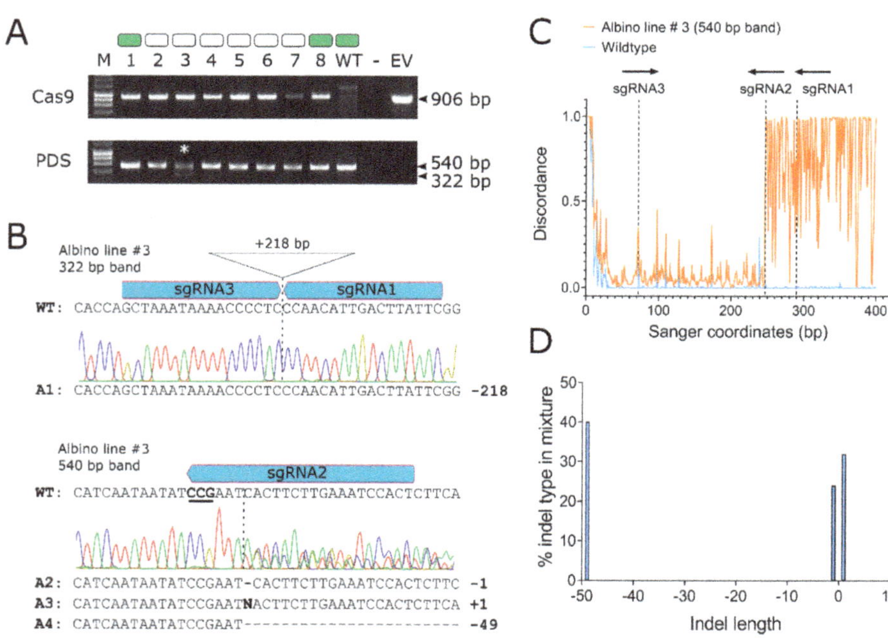

Figure 3. Detecting the presence of the transgene and CRISPR/Cas9-induced indels, by PCR and Sanger sequencing. (**A**) T-DNA integration was confirmed by PCR amplification of a 906 bp fragment of the *Streptococcus pyogenes Cas9* gene (upper gel). Edits were detected by PCR amplification of a 540 bp fragment of the *Hieracium PDS* gene (lower gel). Phenotypes of the plants that were examined in this

assay are depicted above the gel as green (wildtype) or white (albino) boxes. The D36 wildtype control (WT), no template control (–) and pDE-Cas9 empty vector positive control (EV) are also shown. (**B**) Sanger sequencing chromatograms of the gel purified 340-bp band (upper) and 540-bp band (lower) shown in lane 3 (asterisk). The position of the sgRNAs within the wildtype (WT) sequence is shown above the chromatograms, PAM is underlined, and the predicted cut sites are shown as vertical dashed lines. Alleles inferred in the edited sequences are shown below the chromatograms; (**C**) visualization of the discordance between Sanger traces of the ~540-bp band derived from the wild type (blue) and albino line # 3 (orange), determined using ICE analysis. The expected cut sites for each sgRNA are shown as vertical dotted lines; (**D**) inferred alleles and their frequencies, detected within the ~540-bp band of albino line 3, by Synthego ICE analysis.

3.3. Amplicon Deep Sequencing Reveals a Wide Range of CRISPR/Cas9-Induced Indels in Hieracium Primary Transformants

To examine in more detail the type and frequency of CRISPR/Cas9-induced edits in the *HPDS* gene, amplicon deep sequencing was performed on 18 independent T0 transformants (6 green, 5 chimeric and 7 albino). The 7 albino lines included those shown in Table 1 and an additional three plants from leaf-disk transformations where phenotypic frequencies were not scored. Edits were identified as variations in the amplicon sequences near the predicted cleavage sites from those of an untransformed D36 control. Analysis of the combined mutations across two amplicons in 18 transgenic lines revealed differing frequencies of insertions (55.9%), deletions (38.2%), substitutions (4.4%) and combined mutations (1.5%) (i.e., >1 mutation type in an allele) (Figure 4A, Table S2). Single base mutations occurred 60.3% of the time (Figure 4B, Table S2). Adenine was the most frequently inserted single base, accounting for more than half of the 1 base insertions (55.3%), while single guanine insertions were not observed in our selection of transgenics (Figure 4C, Table S2). These single base insertion frequencies are similar to those reported in *Brassica napus* [39]. Promisingly, 6% of the editing events in regenerated *Hieracium* transgenics were large deletions between two distant guide RNAs (Figure 4D). The largest mutation that could be detected by NGS was a 51 bp deletion between the guides sgRNA1 and sgRNA2 (Figure 4D, Table S2). As the binding site of the third guide RNA (sgRNA3) was present within a separate amplicon, we did not detect the largest 218 bp deletions between the two most distant guides using the NGS method, but this was readily observed using standard PCR (Figure 3). Interestingly, three independently transformed green plants without a visible phenotype (Lines 4.1, 9.2 and 29.1) harbored CRISPR induced edits in *HPDS* (Table S2). These plants also retained wildtype *HPDS* alleles at relatively high frequencies (Table S2). This result confirms that complete knockout of all *HPDS* alleles is required to observe the albino phenotype in *Hieracium* cells.

3.4. Editing in Chimeric Hieracium Continues through Apomictic Seed Formation and Is Inherited in the Next Generation

The transmission of CRISPR/Cas9-mediated edits in seeds derived from the apomict T0 transgenics was examined. D36 is a self-incompatible facultative apomict. More than 97% of seeds set arise asexually via apomixis and are tetraploid maternal clones [40]. However, if the aposporous initial cell does not form in an ovule then meiosis can proceed forming a sexually derived female gametophyte where chromosomal segregation has occurred. Viable dihaploid (n + 0) seeds form if the *LOP* locus is inherited in meiotically reduced embryo sacs [41]. To test for *Cas9* segregation, 79 T1 progeny from two independent primary transformants (phenotypic chimeras) with confirmed edits were screened by PCR for the presence of the *Cas9*, showing that 98.7% (78/79) of seedlings contained the *Cas9* cassette (Figure S5). Segregation of *Cas9* in 1.3% of seedlings is consistent with a previous study where approximately 2.5% of D36 seeds were found to be meiotically derived [40]. Polyembryony occurred in 2% and 7% of the seed from the T0 *HPDS* lines 1 and 2, respectively (Table 2). These frequencies align with previous observations of polyembryony in D36, measured at embryo sac maturation [42] and germination [43]. These results suggest that most T1 seeds formed in the transgenic plants arose via apomixis.

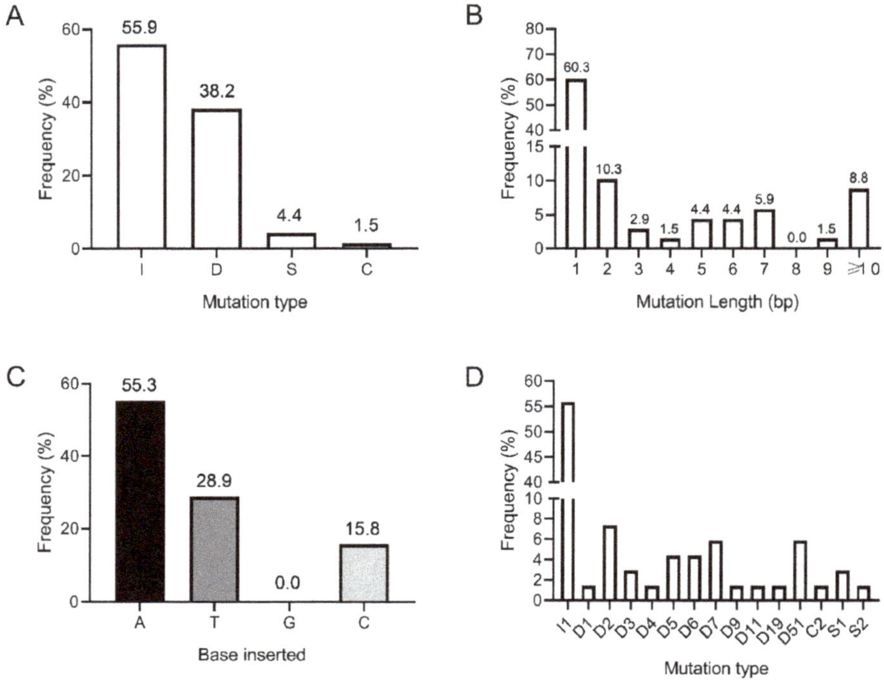

Figure 4. The combined types and frequencies of CRISPR/Cas9-induced mutations in the *Hieracium PDS* gene, mediated by the three different sgRNAs, determined by NGS in 18 T0 independent transformants. (**A**) Frequency of insertions [I], deletions [D], substitutions [S], and combined [C] mutation types; (**B**) Frequency of different mutation lengths regardless of the mutation types; (**C**) Percentage of bases inserted for the 1-bp insertions A = adenine, T = thymine, G = guanine, C = cytosine; (**D**) Frequency of each mutation type for all of the mutations induced by the three sgRNAs. I = insertion; D = deletion; S = substitution; C = combined mutation. Frequencies are shown above the bars in **A–C**.

Table 2. Phenotypes of T1 seedlings from two independent primary (T0) transformants. Seeds derived from chimeric plants.

	HPDS **Line 1**		*HPDS* **Line 2**	
	Progeny	**%**	**Progeny**	**%**
Albino	95	62.9	104	84.6
Chimera	56	37.1	19	15.4
Green	0	0	0	0
Polyembryony	2	1.3	7	5.7

Two independent T0 chimeric parents produced a high percentage of full albino progeny at 62.9% and 84.6%, respectively, while the remaining T1 progeny exhibited albino sectors (Table 2). In *Arabidopsis*, chimeras can be eliminated from the T2 generation by self-fertilizing a chimeric T1 that contains edits in the germline [34]. However, D36 is a self-incompatible apomict, so the high frequency of albino seedlings observed in the T1 generation was likely due to continued activity of the Cas9 nuclease in the T0 parent and/or the parthenogenetic embryo. To test this hypothesis, *HPDS* gene edits were assessed across the T0 and T1 generations using NGS. Sequences from representative individual plants of each generation are illustrated in Figure 5. In the region surrounding sgRNA2, a T0 chimera contained edits in both wildtype alleles, but 39% of sequences remained unedited (Figure 5A). However, a T1 albino progeny derived from this parent contained edits in close to 100% of the amplicon sequence reads, which included a 3 base de novo deletion (Figure 5A). This suggests that the Cas9 remained active in the T0 before or during aposporous embryo sac formation or in the developing T1 embryo. Some alleles that were present in the T0 chimera were absent in the T1 progeny (Figure 5A). This may represent spatial differences in Cas9-mediated edits throughout cells of the chimeric T0 parent.

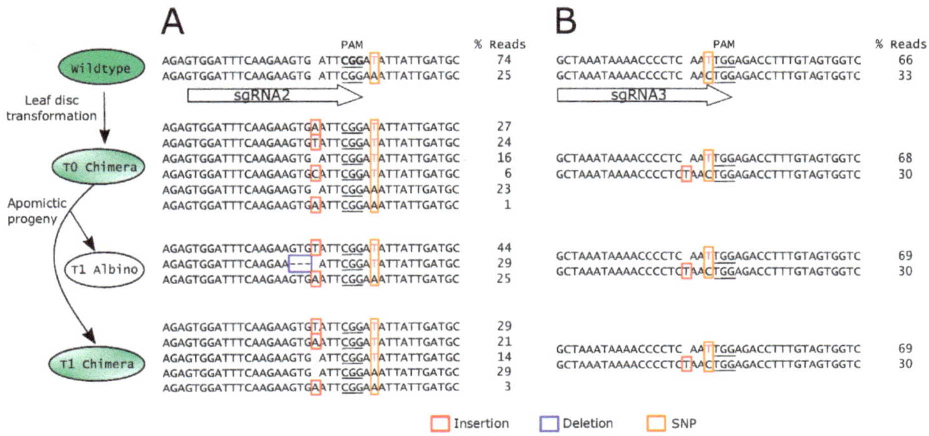

Figure 5. Stability of edits in the *HPDS* gene across different generations of apomictic *Hieracium* is affected by polymorphisms in the guide RNA. A subset of amplicon deep sequencing results surrounding the (**A**) sgRNA2 and (**B**) sgRNA3 binding sites. For clarity, sequences around the sgRNA3 binding site in (**B**) are shown as reverse complement. For each generation and each sgRNA target site, the allelic sequences from one representative plant are shown. Allele frequencies are shown as percentage of total reads. Note that sgRNA2 was 100% complementary to all alleles detected in the wildtype, but sgRNA3 was only 50% complementary due to a natural SNP 1 base upstream of the PAM. SNPs are shown in colored font with yellow boxes. Indels are shown as red boxes (insertion) or blue boxes (deletion). PAM underlined.

Analysis of editing across the generations also revealed allele-specific editing in *Hieracium*. Within one of the *HPDS* alleles, a cytosine/thymine single nucleotide polymorphism (SNP) was present 1 base upstream of the PAM of the sgRNA3 site (Figure 5B). Hence sgRNA3 was complementary to only half of the wildtype *HPDS* alleles (Figure 5B). No edits were mediated by sgRNA3 in the noncomplementary allele in any of the plants analyzed in either the T0 or T1 generations (Figure 5B, Table S2). The allele complementary to sgRNA3 contained a thymine insertion in the T0 (Figure 5B). The frequency of this edited allele remained stable across the T0 and T1 generations, suggesting that an early editing event may have occurred in the T0 (Figure 5B). Collectively, these results show that CRISPR/Cas9 is efficient in *Hieracium* and can be employed to target specific alleles.

4. Discussion

In this study, the combination of *Agrobacterium*-mediated leaf disk transformation together with CRISPR/Cas9-mediated disruption of the *HPDS* gene has shown high frequencies of putative tetra-allelic editing in the T1 generation of isolates examined (62.9% & 84.6%) and, in chimeric plants, albino sectoring continued through floral development. The absence of *Cas9* segregation (Figure S5) and continually active Cas9 activity through the events of apomixis where meiosis is suppressed and fertilization is not necessary to induce seed formation likely facilitated these good editing frequencies.

Recovery of fully albino mutants in the T0 generation occurred at an average frequency of 11.4%, with a maximum observed frequency of 16.7% in Experiment 2 (Table 1). The efficiency of CRISPR/Cas9-mediated editing of the *PDS* gene was variable in other species. For example, 63% of transgenic Cavendish banana plants displayed full albinism [32], while fully albino cassava mutants occurred at frequency of 21% for cultivar 60444 and 52% for cultivar TME 204 [30]. Differences in ploidy may explain these variable efficiencies, as the editing events in cassava (diploid) and Cavendish banana (triploid) would need to occur in fewer alleles to induce a phenotype than in *H. piloselloides* (tetraploid). It would be interesting to test the efficiency of CRISPR/Cas9 within dihaploid (D18) *Hieracium*. The type of explants used may also influence regeneration efficiencies. Like our study, leaf discs were used to generate transgenic apple plants that carried CRISPR/Cas9-induced mutations in PDS [44]. In that study, full and partial albino phenotypes occurred in 32% of regenerated T0 plantlets, which more closely resembles our combined frequency of 42% (Table 1). Different promoters, the total number of sgRNAs used and copy number of target genes, may also influence editing efficiencies.

Results presented here pave the way for the identification of key genes involved in apomixis in *Hieracium* using a genome editing strategy together with a seed screening approach used in a prior γ-mutagenesis screen to identify apomixis loci in *Hieracium* [12]. The *LOA* locus in both apomictic *H. praealtum* (R35) and *H piloselloides* (D36) resides on the long arm of a single hemizygous chromosome. The chromosomal location of the *LOP* locus in *H. praealtum* (R35) remains unknown, and the exact sizes of *LOA* and *LOP* loci are unclear. Gene sequences, SNPs, and molecular markers linked to both loci are known [15,19,22,33] and could be used to design allele-specific distant guide RNAs that initially cause large deletions of multiple gene candidates. Large chromosomal deletions of up to 245 Kb have been observed in rice [45]. In this study, 6% of the editing events in regenerated *Hieracium* transgenics were 51 bp deletions between two guides (Figure 4D), and larger deletions were detected by Sanger sequencing (Figure 3A,B). Future studies in *Hieracium* could therefore deploy pairs of sgRNAs that generate large (Kb) fragment deletions via the NHEJ repair pathway, which would enable the functional analysis of large *cis*-regulatory domains or gene clusters within apomixis loci.

The better characterized *LOA* locus could be a good initial focus for use of CRISPR/Cas9 to target multiple candidate apomixis genes simultaneously in this manner in *Hieracium*. *LOA* controls AI cell formation, sexual suppression and the formation of aposporous embryo sacs. If *LOA* were successfully deleted by CRISPR/Cas9, in tetraploid *H. piloselloides*, then the *LOP* locus would segregate in half of the progeny. Chromosomally reduced gametophytes containing *LOP* would develop into viable dihaploid (n + 0) seeds without fertilization (Figure 6A). If *LOP* were deleted, and *LOA* function remained intact, chromosomally unreduced gametophytes would require fertilization and the resultant progeny would show increased ploidy (Figure 6B). If both *LOA* and *LOP* were deleted, then seed formation would occur via sexual reproduction. These outcomes could be screened initially by seedling and ploidy phenotypes (Figure 6A,B) before progressing to a cytological and molecular assessment to identify functional knockouts in causal apomixis genes [12,46]. Allele specificity would avoid potentially lethal mutations in gene homeologs. A SNP was sufficient to confer allele-specific editing of the *CHLOROPLASTOS ALTERADOS1* (GhCLA1) gene in allotetraploid cotton (*Gossypium hirsutum*) [47]. Similarly, in our study, allele-specific editing was observed at the sgRNA3 target site due to a SNP (Figure 5B). Reversion to meiosis though the inactivation of *LOA* function would enable *Cas9* segregation in the progeny, which would facilitate rapid genetic screening. Candidate apomixis genes identified in this approach could then be transformed into characterized apomictic mutants deficient in *LOA* and *LOP* function to determine whether activity was restored [11].

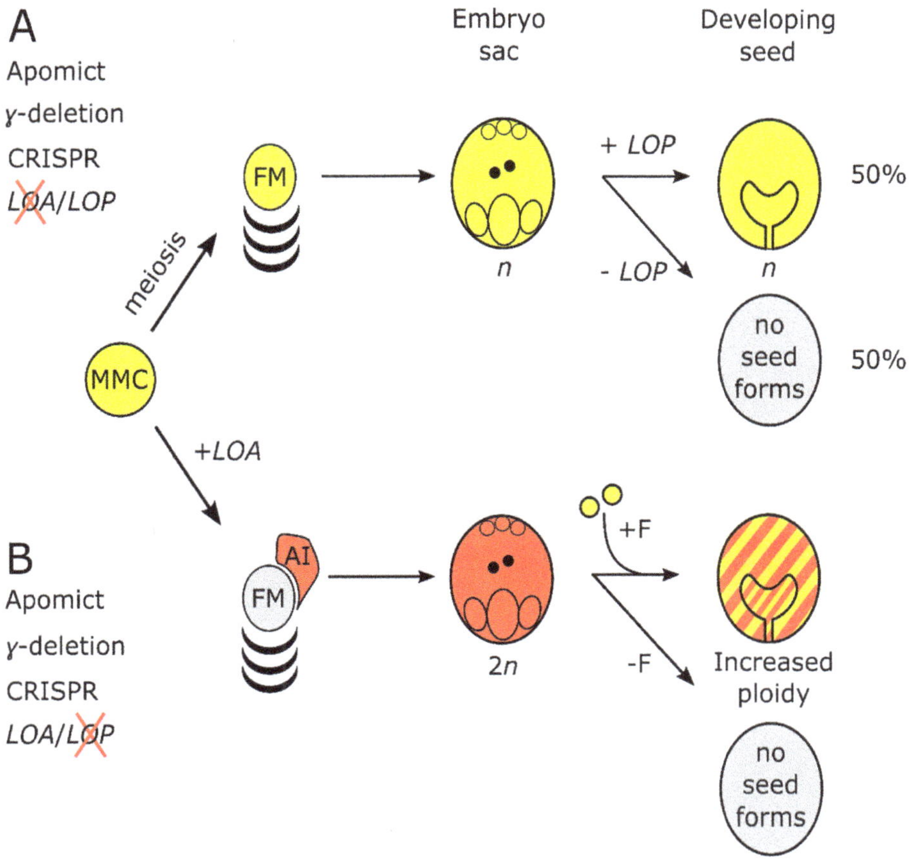

Figure 6. Schematic for identifying apomixis genes in *Hieracium* using functional knockouts and gene editing. (**A**) Known outcome of γ-mutagenesis and expected outcome of CRISPR-mediated disruption of *LOA* in apomictic tetraploid *Hieracium*. *LOA* loss-of-function causes *LOP* to segregate during meiosis. *LOP* activity in the reduced (n) embryo sacs leads to 50% dihaploid progeny and 50% non-viable seeds with aborted embryos. Gene edited progeny, from parents with reduced seed-set, could be screened by flow cytometry, sequencing and cytology; (**B**) γ deletions at *LOP* lead to formation of unreduced embryo sacs which, when fertilized (+F), increase in ploidy, hence the indicated red and yellow shading. Large and targeted CRISPR/Cas9 editing should result in a similar phenotype. Mutations in both loci would result in a sexual phenotype, as in Figure 1A.

Supplementary Materials: The following are available online at http://www.mdpi.com/2073-4425/11/9/1064/s1. Table S1: List of primers used in the study. Table S2: Frequencies of edited *HPDS* alleles identified by next generation sequencing in T0 independent transformants. Figure S1: Schematic of the predicted *Hieracium* D18 *PHYTOENE DESATURASE* gene ortholog, guide RNA-binding sites and the CRISPR/Cas9 construct targeting the *HPDS* gene. Figure S2: Phylogenetic relationships of plant phytoene desaturase proteins. Figure S3: Functional testing of single guide RNAs by an in vitro cleavage assay. Figure S4: Sanger sequencing traces of the edited *HPDS* gene from multiple T0 replicate lines. Figure S5: Detection of the *Cas9* transgene in 78/79 T1 progeny from T0 chimeric *HPDS* mutants. References [48–50] are cited in the supplementary materials.

Author Contributions: Conceptualization, S.T.H. and A.M.G.K.; methodology, S.W.H. and M.G.; formal analysis, S.W.H.; writing—original draft preparation, S.W.H. and A.M.G.K.; writing—review and editing, S.W.H., S.T.H., M.G. and A.M.G.K.; funding acquisition, A.M.G.K. All authors have read and agreed to the published version of the manuscript.

Funding: This research was carried out with funds from the Commonwealth Scientific Industrial Research Organization.

Acknowledgments: We thank Luca Comai (University of California, Davis) for providing the Gateway entry vectors. Johannes Scharwies (Stanford University) for advice on CRISPR vectors. Chloe Henderson for photography in Figure 2J and graphic abstract. Inna Mazonka, Weng Leong and Natasha Bazanova (CSIRO) for technical assistance. Martina Juranić (CSIRO) for advice and discussions. Nial Gursanscky (CSIRO) for advice and assistance with amplicon deep sequencing.

Conflicts of Interest: The authors declare no conflict of interest. The funders had no role in the design of the study; in the collection, analyses or interpretation of data; in the writing of the manuscript or in the decision to publish the results.

References

1. Conner, J.A.; Ozias-Akins, P. Apomixis: Engineering the ability to harness hybrid vigor in crop plants. In *Plant Germline Development. Methods in Molecular Biology*; Schmidt, A., Ed.; Humana Press: New York, NY, USA, 2017; Volume 1669, pp. 17–34.
2. Hand, M.L.; Koltunow, A.M. The genetic control of apomixis: Asexual seed formation. *Genetics* **2014**, *197*, 441–450. [CrossRef]
3. Koltunow, A.M.; Grossniklaus, U. Apomixis: A developmental perspective. *Annu. Rev. Plant Biol.* **2003**, *54*, 547–574. [CrossRef] [PubMed]
4. Sailer, C.; Schmid, B.; Grossniklaus, U. Apomixis allows the transgenerational fixation of phenotypes in hybrid plants. *Curr. Biol.* **2016**, *26*, 331–337. [CrossRef] [PubMed]
5. Hörandl, E.; Temsch, E.M. Introgression of apomixis into sexual species is inhibited by mentor effects and ploidy barriers in the Ranunculus auricomus complex. *Ann. Bot.* **2009**, *104*, 81–89. [CrossRef] [PubMed]
6. Conner, J.A.; Mookkan, M.; Huo, H.; Chae, K.; Ozias-Akins, P. A parthenogenesis gene of apomict origin elicits embryo formation from unfertilized eggs in a sexual plant. *Proc. Natl. Acad. Sci. USA* **2015**, *112*, 11205–11210. [CrossRef]
7. Bortesi, L.; Fischer, R. The CRISPR/Cas9 system for plant genome editing and beyond. *Biotechnol. Adv.* **2015**, *33*, 41–52. [CrossRef]
8. Jia, H.; Wang, N. Targeted genome editing of sweet orange using Cas9/sgRNA. *PLoS ONE* **2014**, *9*, e93806. [CrossRef]
9. Jia, H.; Zhang, Y.; Orbović, V.; Xu, J.; White, F.F.; Jones, J.B.; Wang, N. Genome editing of the disease susceptibility gene CsLOB1 in citrus confers resistance to citrus canker. *Plant Biotechnol. J.* **2017**, *15*, 817–823. [CrossRef]
10. Hand, M.L.; Vit, P.; Krahulcova, A.; Johnson, S.D.; Oelkers, K.; Siddons, H.; Chrtek, J.; Fehrer, J.; Koltunow, A.M.G. Evolution of apomixis loci in Pilosella and Hieracium (Asteraceae) inferred from the conservation of apomixis-linked markers in natural and experimental populations. *Heredity* **2015**, *114*, 17–26. [CrossRef]
11. Koltunow, A.M.G.; Johnson, S.D.; Rodrigues, J.C.M.; Okada, T.; Hu, Y.; Tsuchiya, T.; Wilson, S.; Fletcher, P.; Ito, K.; Suzuki, G.; et al. Sexual reproduction is the default mode in apomictic *Hieracium* subgenus *Pilosella*, in which two dominant loci function to enable apomixis. *Plant J.* **2011**, *66*, 890–902. [CrossRef]
12. Catanach, A.S.; Erasmuson, S.K.; Podivinsky, E.; Jordan, B.R.; Bicknell, R. Deletion mapping of genetic regions associated with apomixis in *Hieracium*. *Proc. Natl. Acad. Sci. USA* **2006**, *103*, 18650–18655. [CrossRef] [PubMed]
13. Henderson, S.T.; Johnson, S.D.; Eichmann, J.; Koltunow, A.M.G. Genetic analyses of the inheritance and expressivity of autonomous endosperm formation in Hieracium with different modes of embryo sac and seed formation. *Ann. Bot.* **2017**, *119*, 1001–1010. [CrossRef] [PubMed]
14. Ogawa, D.; Johnson, S.D.; Henderson, S.T.; Koltunow, A.M. Genetic separation of autonomous endosperm formation (AutE) from the two other components of apomixis in Hieracium. *Plant Reprod.* **2013**, *26*, 113–123. [CrossRef] [PubMed]
15. Okada, T.; Ito, K.; Johnson, S.D.; Oelkers, K.; Suzuki, G.; Houben, A.; Mukai, Y.; Koltunow, A.M. Chromosomes carrying meiotic avoidance loci in three apomictic eudicot *Hieracium* subgenus *Pilosella* species share structural features with two monocot apomicts. *Plant Physiol.* **2011**, *157*, 1327–1341. [CrossRef]

16. Akiyama, Y.; Conner, J.A.; Goel, S.; Morishige, D.T.; Mullet, J.E.; Hanna, W.W.; Ozias-Akins, P. High-resolution physical mapping in *Pennisetum squamulatum* reveals extensive chromosomal heteromorphism of the genomic region associated with apomixis. *Plant Physiol.* **2004**, *134*, 1733–1741. [CrossRef]
17. Akiyama, Y.; Hanna, W.W.; Ozias-Akins, P. High-resolution physical mapping reveals that the apospory-specific genomic region (ASGR) in Cenchrus ciliaris is located on a heterochromatic and hemizygous region of a single chromosome. *Theor. Appl. Genet.* **2005**, *111*, 1042–1051. [CrossRef]
18. Bräuning, S.; Catanach, A.; Lord, J.M.; Bicknell, R.; Macknight, R.C. Comparative transcriptome analysis of the wild-type model apomict *Hieracium praealtum* and its *loss of parthenogenesis* (*lop*) mutant. *BMC Plant Biol.* **2018**, *18*, 206. [CrossRef]
19. Rabiger, D.S.; Taylor, J.M.; Spriggs, A.; Hand, M.L.; Henderson, S.T.; Johnson, S.D.; Oelkers, K.; Hrmova, M.; Saito, K.; Suzuki, G.; et al. Generation of an integrated *Hieracium* genomic and transcriptomic resource enables exploration of small RNA pathways during apomixis initiation. *BMC Biol.* **2016**, *14*, 86. [CrossRef]
20. Juranić, M.; Tucker, M.R.; Schultz, C.J.; Shirley, N.J.; Taylor, J.M.; Spriggs, A.; Johnson, S.D.; Bulone, V.; Koltunow, A.M. Asexual female gametogenesis involves contact with a sexually-fated megaspore in apomictic *Hieracium*. *Plant Physiol.* **2018**, *177*, 1027–1049. [CrossRef]
21. Okada, T.; Hu, Y.; Tucker, M.R.; Taylor, J.M.; Johnson, S.D.; Spriggs, A.; Tsuchiya, T.; Oelkers, K.; Rodrigues, J.C.M.; Koltunow, A.M.G. Enlarging cells initiating apomixis in *Hieracium praealtum* transition to an embryo sac program prior to entering mitosis. *Plant Physiol.* **2013**, *163*, 216–231. [CrossRef]
22. Kotani, Y.; Henderson, S.T.; Suzuki, G.; Johnson, S.D.; Okada, T.; Siddons, H.; Mukai, Y.; Koltunow, A.M.G. The *LOSS OF APOMEIOSIS* (*LOA*) locus in *Hieracium praealtum* can function independently of the associated large-scale repetitive chromosomal structure. *New Phytol.* **2014**, *201*, 973–981. [CrossRef] [PubMed]
23. Bicknell, R.A.; Borst, N.K. Agrobacterium-mediated transformation of *Hieracium aurantiacum*. *Int. J. Plant Sci.* **1994**, *155*, 467–470. [CrossRef]
24. Bertier, L.D.; Ron, M.; Huo, H.; Bradford, K.J.; Britt, A.B.; Michelmore, R.W. High-resolution analysis of the efficiency, heritability, and editing outcomes of CRISPR/Cas9-induced modifications of *NCED4* in lettuce (*Lactuca sativa*). *G3: Genes|Genomes|Genet.* **2018**, *8*, 1513–1521. [CrossRef]
25. Shan, S.; Mavrodiev, E.V.; Li, R.; Zhang, Z.; Hauser, B.A.; Soltis, P.S.; Soltis, D.E.; Yang, B. Application of CRISPR/Cas9 to *Tragopogon* (Asteraceae), an evolutionary model for the study of polyploidy. *Mol. Ecol. Resour.* **2018**, *18*, 1427–1443. [CrossRef] [PubMed]
26. Kishi-Kaboshi, M.; Aida, R.; Sasaki, K. Generation of gene-edited *Chrysanthemum morifolium* using multicopy transgenes as targets and markers. *Plant Cell Physiol.* **2017**, *58*, 216–226. [CrossRef]
27. Iaffaldano, B.; Zhang, Y.; Cornish, K. CRISPR/Cas9 genome editing of rubber producing dandelion *Taraxacum kok-saghyz* using *Agrobacterium rhizogenes* without selection. *Ind. Crop. Prod.* **2016**, *89*, 356–362. [CrossRef]
28. Qin, G.; Gu, H.; Ma, L.; Peng, Y.; Deng, X.W.; Chen, Z.; Qu, L.J. Disruption of phytoene desaturase gene results in albino and dwarf phenotypes in Arabidopsis by impairing chlorophyll, carotenoid, and gibberellin biosynthesis. *Cell Res.* **2007**, *17*, 471–482. [CrossRef]
29. Zhang, H.; Zhang, J.; Wei, P.; Zhang, B.; Gou, F.; Feng, Z.; Mao, Y.; Yang, L.; Zhang, H.; Xu, N.; et al. The CRISPR/Cas9 system produces specific and homozygous targeted gene editing in rice in one generation. *Plant Biotechnol. J.* **2014**, *12*, 797–807. [CrossRef]
30. Odipio, J.; Alicai, T.; Ingelbrecht, I.; Nusinow, D.A.; Bart, R.; Taylor, N.J. Efficient CRISPR/Cas9 genome editing of phytoene desaturase in cassava. *Front. Plant Sci.* **2017**, *8*. [CrossRef]
31. Howells, R.M.; Craze, M.; Bowden, S.; Wallington, E.J. Efficient generation of stable, heritable gene edits in wheat using CRISPR/Cas9. *BMC Plant Biol.* **2018**, *18*, 215. [CrossRef]
32. Naim, F.; Dugdale, B.; Kleidon, J.; Brinin, A.; Shand, K.; Waterhouse, P.; Dale, J. Gene editing the phytoene desaturase alleles of Cavendish banana using CRISPR/Cas9. *Transgenic Res.* **2018**. [CrossRef]
33. Shirasawa, K.; Hand, M.L.; Henderson, S.T.; Okada, T.; Johnson, S.D.; Taylor, J.M.; Spriggs, A.; Siddons, H.; Hirakawa, H.; Isobe, S.; et al. A reference genetic linkage map of apomictic Hieracium species based on expressed markers derived from developing ovule transcripts. *Ann. Bot.* **2015**, *115*, 567–580. [CrossRef]
34. Fauser, F.; Schiml, S.; Puchta, H. Both CRISPR/Cas-based nucleases and nickases can be used efficiently for genome engineering in *Arab. Thaliana Plant J.* **2014**, *79*, 348–359. [CrossRef]
35. Pinello, L.; Canver, M.C.; Hoban, M.D.; Orkin, S.H.; Kohn, D.B.; Bauer, D.E.; Yuan, G.-C. Analyzing CRISPR genome-editing experiments with CRISPResso. *Nat. Biotechnol.* **2016**, *34*, 695. [CrossRef]

36. Clement, K.; Rees, H.; Canver, M.C.; Gehrke, J.M.; Farouni, R.; Hsu, J.Y.; Cole, M.A.; Liu, D.R.; Joung, J.K.; Bauer, D.E.; et al. CRISPResso2 provides accurate and rapid genome editing sequence analysis. *Nat. Biotechnol.* **2019**, *37*, 224–226. [CrossRef]
37. Edwards, K.; Johnstone, C.; Thompson, C. A simple and rapid method for the preparation of plant genomic DNA for PCR analysis. *Nucleic Acids Res.* **1991**, *19*, 1349. [CrossRef]
38. Hsiau, T.; Maures, T.; Waite, K.; Yang, J.; Kelso, R.; Holden, K.; Stoner, R. Inference of CRISPR edits from Sanger trace data. *bioRxiv* **2018**. [CrossRef]
39. Yang, H.; Wu, J.-J.; Tang, T.; Liu, K.-D.; Dai, C. CRISPR/Cas9-mediated genome editing efficiently creates specific mutations at multiple loci using one sgRNA in *Brassica napus*. *Sci. Rep.* **2017**, *7*, 7489. [CrossRef]
40. Bicknell, R.A.; Lambie, S.C.; Butler, R.C. Quantification of progeny classes in two facultatively apomictic accessions of *Hieracium*. *Hereditas* **2003**, *138*, 11–20. [CrossRef]
41. Bicknell, R.A.; Koltunow, A.M. Understanding apomixis: Recent advances and remaining conundrums. *Plant Cell* **2004**, *16*, S228–S245. [CrossRef]
42. Juranić, M.; Johnson, S.D.; Koltunow, A.M. Phenotypic plasticity of aposporous embryo sac development in *Hieracium praealtum*. *Plant Signal. Behav.* **2019**, 1–4. [CrossRef]
43. Koltunow, A.M.; Johnson, S.D.; Bicknell, R.A. Apomixis is not developmentally conserved in related, genetically characterized *Hieracium* plants of varying ploidy. *Sex Plant Reprod.* **2000**, *12*, 253–266. [CrossRef]
44. Nishitani, C.; Hirai, N.; Komori, S.; Wada, M.; Okada, K.; Osakabe, K.; Yamamoto, T.; Osakabe, Y. Efficient genome editing in apple using a CRISPR/Cas9 system. *Sci. Rep.* **2016**, *6*, 31481. [CrossRef]
45. Zhou, H.; Liu, B.; Weeks, D.P.; Spalding, M.H.; Yang, B. Large chromosomal deletions and heritable small genetic changes induced by CRISPR/Cas9 in rice. *Nucleic Acids Res.* **2014**, *42*, 10903–10914. [CrossRef]
46. Koltunow, A.M.G.; Johnson, S.D.; Okada, T. Apomixis in hawkweed: Mendel's experimental nemesis. *J. Exp. Bot.* **2011**, *62*, 1699–1707. [CrossRef]
47. Wang, P.; Zhang, J.; Sun, L.; Ma, Y.; Xu, J.; Liang, S.; Deng, J.; Tan, J.; Zhang, Q.; Tu, L.; et al. High efficient multisites genome editing in allotetraploid cotton (*Gossypium hirsutum*) using CRISPR/Cas9 system. *Plant Biotechnol. J.* **2018**, *16*, 137–150. [CrossRef]
48. Larkin, M.A.; Blackshields, G.; Brown, N.P.; Chenna, R.; McGettigan, P.A.; McWilliam, H.; Valentin, F.; Wallace, I.M.; Wilm, A.; Lopez, R.; et al. Clustal W and Clustal X version 2.0. *Bioinformatics* **2007**, *23*, 2947–2948. [CrossRef]
49. Shan, S.; Mavrodiev, E.V.; Li, R.; Zhang, Z.; Hauser, B.A.; Soltis, P.S.; Soltis, D.E.; Yang, B. Data from: Application of CRISPR/Cas9 to Tragopogon (Asteraceae), an evolutionary model for the study of polyploidy. *Dryad Digit. Repository* **2018**. [CrossRef]
50. Kumar, S.; Steche, G.; Li, M.; Knyaz, C.; Tamura, K. MEGA X: Molecular evolutionary genetics analysis across computing platforms. *Mol. Biol. Evol.* **2018**, *35*, 1547–1549. [CrossRef]

genes

MDPI

Opinion

Apomixis Technology: Separating the Wheat from the Chaff

Diego Hojsgaard

Department of Systematics, Biodiversity and Evolution of Plants, Albrecht-von-Haller Institute for Plant Sciences, Georg-August-University of Göttingen, Untere Karspüle 2, D-37073-1 Göttingen, Germany; Diego.Hojsgaard@biologie.uni-goettingen.de

Received: 12 March 2020; Accepted: 8 April 2020; Published: 10 April 2020

check for updates

Abstract: Projections indicate that current plant breeding approaches will be unable to incorporate the global crop yields needed to deliver global food security. Apomixis is a disruptive innovation by which a plant produces clonal seeds capturing heterosis and gene combinations of elite phenotypes. Introducing apomixis into hybrid cultivars is a game-changing development in the current plant breeding paradigm that will accelerate the generation of high-yield cultivars. However, apomixis is a developmentally complex and genetically multifaceted trait. The central problem behind current constraints to apomixis breeding is that the genomic configuration and molecular mechanism that initiate apomixis and guide the formation of a clonal seed are still unknown. Today, not a single explanation about the origin of apomixis offer full empirical coverage, and synthesizing apomixis by manipulating individual genes has failed or produced little success. Overall evidence suggests apomixis arise from a still unknown single event molecular mechanism with multigenic effects. Disentangling the genomic basis and complex genetics behind the emergence of apomixis in plants will require the use of novel experimental approaches benefiting from Next Generation Sequencing technologies and targeting not only reproductive genes, but also the epigenetic and genomic configurations associated with reproductive phenotypes in homoploid sexual and apomictic carriers. A comprehensive picture of most regulatory changes guiding apomixis emergence will be central for successfully installing apomixis into the target species by exploiting genetic modification techniques.

Keywords: apomeiosis; clonal seeds; endosperm; heterosis capture; molecular breeding; parthenogenesis

---

1. Introduction

With a world population expected to reach 9.7 billion by 2050 [1] and increasing food demand, plant breeders are expected to create more resilient crops to overcome constraints on cereal production imposed by population growth, climate change and environmental degradation [2].

The use of novel tools in biotechnology and the advances in the characterization of genes and genomes is playing a central role speeding up plant breeding efforts to increase yields and grain quality. However, the creation of a new crop variety takes between 7–20 years to bring to the market, costs from hundreds of thousands to millions of euros, and cannot capture all beneficial gene interactions [3–5]. Selected traits in high-yield hybrid cultivars segregate in the offspring during the (normal) mechanism of (sexual) reproduction and seed formation, and new hybrid seeds must be generated every year from selected parental lines to keep superior plant phenotypes. Projections indicate that the current breeding approaches will be unable to produce the increase in global crop yields needed to address global food security [6]. As long as the development of new hybrid varieties relies entirely in the exploitation of sexuality, breeding programs are inevitably costly and time-demanding.

Apomixis is a disruptive innovation, an alternative to sex that can speed up the time and reduce the cost needed to create a new cultivar. By skipping key steps of sexuality (Figure 1), apomictic plants can produce clonal seeds, and hence, capture heterosis and gene combinations of elite phenotypes transgenerationally [7,8]. The potential of breeding apomixis into hybrid cultivars has been known for a long time [5] and scientists had been looking for ways to introgress, induce, or mimic apomixis in sexual crops with the central concept that pollen of apomicts could transmit dominant apomixis genes in crosses with sexual plants or that apomixis might be induced from altering the sexual development. Harnessing apomixis would accelerate breeding programs and the ability of plant breeders to fix and propagate genetic heterozygosity and the associated hybrid vigor, reducing the time required to produce new varieties as well as the costs associated with seed-production [5,7,8]. The introduction of apomixis into hybrid cultivars is a game-changing innovation in the current plant breeding paradigm. Apomixis breeding can lead to rapid mass production of new elite cultivars better adapted to local environments across the world, reduce the problems connected to monocultures, and contribute to crop resilience by meeting food production goals without restrictions from biotic and abiotic stresses, climate change, or environmental degradation.

Whereas the majority of flowering plants produce seeds following the events of sexual reproduction (Figure 1a), around 400 species (or ca. 0.1% of all angiosperm species) belonging to 293 genera [9,10] had evolved the singular reproductive mechanism called apomixis. The fact that (1) apomixis has recurrently arisen across time from sexual progenitors [11], (2) apomictic species naturally occur in diverse plant families throughout the angiosperm phylogeny [9,10], and (3) a few of those apomictic species are minor crops (such as some forage grasses, fruits, and guayule) [5,12] suggest that apomixis can be introduced and/or engineered into major cereal and legume crops. Thus far, attempts to introgress apomixis from wild relatives into important crop species such as maize, wheat, and pearl millet have failed mainly due to interspecific and ploidy barriers [13–15].

The search of the functional control of apomixis has been a goal of crop scientists for the past 70–100 years and has involved a plethora of different approaches with little or no success [16,17]. Identification of particular genes in sexual model plants associated with apomixis-like features date back to before the Bellagio apomixis declaration [18], and yet, limited progress has been done. In the last years, identification of new genes linked to apomixis-like phenotypes or apomixis components had opened new perspectives but had also found new constraints. Apomixis must coordinate molecular and developmental interactions between one sporophytic and two gametophytic tissues and it is certainly not a single gene trait (see details below), which adds hurdles to apomixis breeding [8,19]. Even if simultaneously modifying several key genes produces plants mimicking apomixis phenotypes, relevant knowledge about their genetic background and possible molecular interactions and responses to regulatory signals in developmental cascades during gametogenesis and embryo and endosperm developments will still be missing. This is likely the reason why characterized apomixis-like mutants often display low penetrance and paltry quality phenotypes (see discussion below and in [20]). In the case of apomixis technology, the central problem behind constraints to apomixis breeding is that the genomic background and molecular mechanisms that initiate apomixis and guide the formation of a functional clonal seed are still poorly understood. Thus, about the road to the creation of sexual plants genetically modified to produce seeds carrying non-recombinant embryos also implies resolving the current blindness about the molecular basis behind natural apomixis. Understanding what genetic mechanism and molecular changes elicits apomixis emergence in natural species will not only benefit current efforts toward synthetic apomixis but is also a central prerequisite to harness its potential use in breeding along with an effective (penetrant) expression in sexual, domesticated crop plants.

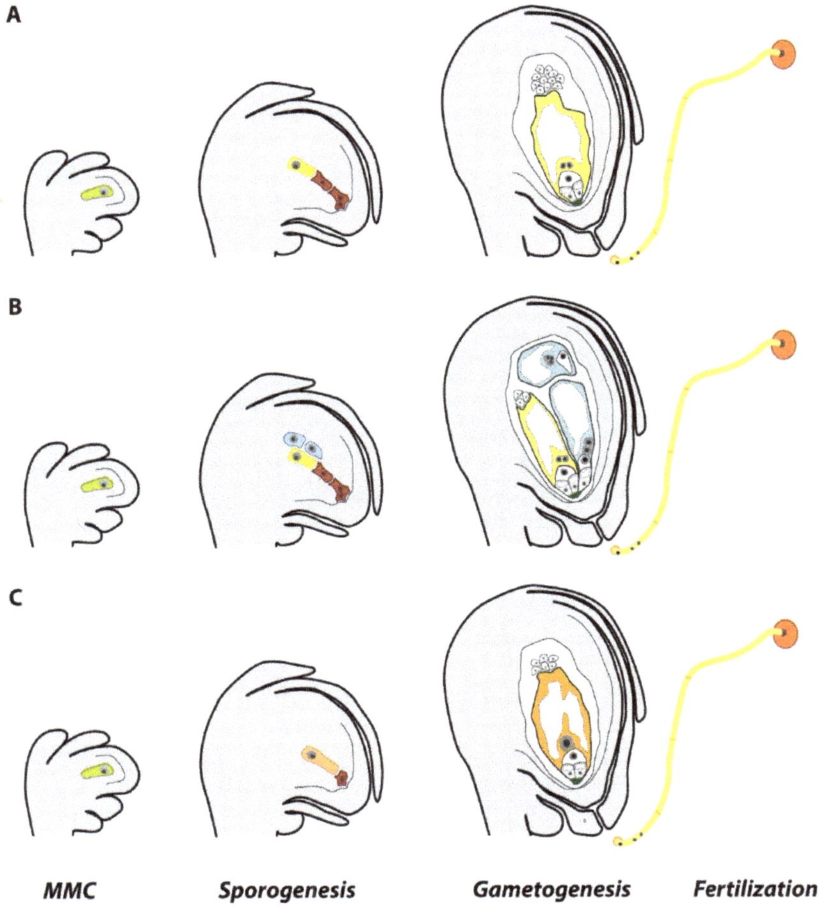

Figure 1. Reproductive alternatives in ovules of angiosperms with gametophytic apomixis. (**A**) Standard events during sexual reproduction, leading to the specification of a megaspore mother cell (MMC), who goes into meiosis and forms a row of three–four megaspores (end of sporogenesis), while the chalazal one (in yellow) develops further into a meiotic female gametophyte carrying haploid nuclei (end of gametogenesis). During fertilization, two haploid sperms fuses to the egg and central cells each to form the embryo and the endosperm of the seed, respectively. (**B**) In aposporous species, simultaneously with the MMC meiotic division, aposporous initial cells (in light blue) appear in the nucellus and acquire a megaspore-like fate entering gametogenesis and developing aposporous (diploid) female gametophytes. During fertilization, one haploid sperm fuse to the central cell to form the endosperm and the egg cell develops parthenogenetically into an embryo. (**C**) In diplosporous species, the MMC goes through a modified meiosis and form a diploid megaspore (in orange) and later a female gametophyte. During fertilization, one sperm fuse to the central cell and develop the endosperm and the embryo is developed by parthenogenesis. The drawing is based on observations in *Paspalum* spp.

2. Developmental Features of Apomixis

Apomictic plants can skip sex and produce seeds carrying clonal embryos. By doing so, ovules of apomictic plants modify key steps of cell specification, female meiosis, gamete fusion, and embryo and endosperm development. Two main types of apomixis had been described: sporophytic and gametophytic apomixis. In sporophytic apomixis (also called adventitious embryony), an embryo is directly formed from a somatic cell (e.g., integumentary or nucellar) in the ovule, which can occupy or attach to the meiotic embryo sac, thus rising seeds with poly-embryos [21,22]. Fertilization of the meiotic embryo sac is needed to produce a functional seed and for the growth of these embryos until maturity [22]. In gametophytic apomixis (from here onwards "apomixis"), the embryo is formed via an unreduced female gametophyte that engenders a mono-embryonic seed in most cases after fertilization of the endosperm. Plants showing gametophytic apomixis modify three components of the standard sexual program of seed formation. First, meiosis is prevented or skipped to form unreduced female gametophytes. The circumvention of meiosis (or apomeiosis) can happen 1) by differentiation of a nucellar cell into a megaspore-like cell independent of the megaspore mother cell (MMC) progression (apospory, Figure 1b) or 2) at the MMC via a failure in chromosome pairing (failing of the formation of the synaptonemal complex) and subsequent conversion of the first meiotic division into a mitosis-like division (diplospory, Figure 1c). In both cases, the unreduced megaspore or megaspore-like cell undergoes gametogenesis and generates an embryo sac [21,23]. The second key step is the lack of egg cell fertilization. In apomeiotic embryo sacs, the egg cell initiates embryogenesis by parthenogenesis, i.e., without fertilization [21]. The third crucial step to produce a functional clonal seed is the development of the endosperm (a nutritious tissue that regulate cellular differentiation and embryonic organogenesis [24]). In most apomictic plants, the endosperm usually develops after fertilization of the unreduced central cell (pseudogamy), which is insensitive or less deterrent to paternal ploidy misbalances than in meiotic embryo sacs [25,26]. Thus, the endosperm primary cell in apomicts tolerates shifts of parental genome dosages (from 2maternal:0paternal contributions in embryo sacs with autonomous endosperm development, and 1m:1p up until 8m:1p in pseudogamous embryo sacs), other than the 2m:1p genomic ratio required for endosperm development during sexual seed formation [27,28].

Furthermore, apomixis is almost exclusively associated with polyploidy in plants, often forming agamic complexes in which diploids are persistently sexual and polyploids arising from diploids are persistently apomictic [29,30]. In these complexes, co-occurring diploids show a tendency to produce low proportions of apomeiotic female gametophytes [31–33], and yet, functional apomixis is not stable except for dihaploids in *Erigeron* [34] and diploid cytotypes in *Boechera* [35,36]. In addition, apomictic plants are facultative, meaning that sexual reproduction is not completely eliminated [37]. Facultative plants keep a low level of sexual seed formation [38–41].

Thus, apomixis is not only a rare biological phenomenon but a developmentally complex trait.

3. Genetic and Genomic Features of Apomixis

The molecular genetics behind apomixis is still unsolved. The inheritance of apomixis has a multifaceted nature. Apomixis inheritance can be explained by Mendelian genetics, but every conceivable complication for genetic analysis seems to accumulate in apomicts [15]. The use of different molecular mapping techniques (comparative mapping, linkage disequilibrium mapping, and deletion mapping) and Sanger sequencing methods helped researchers to uncover details about the complex molecular background associated with apomixis as well as to identify sequences and genes linked to the trait. Studies carried out both in aposporous and diplosporous species (e.g., *Brachiaria*, [42,43]; *Cenchrus/Pennisetum*, [44,45]; *Erigeron*, [46]; *Hieracium*, [47]); *Hypericum* [48]; *Paspalum*, [49–52]; *Ranunculus* [53]; *Taraxacum* [54]; *Trypsacum*, [55]), indicates that apomixis behaves as a dominant trait over sexuality, with apomictic components co-segregating, often inherited as a simplex phenotype and exhibiting segregation distortion. Genetic factors linked to apomixis components had been localized into large chromosomal regions of up to 50 Mb showing suppression of recombination [50,56] and hemizygous linkage groups (i.e., markers do not show hybridization signals in sexuals [56–58]). Apomixis loci of monocots and dicots allocate chromosome structural changes (mainly inversions and translocations; [50,58,59]) and share microsynteny among different apomictic and sexual species [60], a feature that may relay—at least in grasses—upon recombinogenic (euchromatic) or nonrecombinogenic (heterochromatic) regions [61]. Mechanisms operating to reduce recombination and retain linkage disequilibrium could be ascribed to allele sequence divergence, chromosomal rearrangements, and chromatin remodeling to enhance the formation of heterochromatin. BAC clones carrying markers linked to apomixis show presence of numerous rearrangements due to insertions or deletions of transposable elements interrupting gene sequences [57,62–64]. Increased allelic sequence divergence, retrotransposon activity and mutational degradation [65,66] are expected evolutionary consequences of chromosomal regions subjected to repression of meiotic recombination. The evolution of the heteromorphic chromosomal region in organisms with sexual chromosomes is a good example [67,68]. These features reduce the efficiency of map-based cloning strategies and the genetic dissection of apomixis.

Attempts to identify components of apomixis by transcriptional profiling of reproductive organs using Differential Display PCR (DD-PCR), SuperSAGE (serial analysis of gene expression), and high-throughput sequencing on microdissected ovules revealed 1) differentially expressed genes in reproductive tissues of apomictic and sexual relatives from different plant systems (e.g., *Pennisetum*, [69,70]; *Brachiaria*, [71,72]; *Panicum*, [73,74]; *Poa*, [75], *Eragrostis*, [76–78]; *Paspalum*, [42,64,79–81]; *Hieracium*, [82,83]; *Hypericum* [84,85]) and 2) an overall shift in gene regulation at the MMC stage and a global heterochronic gene expression between the sexual and apomeiotic ovules (e.g., *Boechera*, [86–89]; *Paspalum* [79,90,91]; *Pennisetum*, [92]; *Hieracium*, [83,93,94]; *Hypericum* [85,95]; *Ranunculus* [96,97]). Such wide-ranging de-regulation on gene expression levels between sexual and apomictic ovules affect genes encoding varied biological functions (GO classes) and regulatory pathways, including key genes of the sexual pathway, RNA-directed DNA methylation and transcription regulation, hormonal signaling, and cell cycle control (see details in the next section). All these changes in gene expression, modulation of gene networks, cellular metabolism, and communication are associated with temporal and spatial ovule developmental asynchronies [40,88,98,99], incomplete penetrance of the character, and variable expressivity partially modulated by environmental conditions [100–105].

Even when these approaches allowed to identify several candidate genes and sequences associated with apomixis [64,73,74,80,90,106–110], most of them had no clear function assigned yet (however, see details in the next sections and under The Sexual Machinery and Apomixis-Like Phenotypes). Moreover, analyses of transcriptomes are complicated in most apomictic species by the high complexity, redundancy and polyploid nature of their genomes. The limitations of *de novo* transcriptomic assemblies in the absence of genomic information frequently lead to the compression of multiple gene copies into single transcriptome contigs, effectively reducing the gene space, but with potential loss of informative alleles [94,111].

4. Molecular Control of Apomixis in Apomictic Plants

Genetic and genomic studies were successful at identifying genes with a potential role in key steps of apomictic reproduction, especially genes involved in induction of fertilization-independent embryo development (for a detailed review, see [112–114]). Simultaneously, efforts for functional characterization of apomixis related genes and regulatory sequences had revealed an interesting diversity of proteins and small RNAs and long non-coding RNAs most likely being part of the apomixis regulatory cascade, yet their overall role and interconnection needs further evaluation.

Among these findings, particularly diverse are those in *Paspalum*, such as the PnTgs1-like protein (a trimethylguanosine synthase-like protein), whose function has been associated gametophyte and possibly embryo development [115,116], *QUI-GON JINN* (a gene showing homology with mitogen-activated protein kinase kinase kinases (MAP3K/MAPKKK/MEKK)) having probable roles in the acquisition of a gametophytic cell fate by AIs, and the development of aposporous embryo sacs [110], *PsORC3a* (a pseudogene with homology to subunit 3 of the ORIGIN RECOGNITION COMPLEX (ORC3)) with a probable role regulating expression of its functional homolog and with the development of apomictic endosperm [64], *PN_LNC_N13* (a member of a family of long non-coding RNAs (lncRNAs)) involved in splicing regulation [117], or a number of small RNAs differentially represented in sexual and apomictic flowers likely involved in diverse regulatory pathways—including auxin signaling—important in promoting apomixis [118].

In *Boechera*, Corral et al. [108] and Mau et al. [109] functionally characterized genes associated with apomixis and formation of unreduced pollen (see details below). In addition, using laser-assisted microdissection to analyze transcriptomes of MMC and AIC, Schmidt et al. [88] found significant enrichment of different molecular functions, and differential expression of several core cell cycle genes, meiotic genes, and genes involved in epigenetic pathways. More recently, F-box genes and E3 ligases were identified among differentially expressed genes as probable regulators important for germline development in sexual versus apomictic ovular tissues [89]. Amiyete et al. [119,120] exposed several microRNAs, including one targeting a Squamosa promoter binding protein like (SPL11) differentially up-regulated at the MMC stage of ovule development in apomictic genotypes and small RNAs with potential binding sites in exonic regions, indicating a probable role in post-transcriptional gene regulation.

In *Eragrostis*, tissue-specific expression differences were detected for genes *AGO104* and *DMT102*, the first one is relevant for cell fate specificity, which may be preventing entry into meiosis and promoting gametophytic development in the diplosporous ovules, the second one is required for cytosine methylation at CNG sites likely involved in maintenance functions, which might be promoting the establishment of gametophytic specificities (i.e., four-celled embryo sacs) observed in apomictic ovules [121]. Analysis of small RNA families identified two genes, a MADS-box transcription factor gene, and a transposon, specifically repressed in the sexual genotype [122].

In *Hieracium*, using dihaploid and tetraploid apomictic plants and deletion mutants for apomixis (LOA) and parthenogeneisis (LOP) loci, Rabiger et al. [94] identified ovary-expressed ARGONAUTE genes and differentially expressed genes enriched for processes involved in small RNA biogenesis and chromatin silencing in apomicts, plus a small number of putative differentially targeted genes within the mutants and potential candidates in the specification and initiation of aposporous initial (AI) cells. Diverse Arabinogalactan-proteins (AGPs) were identified in AI and FM cells, expressed during early sexual and aposporous gametophyte development, which suggest a role in communication and signal transduction events leading to megaspore death and AI cell differentiation and progression toward gametogenesis [123].

In *Hypericum*, a truncated gene ARI showing homology to *AtARIADNE7* (encoding a ring finger E3 ligase protein involved in regulatory processes and ubiquitin-mediated protein) degradation, might have a role on protein populations altering gametophyte development [48]. In transcriptomic analyses, Galla et al. [85] found evidence that differentiation of the AI cell may be related to the misregulation of RNA-directed DNA methylation (RdDM) pathways (through *MEE57*, *CMT3*, and *IND2*, genes involved in the maintenance of DNA methylation, and AGO9, a component of RNA silencing complexes), chromatin-remodeling proteins (through HPCHC1, a MORC-like CW-type Zinc finger protein), and hormonal homeostasis (through HPPIN8, an auxin efflux carrier involved in ovule intercellular auxin gradients).

Studies on different apomictic systems indicate that epigenetic changes play a relevant role modulating apomixis progression (e.g., [124]) and controlling parthenogenesis. The methylation state of the genomic region for apomixis in *Paspalum* spp. controls the activation/repression of parthenogenesis [125], while in *Boechera* spp., aberrant imprinting and locus-specific DNA methylation changes are responsible for the maternal activation of a MADS-box gene encoding homologs of the transcription factor PHERES1 (PHE1), which promote embryo growth and is paternally expressed in sexual *Arabidopsis thaliana* [126]. In *Hieracium*, a paternally-inherited *PHE1* gene was identified as transcriptionally active in both sexually and apomictic derived seeds, suggesting the imprinting system may be modified [83]. Thus far, several proteins acting as putative inducers of parthenogenesis had been identified using differential transcriptomics (e.g., in *Poa pratensis* the SERK (for SOMATIC EMBRYOGENESIS RECEPTOR-LIKE KINASE); in *Pennisetum squamulatum*, the ASGR-BBML (for apospory-specific genomic region-BABY BOOM-like), and in the *Boechera holboelli*, complex APOLLO (for apomixis linked locus).

Albertini et al. [75] found a group of genes involved in signaling and trafficking events during reproductive development in *Poa pratensis*. The authors were able to identify *Pp*SERK as putative candidates for the induction of apomixis. Since SERK encodes a leucine-rich repeat containing a receptor-like kinase highly expressed during somatic and sexual embryogenesis *in vitro* [127], it was suggested that *Pp*SERK may promote autonomous embryo development in apomictic species [107]. A recent screen for members of the SERK family in *Paspalum notatum* resulted in the identification of two paralogs, *PnSERK1* and *PnSERK2*, the latter displaying a strong differential spatial expression pattern in ovules of apomict and sexual genotypes [128].

Another gene involved in the induction of parthenogenesis has been recently identified in the *Pennisetum/Cenchrus* species aggregate. The *PsASGR-BABY BOOM-like* (*PsASGR-BBML*) gene is expressed in egg cells before fertilization and can induce ectopic parthenogenetic embryos [129,130]. *PsASGR-BBML-like* is a member of the BBM-like clade of APETALA 2 transcription factors, which are conservatively widespread in the plant kingdom and functionally diverse [131]. Similar genes encoding BBM or BBM-like proteins were found to induce ectopic embryo formation in sexual model systems such as *Brassica* and *Arabidopsis* [132]. A functional test of the *PsASGR-BBML* transgene showed it promotes parthenogenesis and the production of haploid offspring in transgenic sexual pearl millet, can induce haploid embryo development in maize and rice at variable rates, but failed to induce haploid seed development in *Arabidopsis thaliana* [129,130,133]. These studies represent a good progress toward understanding of parthenogenetic development. Moreover, the ASGR region and the *PsASGR-BBML* gene sequences are highly conserved across the Paniceae in *Brachiaria* and *Panicum* species having different chromosomal backgrounds, suggesting a relevant role for the parthenogenesis component of apomixis [59,134]. However, the variable penetrance of the trait between transgenic lines and siblings of the same line, the complexity of embryo development observed in rice lines with shifts to vivipary and absence of endosperm development [130] plus the lack of a reference genome of the apomictic species highly restrict the options to deeply characterize the *PsASGR-BBML* gene and the network of genes and/or protein interactions which may further promote parthenogenesis. This limits the chances of manipulating plants with enhanced transcriptional levels (see [20] in this issue).

In *Boechera*, Sharbel et al. [86,87] were able to identify and characterize an apomixis-related gene (APOLLO; [108]). The APOLLO gene encodes an Aspartate Glutamate Aspartate Aspartate histidine exonuclease whose transcripts are down-regulated in sexual ovules entering meiosis while being up-regulated in apomeiotic ovules at the same stage of development in plants of the genus *Boechera* [108]. The gene has apo- and sex-specific alleles, both highly polymorphic (13 apoalleles and 21 sexalleles characterized out of 18 genotypes). All apomictic *Boechera* spp. accessions evaluated had at least one apoallele and one sexallele, i.e., they are heterozygous for the APOLLO gene, while all sexual genotypes were homozygous for sexalleles. Compared to the sexallele consensus sequence, dominant apoalleles are characterized by a set of linked apomixis-specific polymorphisms, of which the most relevant is a 20- nucleotide polymorphism present in the 5′ untranslated region (UTR) that contains specific transcription factor-binding sites for a number of known regulatory factors (e.g., ARABIDOPSIS THALIANA HOMEOBOX PROTEIN5; [108]). The authors suggested that the expression of a deregulated apoallele could induce the cascade of events leading to asexual female gamete formation in apomictic *Boechera* plants. However, thus far, apoalleles have not been validated and specific targets of apoalleles have not been identified yet. The high number of putative alleles (the lack of a consensus apomictic allele), the variation for APOLLO copy number (i.e., CNV) observed among genotypes, the lack of a similar genetic background in flanking regions of both apo- and sexalleles analyzed in three BACs (for details see [108]), and the complex evolutionary history of *Boechera* spp. [135] hamper clarification of apomixis initiation and development in this group.

5. The Sexual Machinery and Apomixis-Like Phenotypes

Over the last years, substantial work has been done in flowering plants on regards of gain- and loss-of-function mutants in different model systems, particularly in Arabidopsis, Petunia, and rice. These mutants had led to the identification of genes involved in different molecular processes during sexual seed development, including primordium differentiation, ovule development, gametophyte differentiation, and reproductive cell identity [136–139]. Several of such genes are associated with key steps of meiosis, embryo, and endosperm development and had shown phenotypes resembling elements of apomixis development [139–144]. However, thus far, all attempts of *de novo* engineering apomixis by, e.g., overexpression of candidate genes, has failed and today not a single fully (displaying high expressivity) apomictic mutant has been recovered from sexual species. I will mention here some relevant genes as an overview of the complexity behind the emergence of apomixis, as similar apomixis-like phenotypes can be caused by diverse genes, and all these mutants represent independent molecular events which should be or are expected to be developmentally "coordinated" in a natural apomictic plant.

5.1. Apomeiosis Mutants

MMC competence is restricted to only one cell from the nucellus in angiosperms, which undergo meiosis and give rise to the female gametophyte. Several genes and small RNAs interact to control gamete cell specification and meiosis by restricting the number of meiotic precursors or by enabling the meiosis progression [138]. The homeodomain transcription factor *WUSCHEL* (*WUS*) is a key regulator of stem cell fate, essential for the formation of the integuments and in the specification of MMCs. In *Arabidopsis*, Zhao et al. [145] demonstrated that genes coding KIP-RELATED PROTEIN (KRP) (an inhibitor of cyclin-dependent kinase, CDK) act to restrict the inactivation of the Retinoblastoma homolog RBR1 by CDKA;1. Since RBR1 is a repressor of *WUS*, *krp* and *rbr1* mutants display supernumerary meiocytes and embryo sac-like structures. Similarly, ICK (another family of CDK inhibitors) function to restrict the formation of megaspore mother cells and functional megaspores to one per ovule [146]. Inactivation of all ICK/KRP genes produces ovules having supernumerary MMCs, FMs and embryo sacs.

Small-RNA pathways have a relevant role in the regulation of female germline specification. Su et al. [147] found that THO non-cell complex autonomously restricts the megaspore mother cell fate to a single cell by preventing ectopic MMC formation via trans-acting small interfering RNA (ta-siRNA). The TEX1 protein of the THO/TREX complex, present only in epidermal cells, restricted the expression of ARF3 to the medio domain of ovule primordia through the biogenesis of TAS3-derived ta-siRNA. *TAS* family genes are transcripts targeted for cleavage by different miRNAs and ARF3 (AUXIN RESPONSE FACTOR) family members regulate leaf polarity, floral stem cell maintenance, and lateral root growth [148]. Mutations in components of the THO/TREX complex led to ovules with multiple megaspore mother cells, as well as a TAS3 ta-siRNA-insensitive mutant.

In another example of non-cell autonomous restriction of MMC fate, Zhao et al. [149] demonstrated that the *Arabidopsis* cytochrome P450 gene *KLU* (expressed in inner integument) produces a mobile signal that recruits the ATP-dependent chromatin, remodeling complex SWR1 to *WRKY28* (a transcription factor) in ovule primordia and promote its expression by the deposition of specific histone variants. *WRKY28* is expressed in somatic cells surrounding the MMC and is required to inhibit the acquisition of MMC-like cell fate.

Members of the ARGONAUTE (AGO) protein family interact with sRNAs to regulate transcriptional and posttranscriptional gene expression through RNA-directed DNA methylation (RdDM) or ta-siRNA pathways. In rice the *MEIOSIS ARRESTED AT LEPTOTENE1* (*MEL1*) (an ortholog of *AtAGO5*) and the *MULTIPLE SPOROCYTE* (*MSP1*) genes initiate the sporogenous development, which regulates the completion of meiosis and the repression of germ cell fate in somatic tissues in rice ovules [150,151]. Mutants of those genes show an increased number of sporocytes, and multiple disorganized, but occasionally viable, female gametophytes [152,153] resembling the formation of multiple embryo sacs from nucellar cells in aposporous grasses [33].

Another case of apomixis-like mutants comes from *Arabidopsis* AGO9, a protein present in the cytoplasm of cells from the epidermal layer and accumulates in the nucleus of the MMC in ovules. AGO9 specifically binds 24-nt sRNAs derived from transposable elements, and together with RNA-DEPENDENT RNA POLYMERASE 6 (RDR6) involved in the biogenesis of siRNAs, they restrict the formation of multiple MMC cells in the ovule [152]. The archesporial cell of *ago9* mutants can differentiate directly into a functional megaspore without undergoing meiosis and produces aposporous-like embryo sacs [152]. This suggests that the transition from a sporophytic to a reproductive fate in cells surrounding the germline is suppressed by factors like AGO9. Another ARGONAUTE protein (AGO104) found in maize is coded by the *Dominant non-reduction4* (*Dnr4*) gene. Mutants of this protein produce viable unreduced female gametophytes due to defects in chromatin condensation during meiosis that finally fails to segregate chromosomes and creates a diplosplory-like phenotype. AGO104 is functionally related to the *At*AGO9 and is involved in gene silencing via methylation [144]. Diplospory is the type of apomixis found in *Tripsacum dactyloides*, a maize relative [154]. Interestingly, the AGO104 locus is located in a region on chromosome 6 of maize that is syntenic to the *Tripsacum* apomixis locus [153].

Maize DNA methyltransferases (DMT) are involved in RNA directed DNA methylation (RdDM). The inactivation of two DMT102 and DMT103 results in phenotypes evocative of the *At*AGO9 mutant, producing unreduced gametes and multiple embryo sacs in the ovule. Comparative analysis between the sexual maize biotype, *dmt* mutants, and apomictic hybrids exposed similar chromatin states in archesporial tissues of both the *dmt102* and apomicts [144]. Using *Arabidopsis* intraspecific hybrids between ecotypes and *ago9* and *rdr6* mutants, Rodriguez-Leal et al. [155] showed that multiple loci may control cell specification at the onset of female meiosis, and that variations in transcriptional regulation and localization of AGO9 and abnormal gamete precursors in *rdr6* ovules are reminiscent of natural phenotypic variation during megasporogenesis. In agreement with studies on apomictic plants (see previous section), these results suggest that epigenetic regulations play a role in the development and differentiation between apomixis and sexuality, and many plants will have the ability to reproduce asexually by short-circuiting the appropriate signals [152,156–158].

Studies in *Arabidopsis* identified the DYAD/SWITCH1 (SWI1) gene required for megasporogenesis (and microsporogenesis). SWI1 is expressed in female and male meiotic cells and have a role in sister chromatid cohesion and centromere organization [159]. SWI1 mutants show defects in female meiosis progression that led to a switch from meiotic to mitotic division and the formation of unreduced gametes in very low frequency [159–161]. The disruption of a single gene resulted in the bypass of meiosis thus resembling diplospory (a key component of apomixis). Another case of mutants showing apomixis-like phenotypes is the triple mutant called *MiMe* ("mitosis instead meiosis") [162]. In *MiMe* plants, the combination of three mutants (*osd1*: omits the second meiosis; *Atspo11*-1: eliminates recombination and pairing; and *Atrec8*: modifies chromatid segregation) replace meiosis by a mitosis, and hence, apomeiotic-like embryo sacs are produced together with diploid gametes genetically identical to the mother plant [162,163].

Several mutants involved in polarity and cell fate determination in the *Arabidopsis thaliana* female gametophyte had been described (including cell cycle regulators, components of the cytokinin signaling pathway or MYB transcription factors) [114,139] showing arrested development at different stages, uncellularized gametophytes and cell identity loss, phenotypes similar to those observed in mature ovules of some apomicts carrying multiple embryo sacs. However, female gametogenesis in apomicts do not generally show drastic structural alterations, except perhaps for the ploidy change in nuclei and the absence of antipodal cells in embryo sacs of grass species.

5.2. Parthenogenetic Mutants

In both animals and plants, the egg cell activation event during sexual reproduction is highly reliant on pulse-signals that increase intracellular calcium ions [164] and can trigger in vitro parthenogenetic embryo development in animals [165]. In plants, however, similar experiments were insufficient to activate parthenogenesis in absence of fertilization [166]. Still, in the *multicopy suppressor of ira 1* (*msi1*) mutant in *Arabidopsis* the egg cell displays the capacity to start developing a parthenogenetic embryo without fertilization, undergoing abortion at later stages of development [143,167]. The *MSI1* gene in *Hieracium* (*HMSI1*) seemingly triggers the initiation of autonomous seed development [168]. Other potential candidate genes are those found to trigger somatic embryogenesis when expressed ectopically. These genes, identified mainly in *Arabidopsis* include *BABY BOOM* (*BBM*) [132] and *WUSCHEL* (*WUS*) [169] (both discussed above), *LEAFY COLTYLEDON 1* (*LEC1*) [170] and *LEAFY COTYLEDON 2* (*LEC2*) [171].

In maize a frame-shift mutation in *MATRILINEAL* (*MTL*), a pollen-specific phospholipase exclusively localized to sperm cytoplasm, triggered an increase in the haploid induction rate ([172]; see also [173,174]), with similar results in Indica rice [175] and wheat [176]. Even though *MTL* is not directly involved in parthenogenesis, it shows that pollen-specific genes may mediate the formation of haploid seeds. Haploid induction is triggered by nuclear-cytoplasmic interactions in durum wheat [177] and other species, and is spontaneously (and rarely) observed in natural apomicts [21] highlighting that specific combination of non-nuclear sperm components may provide fertilization-like signals promoting parthenogenetic egg cell progression.

In Bryophytes, the homeobox gene *BELL1* and *RKD* family of transcription factors might be of relevance for ectopic induction of embryogenesis in angiosperms. Overexpression of *BELL1* induces embryo formation in the moss *Physcomitrella patens* and reproductive diploid sporophytes from gametophytic cells without fertilization [178]. While *BELL1* is a central molecular trigger for the gametophyte-to-sporophyte transition in *P. patens*, in rice *BELL1*-like homeobox genes regulate inflorescence architecture [179] and in *A. thaliana BELL1* interacts MADS box factors and repress *WUS* for proper differentiation of the ovule [180]. In the liverwort *Marchantia polymorpha*, transformants using artificial microRNAs (amiRNA) constructs to downregulate endogenous *MpRKD* transcript levels show severe defects in gemma cup formation and egg-like cells underwent cell divisions in the absence of fertilization before the archegonia reached maturity [181]. However, the pattern of such divisions differed from those of the wild-type embryo, and failed to form viable, normally patterned embryos [181]. Thus, *MpRKD* has a crucial role in the formation of the gemma cup and in establishing and/or maintaining the quiescent state of the egg cell prior to fertilization, preventing it from undergoing parthenogenetic-like grow by cell divisions. Whereas *MpRKD* appears to be primarily active in the control of gametophyte development, in flowering plants RKD function is recruited to early phases of embryogenesis but loss-of-function experiments in *A. thaliana* have not yet produced phenotypes affecting the egg cell [182].

Differences observed among plants in haploid and diploid parthenogenesis likely underly a combination of genetic redundancy and divergent evolution in regulatory factors ensuring egg cell quiescence prior fertilization [183].

5.3. Mutants Associated with Endosperm Development

Despite the biological and economic relevance of the endosperm, the molecular events underlying its development were largely unexplored. In recent years, this has changed with the combined use of mutants, cell specific markers, and plant hormone sensing reporters [184]. Additional loss-of-function *Arabidopsis* mutants associated with apomixis-like phenotypes are those of the Polycomb group complex 2 (PRC2) that repress endosperm formation and seed development in the absence of fertilization. The genes of this complex encode several proteins, such as MEDEA (MEA) [185], FERTILIZATION INDEPENDENT SEED 2 (FIS2) [142,186], FERTILIZATION INDEPENDENT ENDOSPERM (FIE) [141], and the mentioned MULTICOPY SUPPRESSOR OF IRA 1 (MSI1) [143]. *FIS* class genes are expressed in the central cell of the female gametophyte and encode protein subunits of the PRC2 that control the expression of downstream genes promoting endosperm growth (e.g., *MADS-box PHERES1* or *PHE1*) by silencing paternal or maternal alleles [88,187–189]. FIS-PRC2 mutants lead to mitotic activation of the central cell nucleus division in absence of fertilization and the autonomous development of a diploid endosperm [189]. However, like in the majority of mutant phenotypes, in most *fie* seeds the endosperm can only growth until the cellularized phase and the development of the embryo is arrested at globular-heart stages [141,167,190]. Unexpectedly, specific downregulation of *FIE* in sexual *Hieracium* failed to stimulate the autonomous proliferation of the central cell, and pollination was required to activate seed initiation but led to subsequent seed abortion [191]. Similarly, in *FIE* downregulated apomictic *Hieracium*, autonomous embryo and endosperm initiation were also inhibited and seed development was arrested, with most autonomous seeds displaying defective embryo and endosperm growths [191]. Thus, FIE might not be directly involved in endosperm progression in apomicts (and hence is likely to be a poor candidate for the synthesis of apomixis in crop plants [191]); alternatively, it might require interactions with other factors and/or molecular signals to stimulate endosperm progression without fertilization.

6. Overall Features, Hypotheses in Conflict, and the Quest for the Origin of Apomixis

The characterization of a number of genes with likely relevant roles in the activation and/or progression of apomeiosis and/or parthenogenesis in different plant groups and the observed potential of many taxa to express apomixis-like traits indicates that the induction of apomeiosis and/or parthenogenesis by genetic and epigenetic manipulation is feasible. However, the evidence also indicates that apomixis is caused by a cluster of genes—physically linked or not—most likely functionally related that act coordinately in the activation of apomixis. Thus, while the use of different genetic modification techniques (e.g., Transgene-directed Mutagenesis, RNA-directed DNA-Methylation, or new restriction enzyme techniques (gene-editing); [192]) in attempting to induce apomixis in a sexual plant is likely the best strategy, expectations about obtaining a neoapomict may not be realistic. The association of apomictic sequences to accumulation of transposable elements, gene degeneration, and gene-specific silencing mechanisms (likely based on chromatin remodeling factors or transacting and heterochromatic interfering RNAs involved in both transcriptional and post-transcriptional gene regulation, see [139,193,194]) suggests that inducing apomixis *sensu stricto* by simply knocking out meiotic genes is unfeasible. A synthetic approach to engineering apomixis by targeting key regulatory steps has been considered for a long time while many of the above discussed genes were discovered [195], and yet, achievements on this goal progressed slowly (for details see [113]). Examples are the feeble penetrance attained by synthetic apomixis-like mutants (see for example [162]), which expose the fact that changing radically key developmental steps likely require the coordination of multiple metabolic pathways, molecular signals, and cellular players. Unless a proper understanding of the genetic architecture of apomixis exists, the prospect for a genetic engineered apomictic crop remains equivocal (but see [20] other reviews in this issue).

Thus far, three main hypotheses (HA-C) are usually referred to for the cause of apomixis when trying to explain the varied facts and features around the formation of clonal seeds:

H**A**) it is caused by developmental asynchronies (de-regulation of the sexual developmental pathway) resulting as a consequence of hybridization and/or polyploidization [9,195–197]; this hypothesis is supported by ovule development analyses and transcriptional profiling (see discussion above).

H**B**) it is a mutation-based anomaly that involves a simple or (conceivably) complex genetic locus [15,198]; this hypothesis is supported by natural and induced mutants (see discussion above).

H**C**) it is an ancient state from which sexual reproduction arose as an alternative, epigenetically regulated and with remnant capacities relatively conserved across eukaryotes [199,200], supported by mutants affecting methylation pathways (see discussion above).

Functional apomixis needs the coordinated concurrence of three reproductive elements (i.e., apomeiosis or the formation of unreduced non-recombinant gametes, embryogenesis without fertilization, and endosperm development), each with a complex developmental background in sexual plants (e.g., involving fundamental changes in cell specification and fate during ovule and seed developments [201,202]) and several genes involved (see above). Even though these hypotheses are not mutually exclusive, they fail to provide an appropriate frame and a unified explanation covering all empirical data about apomixis. While the simultaneous deregulation in multiple loci (H**B**) can be a downstream effect initiated by mutations on genes controlling apomeiosis, parthenogenesis, or endosperm development (H**A**), it is highly unlikely that these three changes on the sexual reproductive program could evolve together in a sexual ancestor through randomly occurring mutations. As any single step has a negative effect on the fitness of its carrier [11,203,204], the needed genetic changes (and basis) for apomixis must appear concurrently and have a concerted expression to be functional and produce a viable seed (details about the first stages of emergence and stabilization of apomixis in natural populations can be found in Hojsgaard [205] and Hojsgaard and Hoerandl [30]). Likewise, as apomixis can appear in the F_1 progeny from sexual parents after hybridization and polyploidization [99,206], the emergence of apomixis may not require long-term evolutionary changes or separate evolution of mutants for each reproductive step involved in apomixis. This is a relevant point when considering apomixis breeding. Similarly, apomixis is not simply a consequence of polyploidization since 1) most known polyploid plants are sexual; 2) polyploid apomictic species do not exhibit recombinant sibs able to reproduce only by sexuality in nature (as expected if polyploidization would separately affect genes related with apomeiosis, parthenogenesis, and endosperm formation); and 3) polyploid sexuals can be recovered by inducing chromosome doubling in sexual diploids from agamic complexes [207]. Thus, the emergence of apomixis in a plant is likely based on a specific molecular mechanism that has cytological consequences and not *vice versa*. In sexual-apomictic comparisons, sexuals display a normal transcriptional profile ascribed to (stable) meiosis + syngamy (in other words, non-functional apomixis) and apomicts display a general de-regulation of genes and several sequence-level changes, ascribed to the abnormal meiosis and irregular syngamy usually accompanied by high rates of apomeiosis + parthenogenesis (or functional apomixis) [38,40].

This "all-or-nothing" pattern observed at developmental and gene expression levels in different apomicts—together with the other mentioned features—implies that apomixis in plants is triggered by a (likely common) single-event molecular mechanism that must implicate simultaneous changes on genes "masterly" associated with (at least three) critical steps of the sexual program (H**A** and H**C**), and involve a corresponding de-regulation of many more loci (H**B**) leading to the reproductive switch (and associated shift of ploidy) that deplete sex (meiosis and syngamy) and activates apomixis (apomeiosis and parthenogenesis) during seed formation.

Only by unveiling the molecular nature of such allegedly single event that prelude the emergence of apomixis in natural plant species will we be able to have a chance to effectively manipulate genetic sequences and sexuality towards a penetrant apomixis-like development in plants.

7. Current View and Prospects

Disentangling the molecular basis and regulatory mechanisms underlying the emergence of apomixis requires species-specific knowledge about the (epi)genetic and genomic contexts. For doing so, in general terms, it is necessary to compare apomictic genomes against sexual genomes, preferentially of the same ploidy and with no hybrid origin to enable the identification of all changes in gene sequences and genome structure required to activate apomixis. Knowledge about the genomic background interacting with an apomixis locus will provide indirect but valuable information about (primary and secondary) variants of metabolic relevance and interrelation of regulatory control mechanisms in the expression of apomixis (like cell cycle control, protein turnover, signal transduction, and hormonal regulatory pathways). Mainly due to the cost and complexity of such projects (almost all apomicts are polyploids and/or hybrids), no genome of an apomictic species has been yet sequenced. The publication of the perl millet genome [*Cenchrus americanus* (L.) Morrone] (genome size ~1.79 Gb; [208]), a minor crop and diploid sexual relative of tetraploid apomictic *C. ciliaris* L. (Buffelgrass), as well as of the weeping lovegrass genome [*Eragrostis curvula* (Schrad.) Nees] (genome size ~660 Mb; [209]), a diploid sexual forage grass having apomictic polyploid cytotypes, and the genomes of *Boechera stricta* (genome size 216 Mb) and *B. retrofracta* (genome size 200 Mb) [210–212], two sexual diploid herbs relative to apomictic *B. holboellii* and *B. divaricarpa*, represent an effort in that direction. Two recently-initiated projects aiming at sequencing the genome of several sexual and apomictic *Boechera* species in Canada [213] and Switzerland [214], and the generation of an apomictic *Hieracium* (predicted genome size 3.6 Gb) genomic and transcriptomic resource [94], may provide the first datasets for sexual-apomictic pair genomic comparisons.

However, the NGS market has rapidly evolved and a large number of developments have gained accuracy and speed, have partially solved technically complex issues (e.g., genome phasing), and reduced the cost of sequencing [215–217]. Hence, it is timely to progress the state-of-the-art and join efforts in sequencing high quality genomes using long sequence reads to cover repetitive regions (present in some apomixis loci) and linked-read strategies to improve genome contiguity and assembly quality. Different cost-effective and scalable strategies are offered in the market. Convergence between assembled sequences and chromosome number can be achieved through different chromosome conformation capture techniques [218] and scaffolding strategies enhancing sequence continuity and the integration of fragmentary sequences; however, it needs to be decided on a case-by-case basis. Even when genome phasing might not be strictly necessary for the identification of apomixis genes (the simplex haplotype of an apomixis locus is expected to be sufficiently divergent to other sexual haplotypes), defining the most likely phase within a set of connected variants of an allele and appropriate read population phasing [219,220] will improve the ability to correctly identify compound heterozygosity at rare variants in either *trans* or *cis* interactions relevant for apomixis expressivity in both wild and mutant plants. The causality for the observed variation in penetrance and expressivity among natural apomicts and apomixis-like mutants is widely undetermined. Integration of genetic maps and methylomes, when available, into quality genomes will certainly facilitate linking phenotype variability to genotypes and identifying genetic modifiers. Despite that complex, quantitative phenotypes are stably inherited across generations [8], phenotypic instability, likely due to gene-environment interactions or genetic compensation to transgenes, is known for natural apomicts and mutants [8,181]. Hence, combined evaluations of transgenerational phenotype stability and genomic data on apomictic plants and apomixis-like mutants, including possible off-target effects in genome editing, will be required to unclose molecular interactions toward penetrant apomixis phenotypes.

Yet, apomixis research is underfunded, and not every newly available technique is a necessary improvement for solving the molecular control of apomixis in each plant system. Comparative studies (such as coexpression or GWA studies) can help identifying shared elements in expression networks and coordinated activity of gene sets in complex phenotypes [81]. However, methodological differences often neglect integration of multiple data sets from different sources due to incompatibility. Perhaps, given the relevance and remaining challenges toward apomixis technology, thinking of an international consortium between public and private institutions enabling synergies between qualified researchers, know-how and access to funding to implement apomixis technology in main crops is not utopian. By stating clear rules and goals among participants, such an enterprise will facilitate financing and speed up the collection of data, analysis, and results validation while reducing individual efforts and maximizing data compatibility.

Even though the endeavor to get an apomictic crop is still technically demanding and challenging (e.g., underlying gene networks and regulatory pathways differ among apomicts, and hence, manipulating those to shift developmental pathways will present distinct hurdles [144,183]), it will help us realize whether implementing apomixis breeding in agroscience is a feasible task or not, a pending topic that is becoming critical and will help humanity resolving current and future needs on food supply, landscape management, and conservation of biodiversity areas.

Author Contributions: Conceptualization, D.H.; investigation, D.H.; writing—original draft preparation, D.H.; writing—review and editing, D.H. All authors have read and agreed to the published version of the manuscript.

Funding: This research received no external funding.

Acknowledgments: Most of the present manuscript was written by the author during the preparation of an ERC Starting grant application (CloSeToBreedFuture) focused on solving the molecular basis behind apomixis.

Conflicts of Interest: The authors declare no conflict of interest.

References

1. United Nations, Department of Economic and Social Affairs, P.D. *World Population Prospects 2019. Volume II: Demographic Profiles*; United Nations Press: New York, NY, USA, 2019.
2. FAO. *The State of Food and Agriculture*; Food and Agriculture Organization of the United Nations Press: Rome, Italy, 2009; ISBN 978-92-5-106215-9.
3. Sharma, H.C.; Crouch, J.H.; Sharma, K.K.; Seetharama, N.; Hash, C.T. Applications of biotechnology for crop improvement: Prospects and constraints. *Plant Sci.* **2002**, *163*, 381–395. [CrossRef]
4. Acquaah, G. *Principles of Plant Genetics and Breeding*, 2nd ed.; Wiley- Blackwell: Chichester, UK, 2012; ISBN 978-0-470-66476-6.
5. Van Dijk, P.J.; Rigola, D.; Schauer, S.E. Plant breeding: Surprisingly, less sex is better. *Curr. Biol.* **2016**, *26*, R122–R124. [CrossRef] [PubMed]
6. Ray, D.K.; Mueller, N.D.; West, P.C.; Foley, J.A. Yield trends are insufficient to double global crop production by 2050. *PLoS ONE* **2013**, *8*, e66428. [CrossRef] [PubMed]
7. Spillane, C.; Curtis, M.D.; Grossniklaus, U. Apomixis technology development—virgin births in farmers' fields? *Nat. Biotechnol.* **2004**, *22*, 687–691. [CrossRef] [PubMed]
8. Sailer, C.; Schmid, B.; Grossniklaus, U. Apomixis allows the transgenerational fixation of phenotypes in hybrid plants. *Curr. Biol.* **2016**, *26*, 331–337. [CrossRef]
9. Carman, J.G. Asynchonous expression of duplicate genes in angiosperms may cause apomixis, bispory, tetraspory, and polyembryony. *Biol. J. Linn. Soc.* **1997**, *61*, 51–94. [CrossRef]
10. Hojsgaard, D.; Klatt, S.; Baier, R.; Carman, J.G.; Hörandl, E. Taxonomy and biogeography of apomixis in angiosperms and associated biodiversity characteristics. *CRC Crit. Rev. Plant Sci.* **2014**, *33*, 414–427. [CrossRef]
11. van Dijk, P.J.; Vijverberg, K. The significance of apomixis in the evolution of the angiosperms: A reappraisal. In *Plant Species-Level Systematics: New Perspectives on Pattern & Process*; Bakker, F.T., Chatrou, L.W., Gravendeel, B., Pelser, P.B., Eds.; Gantner Verlag: Ruggell, Lichtenstein, 2005; Volume 143, pp. 101–116. ISBN 0080-0694.
12. Ray, D.T.; Coffelt, T.A.; Dierig, D.A. Breeding guayule for commercial production. *Ind. Crops Prod.* **2005**, *22*, 15–25. [CrossRef]
13. Savidan, Y. Transfer of apomixis through wide crosses. In *The Flowering of Apomixis: From Mechanisms to Genetic Engineering*; Savidan, Y., Carman, J.G., Dresselhaus, T., Eds.; CIMMYT, IRD, European Commission OC VI (FAIR): Mexico City, Mexico, 2001; pp. 153–167. ISBN 970-648-074-9.
14. Ozias-Akins, P.; Akiyama, Y.; Hanna, W.W. Molecular characterization of the genomic region linked with apomixis in *Pennisetum/Cenchrus*. *Funct. Integr. Genomics* **2003**, *3*, 94–104. [CrossRef]
15. Ozias-Akins, P.; van Dijk, P.J. Mendelian genetics of apomixis in plants. *Annu. Rev. Genet.* **2007**, *41*, 509–537. [CrossRef]
16. Barcaccia, G.; Albertini, E. Apomixis in plant reproduction: A novel perspective on an old dilemma. *Plant Reprod.* **2013**, *26*, 159–179. [CrossRef] [PubMed]
17. Cardwell, J.H. *A review of apomixis and differential expression analyses using microarrays*; Utah State University Press: Logan, UT, USA, 2013.
18. Grossniklaus, U.; Koltunow, A.; Van Lookeren Campagne, M. A bright future for apomixis. *Trends Plant Sci.* **1998**, *3*, 415–416. [CrossRef]
19. Leblanc, O.; Grimanelli, D.; Hernandez-Rodriguez, M.; Galindo, P.A.; Soriano-Martinez, A.M.; Perotti, E. Seed development and inheritance studies in apomictic maize-*Tripsacum* hybrids reveal barriers for the transfer of apomixis into sexual crops. *Int. J. Dev. Biol.* **2009**, *53*, 585–596. [CrossRef] [PubMed]

20. Scheben, A.; Hojsgaard, D.H. Can we use gene-editing to induce apomixis in sexual plants? *Genes (Basel)* **2020**. Submitted.
21. Asker, S.E.; Jerling, L. *Apomixis in Plants*; CRC Press: Boca Raton, FL, USA, 1992; ISBN 0-8493-4545-6.
22. Naumova, T.N. *Apomixis in Angiosperms-Nucellar and Integumentary embryony*; CRC Press: Boca Raton, FL, USA, 1993; ISBN 13: 978-1-315-89065-4.
23. Nogler, G.A. Gametophytic apomixis. In *Embryology of Angiosperms*; Johri, B.M., Ed.; Springer-Verlag Press: Berlin/Heidelberg, Germany, 1984; pp. 475–518. ISBN 3-540-12739-9.
24. Olsen, O.-A. (Ed.) *Plant Cell Monographs*; Plant Cell; Springer: Berlin/Heidelberg, Germany, 2007; Volume 8, ISBN 978-3-540-71234-3.
25. Quarin, C.L. Effect of pollen source and pollen ploidy on endosperm formation and seed set in pseudogamous apomictic *Paspalum notatum*. *Sex. Plant Rep.* **1999**, *11*, 331–335. [CrossRef]
26. Talent, N.; Dickinson, T.A. Endosperm formation in aposporous *Crataegus* (Rosaceae, Spiraeoideae, tribe Pyreae): Parallels to Ranunculaceae and Poaceae. *New Phytol.* **2007**, *173*, 231–249. [CrossRef]
27. Köhler, C.; Mittelsten Scheid, O.; Erilova, A. The impact of the triploid block on the origin and evolution of polyploid plants. *Trends Genet.* **2010**, *26*, 142–148. [CrossRef]
28. Schatlowski, N.; Wolff, P.; Santos-González, J.; Schoft, V.; Siretskiy, A.; Scott, R.; Tamaru, H.; Köhler, C. Hypomethylated pollen bypasses the interploidy hybridization barrier in *Arabidopsis*. *Plant Cell* **2014**, *26*, 3556–3568. [CrossRef]
29. Hojsgaard, D.; Hörandl, E. Apomixis as a facilitator of range expansion and diversification in plants. In *Evolutionary Biology: Biodiversification from Genotype to Phenotype*; Pontarotti, P., Ed.; Springer International Press: Marseille, France, 2015; pp. 305–327. ISBN 10.1007/978-3-319-19932-0.
30. Hojsgaard, D.; Hörandl, E. The rise of apomixis in natural plant populations. *Front. Plant Sci.* **2019**, *10*, 358. [CrossRef]
31. Norrmann, G.A.; Quarin, C.L.; Burson, B.L. Cytogenetics and reproductive behavior of different chromosome races in six *Paspalum* species. *J. Hered.* **1989**, *80*, 24–28. [CrossRef]
32. Naumova, N.; Osadtchiy, J.V.; Sharma, V.K.; Dijkhuis, P.; Ramulu, K.S. Apomixis in plants: Structural and functional aspects of diplospory in *Poa nemoralis* and *P. palustris*. *Protoplasma* **1999**, *208*, 186–195.
33. Hojsgaard, D.; Schegg, E.; Valls, J.F.M.; Martínez, E.J.; Quarin, C.L. Sexuality, apomixis, ploidy levels, and genomic relationships among four *Paspalum* species of the subgenus *Anachyris* (Poaceae). *Flora Morphol. Distrib. Funct. Ecol. Plants* **2008**, *203*, 535–547. [CrossRef]
34. Noyes, R.D.; Wagner, J.D. Dihaploidy yields diploid apomicts and parthenogens in *Erigeron* (Asteraceae). *Am. J. Bot.* **2014**, *101*, 865–874. [CrossRef] [PubMed]
35. Böcher, T. Cytological and embryological studies in the amphi– apomictic *Arabis holboellii* complex. *Biol. Skr.* **1951**, *6*, 1–58.
36. Kantama, L.; Sharbel, T.F.; Schranz, M.E.; Mitchell-Olds, T.; de Vries, S.; de Jong, H. Diploid apomicts of the *Boechera holboellii* complex display large-scale chromosome substitutions and aberrant chromosomes. *Proc. Natl. Acad. Sci. USA* **2007**, *104*, 14026–14031. [CrossRef]
37. Tucker, M.R.; Koltunow, A.M. Traffic monitors at the cell periphery: The role of cell walls during early female reproductive cell differentiation in plants. *Curr. Opin. Plant Biol.* **2014**, *17*, 137–145. [CrossRef]
38. Aliyu, O.M.; Schranz, M.E.; Sharbel, T.F. Quantitative variation for apomictic reproduction in the genus *Boechera* (Brassicaceae). *Am. J. Bot.* **2010**, *97*, 1719–1731. [CrossRef]
39. Rebozzio, R.N.; Sartor, M.E.; Quarin, C.L.; Espinoza, F. Residual sexuality and its seasonal variation in natural apomictic *Paspalum notatum* accessions. *Biol. Plant.* **2011**, *55*, 391–395. [CrossRef]
40. Hojsgaard, D.H.; Martínez, E.J.; Quarin, C.L. Competition between meiotic and apomictic pathways during ovule and seed development results in clonality. *New Phytol.* **2013**, *197*, 336–347. [CrossRef]
41. Hand, M.L.; Vít, P.; Krahulcová, A.; Johnson, S.D.; Oelkers, K.; Siddons, H.; Chrtek, J.; Fehrer, J.; Koltunow, A.M.G. Evolution of apomixis loci in *Pilosella* and *Hieracium* (Asteraceae) inferred from the conservation of apomixis-linked markers in natural and experimental populations. *Heredity (Edinb.)* **2015**, *114*, 17–26. [CrossRef]
42. Pessino, S.C.; Ortiz, J.P.A.; Leblanc, O.; do Valle, C.B.; Evans, C.; Hayward, M.D. Identification of a maize linkage group related to apomixis in *Brachiaria*. *Theor. Appl. Genet.* **1997**, *94*, 439–444. [CrossRef]

43. Pessino, S.; Evans, C.; Ortiz, J.; Armstead, I.; do Valle, C.; Hayward, M. A genetic map of the apospory-region in *Brachiaria* hybrids: Identification of two markers closely associated with the trait. *Hereditas* **1998**, *128*, 153–158. [CrossRef]
44. Gustine, D.; Sherwood, R.; Huff, D. Apospory-linked molecular markers in buffelgrass. *Crop Sci.* **1997**, *37*, 947–951. [CrossRef]
45. Ozias-Akins, P.; Roche, D.; Hanna, W. Tight clustering and hemizygosity of apomixis-linked markers in *Pennisetum squamulatum* implies genetic control of apospory by a divergent locus that may have no allelic form in sexual genotypes. *Proc. Natl. Acad. Sci. USA* **1998**, *95*, 5127–5132. [CrossRef]
46. Noyes, R.D.; Rieseberg, L.H. Two independent loci control agamospermy (apomixis) in the triploid flowering plant *Erigeron annuus*. *Genetics* **2000**, *155*, 379–390.
47. Catanach, A.S.; Erasmuson, S.K.; Podivinsky, E.; Jordan, B.R.; Bicknell, R. Deletion mapping of genetic regions associated with apomixis in *Hieracium*. *Proc. Natl. Acad. Sci. USA* **2006**, *103*, 18650–18655. [CrossRef]
48. Schallau, A.; Arzenton, F.; Johnston, A.J.; Hähnel, U.; Koszegi, D.; Blattner, F.R.; Altschmied, L.; Haberer, G.; Barcaccia, G.; Bäumlein, H. Identification and genetic analysis of the *APOSPORY* locus in *Hypericum perforatum* L. *Plant J.* **2010**, *62*, 773–784. [CrossRef]
49. Labombarda, P.; Busti, A.; Caceres, M.E.; Pupilli, F.; Arcioni, S. An AFLP marker tightly linked to apomixis reveals hemizygosity in a portion of the apomixis-controlling locus in *Paspalum simplex*. *Genome* **2002**, *45*, 513–519. [CrossRef]
50. Pupilli, F.; Martinez, E.J.; Busti, A.; Calderini, O.; Quarin, C.L.; Arcioni, S. Comparative mapping reveals partial conservation of synteny at the apomixis locus in *Paspalum* spp. *Mol. Genet. Genomics* **2004**, *270*, 539–548. [CrossRef]
51. Stein, J.; Quarin, C.L.; Martínez, E.J.; Pessino, S.C.; Ortiz, J.P.A. Tetraploid races of *Paspalum notatum* show polysomic inheritance and preferential chromosome pairing around the apospory-controlling locus. *Theor. Appl. Genet.* **2004**, *109*, 186–191. [CrossRef]
52. Hojsgaard, D.H.; Martínez, E.J.; Acuña, C.A.; Quarin, C.L.; Pupilli, F. A molecular map of the apomixis-control locus in *Paspalum procurrens* and its comparative analysis with other species of *Paspalum*. *Theor. Appl. Genet.* **2011**, *123*, 959–971. [CrossRef] [PubMed]
53. Nogler, G.A. Genetics of apospory in apomictic *Ranunculus auricomus*. V. Conclusion. *Bot. Helvet.* **1984**, *94*, 411–422.
54. Van Dijk, P.J.; Bakx-Schotman, J.M.T. Formation of unreduced megaspores (diplospory) in apomictic dandelions (*Taraxacum officinale, s.l.*) is controlled by a sex-specific dominant locus. *Genetics* **2004**, *166*, 483–492. [CrossRef]
55. Grimanelli, D.; Leblanc, O.; Espinosa, E.; Perotti, E.; Gonzalez de Leon, D.; Savidan, Y. Mapping diplosporous apomixis in tetraploid *Tripsacum*: One gene or several genes? *Heredity (Edinb)* **1998**, *80*, 33–39. [CrossRef]
56. Goel, S.; Chen, Z.; Conner, J.A.; Akiyama, Y.; Hanna, W.W.; Ozias-Akins, P. Delineation by fluorescence *in situ* hybridization of a single hemizygous chromosomal region associated with aposporous embryo sac formation in *Pennisetum squamulatum* and *Cenchrus ciliaris*. *Genetics* **2003**, *163*, 1069–1082. [PubMed]
57. Calderini, O.; Chang, S.B.; De Jong, H.; Busti, A.; Paolocci, F.; Arcioni, S.; De Vries, S.C.; Abma-Henkens, M.H.C.; Lankhorst, R.M.K.; Donnison, I.S.; et al. Molecular cytogenetics and DNA sequence analysis of an apomixis-linked BAC in *Paspalum simplex* reveal a non pericentromere location and partial microcolinearity with rice. *Theor. Appl. Genet.* **2006**, *112*, 1179–1191. [CrossRef] [PubMed]
58. Okada, T.; Ito, K.; Johnson, S.D.; Oelkers, K.; Suzuki, G.; Houben, A.; Mukai, Y.; Koltunow, A.M. Chromosomes carrying meiotic avoidance loci in three apomictic eudicot *Hieracium* subgenus *Pilosella* species share structural features. *Plant Physiol.* **2011**, *157*, 1327–1341. [CrossRef] [PubMed]
59. Worthington, M.; Ebina, M.; Yamanaka, N.; Heffelfinger, C.; Quintero, C.; Zapata, Y.P.; Perez, J.G.; Selvaraj, M.; Ishitani, M.; Duitama, J.; et al. Translocation of a parthenogenesis gene candidate to an alternate carrier chromosome in apomictic *Brachiaria humidicola*. *BMC Genomics* **2019**, *20*, 1–18. [CrossRef]
60. Galla, G.; Siena, L.A.; Ortiz, J.P.A.; Baumlein, H.; Barcaccia, G.; Pessino, S.C.; Bellucci, M.; Pupilli, F. A Portion of the apomixis locus of *Paspalum simplex* is microsyntenic with an unstable chromosome segment highly conserved among Poaceae. *Sci. Rep.* **2019**, *9*, 3271. [CrossRef]
61. Bowers, J.E.; Arias, M.A.; Asher, R.; Avise, J.A.; Ball, R.T.; Brewer, G.A.; Buss, R.W.; Chen, A.H.; Edwards, T.M.; Estill, J.C.; et al. Comparative physical mapping links conservation of microsynteny to chromosome structure and recombination in grasses. *Proc. Natl. Acad. Sci. USA* **2005**, *102*, 13206–13211. [CrossRef]

62. Akiyama, Y.; Hanna, W.W.; Ozias-Akins, P. High-resolution physical mapping reveals that the apospory-specific genomic region (ASGR) in *Cenchrus ciliaris* is located on a heterochromatic and hemizygous region of a single chromosome. *Theor. Appl. Genet.* **2005**, *111*, 1042–1051. [CrossRef] [PubMed]
63. Conner, J.A.; Goel, S.; Gunawan, G.; Cordonnier-Pratt, M.M.; Johnson, V.E.; Liang, C.; Wang, H.; Pratt, L.H.; Mullet, J.E.; DeBarry, J.; et al. Sequence analysis of bacterial artificial chromosome clones from the apospory-specific genomic region of *Pennisetum* and *Cenchrus*. *Plant Physiol.* **2008**, *147*, 1396–1411. [CrossRef] [PubMed]
64. Siena, L.A.; Ortiz, J.P.A.; Calderini, O.; Paolocci, F.; Cáceres, M.E.; Kaushal, P.; Grisan, S.; Pessino, S.C.; Pupilli, F. An apomixis-linked ORC3-like pseudogene is associated with silencing of its functional homolog in apomictic *Paspalum simplex*. *J. Exp. Bot.* **2016**, *67*, 1965–1978. [CrossRef] [PubMed]
65. Muller, H.J. The relation of recombination to mutational advance. *Mutat. Res. Mol. Mech. Mutagen.* **1964**, *1*, 2–9. [CrossRef]
66. Kondrashov, A.S. Selection against harmful mutations in large sexual and asexual populations. *Genet. Res.* **1982**, *40*, 325–332. [CrossRef] [PubMed]
67. Charlesworth, D. Evolution of recombination rates between sex chromosomes. *Philos. Trans. R. Soc. B Biol. Sci.* **2017**, *372*, 20160456. [CrossRef] [PubMed]
68. Pupilli, F.; Barcaccia, G. Cloning plants by seeds: Inheritance models and candidate genes to increase fundamental knowledge for engineering apomixis in sexual crops. *J. Biotechnol.* **2012**, *159*, 291–311. [CrossRef]
69. Vielle Calzada, J.P.; Craneand, C.F.; Stelly, D.M. Apomixis: The asexual revolution. *Science (80-)* **1996**, *274*, 1322–1323. [CrossRef]
70. Singh, M.; Burson, B.L.; Finlayson, S.A. Isolation of candidate genes for apomictic development in buffelgrass (*Pennisetum ciliare*). *Plant Mol. Biol.* **2007**, *64*, 673–682. [CrossRef]
71. Leblanc, O.; Armstead, I.; Pessino, S.; Ortiz, J.P.A.; Evans, C.; do Valle, C.; Hayward, M.D. Non-radioactive mRNA fingerprinting to visualise gene expression in mature ovaries of *Brachiaria* hybrids derived from *B. brizantha*, an apomictic tropical forage. *Plant Sci.* **1997**, *126*, 49–58. [CrossRef]
72. Rodrigues, J.C.M.; Cabral, G.B.; Dusi, D.M.A.; de Mello, L.V.; Rigden, D.J.; Carneiro, V.T.C. Identification of differentially expressed cDNA sequences in ovaries of sexual and apomictic plants of *Brachiaria brizantha*. *Plant Mol. Biol.* **2003**, *53*, 745–757. [CrossRef] [PubMed]
73. Chen, L.; Miyazaki, C.; Kojimai, A.; Saito, A.; Adachi, T. Isolation and characterization of a gene expressed during early embryo sac development in apomictic Guinea grass (*Panicum maximum*). *J. Plant Physiol.* **1999**, *154*, 55–62. [CrossRef]
74. Yamada-Akiyama, H.; Akiyama, Y.; Ebina, M.; Xu, Q.; Tsuruta, S.; Yazaki, J.; Kishimoto, N.; Kikuchi, S.; Takahara, M.; Takamizo, T.; et al. Analysis of expressed sequence tags in apomictic guineagrass (*Panicum maximum*). *J. Plant Physiol.* **2009**, *166*, 750–761. [CrossRef] [PubMed]
75. Albertini, E.; Marconi, G.; Barcaccia, G.; Raggi, L.; Falcinelli, M. Isolation of candidate genes for apomixis in *Poa pratensis* L. *Plant Mol. Biol.* **2004**, *56*, 879–894. [CrossRef] [PubMed]
76. Cervigni, G.D.L.; Paniego, N.; Díaz, M.; Selva, J.P.; Zappacosta, D.; Zanazzi, D.; Landerreche, I.; Martelotto, L.; Felitti, S.; Pessino, S.; et al. Expressed sequence tag analysis and development of gene associated markers in a near-isogenic plant system of *Eragrostis curvula*. *Plant Mol. Biol.* **2008**, *67*, 1–10. [CrossRef]
77. Rodrigo, J.M.; Zappacosta, D.C.; Selva, J.P.; Garbus, I.; Albertini, E.; Echenique, V. Apomixis frequency under stress conditions in weeping lovegrass (*Eragrostis curvula*). *PLoS ONE* **2017**, *12*, 1–17. [CrossRef]
78. Garbus, I.; Romero, J.R.; Selva, J.P.; Pasten, M.C.; Chinestra, C.; Carballo, J.; Zappacosta, D.C.; Echenique, V. *De novo* transcriptome sequencing and assembly from apomictic and sexual *Eragrostis curvula* genotypes. *PLoS ONE* **2017**, *12*, 1–22. [CrossRef]
79. Laspina, N.V.; Vega, T.; Seijo, J.G.; González, A.M.; Martelotto, L.G.; Stein, J.; Podio, M.; Ortiz, J.P.A.; Echenique, V.C.; Quarin, C.L.; et al. Gene expression analysis at the onset of aposporous apomixis in *Paspalum notatum*. *Plant Mol. Biol.* **2008**, *67*, 615–628. [CrossRef]
80. Pessino, S.C.; Espinoza, F.; Martínez, E.J.; Ortiz, J.P.A.; Valle, E.M.; Quarin, C.L. Isolation of cDNA clones differentially expressed in flowers of apomictic and sexual *Paspalum notatum*. *Hereditas* **2001**, *134*, 35–42. [CrossRef]

81. De Oliveira, F.A.; Vigna, B.B.Z.; Da Silva, C.C.; Fávero, A.P.; De Matta, F.P.; Azevedo, A.L.S.; De Souza, A.P. Coexpression and transcriptome analyses identify active apomixis-related genes in *Paspalum notatum* leaves. *BMC Genomics* **2020**, *21*, 1–15. [CrossRef]
82. Guerin, J.; Rossel, J.B.; Robert, S.; Tsuchiya, T.; Koltunow, A. A *DEFICIENS* homologue is down-regulated during apomictic initiation in ovules of Hieracium. *Planta* **2000**, *210*, 914–920. [CrossRef] [PubMed]
83. Bräuning, S.; Catanach, A.; Lord, J.M.; Bicknell, R.; Macknight, R.C. Comparative transcriptome analysis of the wild-type model apomict *Hieracium praealtum* and its *loss of parthenogenesis* (*lop*) mutant. *BMC Plant Biol.* **2018**, *18*, 1–21. [CrossRef] [PubMed]
84. Galla, G.; Vogel, H.; Sharbel, T.F.; Barcaccia, G. *De novo* sequencing of the *Hypericum perforatum* L. flower transcriptome to identify potential genes that are related to plant reproduction *sensu lato*. *BMC Genomics* **2015**, *16*, 1–22. [CrossRef] [PubMed]
85. Galla, G.; Zenoni, S.; Avesani, L.; Altschmied, L.; Rizzo, P.; Sharbel, T.F.; Barcaccia, G. Pistil transcriptome analysis to disclose genes and gene products related to aposporous apomixis in *Hypericum perforatum* L. *Front. Plant Sci.* **2017**, *8*, 79. [CrossRef]
86. Sharbel, T.F.; Voigt, M.-L.; Corral, J.M.; Thiel, T.; Varshney, A.; Kumlehn, J.; Vogel, H.; Rotter, B. Molecular signatures of apomictic and sexual ovules in the *Boechera holboellii* complex. *Plant J.* **2009**, *58*, 870–882. [CrossRef]
87. Sharbel, T.F.; Voigt, M.-L.; Corral, J.M.; Galla, G.; Kumlehn, J.; Klukas, C.; Schreiber, F.; Vogel, H.; Rotter, B. Apomictic and sexual ovules of *Boechera* display heterochronic global gene expression patterns. *Plant Cell* **2010**, *22*, 655–671. [CrossRef]
88. Schmidt, A.; Schmid, M.W.; Klostermeier, U.C.; Qi, W.; Guthörl, D.; Sailer, C.; Waller, M.; Rosenstiel, P.; Grossniklaus, U. Apomictic and sexual germline development differ with respect to cell cycle, transcriptional, hormonal and epigenetic regulation. *PLoS Genet.* **2014**, *10*, e1004476. [CrossRef]
89. Zühl, L.; Volkert, C.; Ibberson, D.; Schmidt, A.; Wilson, Z. Differential activity of F-box genes and E3 ligases distinguishes sexual versus apomictic germline specification in *Boechera*. *J. Exp. Bot.* **2019**, *70*, 5643–5657. [CrossRef]
90. Polegri, L.; Calderini, O.; Arcioni, S.; Pupilli, F. Specific expression of apomixis-linked alleles revealed by comparative transcriptomic analysis of sexual and apomictic *Paspalum simplex* Morong flowers. *J. Exp. Bot.* **2010**, *61*, 1869–1883. [CrossRef]
91. Ortiz, J.P.A.; Revale, S.; Siena, L.A.; Podio, M.; Delgado, L.; Stein, J.; Leblanc, O.; Pessino, S.C. A reference floral transcriptome of sexual and apomictic *Paspalum notatum*. *BMC Genomics* **2017**, *18*, 318. [CrossRef]
92. Sahu, P.P.; Gupta, S.; Malaviya, D.R.; Roy, A.K.; Kaushal, P.; Prasad, M. Transcriptome analysis of differentially expressed genes during embryo sac development in apomeiotic non-parthenogenetic interspecific hybrid of *Pennisetum glaucum*. *Mol. Biotechnol.* **2012**, *51*, 262–271. [CrossRef] [PubMed]
93. Okada, T.; Hu, Y.; Tucker, M.R.; Taylor, J.M.; Johnson, S.D.; Spriggs, A.; Tsuchiya, T.; Oelkers, K.; Rodrigues, J.C.M.; Koltunow, A.M.G. Enlarging cells initiating apomixis in *Hieracium praealtum* transition to an embryo sac program prior to entering mitosis. *Plant Physiol.* **2013**, *163*, 216–231. [CrossRef] [PubMed]
94. Rabiger, D.S.; Taylor, J.M.; Spriggs, A.; Hand, M.L.; Henderson, S.T.; Johnson, S.D.; Oelkers, K.; Hrmova, M.; Saito, K.; Suzuki, G.; et al. Generation of an integrated *Hieracium* genomic and transcriptomic resource enables exploration of small RNA pathways during apomixis initiation. *BMC Biol.* **2016**, *14*, 1–21. [CrossRef] [PubMed]
95. Galla, G.; Basso, A.; Grisan, S.; Bellucci, M.; Pupilli, F.; Barcaccia, G. Ovule gene expression analysis in sexual and aposporous apomictic *Hypericum perforatum* L. (Hypericaceae) accessions. *Front. Plant Sci.* **2019**, *10*, 654. [CrossRef] [PubMed]
96. Pellino, M. *Gene Expression Analysis and Transcriptome Evolution in Apomicts: A Case Study in Boechera and Ranunculus*; Heidelberg University: Heidelberg, Germany, 2015.
97. Pellino, M.; Hojsgaard, D.; Hörandl, E.; Sharbel, T. Chasing the apomictic factors in the *Ranunculus auricomus* complex: The effects of hybridization, parent of origin and ploidy in comparative ovule specific gene expression analysis. *Genes (Basel)* **2020**. submitted.
98. Grimanelli, D.; García, M.; Kaszas, E.; Perotti, E.; Leblanc, O. Heterochronic expression of sexual reproductive programs during apomictic development in *Tripsacum*. *Genetics* **2003**, *165*, 1521–1531.

99. Hojsgaard, D.; Greilhuber, J.; Pellino, M.; Paun, O.; Sharbel, T.F.; Hörandl, E. Emergence of apospory and bypass of meiosis via apomixis after sexual hybridisation and polyploidisation. *New Phytol.* **2014**, *204*, 1000–1012. [CrossRef]
100. Quarin, C.L. Seasonal changes in the incidence of apomixis of diploid, triploid, and tetraploid plants of *Paspalum cromyorrhizon. Euphytica* **1986**, *35*, 515–522. [CrossRef]
101. Quarin, C.L.; Espinoza, F.; Martinez, E.J.; Pessino, S.C.; Bovo, O.A. A rise of ploidy level induces the expression of apomixis in *Paspalum notatum. Sex. Plant Reprod.* **2001**, *13*, 243–249. [CrossRef]
102. Rios, E.F.; Blount, A.; Kenworthy, K.E.; Acuña, C.A.; Quesenberry, K.H. Seasonal expression of apospory in bahiagrass. *Trop. Grasslands* **2013**, *1*, 116. [CrossRef]
103. Shah, J.N.; Kirioukhova, O.; Pawar, P.; Tayyab, M.; Mateo, J.L.; Johnston, A.J. Depletion of key meiotic genes and transcriptome-wide abiotic stress reprogramming mark early preparatory events ahead of apomeiotic transition. *Front. Plant Sci.* **2016**, *7*, 1539. [CrossRef] [PubMed]
104. Karunarathne, P.; Reutemann, A.V.; Schedler, M.; Gluecksberg, A.; Martinez, E.J.; Honfi, A.I.; Hojsgaard, D.H. Sexual modulation in a polyploid grass: a reproductive contest between environmentally inducible sexual and genetically dominant apomictic pathways. *Sci. Rep.* **2020**. *under review*.
105. Klatt, S.; Hadacek, F.; Hodač, L.; Brinkmann, G.; Eilerts, M.; Hojsgaard, D.; Hörandl, E. Photoperiod extension enhances sexual megaspore formation and triggers metabolic reprogramming in facultative apomictic *Ranunculus auricomus. Front. Plant Sci.* **2016**, *7*, 278. [CrossRef] [PubMed]
106. Chen, L.; Guan, L.; Seo, M.; Hoffmann, F.; Adachi, T. Developmental expression of *ASG-1* during gametogenesis in apomictic guinea grass (*Panicum maximum*). *J. Plant Physiol.* **2005**, *162*, 1141–1148. [CrossRef]
107. Albertini, E.; Marconi, G.; Reale, L.; Barcaccia, G.; Porceddu, A.; Ferranti, F.; Falcinelli, M. *SERK* and *APOSTART*. Candidate genes for apomixis in *Poa pratensis. Plant Physiol.* **2005**, *138*, 2185–2199. [CrossRef]
108. Corral, J.M.; Vogel, H.; Aliyu, O.M.; Hensel, G.; Thiel, T.; Kumlehn, J.; Sharbel, T.F. A conserved apomixis-specific polymorphism is correlated with exclusive exonuclease expression in premeiotic ovules of apomictic *Boechera* species. *Plant Physiol.* **2013**, *163*, 1660–1672. [CrossRef]
109. Mau, M.; Corral, J.M.; Vogel, H.; Melzer, M.; Fuchs, J.; Kuhlmann, M.; de Storme, N.; Geelen, D.; Sharbel, T.F. The conserved chimeric transcript *UPGRADE2* is associated with unreduced pollen formation and is exclusively found in apomictic *Boechera* species. *Plant Physiol.* **2013**, *163*, 1640–1659. [CrossRef]
110. Mancini, M.; Permingeat, H.; Colono, C.; Siena, L.; Pupilli, F.; Azzaro, C.; de Alencar Dusi, D.M.; de Campos Carneiro, V.T.; Podio, M.; Seijo, J.G.; et al. The MAP3K-coding *QUI-GON JINN* (*QGJ*) gene is essential to the formation of unreduced embryo sacs in *Paspalum. Front. Plant Sci.* **2018**, *871*, 1547. [CrossRef]
111. Krasileva, K.V.; Buffalo, V.; Bailey, P.; Pearce, S.; Ayling, S.; Tabbita, F.; Soria, M.; Wang, S.; Akhunov, E.; Uauy, C.; et al. Separating homeologs by phasing in the tetraploid wheat transcriptome. *Genome Biol.* **2013**, *14*, R66. [CrossRef]
112. Vijverberg, K.; Ozias-Akins, P.; Schranz, M.E. Identifying and engineering genes for parthenogenesis in plants. *Front. Plant Sci.* **2019**, *10*, 128. [CrossRef]
113. Ozias-Akins, P.; Conner, J.A. Clonal reproduction through seeds in sight for crops. *Trends Genet.* **2020**, *36*, 215–226. [CrossRef] [PubMed]
114. Schmidt, A. Controlling apomixis: Shared features and distinct characteristics of gene regulation. *Genes (Basel)* **2020**, *11*, 329. [CrossRef] [PubMed]
115. Siena, L.A.; Ortiz, J.P.A.; Leblanc, O.; Pessino, S. *PnTgs1-like* expression during reproductive development supports a role for RNA methyltransferases in the aposporous pathway. *BMC Plant Biol.* **2014**, *14*, 297. [CrossRef] [PubMed]
116. Colono, C.; Ortiz, J.P.A.; Permingeat, H.R.; Bicknell, R.A. A Plant-Specific *TGS1* homolog influences gametophyte development in sexual tetraploid *Paspalum notatum* ovules. *Front. Plant Sci.* **2019**, *10*, 1566. [CrossRef] [PubMed]
117. Ochogavía, A.; Galla, G.; Seijo, J.G.; González, A.M.; Bellucci, M.; Pupilli, F.; Barcaccia, G.; Albertini, E.; Pessino, S. Structure, target-specificity and expression of *PN_LNC_N13*, a long non-coding RNA differentially expressed in apomictic and sexual *Paspalum notatum. Plant Mol. Biol.* **2018**, *96*, 53–67. [CrossRef] [PubMed]
118. Ortiz, J.P.A.; Leblanc, O.; Rohr, C.; Grisolia, M.; Siena, L.A.; Podio, M.; Colono, C.; Azzaro, C.; Pessino, S.C. Small RNA-seq reveals novel regulatory components for apomixis in *Paspalum notatum. BMC Genom.* **2019**, *20*, 487. [CrossRef] [PubMed]

119. Amiteye, S.; Corral, J.M.; Vogel, H.; Sharbel, T.F. Analysis of conserved microRNAs in floral tissues of sexual and apomictic *Boechera* species. *BMC Genomics* **2011**, *12*, 500. [CrossRef]
120. Amiteye, S.; Corral, J.M.; Vogel, H.; Kuhlmann, M.; Mette, M.F.; Sharbel, T.F. Novel microRNAs and microsatellite-like small RNAs in sexual and apomictic *Boechera* species. *MicroRNA* **2013**, *2*, 45–62. [CrossRef]
121. Selva, J.P.; Siena, L.; Rodrigo, J.M.; Garbus, I.; Zappacosta, D.; Romero, J.R.; Ortiz, J.P.A.; Pessino, S.C.; Leblanc, O.; Echenique, V. Temporal and spatial expression of genes involved in DNA methylation during reproductive development of sexual and apomictic *Eragrostis curvula*. *Sci. Rep.* **2017**, *7*, 15092. [CrossRef]
122. Garbus, I.; Selva, J.P.; Pasten, M.C.; Bellido, A.M.; Carballo, J.; Albertini, E.; Echenique, V. Characterization and discovery of miRNA and miRNA targets from apomictic and sexual genotypes of *Eragrostis curvula*. *BMC Genomics* **2019**, *20*, 839. [CrossRef]
123. Juranić, M.; Tucker, M.R.; Schultz, C.J.; Shirley, N.J.; Taylor, J.M.; Spriggs, A.; Johnson, S.D.; Bulone, V.; Koltunow, A.M. Asexual female gametogenesis involves contact with a sexually fated megaspore in apomictic *Hieracium*. *Plant Physiol.* **2018**, *177*, 1027–1049. [CrossRef] [PubMed]
124. Bocchini, M.; Galla, G.; Pupilli, F.; Bellucci, M.; Barcaccia, G.; Ortiz, J.P.A.; Pessino, S.C.; Albertini, E. The vesicle trafficking regulator *PN-SCD1* is demethylated and overexpressed in florets of apomictic *Paspalum notatum* genotypes. *Sci. Rep.* **2018**, *8*, 3030. [CrossRef] [PubMed]
125. Podio, M.; Cáceres, M.E.; Samoluk, S.S.; Seijo, J.G.; Pessino, S.C.; Ortiz, J.P.A.; Pupilli, F. A methylation status analysis of the apomixis-specific region in *Paspalum* spp. suggests an epigenetic control of parthenogenesis. *J. Exp. Bot.* **2014**, *65*, 6411–6424. [CrossRef] [PubMed]
126. Kirioukhova, O.; Shah, J.N.; Larsen, D.S.; Tayyab, M.; Mueller, N.E.; Govind, G.; Baroux, C.; Federer, M.; Gheyselinck, J.; Barrell, P.J.; et al. Aberrant imprinting may underlie evolution of parthenogenesis. *Sci. Rep.* **2018**, *8*, 10626. [CrossRef]
127. Thomas, C.; Meyer, D.; Himber, C.; Steinmetz, A. Spatial expression of a sunflower *SERK* gene during induction of somatic embryogenesis and shoot organogenesis. *Plant Physiol. Biochem.* **2004**, *42*, 35–42. [CrossRef]
128. Podio, M.; Felitti, S.A.; Siena, L.A.; Delgado, L.; Mancini, M.; Seijo, J.G.; González, A.M.; Pessino, S.C.; Ortiz, J.P.A. Characterization and expression analysis of *SOMATIC EMBRYOGENESIS RECEPTOR KINASE* (*SERK*) genes in sexual and apomictic *Paspalum Notatum*. *Plant Mol. Biol.* **2014**, *84*, 479–495. [CrossRef]
129. Conner, J.A.; Mookkan, M.; Huo, H.; Chae, K.; Ozias-Akins, P. A parthenogenesis gene of apomict origin elicits embryo formation from unfertilized eggs in a sexual plant. *Proc. Natl. Acad. Sci. USA* **2015**, *112*, 11205–11210. [CrossRef]
130. Conner, J.A.; Podio, M.; Ozias-Akins, P. Haploid embryo production in rice and maize induced by *PsASGR-BBML* transgenes. *Plant Rep.* **2017**, *30*, 41–52. [CrossRef]
131. Licausi, F.; Ohme-Takagi, M.; Perata, P. APETALA2/Ethylene Responsive Factor (AP2/ERF) transcription factors: Mediators of stress responses and developmental programs. *New Phytol.* **2013**, *199*, 639–649. [CrossRef]
132. Boutilier, K.; Offringa, R.; Sharma, V.K.; Kieft, H.; Ouellet, T.; Zhang, L.; Hattori, J.; Liu, C.-M.; van Lammeren, A.A.M.; Miki, B.L.A.; et al. Ectopic expression of *BABY BOOM* triggers a conversion from vegetative to embryonic growth. *Plant Cell* **2002**, *14*, 1737–1749. [CrossRef]
133. Khanday, I.; Skinner, D.; Yang, B.; Mercier, R.; Sundaresan, V. A male-expressed rice embryogenic trigger redirected for asexual propagation through seeds. *Nature* **2019**, *565*, 91–95. [CrossRef] [PubMed]
134. Worthington, M.; Heffelfinger, C.; Bernal, D.; Quintero, C.; Zapata, Y.P.; Perez, J.G.; De Vega, J.; Miles, J.; Dellaporta, S.; Tohme, J. A parthenogenesis gene candidate and evidence for segmental allopolyploidy in apomictic *Brachiaria decumbens*. *Genetics* **2016**, *203*, 1117–1132. [CrossRef] [PubMed]
135. Beck, J.B.; Alexander, P.J.; Allphin, L.; Al-Shehbaz, I.A.; Rushworth, C.; Bailey, C.D.; Windham, M.D. Does hybridization drive the transition to asexuality in diploid *Boechera*? *Evolution* **2012**, *66*, 985–995. [CrossRef] [PubMed]
136. Colombo, L.; Battaglia, R.; Kater, M.M. *Arabidopsis* ovule development and its evolutionary conservation. *Trends Plant Sci.* **2008**, *13*, 444–450. [CrossRef]
137. Wijnker, E.; Schnittger, A. Control of the meiotic cell division program in plants. *Plant Reprod.* **2013**, *26*, 143–158. [CrossRef]
138. Drews, G.N.; Koltunow, A.M. The female gametophyte. *Arab. B.* **2011**, *9*, e0155. [CrossRef]

139. Schmidt, A.; Schmid, M.W.; Grossniklaus, U. Plant germline formation: Common concepts and developmental flexibility in sexual and asexual reproduction. *Development* **2015**, *142*, 229–241. [CrossRef]
140. Grossniklaus, U.; Schneitz, K. The molecular and genetic basis of ovule and megagametophyte development. *Semin. Cell Dev. Biol.* **1998**, *9*, 227–238. [CrossRef]
141. Ohad, N.; Yadegari, R.; Margossian, L.; Hannon, M.; Michaeli, D.; Harada, J.J.; Goldberg, R.B.; Fischer, R.L. Mutations in *FIE*, a WD Polycomb group gene, allow endosperm development without fertilization. *Plant Cell* **1999**, *11*, 407–415. [CrossRef]
142. Luo, M.; Bilodeau, P.; Dennis, E.S.; Peacock, W.J.; Chaudhury, A. Expression and parent-of-origin effects for *FIS2*, *MEA*, and *FIE* in the endosperm and embryo of developing *Arabidopsis* seeds. *Proc. Natl. Acad. Sci. USA* **2000**, *97*, 10637–10642. [CrossRef]
143. Köhler, C.; Hennig, L.; Spillane, C.; Pien, S.; Gruissem, W.; Grossniklaus, U. The *Polycomb*-group protein MEDEA regulates seed development by controlling expression of the MADS-box gene *PHERES1*. *Genes Dev.* **2003**, *17*, 1540–1553. [CrossRef] [PubMed]
144. Garcia-Aguilar, M.; Michaud, C.; Leblanc, O.; Grimanelli, D. Inactivation of a DNA methylation pathway in Maize reproductive organs results in apomixis-like phenotypes. *Plant Cell* **2010**, *22*, 3249–3267. [CrossRef] [PubMed]
145. Zhao, X.; Bramsiepe, J.; Van Durme, M.; Komaki, S.; Prusicki, M.A.; Maruyama, D.; Forner, J.; Medzihradszky, A.; Wijnker, E.; Harashima, H.; et al. RETINOBLASTOMA RELATED1 mediates germline entry in *Arabidopsis*. *Science* **2017**, *356*, eaaf6532. [CrossRef] [PubMed]
146. Cao, L.; Wang, S.; Venglat, P.; Zhao, L.; Cheng, Y.; Ye, S.; Qin, Y.; Datla, R.; Zhou, Y.; Wang, H. *Arabidopsis* ICK/KRP cyclin-dependent kinase inhibitors function to ensure the formation of one megaspore mother cell and one functional megaspore per ovule. *PLOS Genet.* **2018**, *14*, e1007230. [CrossRef] [PubMed]
147. Su, Z.; Zhao, L.; Zhao, Y.; Li, S.; Won, S.; Cai, H.; Wang, L.; Li, Z.; Chen, P.; Qin, Y.; et al. The THO complex non-cell-autonomously represses female germline specification through the TAS3-ARF3 module. *Curr. Biol.* **2017**, *27*, 1597–1609.e2. [CrossRef]
148. Rajagopalan, R.; Vaucheret, H.; Trejo, J.; Bartel, D.P. A diverse and evolutionarily fluid set of microRNAs in *Arabidopsis thaliana*. *Genes Dev.* **2006**, *20*, 3407–3425. [CrossRef]
149. Zhao, L.; Cai, H.; Su, Z.; Wang, L.; Huang, X.; Zhang, M.; Chen, P.; Dai, X.; Zhao, H.; Palanivelu, R.; et al. KLU suppresses megasporocyte cell fate through SWR1-mediated activation of *WRKY28* expression in *Arabidopsis*. *Proc. Natl. Acad. Sci. USA* **2018**, *115*, E526–E535. [CrossRef]
150. Nonomura, K.-I.; Miyoshi, K.; Eiguchi, M.; Suzuki, T.; Miyao, A.; Hirochika, H.; Kurata, N. The *MSP1* gene is necessary to restrict the number of cells entering into male and female sporogenesis and to initiate anther wall formation in rice. *Plant Cell* **2003**, *15*, 1728–1739. [CrossRef]
151. Nonomura, K.-I.; Morohoshi, A.; Nakano, M.; Eiguchi, M.; Miyao, A.; Hirochika, H.; Kurata, N. A germ cell-specific gene of the ARGONAUTE family is essential for the progression of premeiotic mitosis and meiosis during sporogenesis in rice. *Plant Cell* **2007**, *19*, 2583–2594. [CrossRef]
152. Olmedo-Monfil, V.; Durán-Figueroa, N.; Arteaga-Vázquez, M.; Demesa-Arévalo, E.; Autran, D.; Grimanelli, D.; Slotkin, R.K.; Martienssen, R.A.; Vielle-Calzada, J.-P. Control of female gamete formation by a small RNA pathway in *Arabidopsis*. *Nature* **2010**, *464*, 628–632. [CrossRef]
153. Singh, M.; Goel, S.; Meeley, R.B.; Dantec, C.; Parrinello, H.; Michaud, C.; Leblanc, O.; Grimanelli, D. Production of viable gametes without meiosis in maize deficient for an ARGONAUTE protein. *Plant Cell* **2011**, *23*, 443–458. [CrossRef] [PubMed]
154. Leblanc, O.; Dueñas, M.; Hernández, M.; Bello, S.; Garcia, V.; Berthaud, J.; Savidan, Y. Chromosome doubling in *Tripsacum*: The production of artificial, sexual tetraploid plants. *Plant Breed.* **1995**, *114*, 226–230. [CrossRef]
155. Rodríguez-Leal, D.; León-Martínez, G.; Abad-Vivero, U.; Vielle-Calzada, J.P. Natural variation in epigenetic pathways affects the specification of female gamete precursors in *Arabidopsis*. *Plant Cell* **2015**, *27*, 1034–1045. [CrossRef] [PubMed]
156. Baroux, C.; Raissig, M.T.; Grossniklaus, U. Epigenetic regulation and reprogramming during gamete formation in plants. *Curr. Opin. Genet. Dev.* **2011**, *21*, 124–133. [CrossRef] [PubMed]
157. Grimanelli, D. Epigenetic regulation of reproductive development and the emergence of apomixis in angiosperms. *Curr. Opin. Plant Biol.* **2012**, *15*, 57–62. [CrossRef]
158. Rodriguez-Leal, D.; Vielle-Calzada, J.P. Regulation of apomixis: Learning from sexual experience. *Curr. Opin. Plant Biol.* **2012**, *15*, 549–555. [CrossRef]

159. Mercier, R. SWITCH1 (SWI1): A novel protein required for the establishment of sister chromatid cohesion and for bivalent formation at meiosis. *Genes Dev.* **2001**, *15*, 1859–1871. [CrossRef]
160. Siddiqi, I.; Ganesh, G.; Grossniklaus, U.; Subbiah, V. The dyad gene is required for progression through female meiosis in *Arabidopsis*. *Development* **2000**, *127*, 197–207.
161. Ravi, M.; Marimuthu, M.P.A.; Siddiqi, I. Gamete formation without meiosis in *Arabidopsis*. *Nature* **2008**, *451*, 1121–1124. [CrossRef]
162. d'Erfurth, I.; Jolivet, S.; Froger, N.; Catrice, O.; Novatchkova, M.; Mercier, R. Turning meiosis into mitosis. *PLoS Biol.* **2009**, *7*, e1000124. [CrossRef]
163. d'Erfurth, I.; Cromer, L.; Jolivet, S.; Girard, C.; Horlow, C.; Sun, Y.; To, J.P.C.; Berchowitz, L.E.; Copenhaver, G.P.; Mercier, R. The CYCLIN-A CYCA1;2/TAM is required for the meiosis I to meiosis II transition and cooperates with OSD1 for the prophase to first meiotic division transition. *PLoS Genet.* **2010**, *6*, e1000989. [CrossRef] [PubMed]
164. Miyazaki, S.; Ito, M. Calcium signals for egg activation in mammals. *J. Pharmacol. Sci.* **2006**, *100*, 545–552. [CrossRef] [PubMed]
165. Uranga, J.A.; Pedersen, R.A.; Arechaga, J. Parthenogenetic activation of mouse oocytes using calcium ionophores and protein kinase C stimulators. *Int. J. Dev. Biol.* **1996**, *40*, 515–519. [PubMed]
166. Antoine, A.-F.; Faure, J.-E.; Dumas, C.; Feijó, J.A. Differential contribution of cytoplasmic Ca2+ and Ca2+ influx to gamete fusion and egg activation in maize. *Nat. Cell Biol.* **2001**, *3*, 1120–1123. [CrossRef] [PubMed]
167. Guitton, A.-E.; Berger, F. Loss of function of *MULTICOPY SUPPRESSOR OF IRA 1* produces nonviable parthenogenetic embryos in *Arabidopsis*. *Curr. Biol.* **2005**, *15*, 750–754. [CrossRef] [PubMed]
168. Rodrigues, J.C.M.; Luo, M.; Berger, F.; Koltunow, A.M.G. Polycomb group gene function in sexual and asexual seed development in angiosperms. *Sex. Plant Reprod.* **2010**, *23*, 123–133. [CrossRef]
169. Zuo, J.; Niu, Q.-W.; Frugis, G.; Chua, N.-H. The *WUSCHEL* gene promotes vegetative-to-embryonic transition in *Arabidopsis*. *Plant J.* **2002**, *30*, 349–359. [CrossRef]
170. Lotan, T.; Ohto, M.; Yee, K.M.; West, M.A.; Lo, R.; Kwong, R.W.; Yamagishi, K.; Fischer, R.L.; Goldberg, R.B.; Harada, J.J. *Arabidopsis LEAFY COTYLEDON1* is sufficient to induce embryo development in vegetative cells. *Cell* **1998**, *93*, 1195–1205.
171. Stone, S.L.; Kwong, L.W.; Yee, K.M.; Pelletier, J.; Lepiniec, L.; Fischer, R.L.; Goldberg, R.B.; Harada, J.J. *LEAFY COTYLEDON2* encodes a B3 domain transcription factor that induces embryo development. *Proc. Natl. Acad. Sci. USA* **2001**, *98*, 11806–11811. [CrossRef]
172. Kelliher, T.; Starr, D.; Richbourg, L.; Chintamanani, S.; Delzer, B.; Nuccio, M.L.; Green, J.; Chen, Z.; McCuiston, J.; Wang, W.; et al. MATRILINEAL, a sperm-specific phospholipase, triggers maize haploid induction. *Nature* **2017**, *542*, 105–109. [CrossRef]
173. Gilles, L.M.; Khaled, A.; Laffaire, J.; Chaignon, S.; Gendrot, G.; Laplaige, J.; Bergès, H.; Beydon, G.; Bayle, V.; Barret, P.; et al. Loss of pollen-specific phospholipase NOT LIKE DAD triggers gynogenesis in maize. *EMBO J.* **2017**, *36*, 707–717. [CrossRef] [PubMed]
174. Liu, C.; Li, X.; Meng, D.; Zhong, Y.; Chen, C.; Dong, X.; Xu, X.; Chen, B.; Li, W.; Li, L.; et al. A 4-bp insertion at *ZmPLA1* encoding a putative phospholipase A generates haploid induction in maize. *Mol. Plant* **2017**, *10*, 520–522. [CrossRef] [PubMed]
175. Yao, L.; Zhang, Y.; Liu, C.; Liu, Y.; Wang, Y.; Liang, D.; Liu, J.; Sahoo, G.; Kelliher, T. *OsMATL* mutation induces haploid seed formation in *indica* rice. *Nat. Plants* **2018**, *4*, 530–533. [CrossRef] [PubMed]
176. Liu, C.; Zhong, Y.; Qi, X.; Chen, M.; Liu, Z.; Chen, C.; Tian, X.; Li, J.; Jiao, Y.; Wang, D.; et al. Extension of the in vivo haploid induction system from diploid maize to hexaploid wheat. *Plant Biotechnol. J.* **2020**, *18*, 316–318. [CrossRef]
177. Matzk, F.; Meyer, H.-M.; Bäumlein, H.; Balzer, H.-J.; Schubert, I. A novel approach to the analysis of the initiation of embryo development in gramineae. *Sex. Plant Rep.* **1995**, *8*, 266–272. [CrossRef]
178. Horst, N.A.; Katz, A.; Pereman, I.; Decker, E.L.; Ohad, N.; Reski, R. A single homeobox gene triggers phase transition, embryogenesis and asexual reproduction. *Nat. Plants* **2016**, *2*, 15209. [CrossRef]
179. Ikeda, T.; Tanaka, W.; Toriba, T.; Suzuki, C.; Maeno, A.; Tsuda, K.; Shiroishi, T.; Kurata, T.; Sakamoto, T.; Murai, M.; et al. *BELL 1*-like homeobox genes regulate inflorescence architecture and meristem maintenance in rice. *Plant J.* **2019**, *98*, 465–478. [CrossRef]

180. Brambilla, V.; Battaglia, R.; Colombo, M.; Masiero, S.; Bencivenga, S.; Kater, M.M.; Colombo, L. Genetic and molecular interactions between BELL1 and MADS box factors support ovule development in *Arabidopsis*. *Plant Cell* **2007**, *19*, 2544–2556. [CrossRef]
181. Rövekamp, M.; Bowman, J.L.; Grossniklaus, U. *Marchantia* Mp*RKD* regulates the gametophyte-sporophyte transition by keeping egg cells quiescent in the absence of fertilization. *Curr. Biol.* **2016**, *26*, 1782–1789. [CrossRef]
182. Köszegi, D.; Johnston, A.J.; Rutten, T.; Czihal, A.; Altschmied, L.; Kumlehn, J.; Wüst, S.E.J.; Kirioukhova, O.; Gheyselinck, J.; Grossniklaus, U.; et al. Members of the RKD transcription factor family induce an egg cell-like gene expression program. *Plant J.* **2011**, *67*, 280–291. [CrossRef]
183. León-Martínez, G.; Vielle-Calzada, J.P. Apomixis in flowering plants: Developmental and evolutionary considerations. *Curr. Top. Dev. Biol.* **2019**, *131*, 565–604. [PubMed]
184. Hands, P.; Rabiger, D.S.; Koltunow, A. Mechanisms of endosperm initiation. *Plant Rep.* **2016**, *29*, 215–225. [CrossRef] [PubMed]
185. Grossniklaus, U.; Vielle-Calzada, J.; Hoeppner, M.; Gagliano, W. Maternal control of embryogenesis by *MEDEA*, a *polycomb* group gene in *Arabidopsis*. *Science* **1998**, *280*, 446–450. [CrossRef] [PubMed]
186. Chaudhury, A.M.; Ming, L.; Miller, C.; Craig, S.; Dennis, E.S.; Peacock, W.J. Fertilization-independent seed development in *Arabidopsis thaliana*. *Proc. Natl. Acad. Sci. USA* **1997**, *94*, 4223–4228. [CrossRef] [PubMed]
187. Schwartz, Y.B.; Pirrotta, V. Polycomb silencing mechanisms and the management of genomic programmes. *Nat. Rev. Genet.* **2007**, *8*, 9–22. [CrossRef]
188. Makarevich, G.; Villar, C.B.R.; Erilova, A.; Kohler, C. Mechanism of *PHERES1* imprinting in Arabidopsis. *J. Cell Sci.* **2008**, *121*, 906–912. [CrossRef]
189. Grossniklaus, U.; Paro, R. Transcriptional silencing by polycomb-group proteins. *Cold Spring Harb. Perspect. Biol.* **2014**, *6*, a019331. [CrossRef]
190. Vinkenoog, R.; Spielman, M.; Adams, S.; Fischer, R.L.; Dickinson, H.G.; Scott, R.J. Hypomethylation promotes autonomous endosperm development and rescues postfertilization lethality in fie mutants. *Plant Cell* **2000**, *12*, 2271–2282. [CrossRef]
191. Rodrigues, J.C.M.; Tucker, M.R.; Johnson, S.D.; Hrmova, M.; Koltunow, A.M.G. Sexual and apomictic seed formation in *Hieracium* requires the plant Polycomb-group gene *FERTILIZATION INDEPENDENT ENDOSPERM*. *Plant Cell* **2008**, *20*, 2372–2386. [CrossRef]
192. Vogel, B. *New Plant Breeding Techniques*; Commissioned by the Swiss Federal Office for the Environment [BAFU/FOEN]: Bern, Switzerland, 2016.
193. Pillot, M.; Baroux, C.; Vazquez, M.A.; Autran, D.; Leblanc, O.; Vielle-Calzada, J.P.; Grossniklaus, U.; Grimanelli, D. Embryo and endosperm inherit distinct chromatin and transcriptional states from the female gametes in *Arabidopsis*. *Plant Cell* **2010**, *22*, 307–320. [CrossRef]
194. She, W.; Grimanelli, D.; Rutowicz, K.; Whitehead, M.W.J.; Puzio, M.; Kotliński, M.; Jerzmanowski, A.; Baroux, C. Chromatin reprogramming during the somatic-to-reproductive cell fate transition in plants. *Development* **2013**, *140*, 4008–4019. [CrossRef] [PubMed]
195. Grossniklaus, U. From sexuality to apomixis: Molecular and genetic approaches. In *The flowering of Apomixis: From Mechanisms to Genetic Engineering*; Savidan, Y., Carman, J.G., Dresselhaus, T., Eds.; CIMMYT, IRD, European Commission OC VI (FAIR): Mexico City, Mexico, 2001; pp. 168–211. ISBN 970-648-074-9.
196. Koltunow, A.M. Apomixis: Embryo sacs and embryos formed without meiosis or fertilization in ovules. *Plant Cell* **1993**, *5*, 1425–1437. [CrossRef] [PubMed]
197. Grimanelli, D.; Leblanc, O.; Perotti, E.; Grossniklaus, U. Developmental genetics of gametophytic apomixis. *Trends Genet.* **2001**, *17*, 597–604. [CrossRef]
198. Richards, A.J. Apomixis in flowering plants: An overview. *Philos. Trans. R. Soc. London. Ser. B Biol. Sci.* **2003**, *358*, 1085–1093. [CrossRef] [PubMed]
199. Carman, J.G.; Jamison, M.; Elliott, E.; Dwivedi, K.K.; Naumova, T.N. Apospory appears to accelerate onset of meiosis and sexual embryo sac formation in *Sorghum* ovules. *BMC Plant Biol.* **2011**, *11*, 9. [CrossRef]
200. Albertini, E.; Barcaccia, G.; Carman, J.G.; Pupilli, F. Did apomixis evolve from sex or was it the other way around? *J. Exp. Bot.* **2019**, *70*, 2951–2964. [CrossRef]
201. Wilson, Z.A.; Yang, C. Plant gametogenesis: Conservation and contrasts in development. *Reproduction* **2004**, *128*, 483–492. [CrossRef]

202. Hisanaga, T.; Yamaoka, S.; Kawashima, T.; Higo, A.; Nakajima, K.; Araki, T.; Kohchi, T.; Berger, F. Building new insights in plant gametogenesis from an evolutionary perspective. *Nat. Plants* **2019**, *5*, 663–669. [CrossRef]
203. Mogie, M. *The Evolution of Asexual Reproduction in Plants*; Chapman & Hall: London, UK, 1992; ISBN 13: 0-412-44220-5.
204. Savidan, Y. Apomixis: Genetics and Breeding. In *Plant Breeding Reviews*; John Wiley & Sons, Inc.: Oxford, UK, 2000; pp. 13–86.
205. Hojsgaard, D. Transient activation of apomixis in sexual neotriploids may retain genomically altered states and enhance polyploid establishment. *Front. Plant Sci.* **2018**, *9*, 230. [CrossRef]
206. Barke, B.H.; Daubert, M.; Hörandl, E. Establishment of apomixis in diploid F2 hybrids and inheritance of apospory from F1 to F2 hybrids of the *Ranunculus auricomus* complex. *Front. Plant Sci.* **2018**, *9*, 1111. [CrossRef]
207. Sartor, M.E.; Quarin, C.L.; Espinoza, F. Mode of reproduction of colchicine-induced *Paspalum plicatulum* tetraploids. *Crop Sci.* **2009**, *49*, 1270–1276. [CrossRef]
208. Varshney, R.K.; Shi, C.; Thudi, M.; Mariac, C.; Wallace, J.; Qi, P.; Zhang, H.; Zhao, Y.; Wang, X.; Rathore, A.; et al. Pearl millet genome sequence provides a resource to improve agronomic traits in arid environments. *Nat. Biotechnol.* **2017**, *35*, 969–976. [CrossRef] [PubMed]
209. Carballo, J.; Santos, B.A.C.M.; Zappacosta, D.; Garbus, I.; Selva, J.P.; Gallo, C.A.; Díaz, A.; Albertini, E.; Caccamo, M.; Echenique, V. A high-quality genome of *Eragrostis curvula* grass provides insights into Poaceae evolution and supports new strategies to enhance forage quality. *Sci. Rep.* **2019**, *9*, 1–15. [CrossRef] [PubMed]
210. Anderson, J.T.; Willis, J.H.; Mitchell-Olds, T. Evolutionary genetics of plant adaptation. *Trends Genet.* **2011**, *27*, 258–266. [CrossRef] [PubMed]
211. Lee, C.R.; Wang, B.; Mojica, J.P.; Mandáková, T.; Prasad, K.V.S.K.; Goicoechea, J.L.; Perera, N.; Hellsten, U.; Hundley, H.N.; Johnson, J.; et al. Young inversion with multiple linked QTLs under selection in a hybrid zone. *Nat. Ecol. Evol.* **2017**, *1*, 0119. [CrossRef]
212. Kliver, S.; Rayko, M.; Komissarov, A.; Bakin, E.; Zhernakova, D.; Prasad, K.; Rushworth, C.; Baskar, R.; Smetanin, D.; Schmutz, J.; et al. Assembly of the *Boechera retrofracta* genome and evolutionary analysis of apomixis-associated genes. *Genes (Basel)* **2018**, *9*, 185. [CrossRef]
213. Sharbel, T.F.; Rigault, P. Comparative Genomics of Apomictic Plants: Advancing Novel Tools for Niche Breeding. Available online: https://www.researchgate.net/project/Comparative-Genomics-of-Apomictic-Plants-Advancing-Novel-Tools-for-Niche-Breeding (accessed on 11 March 2020).
214. Brukhin, V.; Grossniklaus, U. Sequencing, Assembly, and Annotation of the Highly Heterozygous Genome of the Apomictic Plant Boechera divaricarpa. Available online: https://www.researchgate.net/project/Sequencing-Assembly-and-Annotation-of-the-Highly-Heterozygous-Genome-of-the-Apomictic-Plant-Boechera-divaricarpa (accessed on 11 March 2020).
215. Muir, P.; Li, S.; Lou, S.; Wang, D.; Spakowicz, D.J.; Salichos, L.; Zhang, J.; Weinstock, G.M.; Isaacs, F.; Rozowsky, J.; et al. The real cost of sequencing: Scaling computation to keep pace with data generation. *Genome Biol.* **2016**, *17*, 53. [CrossRef]
216. Jiao, W.-B.; Schneeberger, K. The impact of third generation genomic technologies on plant genome assembly. *Curr. Opin. Plant Biol.* **2017**, *36*, 64–70. [CrossRef]
217. Kyriakidou, M.; Tai, H.H.; Anglin, N.L.; Ellis, D.; Strömvik, M.V. Current strategies of polyploid plant genome sequence assembly. *Front. Plant Sci.* **2018**, *9*, 1660. [CrossRef]
218. Kadota, M.; Nishimura, O.; Miura, H.; Tanaka, K.; Hiratani, I.; Kuraku, S. Multifaceted Hi-C benchmarking: What makes a difference in chromosome-scale genome scaffolding? *Gigascience* **2020**, *9*, 1–15. [CrossRef]
219. Blackburn, A.N.; Blondell, L.; Kos, M.Z.; Blackburn, N.B.; Peralta, J.M.; Stevens, P.T.; Lehman, D.M.; Blangero, J.; Göring, H.H.H. Genotype phasing in pedigrees using whole-genome sequence data. *Eur. J. Hum. Genet.* **2020**. [CrossRef] [PubMed]
220. Castel, S.E.; Mohammadi, P.; Chung, W.K.; Shen, Y.; Lappalainen, T. Rare variant phasing and haplotypic expression from RNA sequencing with phASER. *Nat. Commun.* **2016**, *7*, 1–6. [CrossRef] [PubMed]

genes

MDPI

Discussion

Can We Use Gene-Editing to Induce Apomixis in Sexual Plants?

Armin Scheben [1] and Diego Hojsgaard [2,*]

1 Simons Center for Quantitative Biology, Cold Spring Harbor Laboratory, Cold Spring Harbor, NY 11724, USA; scheben@cshl.edu

2 Department of Systematics, Biodiversity and Evolution of Plants, Albrecht-von-Haller Institute for Plant Sciences, University of Goettingen, Untere Karspuele 2, 37073 Goettingen, Germany

* Correspondence: Diego.Hojsgaard@biologie.uni-goettingen.de

Received: 17 June 2020; Accepted: 9 July 2020; Published: 12 July 2020

Abstract: Apomixis, the asexual formation of seeds, is a potentially valuable agricultural trait. Inducing apomixis in sexual crop plants would, for example, allow breeders to fix heterosis in hybrid seeds and rapidly generate doubled haploid crop lines. Molecular models explain the emergence of functional apomixis, i.e., apomeiosis + parthenogenesis + endosperm development, as resulting from a combination of genetic or epigenetic changes that coordinate altered molecular and developmental steps to form clonal seeds. Apomixis-like features and synthetic clonal seeds have been induced with limited success in the sexual plants rice and maize by using gene editing to mutate genes related to meiosis and fertility or via egg-cell specific expression of embryogenesis genes. Inducing functional apomixis and increasing the penetrance of apomictic seed production will be important for commercial deployment of the trait. Optimizing the induction of apomixis with gene editing strategies that use known targets as well as identifying alternative targets will be possible by better understanding natural genetic variation in apomictic species. With the growing availability of genomic data and precise gene editing tools, we are making substantial progress towards engineering apomictic crops.

Keywords: apomeiosis; character segregation; crop biotechnology; heterosis; meiosis; molecular breeding; recombination

---

1. Introduction

The formation of clonal offspring through parthenogenesis is a well-known feature of many phylogenetically distant organisms [1–5], with shared developmental features at least in flowering plants and vertebrate animals [6]. Yet, in most cases, details of the genetic basis and molecular coordination of parthenogenetic development are lacking.

In plants, parthenogenesis holds a gargantuan economic dimension because of its probable impact on breeding. Sexuality imposes critical disadvantages to genetic improvement, making the exploitation of hybrid vigor and high yield only accessible in the short time frame of a single F_1 generation. This raises the need for costly and laborious methods to identify parental lines with optimal 'combining ability,' i.e., inbred lines which complement each other for desired traits in the hybrid [7]. In F_1 hybrid breeding procedures, such a testing of inbred lines for their combining ability is the most limiting factor [7]. Since clonal seed embryos can be created through parthenogenesis (i.e., embryo development from an unfertilized egg cell), engineering parthenogenesis in sexual crops has long been a goal for many researchers around the world. Introducing parthenogenesis into plant breeding would both simplify the selection of parental lines and extend over time the exploitation of the desired F_1 hybrid (expectedly through many generations) breaking the breeding loop in traditional schemes.

However, unlike in animals, in plants the formation of parthenogenetic individuals involves a higher level of complexity as seeds are formed literally by two individuals, the embryo per se and the endosperm, and a second individual that has evolved to have an acquiescent temporary role nourishing the embryo [8]. Thus, the formation of a seed involves twice the number of gametes, a double fertilization step to initiate the embryo and the endosperm developments and more complex molecular controls to produce a functional, viable seed. First, a megaspore mother cell (MMC) in the ovule goes through a meiotic division and produces four (sometimes three) reduced (haploid or n) megaspores. Only one megaspore grows and differentiates by mitosis into a multicellular female gametophyte carrying two gametes with different ploidy, a haploid egg cell (n) and a diploid central cell (two haploid nuclei, $n + n$). Both female gametes are fertilized by haploid sperms delivered by the pollen tube to produce a diploid zygote and a triploid primary endosperm, which develop into the embryo and endosperm tissues of the sexual seed, respectively. Hence, along such reproductive development in ovules of diverse species of flowering plants, there are a few critical developmental steps which can be divided into defined reproductive modules or developmental programs, each having specific molecular controls and functional roles during seed formation (Figure 1).

Some plants can alter these controls along different reproductive steps and develop parthenogenetic embryos through apomixis. Thus, apomixis exploits natural developmental programs for the formation of parthenogenetic embryos and clonal seeds. In plants with sporophytic apomixis, the formation of the female gametophyte and the successive developmental steps are not altered, but extra somatic embryos are attached to the gametophyte usually constraining the development of the zygotic embryo.

In plants with gametophytic apomixis, the developed female gametophytes are unreduced ($2n$), produced either from the germline going through a modified meiotic division sidestepping genetic recombination (diplospory) or from a somatic cell in the ovule nucellus without going through a meiotic division (apospory). Like in sexual ovules, unreduced female gametes show different ploidy, the egg cell being diploid ($2n$) and the central cell (usually) being tetraploid ($2n + 2n$). However, in apomictic ovules the egg cell develops by parthenogenesis into an embryo genetically equal to the mother plant, and the central cell is often fertilized by one (rarely more) haploid (n) sperm(s) to produce the endosperm tissue of the apomictic seed.

Thus, the so-called '*elements of apomixis*' [2] are alterations in the functional output of successive sexual reproductive modules (i.e., meiosis, gametogenesis, and fertilization steps) that do not necessarily interfere with each other. However, since the molecular basis of apomixis is unknown, some factual ambiguities in the developmental features and occurrence of these altered reproductive modules cannot yet be explained. First, even though apomictic ovules show global gene de-regulation (on genes affecting varied functions) and heterochronic developments compared to sexual ones (e.g., [9]), there seem to be concerted regulatory changes acting in coordination to achieve the goal of forming a seed. Second, even when developmental programs (in terms of alleles and gene networks) are expectedly the same between diploids and polyploids (at least in autopolyploids), apomixis is transgenerationally stable [10], penetrant, and highly expressed only in polyploids. Third, while apomixis in nature is dominant over sexuality (though it often shows segregation distortion; [11]), apomixis-like mutants in sexual plants have recessive phenotypes.

In the last 30 years, different plant mutants affecting specific developmental steps in those reproductive modules have been described [12,13]. Many of these reproductive mutants display phenotypes resembling alterations to the sexual developmental programs observed in natural apomictic plants. More recently, the arrival of gene editing methods has sped up the use of several of these plant reproduction mutants individually or in combination to simultaneously alter the normal output of each developmental program and reproductive module within plant ovules, with the goal of obtaining plants simulating natural apomixis. Although these studies successfully produced multiple concurrent mutants, they had relatively low success on attaining a synthetic apomictic plant exhibiting high expressivity. Therefore, the question remains whether apomixis per se can be induced in sexual plants.

Here, we address this question by discussing the genetic control of apomixis and how gene editing approaches can be used to induce this complex trait.

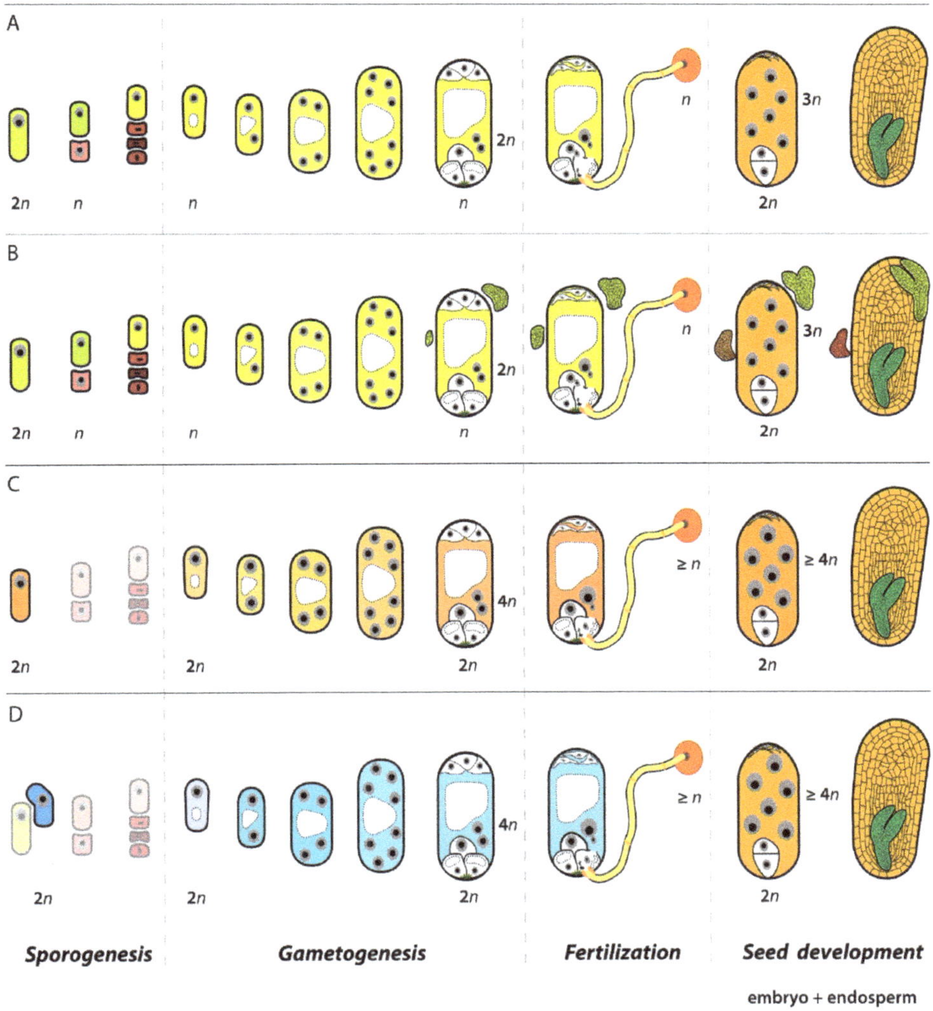

Figure 1. Pathways of seed formation in plants. At least one or two reproductive modules or developmental programs are distinctively altered between sexual (**A**) and apomictic developments (**B–D**). In sporophytic apomixis, a clonal embryo developed from the surrounding somatic tissue replace the zygotic embryo (**B**). In gametophytic apomixis, sporogenesis, or the acquisition of a megaspore mother cell identity and progression through meiosis is the first altered developmental program. In diplospory (**C**) meiosis is bypassed, while in apospory (**D**) it is highly depleted and surrounding somatic cells acquire a megaspore-like fate. Gametogenesis, or the formation of the female gametophyte, is generally conserved in the three pathways. The second developmental step that is altered is fertilization, i.e., the delivery of male gametes and fusion with the female ones. While in sexual plants double fertilization happens (**A**), in most apomicts only one fertilization event is possible due to parthenogenesis and this fertilization is required for developing the endosperm (**C**,**D**).

2. Precise Gene-Editing of Complex Traits

Although most genetic trait modifications in crops still rely on gene knockouts, in recent years the gene editing toolbox has expanded to facilitate many other modifications of functional sequences [14,15]. Gene knockouts nevertheless remain a powerful approach and have induced several apomixis-related traits in plants (see details in Section 5 on *de novo* apomixis). Complex traits result from the interaction among multiple genes and their environment. Induction of a complex trait such as apomixis will likely require a combination of different types of gene editing. Up to now, attempts to induce synthetic apomixis are limited by low expressivity of the trait in modified sexual plants. To efficiently introduce the apomictic pathway into sexual plants, researchers can use modification of gene expression, precise base editing and trait optimization using allelic series.

2.1. Modulating Gene Expression

Gene expression can be modulated by editing regulatory machinery such as promotors and transcription factors. Particularly in polyploids, an additional strategy to reduce gene expression can also be the knockout of redundant paralogs. Editing approaches to increase expression include promoter knockin and upstream open reading frame editing [16]. For example, a promoter swap targeting the *ARGOS8* gene in maize induced overexpression of the gene in multiple tissues and during different developmental stages [17]. This overexpression was associated with a gain-of-function that led to increased drought tolerance.

Complementing nucleotide-level editing to modify gene expression, catalytically inactive 'dead' Cas (dCas) proteins can also act as a recruiting platform for repressors/activators [18] and epigenetic modifiers. In recent years, several approaches have been optimized in plant cells. By joining transcriptional activators such as VP64 and TV to dCas9, expression of a targeted luciferase reporter gene could be increased 2.4-fold to 215-fold, respectively [19]. A combination of VP64 with a modified guide RNA can increase the effectiveness of this activator and a system based on transcription activator-like effector nucleases (TALENs) may be even more effective [20]. The level of activation possible using these systems can differ by gene and further optimization may be achieved by testing combinations of single and multiple activators with a targeting system.

A further approach to modulate expression is to make modifications at the epigenome level. Epigenomic features such as DNA methylation and histone modification play important roles in gene regulation. By fusing dCas9 with histone acetyltransferases, methyltransferases or demethylases, gene expression can be activated or repressed by changing epigenomic states. Site-specific methylation and demethylation using dCas9 together with the SunTag system has been demonstrated to be effective in *Arabidopsis* [21–23]. The SunTag system links the Cas molecule with an GCN4 peptide array that can recruit specific antibodies with attached epigenetic modifiers or transcription activators/repressors. This approach could be used to down- or upregulate the expression of target reproduction genes. However, off-target effects can occur, including hypermethylation in large regions of the genome [21]. The specificity of epigenome editing therefore needs to be increased, particularly when stably modifying complex pathways in crops for commercial use. A further challenge when using this epigenome editing approach in crops is that the CRISPR/Cas transgenes must remain in the plant, triggering governmental regulation in many markets. Although achieving high specificity and efficiency remain challenging, the modulation of gene expression on the genomic and epigenomic level can help reprogram complex reproductive gene networks.

2.2. Targeting Gene Sequence with Base Editing and Prime Editing

Gain-of-function mutations that modify how a protein works can be induced by point mutations, precise indels in functional domains, or gene replacement. This type of mutation can be induced in plants with moderate efficiency (<10%) using gene targeting that relies on double strand breaks and DNA template [24]. However, this gene targeting approach mostly generates imprecise indels.

As an alternative, the more precise base editing can be used to generate point mutations, without requiring DNA template or induction of double strand breaks [25]. The critical component of base editing is a fusion of a cytidine deaminase [26] or adenosine deaminase [27] to a Cas9 nickase (nCas9), which only breaks a single DNA strand, or dCas9. The fusion proteins enable four types of transition point mutations (C→T, G→A, A→G, and T→C). A drawback of base editors is that they can generate off-target point mutations within the deamination window spanning several bases. In rice, cytosine base editors have also been found to introduce off-target mutations outside of the target region [28]. Finally, a broader limitation of base editing is that it cannot be used to perform all possible base-to-base conversion or to introduce indels.

A substantial advance in gene targeting was the recent introduction of prime editing [29], which has increased gene targeting efficiency in plants to >20% [30]. The first component of prime editing is a fusion protein consisting of an nCas9 and a reverse transcriptase. The second component is a prime editing guide RNA that includes both a guide that is homologous to the genomic target and an RNA template with the desired edit. The nCas9 nicks the target location and the RNA edit is reverse transcribed to DNA and written into the genome. Prime editing is now the most versatile and efficient gene targeting approach in most situations, enabling the introduction of all 12 base-to-base conversions as well as indels [29]. Transition point mutations can, however, still be introduced more efficiently using base editors [30]. Prime editing is also less limited by the availability of protospacer adjacent motif (PAM) sequences than base editing or other editing methods [29], making it possible to target more regions of the genome. Prime editing efficiency, however, decreases with the length of the introduced indel and remains restricted to small indels <20 nt [30]. Importantly, the relative efficiencies of different gene targeting approaches vary by target site [30], suggesting that case-by-case optimization for each target will be important when tackling complex traits such as apomixis.

2.3. Trait Optimization with Allelic Series

When an edit that affects the target trait has been identified, the trait can be optimized by inducing a range of allelic variants and combinations of these variants [31]. Such an allelic series can identify functional variants with diverse impacts on a trait. For example, a series of cis-regulatory alleles of the *CLAVATA3* gene in tomato generated step-wise variation in fruit size [32]. In this case different expression levels of the target gene were associated with the trait variation, but different allelic variants can also impact gene function in different ways such as loss or gain of functions [33]. Prior information on target genes and pathways, including on epistasis, protein structure, and natural variation, can help select candidate edits to use for optimization of a trait.

Multiplexed gene editing is a useful tool to accelerate the combination of different edited alleles. In plants, up to 107 simultaneous targeting events have been reported [34]. A simple case of multiplexed gene editing is the knockout of functionally redundant paralogs or members of a gene family to disrupt a function. However, guide RNAs can also be targeted to a complex combination of regulatory regions and coding sequences in less similar genes. Mixing and matching edits in this way can facilitate the manipulation of complex plant traits such as apomixis. Provided the molecular nature of apomixis is known, a certain level of confidence can be assumed that inducing *de novo* apomixis would be a technically demanding but feasible task.

3. The Molecular Basis of Apomixis: Three Models to Explain Empirical Data

One of the main challenges for manipulating apomixis in plant breeding is the lack of a molecular model able to suit all empirical data collected today about apomixis (see Hojsgaard [35] in this issue). In 1990, Savidan wrote a an article [36] summarizing all available information about the genetic control of apomixis and pointed out that around 95% of all data was inconclusive. Today, 30 years after Savidan's review, we have a lot more information, particularly on the genetic regulation of sexual and apomixis-like developments (see e.g., Vijverberg et al. [37]; Ozias-Akins et al. [38]), and yet no unified hypothesis explaining the origin of apomixis exists. Three hypotheses based on different molecular mechanisms are often used to explain the observed diversity of results about asexual seed formation coming from different plant systems.

3.1. Apomixis is a Consequence of Developmental Asynchronies

The first hypothesis is—to some extent—related to the hybridization theory discussed by Juel [39], Murbeck [40], and Ostenfeld [41] at the beginning of the 20th century based on the observation that many *Hieracium* apomictic species were hybrids. Sexual parental types were considered to have apomictic tendencies that would reveal in the hybrids. Later, Rosenberg [42], Winge [43], and Ernst [44] added polyploidy as another prerequisite for apomixis. Of course, back then the DNA molecule was yet to be discovered, and the Mendelian theory of inheritance was just starting to be examined and accepted in a variety of organisms.

More recently, diverse studies on ovule development in related sexual and apomictic species have exposed obvious temporal and spatial developmental asynchronies during seed formation in apomicts pointing to apomixis as a consequence of de-regulation of the sexual program, likely due to hybridization and/or polyploidization [45–50]. According to this hypothesis, apomixis in hybrids (mainly allopolyploids) arises as a consequence of the evolutionary divergence of regulatory sequences controlling sporogenesis and gametogenesis in parental species, and in polyploids (mainly autopolyploids) arises as a consequence of dosage effects and stoichiometric disbalances of macromolecular complexes. This idea has nowadays been reinforced by gene expression and transcriptomic analyses on different plant species exposing significant changes in expression levels (up- and down-regulations) of many genes in apomictic compared to sexual ovules [51–56]. Such regulatory alterations might well support weaker controls on cell fate and key developmental steps allowing, e.g., nucellar cells to acquire a gametogenesis fate (in apospory), or the primary endosperm to develop under imbalanced paternal to maternal genome ratios. However, this hypothesis cannot explain observations on genetic inheritance studies showing reproductive modules in apomicts had independent Mendelian segregation, in some cases showing the expected segregation pattern [57] and in others showing patterns distorted by secondary genetic effects including epistasis, segregation distortion, polyploidy, and lack of recombination (reviewed in [11]).

3.2. Apomixis is a Mutation-Based Phenomenon

For a long time, apomixis components were assumed to have independent inheritance supported by the rare occurrence of B_{III} hybrids due to spontaneous uncoupling of reproductive modules (meiosis and fertilization) [2]. Studies focused on understanding the genetic control of apomixis used experimental crosses between sexual and apomictic plants and their segregating progeny [2,11]. Such studies usually provided inconclusive results and conflicting ideas [36], from models postulating apomixis was a delicate balance of many recessive genes [58] or the result of the action of three recessive genes [59] to models in which apomixis is controlled by a single dominant gene [60–62]. The introduction of different methods of molecular biology, especially in genetics and computing, facilitated the use of larger progenies as well as inheritance and genetic mapping analysis of higher complexity. Today, most apomictic species studied show Mendelian segregation of one or two dominant genes, often modulated by segregation distortion, modifiers genes, epistasis, polyploidy, aneuploidy, etc. (see details in Ozias-Akins and van Dijk [11]), although in some cases more genes had been suggested to regulate apomixis expression [63]. Thus, the overall evidence suggests apomixis is a mutation-based anomaly that involves a relatively simple genetic locus or two loci [11,64] carrying genes for apomeiosis and parthenogenesis. Even though this hypothesis is supported by diverse mutants genes showing diplosporous-like or aposporous-like and parthenogenetic phenotypes [65–67], it is tricky to explain an independent evolution of mutants for each apomixis component in an ancestral population. A possible explanation may lie in the characteristics of the apomixis locus. In several apomictic species showing monogenic or digenic inheritance, the apomixis locus has been identified with large non-recombinant chromosomal segments likely carrying many more genes [68–71] that are microsyntenic to chromosomal segments conserved among different species, including sexual taxa [72]. These features place the possibility of a single-event mechanism involved in the evolutionary origin of apomixis and may suggest a concerted multigenic activity in the control of reproductive modules [35].

3.3. *Apomixis is an Ancient Switch, Polyphenic to Sex, and Epigentically Regulated*

More recently, Albertini et al. [73] discussed the idea that apomixis may be anciently polyphenic with sex, with both reproductive phenisms involving canalized components of complex molecular processes. According to the polyphenism viewpoint, under different environmental conditions, plants can switch on/off certain genes, change the metabolic status in ovules, and consequently, choose between an apomictic or sexual program for seed formation. Although the same genome would encode both sex and apomixis, according with this view, apomixis fails to occur in obligately sexual eukaryotes because genetic or epigenetic modifications have silenced the primitive sex-apomixis switch and/or disrupted molecular capacities for apomixis. Thus, apomixis would be an ancient character epigenetically regulated with a relictual presence in all eukaryotes [74], a view that has gained interest based on methylation analyses of apomicts and on studies in mutants affecting methylation pathways [75–78].

Each of these hypotheses suggests a different molecular frame for the molecular manipulation of sexuality and induction of apomixis in plants via *heterochrony, de novo* through mutations or by *restitutio* of an ancient polyphenic switch.

4. Can Apomixis *Sensu Stricto* be Induced Through Gene-Editing Approaches?

As of the current state of the art, apomixis *sensu stricto* cannot be induced. Unless the specific molecular basis of apomixis is revealed, scientists will not be able to induce apomixis as we know it from natural plants by simply manipulating a few genes in a sexual plant (however, see the next section on *de novo* apomixis).

However, framing that possibility under the different models of the regulatory control of apomixis is a good exercise to bring up points of relevance for genetic engineering. Each of the different models suggest inducing apomixis per se will be challenging. Exploiting gene editing for induction of apomixis is confronted with different obstacles depending on the type of molecular control.

4.1. *Apomixis Caused by Heterochronic Gene Expression*

In this case apomixis is a consequence of heterochronic gene expression derived from divergent evolution or stoichiometric disbalances of macromolecular complexes; inducing apomixis would be a formidable if not quixotic task unless a few molecular edits at sequence level are enough to mimic global regulatory changes and alter the output of specific reproductive modules.

Even in such a case, there would be several hurdles to induce apomixis-like phenotypes. For example, in hybrid apomicts formed from two putatively diverged parents, no information on sequence divergence at gene level exist for most cases. Retrieving such information from genetic data on specific apomictic individuals might be demanding as dosage of alleles (e.g., AABC, ABBC, ABCC) is difficult to obtain [79,80], and using sequence level information from parental plant materials might provide biased information depending on the apomicts age, recombination rates, and rates of molecular divergence. To sum up, it is not possible to know a priori which genes should be modified, nor the extent of the changes needed to shift molecular interactions enough to create a developmental asynchrony able to induce functional changes in the output of each reproductive module without underrunning or preventing it.

In the case of autopolyploids with higher allele similarity (e.g., $A_1A_1A_2A_2$), creating disbalances in stoichiometry of macromolecular complexes that are sufficient to shift reproductive pathways is hardly an option. Besides the above challenges, scientists must deal with difficulties of allelic bias, overdispersion and outliers when modelling autopolyploids [81]. Thus, lack of information about quantity and quality of allelic deviations required for an operating asynchronous development would be a barrier to inducing apomixis under this regulatory model.

In either case (alloploid or autoploid genomes), integrating genomic data and well-designed studies collecting proteomic and metabolomic data will facilitate exposing gene-protein interactions, as well as recognizing proteins and molecular complexes with relevant functions during apomixis emergence. These studies could also reveal the extent of disturbance in cellular metabolic pathways that is tolerated without triggering apoptosis.

4.2. Apomixis Caused by a Few Genes

In the case that apomixis is controlled by a few genes, we assume the existence of single 'master' genes governing independent developmental programs and shifts in the functional output of each reproductive module. Specifically, one master gene for changing meiosis into an apomeiosis, one for inducing parthenogenetic development of the embryo, and one controlling the initiation/progression of endosperm development (a mechanism not yet clarified but likely regulated epigenetically). Activation of those master genes would be enough to initiate multiple concurrent changes observed at the gene level [37,38]. Thus, inducing apomixis might only require targeting those master genes (see Figure 2 and the next section on *de novo* apomixis). By choosing the right combination of sequence-level changes and regulatory modifications, a genotype could be engineered that holds the level of gene expression needed for the correct interaction between gene networks and macromolecular complexes. Moreover, this genotype would navigate changes in cell cycles and ensure a coordinated progression throughout reproductive modules and altered developmental programs to finally produce an asexual seed.

Although this may appear comprehensive from a theoretical viewpoint, it includes many drawbacks from an empirical and technical perspective. We know chromosomal regions associated with the control of apomixis present high allelic divergence, activity of transposable elements, and mutational degradation [68,72,82–84]. Interruptions of gene sequences located in apomixis loci suggest deregulation of apomixis-related genes may be more complex than expected and involve snRNAs and RNAi players [84–87]. Such changes on individual genes may not be easy to mimic using gene editing.

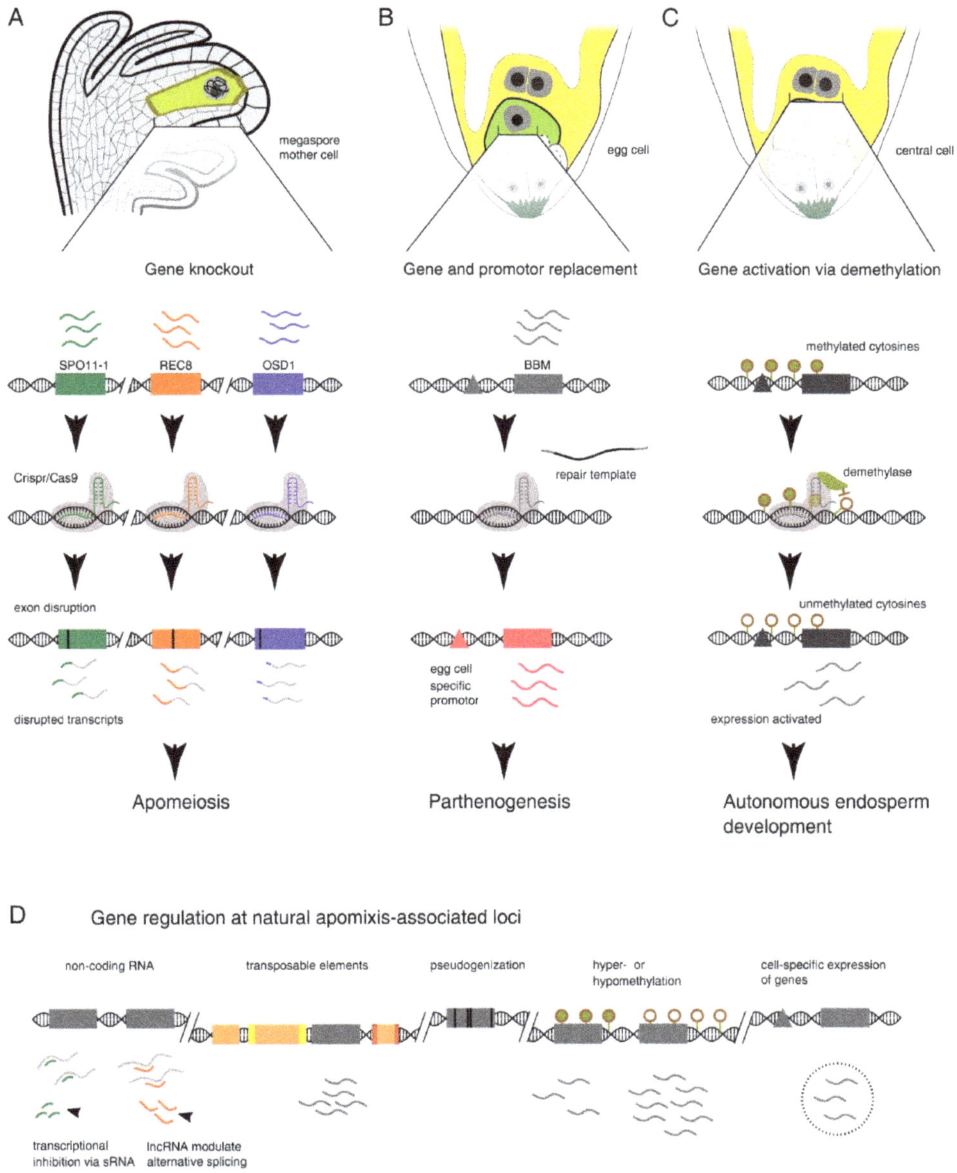

Figure 2. Illustration of gene editing approaches that could be used to target apomixis master genes or multiple genes affecting specific developmental steps and having a master-like effect (for simplicity, we display known gene names associated with reproductive modules and targeted in *de novo* apomixis strategies). (**A**) Gene knockout can be used to convert meiosis to mitosis, as has been shown using the Mitosis instead of Meiosis (MiMe) mutants. (**B**) Tissue-specific promotors and targeted gene replacement for a gene such as *BBM* could induce embryo development without fertilization. (**C**) Using Cas9 as a recruiting platform for demethylase would allow activation of repressed female genes, for instance to erase genomic imprinting and trigger autonomous endosperm development. (**D**) Features of observed genomic and regulatory variation at natural apomixis loci such as those known from *Pennisetum/Cenchrus* [82] and *Paspalum* [86–88].

Another hurdle might relate to the hemizygous condition of sequences linked to apomixis [68,69,89]. Sequences were considered hemizygous, because previous studies often did not detect an allele other than the one associated with apomixis. Hemizygosity may well be considered as an advantage for exploiting gene editing techniques, as modifying only one allele of the polyploid set for each gene would expectedly be enough to express apomixis. However, it also implies that a sexual plant may not carry the sequences required to be modified for apomixis expression. In this case, insertion of the functional alleles using an approach like homology-directed repair or prime editing may be an option to overcome the hurdle of hemizygosity. It remains difficult to insert sequences longer than several kilobases via gene editing, thus, alternative transformation approaches may be required if large sequences need to be inserted.

The *PsASGR-BABY BOOM-LIKE* gene (*BBML*), a gene involved in the induction of parthenogenesis located in an hemizygous region of the apospory-specific genomic region (ASGR) in *Pennisetum/Cenchrus*, is the only trait-associated gene isolated from an apomictic species and tested in intra- and interspecific sexual species (i.e., *Arabidopsis*, rice, maize) [67,90]. Orthologous genes present in sexual *Brassica* and *Arabidopsis* plants can induce embryogenesis [91], meaning the *BBML* family gene has a relevant function enabling embryo development (activating parthenogenesis). However, its specific molecular function and interconnection to other molecular players remains to be solved, and is likely to have a part in the observed low penetrance and its species-specific activity.

Considerable developmental variation has been observed among natural apomicts (diplosporous *Antennaria-*, *Taraxacum-*, *Ixeris-*, *Eragrostis*-types, or aposporous *Hieracium-* and *Panicum*-types of apomixis with other less frequent developmental deviations, plus pseudogamous or autonomous development of the endosperm, [2]) and several genetic mutants resembling apomixis-like phenotypes (see details in the next section). This suggest that apomixis depends on alteration of different species-specific genes controlling and/or coordinating developmental steps within reproductive modules to produce genetically balanced clonal progeny. Editing one or several genes related to each reproductive module, and generating a series of allelic variants, will be useful to induce functional changes in proteins and to modify the output of the reproductive modules. However, the impact of such changes on the coordinated development of reproductive modules within the ovule and the stable formation of apomictic seeds remain to be tested.

4.3. Apomixis Caused by Epigenetic Signals

In the case apomixis is controlled by epigenetic signals which determine cellular metabolic pathways opting between sexual or apomictic developments, manipulating apomixis would be challenging. Canonical or noncanonical signaling may be needed for epigenetically channeling shifts in cellular metabolic programs. The molecular context of each metabolic pathway involved in the steps to sex or apomixis might have substantial relevance, making the induction of apomixis genotype-dependent. The cellular metabolism influences chromatin dynamics and epigenetics and thereby has functional roles in genome regulation [92]. The idea of reproductive phenisms involving anciently canalized components of complex molecular processes [73] might suit Waddington's image of "epigenetic landscapes," depicting valleys for stable cellular states and cell lineage specification [93]. However, quantitatively mapping epigenetic landscapes to provide predictive models of cellular differentiation and to identify optimal routes of cell fate reprogramming is not a simple task [94].

On the other hand, according to metabolic control theory, gene networks and individual metabolic pathway architecture coevolve and constrain evolutionary change through selective partitioning of metabolic fluxes into alternate channels [95]. While a *de novo* induction of apomixis implies placing altered genes in a genome not adapted to their mutated functions and hence lacking *ab initio* a buffering background, a scenario of *restitutio* of an ancient switch provisionally blocked may provide an appropriate background, delivering suitably partitioned metabolic channels. Both sexuality and apomixis are stable and penetrant in those plants in which they occur. This fit well with the hypothesis of an ancient polyphenic switch partitioning sex and apomixis into alternate channels depending upon metabolic status [73]. The absence of apomixis-like mutants displaying high expressivity might well be a consequence of the partial or incomplete partition of metabolic fluxes, and highlight the relevance of understanding protein–protein interaction networks (see discussion in the next section).

While sexuality uses regular developmental programs to form a seed, apomixis changes several regulatory mechanisms of the sexual program to do so. These mechanisms include epigenetic signals, the activity of transcription factors, non-coding RNA signaling and modulation of gene expression, protein turnover, cell-to-cell and hormonal signal transduction [9]. Having all these mechanisms canalized by an ancient switch will clearly ease the induction of apomixis, if only simple internal or external signals are required. For instance, switching between asexuality and sex in lower eukaryotes can be a response to sensed environmental signals. However, in plants such a decision would regulate whether to allocate energy into flowering, i.e., (sexual) reproduction [96]. Between lower and higher eukaryotes (such as plants), evolution has over imposed much more developmental complexity (in size and topology of gene and metabolic networks [97]) on top of the ancient switch between asexual-sexual growth (reproduction). Hence, it might be unlikely that once flowering is decided, one (or more) master-like signal(s) could channel a reproductive switch back to asexual reproduction, without being backed up with genetic redundancy controlling decisions during seed formation. Such redundancy is observed in different plant mutants [98].

Taken together, all collected data on apomixis phenotypes suggest the expression of apomixis involves permanent (genetic) or temporal (epigenetic) modification of several genes and regulatory changes [9,99]. Instead of creating developmental chaos, these modifications seem to provide a coordinated developmental flexibility between reproductive modules, and maintain high levels of expressivity in individual plants (Figure 3). Stable induction of apomixis *sensu stricto* in a sexual plant will likely require knowing most of these changes from the phylogenetically closest apomictic relative. Multiplexed gene editing would need to combine numerous edits optimized to the target genetic background.

5. Closing in on *De Novo* Apomixis: Making Rudimentary Changes in Reproductive Modules to Synthesize Clonal Seeds

Initiating apomixis in a sexual plant through the same molecular mechanism responsible for its occurrence in natural species is not yet a feasible task. Besides the regulatory complexity of sexual seed formation, understanding apomixis as a consequence of specific alterations in reproductive modules has allowed researchers to progress on *de novo* engineering clonal seed formation in otherwise sexual plants. A high number of genes and pseudogenes associated with apomixis or apomixis-like phenotypes in sexual and apomictic plants are enabling the manipulation of meiotic recombination [9,37,98,100–103]. Gene knockouts have induced several apomixis-related traits in plants (Table 1). Promising results were recently published based on attempts to synthesize apomixis in sexual plants by targeted modification and combination of mutants that complement key steps in reproductive modules [104,105].

Table 1. Common genes used to manipulate specific reproductive steps during gamete and seed formation.

Gene	Function	Mutant Phenotype	Reproductive Module	Expressivity	Reference
SWI1/dyad	sister chromatid cohesion	Arrested meiosis	Meiosis	0.99 [1]	[106]
OsPAIR	Homologous chromosome pairing	Arrested meiosis	Meiosis	1.00	[107]
AtSPO11-1	DSBs and initiation of homologous chromosome recombination	Lack recombination	Meiosis	0.97 [2]	[108]
AtREC8	sister chromatid cohesion	Univalents; aberrant chromosomal segregation	Meiosis	1.00 [3]	[109]
AtOSD1	Entry into MII	Lack meiosis II; dyad formation	Meiosis	0.85 [4]	[110,111]
ZmMATL [5]	Sperm-specific phospholipase	Haploid induction, haploid seeds	Fertilization	0.07 [6]	[112]

Table 1. *Cont.*

Gene	Function	Mutant Phenotype	Reproductive Module	Expressivity	Reference
AtCENH3	Centromere-mediated genome elimination	Haploid seeds	Embryogenesis	0.08 [7]	[113]
RKD	Transcription factor	Somatic embryogenesis	Embryogenesis	1 [8]	[114]
BBM	Transcription factor; embryo development	Somatic embryogenesis	Embryogenesis	! [9]	[91]

[1] in a very low proportion (0.24%), one dyad cell can develop into a mature embryo sac and produce a seed [106]; [2] mutants are semi-sterile with females producing 3% mature female gametophytes and three seeds per silique when female mutants were used in crosses to the wild type ecotype (wild type seed set per fruit = 45 ± 5; [108]; [3] mutants show complete male and female sterility [109]; [4] dyads in *osd1* mutants produce on average 35 ± 6 polyploid (3× and 4×) seeds per fruit compared to 38 ± 11 of the wild type [110]; [5] also named *NLD* [115] and *ZmPLA1* [116]; [6] haploid induction rates of 4–12.5% (average 6.65%) in maize; haploid induction rates of 2–6% in rice [117], seed-setting rate in rice 11.5% [105]; [7] estimated from circa 12% fertile ovules multiplied 62.5% of haploid seeds; crosses between GFP-tailwasp X wild-type plants produced 25–45% viable haploid offspring (the rest corresponded to diploids and aneuploids); no clear information on ovule abortion or seed-set is provided; [8] loss of *RKD1* function by antisense overexpression abolished somatic embryogenesis in transgenic *Citrus* and the transgenic T1 plants were derived from self-pollinated zygotic embryos [114]; [9] overexpression using semiconstitutive promoters induces ectopic embryo formation in leaf tissues and other pleiotropic effects; thus, there are no reproductive units per se, and neither penetrance nor expressivity can be estimated as in the other cases.

5.1. Mimicking Sporophytic Apomixis

The simplest road to inducing production of clonal seeds is to mimic sporophytic apomixis. Inducing an ectopic embryo within the ovule while arresting or delaying egg cell progression or zygote development in the fertilized meiotic female gametophyte may involve as few as two genes.

A simple development like this could exploit knowledge of somatic embryogenesis (e.g., *RKW* or *BBM* gene families [99,102]), and a gene (partly) arresting egg cell progression. Here, however, fertilization of the gametophyte must not be avoided, as the endosperm is crucial to the development of a viable seed. While complete arrest of zygote development will impose arrestment of the endosperm and seed failure due to embryo-endosperm signaling and communication [118,119], a temporal interruption of the development of the sexual embryo might well give advantages for somatic embryos to hoard resources from the nucellus and endosperm, as usually observed in plant exhibiting sporophytic apomixis [120].

Studies in *Citrus* had shown somatic embryogenesis is likely regulated by *CiRKD1*, a gene encoding an RWP-RK domain-containing transcription factor [121]. *RKD* genes are expressed in egg and synergid cells of different species, including sexual *Arabidopsis thaliana* and apomictic *Boechera gunnisoniana* [122–125] and likely play a role in cell fate, cell identity in absence of fertilization, and acquisition of embryogenic competence. In *Citrus*, one of the two characterized *CiRKD1* alleles carries an upstream miniature inverted-repeat transposable element (MITE)-like insertion, which may be responsible for its increased expression in tissues with somatic embryogenesis. Antisense silencing of *CiRKD1* genes in transgenic tissues leads to a complete loss of somatic embryogenesis [114]. However, the presence of multiple *CiRKD1* gene copies and its location in a genomic region of about 80 kb including other 11 genes [121,126] may render embryogenesis activation by gene-editing approaches more complex.

Another gene controlling somatic embryogenesis is the *BABY BOOM* gene (*BBM*), which is able to induce embryos in microspores of *Brassica napus* and somatic cells of *A. thaliana* [91,127]. *BBM* genes belong to the AP2/ERF family of transcription factors encoding an AINTEGUMENTA-LIKE (AIL) APETALA2/ethylene responsive element-binding factor carrying two APETALA2 (AP2) DNA-binding domains [128], and have a crucial role regulating totipotency and embryonic identity [102].

Regulatory acquisition of embryogenic competence through genetic modification of a single gene (either *CiRKD1* or *BBM* genes) in somatic cells within the ovule is feasible. Yet, the challenge for mimicking sporophytic apomixis is induction of embryogenesis in somatic cells while simultaneously postponing zygotic development in the gametophytic tissue. Using genes acting in the post-fertilization processes dependent on the paternal allele might be a plausible option to delay or arrest the progression of the fertilized egg-cell. In rice, paternal expression of *BBM1* and its paralogues is required to bypass the fertilization checkpoint and transit to zygotic embryogenesis [104]. However, the zygotic transition postfertilization is initiated by asymmetric activation of parental genomes, with most genes playing major roles in the early development of plant embryos being maternally expressed [129,130]. Yet, *de novo* post-fertilization epigenetic reprogramming and transcriptional silencing of the paternal genome is controlled by *DNMT3A* in mice, a DNA methyltransferase highly expressed in oocytes [131]. *Dnmt3a* maternal knockout embryos die during post-implantation development [131]. Thus, using a plant ortholog of *DNMT3A* might help to unblock paternal gene expression in the early embryo and possibly result in zygotic embryo lethality without affecting the development of somatic embryos.

5.2. Mimicking Gametophytic Apomixis

As few as three genes may be needed to mimic gametophytic apomixis, provided individual genes can create the desired changes in each reproductive module. Under such a simple model, this means (1) producing unreduced non-recombinant gametes (either by skipping meiosis or by blocking the reductional division of meiosis), (2) developing an embryo parthenogenetically, and (3) promoting endosperm development to complete the formation of a seed.

For mimicking gametophytic apomixis, researchers have a wide collection of meiosis related mutants to work with (e.g., [103,132]), though most of these show feeble phenotypes and low expressivity (see discussion below).

While the formation of unreduced gametes might implicate changing the regulation of simple genes related to RNA-directed DNA methylation pathways inducing either aposporous-like (e.g., through *AGO9*, [66]) or diplosporous-like (through *AGO104*, *SWI1*, or *DMT* genes, [133–135]) ovule progression, other strategies like annulling the main features of the meiotic division would require changes in more genes. A number of mutants affecting specific steps of the meiotic prophase and both meiotic cell divisions can be combined to convert meiosis into a mitosis-like division. For example, a triple knockout of the meiotic genes *SPO11-1*, *REC8* and *OSD1* can be used to generate MiMe (Mitosis instead of Meiosis) *Arabidopsis* plants. The resulting meiotic mutant phenotypes eliminate DNA double-strand breaks, meiotic recombination and chromosomal pairing (*Atspo11-1*, [108]) and destabilize centromeric cohesion and thus modifies chromatid segregation by impeding monopolar orientation of the sister kinetochores at metaphase I (*Atrec8*, [136]). Finally, the second division is omitted, likely by modulating the anaphase promoting complex/cyclin levels at the end of first meiosis (*Atosd1*, [110]).

Using a similar MiMe strategy in rice, Mieulet et al. [111] combined knockouts of *REC8* and *OSD1* genes with *PAIR1*, a gene controlling homologous chromosome pairing, and suggested another three genes to be used instead of *SPO11-1*. Like in *Arabidopsis*, the triple rice mutant *pair1*, *Osrec8*, and *Ososd1* produced unreduced diploid gametes. Induction of a certain phenotype between different plant species and the use of a particular gene editing technology will depend upon synteny, protein sequence similarity and interactions with regulatory networks. For example, *OSD1* orthologs are single genes in *Hordeum vulgare* and *Brachypodium distachyon*, but are tandem duplications in *Zea mays*, *Sorghum bicolor*, and *Setaria italica* [137]. Thus, exploitation of such a gene for induction of unreduced gametes will require the use of RNAi or additional disruption of redundant gene copies to deal with duplications.

The next step toward clonal offspring needs to either skip fertilization or the incorporation of male chromosomes into the diploid female gamete. This has been tested using haploid induction genes that promote gynogenesis, i.e., the development of a fertilized egg-cell carrying only maternal chromosomes. One of these genes is a patatin-like phospholipase A restricted to the pollen tube, which might cause sperm chromosome fragmentation and paternal genome elimination in the fertilized egg cell [116]. The gene was characterized almost simultaneously by three research groups and called *MATRILINEAL* (*MTL*; [112]), *NOT LIKE DAD* (*NLD*; [115]), and *ZmPLA1* [116]. Haploid induction can be induced in rice by knocking out the gene *OsMATL* [117], and *TaMTL* in wheat [138]. Combining either the MiMe strategy with *MATL* successfully produced clonal seed, but at exceptionally low frequencies (Table 2; [105,139]). Similarly, adding a modified CENH3 to the MiMe or *dyad* phenotypes also eliminates the paternal genome postfertilization, creating clonal offspring in *Arabidopsis*, but at exceedingly low levels (Table 2; [140]). Alternatively, formation of unreduced gametes can be combined with genetic modification of *BBM/BBM*-like genes. The ectopic expression of these genes in egg cells before fertilization induces parthenogenetic embryos in pearl millet, maize, and rice, though not in *Arabidopsis*, at variable but overall low rates [67,104,141]. Loss-of-function mutations, such as those used for haploid induction, or spatial-temporal regulatory changes can be achieved by introducing indels in coding regions [112,115] or carrying out gene and promoter swaps with the CRISPR/Cas system. However, even when unreduced gametes can be produced at almost wild-type levels, apparent ineffective molecular coordination between distinct reproductive modules (i.e., sporogenesis and gametogenesis to form unreduced gametes, egg-cell parthenogenesis, and endosperm progression to produce a seed) fail to deliver plants with high expressivity of synthetic apomixis, with plants showing drastically reduced fertility in all cases (Table 2). The induction of synthetic apomixis, therefore, remains limited by low efficiency. Although the search for 'silver bullet' inducer genes seems momentarily appropriate, exploiting *de novo* apomixis will rely more on understanding the molecular interaction and background responsible for the low expressivity of combined targeted modifications.

Table 2. Combination of mutants used to create unreduced nonrecombinant gametes and clonal progeny.

Gene Combination	Reproductive Phenotype	Expressivity	Fertility [1]	Reference
AtSPO11-1 + AtREC8 + AtOSD1	Unreduced nonrecombinant gametes	1.00	0.66 [2]	[110]
OsPAIR1 + OsREC8 + OsOSD1	Unreduced nonrecombinant gametes	1.00	0.74 [3]	[111]
AtSPO11-1 + AtREC8 + AtOSD1 + GEM [4]	Clonal offspring (mixed) [5]	0.33	0.3 [6]	[140]
dyad + GEM [4]	Clonal offspring (mixed) [5]	0.13	0.0018 [7]	[140]
AtSPO11-1 + AtREC8 + AtOSD1 + BBM1	Clonal offspring (mixed) [5]	0.11–0.29	? [8]	[104]
OsPAIR1 + OsREC8 + OsOSD1 + OsMATL	Clonal offspring (mixed) [5]	0.02–0.04	0.045	[105]
OsSPO11-1 + OsREC8 + OsOSD1 + OsMATL	Clonal offspring	? [9]	? [9]	[139]

[1] fertility is considered as a seed set or number of seeds formed from the total number of ovules; the data were collected from each study without considering germinability (which ranged between 73–92% among studies). When the number of ovules per fruit were not provided, the number of seeds were compared to those of the wild type plant; [2] dyads in this triple mutant produce on average 25 ± 6 polyploid (3× and 4×) seeds per silique compared to 38 ± 11 of the wild type [110]. MiMe rice plants produce 81.2% of seeds (all 4x derived from unreduced gametes) compared to 79.2% in the wild type [105]; [3] from a total of 1012 seeds from 1370 ovules (flowers) analyzed [111]; [4] GEM line called for Genome Elimination caused by a Mix (GEM) of CENH3 variants (Marimuthu et al. [140], supporting online material page 5); [5] clonal offspring were intermixed with polyploid and aneuploid offspring; [6] considering fertility as 15 seeds per silique in comparison to 50 seeds per silique in the wild type (Marimuthu et al. [140], supporting online material pages 12–13); [7] considering fertility as 0.9 seeds per silique in comparison to 50 seeds per silique in the wild type (Marimuthu et al. [140], supporting online material pages 12–13); [8] no proportion of seed set in comparison to total number of ovules/flowers is mentioned; [9] Xie et al. [139] provide cytological evidence of MiMe induction in a gene-edited rice plant, but no data about formation of clonal progeny by the edited *OsMATL* gene, nor about fertility of the gene modified plant.

The last step to form clonal seeds is the development of the endosperm. Even though combining genetically modified genes is strictly necessary to produce unreduced gametes and parthenogenetic embryos, inducing the formation of the endosperm without fertilization is not crucial for developing synthetic apomixis. Although fertilization in MiMe + *BBM1* rice plants limits the formation of clonal offspring to 29%, with most of the unreduced gametes forming tetraploid progeny [104], a proportion of such clonal seed discounting can be reduced in alternative ways. For instance, by targeting polyspermy avoidance mechanisms [142] or mechanisms needed for gamete fusion (such as secreting EC1 protein and further translocation of sperm specific gamete fusion proteins to the egg-cell surface; [143]). Even when the ratio between the endosperm and the embryo shifted from the normal 3:2 to 3:1 after fertilization of the unreduced central cell in MiMe + *BBM1* plants [104], the 2:1 maternal-to-paternal genome ratio in the endosperm required for appropriate development was maintained together with formation of viable seeds.

Engineering the autonomous formation of the endosperm without fertilization to obtain a complete asexual system may be ideal from a biotechnological perspective. From a biological viewpoint, despite studies showing that a single dominant locus is able to induce the autonomous endosperm phenotype in apomictic *Hieracium* [144], inducing autonomous endosperm in sexuals might be hard due to the molecular complexity of its development. Autonomous endosperm development differs between dicots and monocots, and the dynamics underlying this complex process rely on genome balance, epigenetic gene regulation, and parent-of-origin effects founded upon the contribution of the male gamete [145]. In addition, still unknown genetic modifier elements, protein interactions, and regulatory pathways underlying embryo-endosperm developments add to the complexity [101,146,147] and restrain autonomous endosperm expressivity [144]. Regardless of mutations in *fertilization independent endosperm* (*FIE*) and other Polycomb group genes leading to autonomous endosperm development in *Arabidopsis* [148], orthologs reported in both rice (*OsFIE1* and *OsFIE2*) and maize (*ZmFIE1* and *ZmFIE2*; [149]) produce distinct phenotypes. However, most plants (including crops) have hermaphrodite flowers and pseudogamous development of the endosperm via fertilization is possible in natural and synthetic apomictic plants. Considering, in addition, the lack of specific genes to be targeted for autonomous endosperm development, this trait is not essential per se for inducing apomixis.

5.3. Tuning Changes for Complete Penetrance, and High Expressivity and Fertility

Penetrance measures the percentage of individuals in a population who carry a specific gene (genotype) and express the gene-related trait (phenotype). Penetrance is complete (or incomplete) when all (or less than 100%) of the individuals with a specific genotype express the corresponding phenotype. The degree of expression of a trait is called expressivity and is used to describe variation among individuals with a specific gene (genotype) [150]. In the case of plant reproductive traits, and apomixis specifically, any apomictic individual showing the apomictic phenotype will contribute to the penetrance of the apomixis trait in the population, and any individual variation of the proportions of apomictic or sexual flowers will determine the expressivity of the trait on that particular genotype. While the penetrance of apomixis in natural populations is typically 1 (no natural sexuals coexist with apomicts), observed expressivity of apomixis in individual plants is generally high (expressivity = 0.89, Figure 3). In laboratory plants modified to express certain phenotypes, expressivity rather than penetrance can be determined. Plants genetically modified to produce clonal gametes usually exhibit high expressivity, while plants modified for clonal seed formation exhibit low expressivity of the trait (Table 2). The cause for the apparent uncoupling between reproductive modules for unreduced gametes and parthenogenesis in synthetic apomicts remains to be resolved. An important role in the uncoupling is likely played by interaction between modified genes and gene networks. Since MiMe phenotypes display high expressivity, the low expressivity for clonal seeds and the occurrence of higher ploidy progeny (see e.g., Khanday et al. [104]) can likely be attributed either to genes for parthenogenesis having secondary roles in the expression of the phenotype itself, or, most likely, on pleiotropic effects and genetic redundancy in the development of the phenotype.

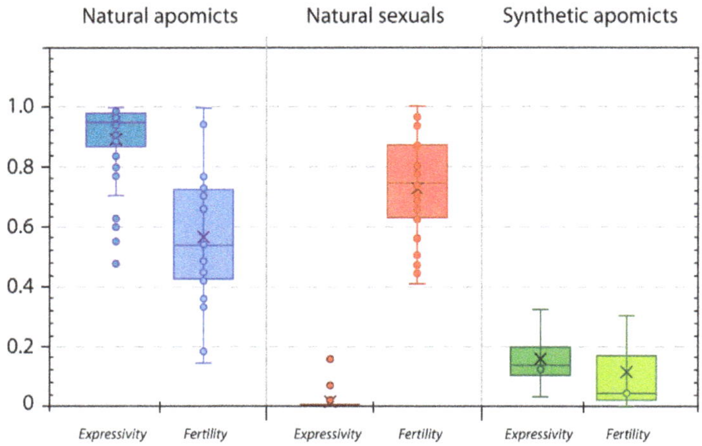

Figure 3. Box and whiskers plot representing observed variability in expressivity and fertility of apomixis in different biological conditions. In natural apomictic genotypes, apomixis expressivity is high ($n = 35$; average = 0.89 ± 0.14, max = 1, min = 0.48) and exhibits medium fertility ($n = 18$; average = 0.57 ± 0.24, max = 1, min = 0,15). In natural sexuals, only traces of apomixis have been recorded in specific cases ($n = 33$; average = 0.007 ± 0.027, max = 0.15, min = 0.0) and these plants exhibit high fertility ($n = 35$; 0.73 ± 0.17, max = 1, min = 0.41). Synthetic apomicts created so far show very low expressivity ($n = 4$; average = 0.16 ± 0.012, max = 0.33, min = 0.03; Table 2) and very low fertility ($n = 3$; average = 0.12 ± 0.16, max = 0.3, min = 0.002; Table 2). The graph is based on data from 42 species showing diplosporous and aposporous apomixis, autonomous and pseudogamous seed development, and their sexual conspecific taxa. Because gametophyte frequencies underestimate apomixis expressivity, all data were collected from studies on flow cytometry on seeds developed under open pollination (except for *Erigeron* in which parthenogenetic development was confirmed through embryology; [151]). Seeds developed through syngamy of meiotic gametes or meiotic plus apomeiotic gametes (i.e., incomplete apomixis and B_{III} hybrids) were classified as sexual. In species where data from multiple genotypes were available, only the mean value was used to avoid sampling bias (e.g., in *Paspalum intermedium*, $n = 18$ genotypes, expressivity = 0.86 ± 0.08, fertility = 0.19 ± 0.11; [152]). In some cases, fertility data were missing because of technical constraints (e.g., *Hypericum* forms several thousand seeds *per* fruit). Species included belong to the genus *Boechera* ($n = 8$, [153–155]), *Boehmeria* ($n = 1$, [156]), *Erigeron* ($n = 2$, [151]), *Hieracium* ($n = 10$, [157,158]), *Hypericum* ($n = 1$, [159]), *Paspalum* ($n = 9$, [152,160–165]), *Pilosella* ($n = 2$, [166]), *Poa* ($n = 3$, [167]), *Ranunculus* ($n = 4$, [168,169]), *Taraxacum* ($n = 2$, [170]).

To some extent, gene editing can help tune and increase the levels of parthenogenesis and formation of clonal seeds. For instance, after a comprehensive genetic analysis of haploid induction in maize [171], distinct studies identified the *MTL/ZmPLA1/NLD* gene and took advantage of gene editing to obtain plants with 6.7% haploid induction rate [112], or to combine *mtl/zmpla1/nld* with a *zmdmp* mutant to increase the haploid induction rate 5–6-fold [172]. These works were a step forward to accelerating crop breeding through in vivo haploid induction systems. There is now a strong opportunity to search for new genes and explore alternative combinations of gene edits to create MiMe-like phenotypes that better interact with parthenogenetic genes and vice versa. A number of meiotic related genes are known to play relevant functions in specific steps, e.g., for recombination initiation [173]. In addition, various genes responsible for acquisition of an embryogenic cell fate [101,102] are available for being tested in different species and under distinct gene-edited configurations. Examples include the *LEAFY COTYLEDON* (*LEC*) family of transcription factors, the *SOMATIC EMBRYOGENESIS RECEPTOR KINASE* (*SERK*), *WUSCHEL* and *AGAMOUS-Like 15* (*AGL15*), *WOUND INDUCED DEDIFFERENTIATION 1* (*WIND1*), and the homeobox gene *BELL1*, among others [174–178].

For efficient expression of an apomixis-like pathway in sexual plants, researchers will need to bring to bear gene expression modification, precise base editing and trait optimization using allelic series on different gene targets to increase trait expressivity. Single genes can be knocked out in different ways, creating specific changes in metabolic pathways and cellular phenotypes. Additionally, different plant mutants can produce similar phenotypes (e.g., multiple MMC or embryo sacs). Therefore, we expect future *de novo* engineering of apomixis to involve appropriately modifying different genes to mimic apomixis-like steps, modulating their expression, and optimizing the fittest combination of mutants toward full trait expressivity. All natural apomicts display complete penetrance and high expressivity (Figure 3). Increasing *de novo* apomixis efficiency will require expanding data collection (see Section 6) and compatibility among apomictic plant systems to provide useful information about the involved metabolic pathways, gene networks and protein interactions.

Another significant reproductive trait in genetically modified plants is fertility, i.e., the proportion of seeds formed in comparison to the total number of developing ovules. While in studies on mutants affecting distinct plant traits, fertility might not be that important, in the present case it is of special interest because (1) we are modifying reproductive development, and such changes will impact the total amount of developing ovules, and (2) inducing phenotypic changes associated with negative effects on plant fertility is undesirable or in opposition to plant breeding goals. Despite the obvious differences between estimates of the proportion of clonal seeds and fertility, in some studies on plants modified for clonal seed formation, there is a lack of clear information about fertility (Table 2). As an example, a plant might produce a total of 10 viable seeds, five being clonal, from a total of 100 developing ovules. If fertility is not considered there might be a misreading of the expressivity of the trait (as 50% instead of the real 5%) in that plant. Here as well, apomicts display higher fertility compared to synthetic apomixis mutants, but lower compared to sexual conspecific taxa (Figure 3). A reduction in fertility is inherently associated with apomixis, and hence, it is a crucial attribute to consider in any attempt to induce apomixis. Increasing data collections about fertility in apomicts, sexuals and mutants will be necessary for understanding changes in fertility caused by modifications of developmental programs and plant reproductive modules, including those observed in natural apomicts and genetically modified plants.

Experimental studies on cultivar developments and wild species have shown apomixis is a dominant trait and is transgenerationally stable [10,179]. Yet, unstable phenotypes occur by spontaneous uncoupling of reproductive modules and formation of segregating sexual offspring (either B_{II} or B_{III}). Phenotypic instability due to gene-environment interactions, genetic heterogeneity or genetic compensation effects has also been observed in mutants [124,180]. Even if complete penetrance of a synthetic phenotype is attained in a natural population or a crop field, trait expressivity will likely vary due to genotype-by-environment interaction. In natural apomicts, experimental assays demonstrated an influence of climatic factors such as daylength, temperature, and salinity on proportions of apomixis and sex [152,181–184]. Environmental influence on sexuality and apomixis was recently exposed in situ on natural populations of facultative apomictic *Paspalum intermedium* [152]. In apomictic grass cultivars, differential rates of residual sexuality may cause loss of productivity due to segregating offspring, impacting cultivar management strategies [163].

The above aspects influencing penetrance, expressivity and fertility should be rigorously screened and evaluated while establishing a synthetic apomictic crop. For wider market release, the aim will be to produce a fertile highly expressive apomict, with no off-target effects of gene editing.

6. The Remaining Challenge of Data Collection for Genomic Dissection of Apomixis Loci

Most of the over 600 plant reference genomes publicly available via GenBank represent model or crop plants and their close wild relatives. Collections of genomic data from apomictic plants are still meagre, and no apomictic plant genome has yet been assembled. By comparing the list of species with sequenced genomes to those genera in which apomixis is present in at least one species [185], we found that a total of 60 species that belong to 33 apomictic-containing genera have a GenBank genome assembly (Table S1). The sequenced species are sexual, mostly diploids. The majority of such species belong to genera containing species with sporophytic apomixis (39 species), with *Citrus* being the most common genus (nine species). Genome assemblies are also available for species belonging to intensively studied apomictic genera, such as *Boechera*, *Boehmeria*, *Cenchrus*, *Eragrostis*, *Erigeron*, *Hypericum*, *Panicum*, and *Setaria* (13 species). Hence, the currently available data represents an interesting collection of sexual species having conspecific apomicts. These data will likely mark a direction and foster future research on apomixis.

We also found another 54 species belonging to six genera in which sporadic occurrence of apomixis or elements of apomixis had been reported (including *Solanum* and *Oryza*, having 20 and 16 species and varieties sequenced, respectively). In these cases, some reports on apomixis related features date back to the 1950s or earlier (e.g., [186,187]), and reported attributes are likely to be genotype-dependent and thus, unless alike genotypes are analyzed, the information about apomixis that could be extracted from these genomes is probably limited.

Although no apomictic genomes are yet available, the dropping cost of genome sequencing is enabling sequencing projects with broad phylogenetic sampling across thousands of non-model plants [188]. Undoubtedly, apomictic genomes will become available, and these genomes can be mined together with those of conspecific sexuals as an invaluable resource to uncover the basis of natural apomixis in plants. Comparative whole genome analysis can detect apomixis-associated regions such as the ASGR locus in *Pennisetum squamulatum* [67,189] or the ACL in *Paspalum simplex* [89,190] and help better understand how sequences related to apomixis are co-evolving within and among plant lineages. A further example of apomixis-related comparative genomics is the discovery of the apomixis candidate gene *CitRKD1*, which relied on *de novo* assembly and sequencing of cultivated *Citrus* and wild relatives [121]. Reference genomes are also useful to align population-level sequencing data and identify genetic differentiation between sexual and apomictic individuals within a species. At broader phylogenetic scales, trends in the evolution of apomictic loci can emerge that pinpoint common apomixis-associated genes, allowing these to be targeted in diverse crop species. Although genome editing is not limited to introducing naturally occurring variation, comparative genomic data can help identify and prioritize editing targets.

Before high-quality apomictic plant genomes become widely available, several technical challenges in their assembly will need to be overcome. *De novo* whole-genome assembly of plants is generally challenging due to large genome sizes, variable ploidy and extensive duplications, repetitive elements, and areas of high GC or AT content. In contrast to most crops, wild species can also harbor high levels of heterozygosity, complicating genome assembly. As mentioned above, apomicts are frequently polyploid and may exhibit apomixis-associated regions with high repetitive element content [88,99,191]. These issues can impede accurate assembly of apomixis-associated genomic regions, especially when relying on short reads. Genomes assembled from short reads are generally draft genomes, comprised of thousands of contigs, with many gaps and errors. Unresolved haplotypes leave us without access to a whole layer of genetic variation information. Long-read sequencing, linked-read strategies, and optical mapping offer significant improvements to genome assembly quality [192], though often at substantially higher costs. However, even when accurate and phased assemblies exist, a single linear reference genome only provides a limited view of within-species sequence diversity. As the apomixis trait can be hemizygous and may thus involve complex structural variation, reference bias could undermine analyses of the trait. One approach to address this issue is by constructing plant pangenomes based on population-level genomic sampling [193]. The pangenome reveals genes with presence/absence variation and can lead to trait discoveries such as the recent tomato pangenome analysis that identified the *TomLoxC* gene as an important player in fruit flavor [194]. Alternatively, genomic sequencing reads can be broken down into k-mers, which can then be associated with phenotypes without requiring computationally expensive and error-prone assembly [195].

Building on a layer of genomic information also allows the integration of functional data on transcripts and proteins to better infer interactions between genes in networks. Substantial advances in this field include analysis of RNA-seq and open chromatin data from bulked and single cells [196–198] as well as in silico methods to infer functions such as binding properties from amino acid sequences [199]. Improving our knowledge on how genes are interconnected in different metabolic pathways will be important for editing different reproductive genes without disrupting the processes that lead to formation of a mature seed. Apomixis-associated genes may interact with other genes requiring multigene editing to alter a trait, as illustrated by genetic redundancy and/or cross-regulation among the five RKD transcription factors that regulate egg cell differentiation [123]. A lineage-specific view of gene regulation will be important when studying regulatory pathways in apomicts because these shows considerable differentiation. Homologous genes can show different regulatory expressions. For instance, CHR106/DDM1 was downregulated in *Tripsacum dactyloides* apomictic plants [66], but showed no difference in *Boechera holboellii* apomictic and sexual plants, and was reported as upregulated in *Eragrostis curvula* [200].

Substantial challenges may lie ahead on the path to induction of penetrant apomixis, but recent progress is encouraging. The more integrated view of genetics enabled by new technologies and data makes it an ideal time to investigate complex traits such as apomixis. A better understanding of the molecular basis of apomixis through exploitation of next-generation sequencing tools in different types of apomicts while guaranteeing data compatibility among studies [35] will be central to implementing gene editing tools in the creation of a fertile synthetic apomict displaying high expressivity. Although most possible alleles can be generated using gene editing, natural allelic variation in apomixis-associated loci can be used to narrow the choice of target allele. Nature has tinkered with alleles over evolutionary time, and we can harness the results rather than reinventing the wheel entirely with editing.

Supplementary Materials: The following are available online at http://www.mdpi.com/2073-4425/11/7/781/s1: Table S1: List of sexual plant species belonging to apomictic-containing genera with public genomes assemblies available in GenBank.

Author Contributions: Conceptualization, D.H.; methodology, A.S. and D.H.; investigation, A.S. and D.H.; writing—original draft preparation, review and editing, A.S. and D.H. All authors have read and agreed to the published version of the manuscript.

Funding: Armin Scheben was supported by NSF grant IOS-1822330. The APC was funded by the Open Access Publication Funds of the Göttingen University.

Conflicts of Interest: The authors declare no conflict of interest. The funders had no role in the design of the study; in the collection, analyses, or interpretation of data; in the writing of the manuscript, or in the decision to publish the results.

References

1. Suomalainen, E.; Saura, A.L.J. *Cytology and Evolution in Parthenogenesis*; CRC Press: Boca Raton, FL, USA, 1987.
2. Asker, S.E.; Jerling, L. *Apomixis in plants*; CRC Press: Boca Raton, FL, USA, 1992.
3. Avise, J.C. *Clonality: The Genetics, Ecology, and Evolution of Sexual Abstinence in Vertebrate Animals*; Oxford University Press: New York, NY, USA, 2008.
4. Schön, I.; Van Dijk, P.; Martens, K. *Lost Sex*; Springer: Dordrecht, The Netherlands, 2009.
5. Hörandl, E.; Grossniklaus, U.; van Dijk, P.; Sharbel, T. *Apomixis: Evolution, Mechanisms and Perspectives*; A. R. G. Gantner: Rugell, Liechtenstein, 2007.
6. Hojsgaard, D.; Schartl, M. Exceptionally rare genomic combinations to skip sex: blending old ideas and new data. *Preprints* **2020**, 2020020363. [CrossRef]
7. Acquaah, G. *Principles of Plant Ggenetics and Breeding*; Wiley-Blackwell: Chichester, UK, 2012.
8. Friedman, E. The evolution of embryogeny in seed plants and the developmental origin and early history of endosperm. *Am. J. Bot.* **1994**, *81*, 1468–1486. [CrossRef]
9. Schmidt, A. Controlling apomixis: shared features and distinct characteristics of gene regulation. *Genes* **2020**, *11*, 329. [CrossRef]
10. Sailer, C.; Schmid, B.; Grossniklaus, U. Apomixis allows the transgenerational fixation of phenotypes in hybrid plants. *Curr. Biol.* **2016**, *26*, 331–337. [CrossRef]

11. Ozias-Akins, P.; van Dijk, P.J. Mendelian genetics of apomixis in plants. *Annu. Rev. Genet.* **2007**, *41*, 509–537. [CrossRef]
12. Drews, G.N.; Koltunow, A.M. The female gametophyte. *Arabidopsis Book* **2011**, *9*, e0155. [CrossRef]
13. Schmidt, A.; Schmid, M.W.; Grossniklaus, U. Plant germline formation: common concepts and developmental flexibility in sexual and asexual reproduction. *Development* **2015**, *142*, 229–241. [CrossRef] [PubMed]
14. Scheben, A.; Wolter, F.; Batley, J.; Puchta, H.; Edwards, D. Towards CRISPR/Cas crops-Bringing together genomics and genome editing. *New Phytol.* **2017**, *216*, 682–698. [CrossRef] [PubMed]
15. Chen, K.; Wang, Y.; Zhang, R.; Zhang, H.; Gao, C. CRISPR/Cas genome editing and precision plant breeding in agriculture. *Annu. Rev. Plant Biol.* **2019**, *70*, 667–697. [CrossRef]
16. Zhang, H.; Si, X.; Ji, X.; Fan, R.; Liu, J.; Chen, K.; Wang, D.; Gao, C. Genome editing of upstream open reading frames enables translational control in plants. *Nat. Biotechnol.* **2018**, *36*, 894–900. [CrossRef] [PubMed]
17. Shi, J.; Gao, H.; Wang, H.; Lafitte, H.R.; Archibald, R.L.; Yang, M.; Hakimi, S.M.; Mo, H.; Habben, J.E. ARGOS8 variants generated by CRISPR-Cas9 improve maize grain yield under field drought stress conditions. *Plant Biotechnol. J.* **2017**, *15*, 207–216. [CrossRef]
18. Tang, X.; Lowder, L.G.; Zhang, T.; Malzahn, A.A.; Zheng, X.; Voytas, D.F.; Zhong, Z.; Chen, Y.; Ren, Q.; Li, Q.; et al. A CRISPR-Cpf1 system for efficient genome editing and transcriptional repression in plants. *Nat. Plants* **2017**, *3*, 17018. [CrossRef]
19. Li, Z.; Zhang, D.; Xiong, X.; Yan, B.; Xie, W.; Sheen, J.; Li, J. A potent Cas9-derived gene activator for plant and mammalian cells. *Nat. Plants* **2017**, *3*, 930–936. [CrossRef] [PubMed]
20. Lowder, L.G.; Zhou, J.; Zhang, Y.; Malzahn, A.; Zhong, Z.; Hsieh, T.F.; Voytas, D.F.; Zhang, Y.; Qi, Y. Robust transcriptional activation in plants using multiplexed CRISPR-Act2.0 and mTALE-Act systems. *Mol. Plant* **2018**, *11*, 245–256. [CrossRef] [PubMed]
21. Papikian, A.; Liu, W.; Gallego-Bartolomé, J.; Jacobsen, S.E. Site-specific manipulation of *Arabidopsis* loci using CRISPR-Cas9 SunTag systems. *Nat. Commun.* **2019**, *10*, 729. [CrossRef]
22. Gallego-Bartolomé, J.; Gardiner, J.; Liu, W.; Papikian, A.; Ghoshal, B.; Kuo, H.Y.; Zhao, J.M.C.; Segal, D.J.; Jacobsen, S.E. Targeted DNA demethylation of the *Arabidopsis* genome using the human TET1 catalytic domain. *Proc. Natl. Acad. Sci. USA* **2018**, *115*, E2125–E2134.
23. Gallego-Bartolomé, J. DNA methylation in plants: mechanisms and tools for targeted manipulation. *New Phytol.* **2020**, *227*, 38–44. [CrossRef] [PubMed]
24. Wolter, F.; Puchta, H. *In planta* gene targeting can be enhanced by the use of CRISPR/Cas12a. *Plant J.* **2019**, *100*, 1083–1094. [CrossRef] [PubMed]
25. Komor, A.C.; Kim, Y.B.; Packer, M.S.; Zuris, J.A.; Liu, D.R. Programmable editing of a target base in genomic DNA without double-stranded DNA cleavage. *Nature* **2016**, *533*, 420–424. [CrossRef] [PubMed]
26. Zong, Y.; Wang, Y.; Li, C.; Zhang, R.; Chen, K.; Ran, Y.; Qiu, J.-L.; Wang, D.; Gao, C. Precise base editing in rice, wheat and maize with a Cas9-cytidine deaminase fusion. *Nat. Biotechnol.* **2017**, *35*, 438–440. [CrossRef]
27. Li, C.; Zong, Y.; Wang, Y.; Jin, S.; Zhang, D.; Song, Q.; Zhang, R.; Gao, C. Expanded base editing in rice and wheat using a Cas9-adenosine deaminase fusion. *Genome Biol.* **2018**, *19*, 59. [CrossRef]
28. Jin, S.; Zong, Y.; Gao, Q.; Zhu, Z.; Wang, Y.; Qin, P.; Liang, C.; Wang, D.; Qiu, J.L.; Zhang, F.; et al. Cytosine, but not adenine, base editors induce genome-wide off-target mutations in rice. *Science* **2019**, *364*, 292–295. [CrossRef] [PubMed]
29. Anzalone, A.V.; Randolph, P.B.; Davis, J.R.; Sousa, A.A.; Koblan, L.W.; Levy, J.M.; Chen, P.J.; Wilson, C.; Newby, G.A.; Raguram, A.; et al. Search-and-replace genome editing without double-strand breaks or donor DNA. *Nature* **2019**, *576*, 149–157. [CrossRef]
30. Lin, Q.; Zong, Y.; Xue, C.; Wang, S.; Jin, S.; Zhu, Z.; Wang, Y.; Anzalone, A.V.; Raguram, A.; Doman, J.L.; et al. Prime genome editing in rice and wheat. *Nat. Biotechnol.* **2020**, *38*, 582–585. [CrossRef] [PubMed]
31. Scheben, A.; Edwards, D. Towards a more predictable plant breeding pipeline with CRISPR/Cas-induced allelic series to optimize quantitative and qualitative traits. *Curr. Opin. Plant Biol.* **2018**, *45*, 218–225. [CrossRef] [PubMed]
32. Rodríguez-Leal, D.; Lemmon, Z.H.; Man, J.; Bartlett, M.E.; Lippman, Z.B. Engineering quantitative trait variation for crop improvement by genome editing. *Cell* **2017**, *171*, 470–480.e8. [CrossRef] [PubMed]
33. Paran, I.; Zamir, D. Quantitative traits in plants: Beyond the QTL. *Trends Genet.* **2003**, *19*, 303–306. [CrossRef]

34. Kannan, B.; Jung, J.H.; Moxley, G.W.; Lee, S.M.; Altpeter, F. TALEN-mediated targeted mutagenesis of more than 100 COMT copies/alleles in highly polyploid sugarcane improves saccharification efficiency without compromising biomass yield. *Plant Biotechnol. J.* **2018**, *16*, 856–866. [CrossRef]
35. Hojsgaard, D. Apomixis technology: Separating the wheat from the chaff. *Genes* **2020**, *11*, 411. [CrossRef]
36. Savidan, Y.H. The genetic control of apomixis. *Apomixis Newsl.* **1990**, *2*, 24–27.
37. Vijverberg, K.; Ozias-Akins, P.; Schranz, M.E. Identifying and engineering genes for parthenogenesis in plants. *Front. Plant Sci.* **2019**, *10*, 128. [CrossRef] [PubMed]
38. Ozias-Akins, P.; Conner, J.A. Clonal reproduction through seeds in sight for crops. *Trends Genet.* **2020**, *36*, 215–226. [CrossRef] [PubMed]
39. Juel, H.O. Vergleichende Untersuchungen über typische und parthenogenetische Fortpflanzung bei der Gattung *Antennaria*. *Kgl. Sven. Vetenskapsakad. Handl.* **1900**, *33*, 1–59.
40. Murbeck, S. Parthenogenetische Embryobildung in der Gattung *Alchemilla*. *Lunds Univ. Årsskrift, avd. 2* **1901**, *38*, 1–41.
41. Ostenfeld, C.H. Further studies on the apogamy and hybridization of the Hieracia. *Z. Indukt. Abstamm. Vererbungsl.* **1910**, *3*, 241–285. [CrossRef]
42. Rosenberg, O. Die Reduktionsteilung und ihre Degeneration in *Hieracium*. *Sven. Bot. Tidskr.* **1917**, *10*, 145–206. [CrossRef]
43. Winge, O. The chromosomes, their number and general importance. *Compt. Rend. Trav. Lab. Carlsb.* **1917**, *13*, 131–275.
44. Ernst, A. *Bastardierung als Ursache der Apogamie im Pflanzenreich.*; Verlag von Gustav Fischer: Jena, Germany, 1918.
45. Koltunow, A.M. Apomixis: Embryo sacs and embryos formed without meiosis or fertilization in ovules. *Plant Cell* **1993**, *5*, 1425–1437. [CrossRef]
46. Carman, J.G. Asynchonous expression of duplicate genes in angiosperms may cause apomixis, bispory, tetraspory, and polyembryony. *Biol. J. Linn. Soc.* **1997**, *61*, 51–94. [CrossRef]
47. Grossniklaus, U. From sexuality to apomixis: Molecular and genetic approaches. In *The Flowering of Apomixis: from Mechanisms to Genetic Engineering*; Savidan, Y., Carman, J.G., Dresselhaus, T., Eds.; International Maize and Wheat Improvement Center: DF, Mexico, 2001; pp. 168–211.
48. Grimanelli, D.; Leblanc, O.; Perotti, E.; Grossniklaus, U. Developmental genetics of gametophytic apomixis. *Trends Genet.* **2001**, *17*, 597–604. [CrossRef]
49. Grimanelli, D.; García, M.; Kaszas, E.; Perotti, E.; Leblanc, O. Heterochronic expression of sexual reproductive programs during apomictic development in *Tripsacum*. *Genetics* **2003**, *165*, 1521–1531. [PubMed]
50. Koltunow, A.M.; Grossniklaus, U. Apomixis: A developmental perspective. *Annu. Rev. Plant Biol.* **2003**, *54*, 547–574. [CrossRef] [PubMed]
51. Polegri, L.; Calderini, O.; Arcioni, S.; Pupilli, F. Specific expression of apomixis-linked alleles revealed by comparative transcriptomic analysis of sexual and apomictic *Paspalum simplex* Morong flowers. *J. Exp. Bot.* **2010**, *61*, 1869–1883. [CrossRef] [PubMed]
52. Sharbel, T.F.; Voigt, M.-L.; Corral, J.M.; Thiel, T.; Varshney, A.; Kumlehn, J.; Vogel, H.; Rotter, B. Molecular signatures of apomictic and sexual ovules in the *Boechera holboellii* complex. *Plant J.* **2009**, *58*, 870–882. [CrossRef] [PubMed]
53. Sharbel, T.F.; Voigt, M.-L.; Corral, J.M.; Galla, G.; Kumlehn, J.; Klukas, C.; Schreiber, F.; Vogel, H.; Rotter, B. Apomictic and sexual ovules of *Boechera* display heterochronic global gene expression patterns. *Plant Cell* **2010**, *22*, 655–671. [CrossRef] [PubMed]
54. Galla, G.; Vogel, H.; Sharbel, T.F.; Barcaccia, G. *De novo* sequencing of the *Hypericum perforatum* L. flower transcriptome to identify potential genes that are related to plant reproduction *sensu lato*. *BMC Genom.* **2015**, *16*, 254. [CrossRef] [PubMed]
55. Galla, G.; Basso, A.; Grisan, S.; Bellucci, M.; Pupilli, F.; Barcaccia, G. Ovule gene expression analysis in sexual and aposporous apomictic *Hypericum perforatum* L. (Hypericaceae) accessions. *Front. Plant Sci.* **2019**, *10*, 654. [CrossRef]
56. Pellino, M.; Hojsgaard, D.; Hörandl, E.; Sharbel, T. Chasing the apomictic factors in the *Ranunculus auricomus* complex: exploring gene expression patterns in microdissected sexual and apomictic ovules. *Genes* **2020**, *11*, 728. [CrossRef]
57. Noyes, R.D.; Baker, R.; Mai, B. Mendelian segregation for two-factor apomixis in *Erigeron annuus* (Asteraceae). *Heredity* **2007**, *98*, 92–98. [CrossRef]

58. Muntzing, A. Further studies on apoximis and sexuality in *Poa*. *Hereditas* **1940**, *27*, 115–190.
59. Powers, L. Fertilization without reduction in Guayule (*Parthenium argentatum* Gray) and a hypothesis as to the evolution of apomixis and polyploidy. *Genetics* **1945**, *30*, 323–346. [PubMed]
60. Harlan, J.R.; Brooks, M.H.; Borgaonkar, D.; de Wet, J.M.J. Nature and inheritance of apomixis in *Bothriochloa* and *Dichanthium*. *Bot. Gaz.* **1964**, *125*, 41–46. [CrossRef]
61. Voigt, P.W.; Bashaw, E.C. Apomixis and sexuality in *Eragrostis curvula*. *Crop Sci.* **1972**, *12*, 843–847. [CrossRef]
62. Nogler, G.A. Genetics of apospory in apomictic *Ranunculus auricomus*. V. Conclusion. *Bot. Helv.* **1984**, *94*, 411–422.
63. Matzk, F.; Prodanovic, S.; Bäumlein, H.; Schubert, I. The inheritance of apomixis in *Poa pratensis* confirms a five locus model with differences in gene expressivity and penetrance. *Plant Cell* **2005**, *17*, 13–24. [CrossRef]
64. Richards, A.J. Apomixis in flowering plants: An overview. *Philos. Trans. R. Soc. London. Biol. Sci.* **2003**, *358*, 1085–1093. [CrossRef] [PubMed]
65. Pillot, M.; Baroux, C.; Vazquez, M.A.; Autran, D.; Leblanc, O.; Vielle-Calzada, J.P.; Grossniklaus, U.; Grimanelli, D. Embryo and endosperm inherit distinct chromatin and transcriptional states from the female gametes in *Arabidopsis*. *Plant Cell* **2010**, *22*, 307–320. [CrossRef] [PubMed]
66. Garcia-Aguilar, M.; Michaud, C.; Leblanc, O.; Grimanelli, D. Inactivation of a DNA methylation pathway in maize reproductive organs results in apomixis-like phenotypes. *Plant Cell* **2010**, *22*, 3249–3267. [CrossRef] [PubMed]
67. Conner, J.A.; Mookkan, M.; Huo, H.; Chae, K.; Ozias-Akins, P. A parthenogenesis gene of apomict origin elicits embryo formation from unfertilized eggs in a sexual plant. *Proc. Natl. Acad. Sci. USA* **2015**, *112*, 11205–11210. [CrossRef]
68. Akiyama, Y.; Hanna, W.W.; Ozias-Akins, P. High-resolution physical mapping reveals that the apospory-specific genomic region (ASGR) in *Cenchrus ciliaris* is located on a heterochromatic and hemizygous region of a single chromosome. *Theor. Appl. Genet.* **2005**, *111*, 1042–1051. [CrossRef]
69. Goel, S.; Chen, Z.; Conner, J.A.; Akiyama, Y.; Hanna, W.W.; Ozias-Akins, P. Delineation by fluorescence *in situ* hybridization of a single hemizygous chromosomal region associated with aposporous embryo sac formation in *Pennisetum squamulatum* and *Cenchrus ciliaris*. *Genetics* **2003**, *163*, 1069–1082. [PubMed]
70. Pupilli, F.; Labombarda, P.; Caceres, M.E.; Quarín, C.L.; Arcioni, S. The chromosome segment related to apomixis in *Paspalum simplex* is homoeologous to the telomeric region of the long arm of rice chromosome 12. *Mol. Breed.* **2001**, *8*, 53–61. [CrossRef]
71. Okada, T.; Ito, K.; Johnson, S.D.; Oelkers, K.; Suzuki, G.; Houben, A.; Mukai, Y.; Koltunow, A.M. Chromosomes carrying meiotic avoidance loci in three apomictic eudicot *Hieracium* subgenus *Pilosella* species share structural features. *Plant Physiol.* **2011**, *157*, 1327–1341. [CrossRef] [PubMed]
72. Galla, G.; Siena, L.A.; Ortiz, J.P.A.; Baumlein, H.; Barcaccia, G.; Pessino, S.C.; Bellucci, M.; Pupilli, F. A portion of the apomixis locus of *Paspalum simplex* is microsyntenic with an unstable chromosome segment highly conserved among Poaceae. *Sci. Rep.* **2019**, *9*, 3271. [CrossRef]
73. Albertini, E.; Barcaccia, G.; Carman, J.G.; Pupilli, F. Did apomixis evolve from sex or was it the other way around? *J. Exp. Bot.* **2019**, *70*, 2951–2964. [CrossRef] [PubMed]
74. Carman, J.G.; Jamison, M.; Elliott, E.; Dwivedi, K.K.; Naumova, T.N. Apospory appears to accelerate onset of meiosis and sexual embryo sac formation in *Sorghum* ovules. *BMC Plant Biol.* **2011**, *11*, 9. [CrossRef]
75. Grimanelli, D. Epigenetic regulation of reproductive development and the emergence of apomixis in angiosperms. *Curr. Opin. Plant Biol.* **2012**, *15*, 57–62. [CrossRef]
76. Podio, M.; Cáceres, M.E.; Samoluk, S.S.; Seijo, J.G.; Pessino, S.C.; Ortiz, J.P.A.; Pupilli, F. A methylation status analysis of the apomixis-specific region in *Paspalum* spp. suggests an epigenetic control of parthenogenesis. *J. Exp. Bot.* **2014**, *65*, 6411–6424. [CrossRef]
77. Kirioukhova, O.; Shah, J.N.; Larsen, D.S.; Tayyab, M.; Mueller, N.E.; Govind, G.; Baroux, C.; Federer, M.; Gheyselinck, J.; Barrell, P.J.; et al. Aberrant imprinting may underlie evolution of parthenogenesis. *Sci. Rep.* **2018**, *8*, 10626. [CrossRef]
78. Bocchini, M.; Galla, G.; Pupilli, F.; Bellucci, M.; Barcaccia, G.; Ortiz, J.P.A.; Pessino, S.C.; Albertini, E. The vesicle trafficking regulator *PN-SCD1* is demethylated and overexpressed in florets of apomictic *Paspalum notatum* genotypes. *Sci. Rep.* **2018**, *8*, 1–11. [CrossRef]

79. Dufresne, F.; Stift, M.; Vergilino, R.; Mable, B.K. Recent progress and challenges in population genetics of polyploid organisms: An overview of current state-of-the-art molecular and statistical tools. *Mol. Ecol.* **2014**, *23*, 40–69. [CrossRef] [PubMed]
80. Meirmans, P.G.; Liu, S. Analysis of molecular variance (AMOVA) for autopolyploids. *Front. Ecol. Evol.* **2018**, *6*, 66. [CrossRef]
81. Gerard, D.; Ferrão, L.F.V.; Garcia, A.A.F.; Stephens, M. Genotyping polyploids from messy sequencing data. *Genetics* **2018**, *210*, 789–807. [CrossRef] [PubMed]
82. Conner, J.A.; Goel, S.; Gunawan, G.; Cordonnier-Pratt, M.M.; Johnson, V.E.; Liang, C.; Wang, H.; Pratt, L.H.; Mullet, J.E.; DeBarry, J.; et al. Sequence analysis of bacterial artificial chromosome clones from the apospory-specific genomic region of *Pennisetum* and *Cenchrus*. *Plant Physiol.* **2008**, *147*, 1396–1411. [CrossRef]
83. Calderini, O.; Chang, S.; De Jong, H.; Bietta, A. Molecular cytogenetics of the apomixis controlling locus in Paspalum simplex. In *Apomixis: Evolution, Mechanisms and Perspectives*; Hörandl, E., Grossniklaus, U., van Dijk, P., Sharbel, T., Eds.; A. R. G. Gantner: Rugell, Liechtenstein, 2007; pp. 125–136.
84. Schallau, A.; Arzenton, F.; Johnston, A.J.; Hähnel, U.; Koszegi, D.; Blattner, F.R.; Altschmied, L.; Haberer, G.; Barcaccia, G.; Bäumlein, H. Identification and genetic analysis of the *APOSPORY* locus in *Hypericum perforatum* L. *Plant J.* **2010**, *62*, 773–784. [CrossRef]
85. Bräuning, S.; Catanach, A.; Lord, J.M.; Bicknell, R.; Macknight, R.C. Comparative transcriptome analysis of the wild-type model apomict *Hieracium praealtum* and its loss of parthenogenesis (lop) mutant. *BMC Plant Biol.* **2018**, *18*, 206. [CrossRef]
86. Ochogavía, A.; Galla, G.; Seijo, J.G.; González, A.M.; Bellucci, M.; Pupilli, F.; Barcaccia, G.; Albertini, E.; Pessino, S. Structure, target-specificity and expression of *PN_LNC_N13*, a long non-coding RNA differentially expressed in apomictic and sexual *Paspalum notatum*. *Plant Mol. Biol.* **2018**, *96*, 53–67. [CrossRef]
87. Ortiz, J.P.A.; Leblanc, O.; Rohr, C.; Grisolia, M.; Siena, L.A.; Podio, M.; Colono, C.; Azzaro, C.; Pessino, S.C. Small RNA-seq reveals novel regulatory components for apomixis in *Paspalum notatum*. *BMC Genom.* **2019**, *20*, 487. [CrossRef]
88. Calderini, O.; Chang, S.B.; De Jong, H.; Busti, A.; Paolocci, F.; Arcioni, S.; De Vries, S.C.; Abma-Henkens, M.H.C.; Lankhorst, R.M.K.; Donnison, I.S.; et al. Molecular cytogenetics and DNA sequence analysis of an apomixis-linked BAC in *Paspalum simplex* reveal a non pericentromere location and partial microcolinearity with rice. *Theor. Appl. Genet.* **2006**, *112*, 1179–1191. [CrossRef]
89. Pupilli, F.; Martinez, E.J.; Busti, A.; Calderini, O.; Quarin, C.L.; Arcioni, S. Comparative mapping reveals partial conservation of synteny at the apomixis locus in *Paspalum* spp. *Mol. Genet. Genom.* **2004**, *270*, 539–548. [CrossRef]
90. Conner, J.A.; Podio, M.; Ozias-Akins, P. Haploid embryo production in rice and maize induced by *PsASGR-BBML* transgenes. *Plant Reprod.* **2017**, *30*, 41–52. [CrossRef] [PubMed]
91. Boutilier, K.; Offringa, R.; Sharma, V.K.; Kieft, H.; Oüellet, T.; Zhang, L.; Hattori, J.; Liu, C.-M.; van Lammeren, A.A.M.; Miki, B.L.A.; et al. Ectopic expression of *BABY BOOM* triggers a conversion from vegetative to embryonic growth. *Plant Cell* **2002**, *14*, 1737–1749. [CrossRef] [PubMed]
92. Reid, M.A.; Dai, Z.; Locasale, J.W. The impact of cellular metabolism on chromatin dynamics and epigenetics. *Nat. Cell Biol.* **2017**, *19*, 1298–1306. [CrossRef] [PubMed]
93. Waddington, C.H. *An Introduction to Modern Genetics.*; G. Allen and Unwin: London, UK, 1939.
94. Bhattacharya, S.; Zhang, Q.; Andersen, M.E. A deterministic map of Waddington's epigenetic landscape for cell fate specification. *BMC Syst. Biol.* **2011**, *5*, 85. [CrossRef] [PubMed]
95. Cork, J.M.; Purugganan, M.D. The evolution of molecular genetic pathways and networks. *BioEssays* **2004**, *26*, 479–484. [CrossRef] [PubMed]
96. Ram, Y.; Hadany, L. Condition-dependent sex: Who does it, when and why? *Philos. Trans. R. Soc. B Biol. Sci.* **2016**, *371*, 20150539. [CrossRef] [PubMed]
97. Stuart, J.M.; Segal, E.; Koller, D.; Kim, S.K. A gene-coexpression network for global discovery of conserved genetic modules. *Science* **2003**, *302*, 249–255. [CrossRef]
98. León-Martínez, G.; Vielle-Calzada, J.P. Apomixis in flowering plants: Developmental and evolutionary considerations. *Curr. Top. Dev. Biol.* **2019**, *131*, 565–604.
99. Hand, M.L.; Koltunow, A.M.G.G. The genetic control of apomixis: Asexual seed formation. *Genetics* **2014**, *197*, 441–450. [CrossRef]
100. Grelon, R.M.M. Meiosis in plants: Ten years of gene discovery. *Cytogenet. Genome Res.* **2008**, *290*, 281–290.

101. Hands, P.; Rabiger, D.S.; Koltunow, A. Mechanisms of endosperm initiation. *Plant Reprod.* **2016**, *29*, 215–225. [CrossRef]
102. Horstman, A.; Li, M.; Heidmann, I.; Weemen, M.; Chen, B.; Muino, J.M.; Angenent, G.C.; Boutiliera, K. The BABY BOOM transcription factor activates the LEC1-ABI3-FUS3-LEC2 network to induce somatic embryogenesis. *Plant Physiol.* **2017**, *175*, 848–857. [CrossRef]
103. Lambing, C.; Franklin, F.C.H.; Wang, C.J.R. Understanding and manipulating meiotic recombination in plants. *Plant Physiol.* **2017**, *173*, 1530–1542. [CrossRef] [PubMed]
104. Khanday, I.; Skinner, D.; Yang, B.; Mercier, R.; Sundaresan, V. A male-expressed rice embryogenic trigger redirected for asexual propagation through seeds. *Nature* **2019**, *565*, 91–95. [CrossRef] [PubMed]
105. Wang, C.; Liu, Q.; Shen, Y.; Hua, Y.; Wang, J.; Lin, J.; Wu, M.; Sun, T.; Cheng, Z.; Mercier, R.; et al. Clonal seeds from hybrid rice by simultaneous genome engineering of meiosis and fertilization genes. *Nat. Biotechnol.* **2019**, *37*, 283–286. [CrossRef] [PubMed]
106. Ravi, M.; Marimuthu, M.P.A.; Siddiqi, I. Gamete formation without meiosis in *Arabidopsis*. *Nature* **2008**, *451*, 1121–1124. [CrossRef] [PubMed]
107. Nonomura, K.; Nakano, M.; Fukuda, T.; Eiguchi, M.; Miyao, A.; Hirochika, H.; Kurata, N. The novel gene *HOMOLOGOUS PAIRING ABERRATION IN RICE MEIOSIS1* of rice encodes a putative coiled-coil protein required for homologous chromosome pairing in meiosis. *Plant Cell* **2004**, *16*, 1008–1020. [CrossRef] [PubMed]
108. Grelon, M.; Vezon, D.; Gendrot, G.; Pelletier, G. AtSPO11-1 is necessary for efficient meiotic recombination in plants. *EMBO J.* **2001**, *20*, 589–600. [CrossRef]
109. Bhatt, A.M.; Lister, C.; Page, T.; Fransz, P.; Findlay, K.; Jones, G.H.; Dickinson, H.G.; Dean, C. The DIF1 gene of *Arabidopsis* is required for meiotic chromosome segregation and belongs to the *REC8/RAD21* cohesin gene family. *Plant J.* **1999**, *19*, 463–472. [CrossRef]
110. D'Erfurth, I.; Jolivet, S.; Froger, N.; Catrice, O.; Novatchkova, M.; Mercier, R. Turning meiosis into mitosis. *PLoS Biol.* **2009**, *7*, e1000124. [CrossRef]
111. Mieulet, D.; Jolivet, S.; Rivard, M.; Cromer, L.; Vernet, A.; Mayonove, P.; Pereira, L.; Droc, G.; Courtois, B.; Guiderdoni, E.; et al. Turning rice meiosis into mitosis. *Cell Res.* **2016**, *26*, 1242–1254. [CrossRef]
112. Kelliher, T.; Starr, D.; Richbourg, L.; Chintamanani, S.; Delzer, B.; Nuccio, M.L.; Green, J.; Chen, Z.; McCuiston, J.; Wang, W.; et al. MATRILINEAL, a sperm-specific phospholipase, triggers maize haploid induction. *Nature* **2017**, *542*, 105–109. [CrossRef] [PubMed]
113. Ravi, M.; Chan, S.W.L. Haploid plants produced by centromere-mediated genome elimination. *Nature* **2010**, *464*, 615–618. [CrossRef]
114. Shimada, T.; Endo, T.; Fujii, H.; Nakano, M.; Sugiyama, A.; Daido, G.; Ohta, S.; Yoshioka, T.; Omura, M. MITE insertion-dependent expression of CitRKD1 with a RWP-RK domain regulates somatic embryogenesis in citrus nucellar tissues. *BMC Plant Biol.* **2018**, *18*, 166. [CrossRef] [PubMed]
115. Gilles, L.M.; Khaled, A.; Laffaire, J.; Chaignon, S.; Gendrot, G.; Laplaige, J.; Bergès, H.; Beydon, G.; Bayle, V.; Barret, P.; et al. Loss of pollen-specific phospholipase NOT LIKE DAD triggers gynogenesis in maize. *EMBO J.* **2017**, *36*, 707–717. [CrossRef] [PubMed]
116. Liu, C.; Li, X.; Meng, D.; Zhong, Y.; Chen, C.; Dong, X.; Xu, X.; Chen, B.; Li, W.; Li, L.; et al. A 4-bp insertion at *ZmPLA1* encoding a putative phospholipase A generates haploid induction in maize. *Mol. Plant* **2017**, *10*, 520–522. [CrossRef] [PubMed]
117. Yao, L.; Zhang, Y.; Liu, C.; Liu, Y.; Wang, Y.; Liang, D.; Liu, J.; Sahoo, G.; Kelliher, T. *OsMATL* mutation induces haploid seed formation in *indica* rice. *Nat. Plants* **2018**, *4*, 530–533. [CrossRef]
118. Berger, F.; Fitz Gerald, J.N.; Ingouff, M. Arabidopsis as a model for understanding the basics of endosperm development. In *Endosperm*; Olsen, O.-A., Ed.; Springer: Berlin/Heidelberg, Germany, 2007; pp. 91–110.
119. Doll, N.M.; Royek, S.; Fujita, S.; Okuda, S.; Chamot, S.; Stintzi, A.; Widiez, T.; Hothorn, M.; Schaller, A.; Geldner, N.; et al. A two-way molecular dialogue between embryo and endosperm is required for seed development. *Science* **2020**, *367*, 431–435. [CrossRef]
120. Naumova, T.N. *Apomixis in Angiosperms-Nucellar and Integumentary Embryony*; CRC Press: Boca Raton, FL, USA, 1993.
121. Wang, X.; Xu, Y.; Zhang, S.; Cao, L.; Huang, Y.; Cheng, J.; Wu, G.; Tian, S.; Chen, C.; Liu, Y.; et al. Genomic analyses of primitive, wild and cultivated citrus provide insights into asexual reproduction. *Nat. Genet.* **2017**, *49*, 765–772. [CrossRef]

122. Tedeschi, F.; Rizzo, P.; Rutten, T.; Altschmied, L.; Bäumlein, H. RWP-RK domain-containing transcription factors control cell differentiation during female gametophyte development in *Arabidopsis*. *New Phytol.* **2017**, *213*, 1909–1924. [CrossRef]
123. Köszegi, D.; Johnston, A.J.; Rutten, T.; Czihal, A.; Altschmied, L.; Kumlehn, J.; Wüst, S.E.J.; Kirioukhova, O.; Gheyselinck, J.; Grossniklaus, U.; et al. Members of the RKD transcription factor family induce an egg cell-like gene expression program. *Plant J.* **2011**, *67*, 280–291. [CrossRef]
124. Rövekamp, M.; Bowman, J.L.; Grossniklaus, U. *Marchantia* Mp*RKD* regulates the gametophyte-sporophyte transition by keeping egg cells quiescent in the absence of fertilization. *Curr. Biol.* **2016**, *26*, 1782–1789. [CrossRef] [PubMed]
125. Koi, S.; Hisanaga, T.; Sato, K.; Shimamura, M.; Yamato, K.T.; Ishizaki, K.; Kohchi, T.; Nakajima, K. An evolutionarily conserved plant RKD factor controls germ cell differentiation. *Curr. Biol.* **2016**, *26*, 1775–1781. [CrossRef]
126. Nakano, M.; Shimada, T.; Endo, T.; Fujii, H.; Nesumi, H.; Kita, M.; Ebina, M.; Shimizu, T.; Omura, M. Characterization of genomic sequence showing strong association with polyembryony among diverse *Citrus* species and cultivars, and its synteny with *Vitis* and *Populus*. *Plant Sci.* **2012**, *183*, 131–142. [CrossRef] [PubMed]
127. Horstman, A. BABY BOOM-induced somatic embryogenesis in *Arabidopsis*. Ph.D. Thesis, Wageningen University, Wageningen, The Netherlands, 2015.
128. Licausi, F.; Ohme-Takagi, M.; Perata, P. APETALA2/Ethylene Responsive Factor (AP2/ERF) transcription factors: mediators of stress responses and developmental programs. *New Phytol.* **2013**, *199*, 639–649. [CrossRef] [PubMed]
129. Pagnussat, G.C.; Yu, H.J.; Ngo, Q.A.; Rajani, S.; Mayalagu, S.; Johnson, C.S.; Capron, A.; Xie, L.F.; Ye, D.; Sundaresan, V. Genetic and molecular identification of genes required for female gametophyte development and function in *Arabidopsis*. *Development* **2005**, *132*, 603–614. [CrossRef] [PubMed]
130. Anderson, S.N.; Johnson, C.S.; Chesnut, J.; Jones, D.S.; Khanday, I.; Woodhouse, M.; Li, C.; Conrad, L.J.; Russell, S.D.; Sundaresan, V. The zygotic transition is initiated in unicellular plant zygotes with asymmetric activation of parental genomes. *Dev. Cell* **2017**, *43*, 349–358.e4. [CrossRef]
131. Albert, J.R.; Yeung, W.K.A.; Toriyama, K.; Kobayashi, H.; Hirasawa, R.; Brind'Amour, J.; Bogutz, A.; Sasaki, H.; Lorincz, M. Maternal DNMT3A-dependent *de novo* methylation of the zygotic paternal genome inhibits gene expression in the early embryo. *BioRxiv* **2020**, 1–35.
132. Ronceret, A.; Vielle-Calzada, J.P. Meiosis, unreduced gametes, and parthenogenesis: Implications for engineering clonal seed formation in crops. *Plant Reprod.* **2015**, *28*, 91–102. [CrossRef]
133. Olmedo-Monfil, V.; Durán-Figueroa, N.; Arteaga-Vázquez, M.; Demesa-Arévalo, E.; Autran, D.; Grimanelli, D.; Slotkin, R.K.; Martienssen, R.A.; Vielle-Calzada, J.-P. Control of female gamete formation by a small RNA pathway in *Arabidopsis*. *Nature* **2010**, *464*, 628–632. [CrossRef]
134. Singh, M.; Goel, S.; Meeley, R.B.; Dantec, C.; Parrinello, H.; Michaud, C.; Leblanc, O.; Grimanelli, D. Production of viable gametes without meiosis in maize deficient for an ARGONAUTE protein. *Plant Cell* **2011**, *23*, 443–458. [CrossRef]
135. Mercier, R. SWITCH1 (SWI1): A novel protein required for the establishment of sister chromatid cohesion and for bivalent formation at meiosis. *Genes Dev.* **2001**, *15*, 1859–1871. [CrossRef] [PubMed]
136. Chelysheva, L.; Diallo, S.; Vezon, D.; Gendrot, G.; Vrielynck, N.; Belcram, K.; Rocques, N.; Márquez-Lema, A.; Bhatt, A.M.; Horlow, C.; et al. AtREC8 and AtSCC3 are essential to the monopolar orientation of the kinetochores during meiosis. *J. Cell Sci.* **2005**, *118*, 4621–4632. [CrossRef] [PubMed]
137. Lloyd, A.H.; Ranoux, M.; Vautrin, S.; Glover, N.; Fourment, J.; Charif, D.; Choulet, F.; Lassalle, G.; Marande, W.; Tran, J.; et al. Meiotic gene evolution: Can you teach a new dog new tricks? *Mol. Biol. Evol.* **2014**, *31*, 1724–1727. [CrossRef] [PubMed]
138. Liu, C.; Zhong, Y.; Qi, X.; Chen, M.; Liu, Z.; Chen, C.; Tian, X.; Li, J.; Jiao, Y.; Wang, D.; et al. Extension of the *in vivo* haploid induction system from diploid maize to hexaploid wheat. *Plant Biotechnol. J.* **2020**, *18*, 316–318. [CrossRef]
139. Xie, E.; Li, Y.; Tang, D.; Lv, Y.; Shen, Y.; Cheng, Z. A strategy for generating rice apomixis by gene editing. *J. Integr. Plant Biol.* **2019**, *61*, 911–916. [CrossRef]
140. Marimuthu, M.P.A.; Jolivet, S.; Ravi, M.; Pereira, L.; Davda, J.N.; Cromer, L.; Wang, L.; Nogué, F.; Chan, S.W.L.; Siddiqi, I.; et al. Synthetic clonal reproduction through seeds. *Science* **2011**, *331*, 876. [CrossRef]

141. Conner, J.A.; Ozias-Akins, P. Apomixis: Engineering the ability to harness hybrid vigor in crop plants. In *Methods in Molecular Biology*; Humana Press: New York, NY, USA, 2017; Volume 1669, pp. 17–34.
142. Tekleyohans, D.G.; Mao, Y.; Kägi, C.; Stierhof, Y.D.; Groß-Hardt, R. Polyspermy barriers: A plant perspective. *Curr. Opin. Plant Biol.* **2017**, *35*, 131–137. [CrossRef]
143. Sprunck, S.; Rademacher, S.; Vogler, F.; Gheyselinck, J.; Grossniklaus, U.; Dresselhaus, T. Egg cell-secreted EC1 triggers sperm cell activation during double fertilization. *Science* **2012**, *338*, 1093–1097. [CrossRef]
144. Henderson, S.T.; Johnson, S.D.; Eichmann, J.; Koltunow, A.M.G. Genetic analyses of the inheritance and expressivity of autonomous endosperm formation in *Hieracium* with different modes of embryo sac and seed formation. *Ann. Bot.* **2017**, *119*, 1001–1010. [CrossRef]
145. Becraft, P.W.; Gutierrez-Marcos, J. Endosperm development: dynamic processes and cellular innovations underlying sibling altruism. *Wiley Interdiscip. Rev. Dev. Biol.* **2012**, *1*, 579–593. [CrossRef] [PubMed]
146. Sabelli, P.A.; Larkins, B.A. The development of endosperm in grasses. *Plant Physiol.* **2009**, *149*, 14–26. [CrossRef]
147. Kuang, Q.; Wang, Y.; Li, S. Detailed observation on expression dynamics of Polycomb group genes during rice early endosperm development in subspecies hybridization reveals their characteristics of parent-of-origin genes. *Rice* **2019**, *12*, 64. [CrossRef] [PubMed]
148. Baroux, C.; Gagliardini, V.; Page, D.R.; Grossniklaus, U. Dynamic regulatory interactions of Polycomb group genes: MEDEA autoregulation is required for imprinted gene expression in *Arabidopsis*. *Genes Dev.* **2006**, *20*, 1081–1086. [CrossRef] [PubMed]
149. Luo, M.; Platten, D.; Chaudhury, A.; Peacock, W.J.; Dennis, E.S. Expression, imprinting, and evolution of rice homologs of the polycomb group genes. *Mol. Plant* **2009**, *2*, 711–723. [CrossRef] [PubMed]
150. Pierce, B. *Genetics: A Conceptual Approach*; W. H. Freeman: New York, NY, USA, 2005.
151. Noyes, R.D.; Givens, A.D. Quantitative assessment of megasporogenesis for the facultative apomicts *Erigeron annuus* and *Erigeron strigosus* (Asteraceae). *Int. J. Plant Sci.* **2013**, *174*, 1239–1250. [CrossRef]
152. Karunarathne, P.; Reutemann, A.V.; Schedler, M.; Gluecksberg, A.; Martinez, E.J.; Honfi, A.I.; Hojsgaard, D.H. Sexual modulation in a polyploid grass: A reproductive contest between environmentally inducible sexual and genetically dominant apomictic pathways. *Sci. Rep.* **2020**, *10*, 8319. [CrossRef]
153. Roy, B.A. The breeding systems of six species of *Arabis* (Brassicaceae). *Am. J. Bot.* **1995**, *82*, 869–877. [CrossRef]
154. Aliyu, O.M.; Schranz, M.E.; Sharbel, T.F. Quantitative variation for apomictic reproduction in the genus *Boechera* (Brassicaceae). *Am. J. Bot.* **2010**, *97*, 1719–1731. [CrossRef]
155. Carman, J.G.; Mateo de Arias, M.; Gao, L.; Zhao, X.; Kowallis, B.M.; Sherwood, D.A.; Srivastava, M.K.; Dwivedi, K.K.; Price, B.J.; Watts, L.; et al. Apospory and diplospory in diploid *Boechera* (brassicaceae) may facilitate speciation by recombination-driven apomixis-to-sex reversals. *Front. Plant Sci.* **2019**, *10*, 724. [CrossRef]
156. Tang, Q.; Zang, G.; Cheng, C.; Luan, M.; Dai, Z.; Xu, Y.; Yang, Z. Diplosporous development in *Boehmeria tricuspis*: Insights from *de novo* transcriptome assembly and comprehensive expression profiling. *Sci. Rep.* **2017**, *7*, 46043. [CrossRef] [PubMed]
157. Hand, M.L.; Vít, P.; Krahulcová, A.; Johnson, S.D.; Oelkers, K.; Siddons, H.; Chrtek, J.; Fehrer, J.; Koltunow, A.M.G. Evolution of apomixis loci in *Pilosella* and *Hieracium* (Asteraceae) inferred from the conservation of apomixis-linked markers in natural and experimental populations. *Heredity* **2015**, *114*, 17–26. [CrossRef] [PubMed]
158. Mráz, P.; Zdvořák, P. Reproductive pathways in *Hieracium* s.s. (Asteraceae): Strict sexuality in diploids and apomixis in polyploids. *Ann. Bot.* **2019**, *123*, 391–403. [CrossRef] [PubMed]
159. Molins, M.P.; Corral, J.M.; Aliyu, O.M.; Koch, M.A.; Betzin, A.; Maron, J.L.; Sharbel, T.F. Biogeographic variation in genetic variability, apomixis expression and ploidy of St. John's wort (*Hypericum perforatum*) across its native and introduced range. *Ann. Bot.* **2014**, *113*, 417–427. [CrossRef]
160. Norrmann, G.A.; Quarin, C.L.; Burson, B.L. Cytogenetics and reproductive behavior of different chromosome races in six *Paspalum* species. *J. Hered.* **1989**, *80*, 24–28. [CrossRef]
161. Hojsgaard, D.; Schegg, E.; Valls, J.F.M.; Martínez, E.J.; Quarin, C.L. Sexuality, apomixis, ploidy levels, and genomic relationships among four *Paspalum* species of the subgenus *Anachyris* (Poaceae). *Flora Morphol. Distrib. Funct. Ecol. Plants* **2008**, *203*, 535–547. [CrossRef]

162. Hojsgaard, D.H.; Martínez, E.J.; Quarin, C.L. Competition between meiotic and apomictic pathways during ovule and seed development results in clonality. *New Phytol.* **2013**, *197*, 336–347. [CrossRef]
163. Hojsgaard, D.H.; Burson, B.L.; Quarin, C.L.; Martínez, E.J. Unravelling the ambiguous reproductive biology of *Paspalum malacophyllum*: a decades old story clarified. *Genet. Resour. Crop Evol.* **2016**, *63*, 1063–1071. [CrossRef]
164. Sartor, M.E.; Quarin, C.L.; Urbani, M.H.; Espinoza, F. Ploidy levels and reproductive behaviour in natural populations of five *Paspalum* species. *Plant Syst. Evol.* **2011**, *293*, 31–41. [CrossRef]
165. Delgado, L.; Galdeano, F.; Sartor, M.E.; Quarin, C.L.; Espinoza, F.; Ortiz, J.P.A. Analysis of variation for apomictic reproduction in diploid *Paspalum rufum*. *Ann. Bot.* **2014**, *113*, 1211–1218. [CrossRef]
166. Krahulcová, A.; Krahulec, F.; Rosenbaumová, R. Expressivity of apomixis in 2n + n hybrids from an apomictic and a sexual parent: Insights into variation detected in *Pilosella* (Asteraceae: Lactuceae). *Sex. Plant Reprod.* **2011**, *24*, 63–74. [CrossRef] [PubMed]
167. Kelley, A.M.; Johnson, P.G.; Waldron, B.L.; Peel, M.D. A survey of apomixis and ploidy levels among poa l. (Poaceae) using flow cytometry. *Crop Sci.* **2009**, *49*, 1395–1402. [CrossRef]
168. Hörandl, E. Evolutionary implications of self-compatibility and reproductive fitness in the apomictic *Ranunculus auricomus* polyploid complex (Ranunculaceae). *Int. J. Plant Sci.* **2008**, *169*, 1219–1228. [CrossRef] [PubMed]
169. Hojsgaard, D.; Greilhuber, J.; Pellino, M.; Paun, O.; Sharbel, T.F.; Elvira, H.; Hörandl, E. Emergence of apospory and bypass of meiosis via apomixis after sexual hybridisation and polyploidisation. *New Phytol.* **2014**, *204*, 1000–1012. [CrossRef] [PubMed]
170. Verduijn, M.H.; Van Dijk, P.J.; Van Damme, J.M.M. The role of tetraploids in the sexual-asexual cycle in dandelions (*Taraxacum*). *Heredity* **2004**, *93*, 390–398. [CrossRef] [PubMed]
171. Hu, H.; Schrag, T.A.; Peis, R.; Unterseer, S.; Schipprack, W.; Chen, S.; Lai, J.; Yan, J.; Prasanna, B.M.; Nair, S.K.; et al. The genetic basis of haploid induction in maize identified with a novel genome-wide association method. *Genetics* **2016**, *202*, 1267–1276. [CrossRef] [PubMed]
172. Zhong, Y.; Liu, C.; Qi, X.; Jiao, Y.; Wang, D.; Wang, Y.; Liu, Z.; Chen, C.; Chen, B.; Tian, X.; et al. Mutation of ZmDMP enhances haploid induction in maize. *Nat. Plants* **2019**, *5*, 575–580. [CrossRef]
173. Lambing, C.; Heckmann, S. Tackling plant meiosis: From model research to crop improvement. *Front. Plant Sci.* **2018**, *9*, 829. [CrossRef]
174. Lotan, T.; Ohto, M.; Yee, K.M.; West, M.A.; Lo, R.; Kwong, R.W.; Yamagishi, K.; Fischer, R.L.; Goldberg, R.B.; Harada, J.J. Arabidopsis LEAFY COTYLEDON1 is sufficient to induce embryo development in vegetative cells. *Cell* **1998**, *93*, 1195–1205.
175. Zuo, J.; Niu, Q.-W.; Frugis, G.; Chua, N.-H. The *WUSCHEL* gene promotes vegetative-to-embryonic transition in *Arabidopsis*. *Plant J.* **2002**, *30*, 349–359. [CrossRef]
176. Horst, N.A.; Katz, A.; Pereman, I.; Decker, E.L.; Ohad, N.; Reski, R. A single homeobox gene triggers phase transition, embryogenesis and asexual reproduction. *Nat. Plants* **2016**, *2*, 15209. [CrossRef]
177. Harding, E.W.; Tang, W.; Nichols, K.W.; Fernandez, D.E.; Perry, S.E. Expression and maintenance of embryogenic potential is enhanced through constitutive expression of AGAMOUS-Like 15. *Plant Physiol.* **2003**, *133*, 653–663. [CrossRef] [PubMed]
178. Iwase, A.; Mitsuda, N.; Ikeuchi, M.; Ohnuma, M.; Koizuka, C.; Kawamoto, K.; Imamura, J.; Ezura, H.; Sugimoto, K. *Arabidopsis* WIND1 induces callus formation in rapeseed, tomato, and tobacco. *Plant Signal. Behav.* **2013**, *8*, e27432. [CrossRef]
179. Miles, J.W. Apomixis for cultivar development in tropical forage grasses. *Crop Sci.* **2007**, *47*, S238–S249. [CrossRef]
180. Wang, S.; Jin, W.; Wang, K. Centromere histone H3- and phospholipase-mediated haploid induction in plants. *Plant Methods* **2019**, *15*, 42. [CrossRef] [PubMed]
181. Quarin, C.L. Seasonal changes in the incidence of apomixis of diploid, triploid, and tetraploid plants of *Paspalum cromyorrhizon*. *Euphytica* **1986**, *35*, 515–522. [CrossRef]
182. Gounaris, E.K.K.; Sherwood, R.T.T.; Gounaris, I.; Hamilton, R.H.H.; Gustine, D.L.L. Inorganic salts modify embryo sac development in sexual and aposporous *Cenchrus ciliaris*. *Sex. Plant Reprod.* **1991**, *4*, 188–192. [CrossRef]

183. Klatt, S.; Hadacek, F.; Hodač, L.; Brinkmann, G.; Eilerts, M.; Hojsgaard, D.; Hörandl, E. Photoperiod extension enhances sexual megaspore formation and triggers metabolic reprogramming in facultative apomictic *Ranunculus auricomus*. *Front. Plant Sci.* **2016**, *7*, 278. [CrossRef]
184. Rodrigo, J.M.; Zappacosta, D.C.; Selva, J.P.; Garbus, I.; Albertini, E.; Echenique, V. Apomixis frequency under stress conditions in weeping lovegrass (*Eragrostis curvula*). *PLoS ONE* **2017**, *12*, e0175852. [CrossRef] [PubMed]
185. Hojsgaard, D.; Klatt, S.; Baier, R.; Carman, J.G.; Hörandl, E. Taxonomy and biogeography of apomixis in angiosperms and associated biodiversity characteristics. *CRC. Crit. Rev. Plant Sci.* **2014**, *33*, 414–427. [CrossRef]
186. Bremer, G. A cytological investigation of some species and species hybrids within the genus *Saccharum*. *Genetica* **1923**, *5*, 97–148, 273–326. [CrossRef]
187. Young, W.J. Potato ovules with two embryo sacs. *Am. J. Bot.* **1922**, *9*, 213–214. [CrossRef]
188. Cheng, S.; Melkonian, M.; Smith, S.A.; Brockington, S.; Archibald, J.M.; Delaux, P.M.; Li, F.W.; Melkonian, B.; Mavrodiev, E.V.; Sun, W.; et al. 10KP: A phylodiverse genome sequencing plan. *Gigascience* **2018**, *7*, giy013. [CrossRef] [PubMed]
189. Sapkota, S.; Conner, J.A.; Hanna, W.W.; Simon, B.; Fengler, K.; Deschamps, S.; Cigan, M.; Ozias-Akins, P. *In silico* and fluorescence in situ hybridization mapping reveals collinearity between the *Pennisetum squamulatum* apomixis carrier-chromosome and chromosome 2 of sorghum and foxtail millet. *PLoS ONE* **2016**, *11*, e0152411. [CrossRef] [PubMed]
190. Calderini, O.; Donnison, I.; Polegri, L.; Panara, F.; Thomas, A.; Arcioni, S.; Pupilli, F. Partial isolation of the genomic region linked with apomixis in *Paspalum simplex*. *Mol. Breed.* **2011**, *28*, 265–276. [CrossRef]
191. Ferreira de Carvalho, J.; de Jager, V.; van Gurp, T.P.; Wagemaker, N.C.A.M.; Verhoeven, K.J.F. Recent and dynamic transposable elements contribute to genomic divergence under asexuality. *BMC Genom.* **2016**, *17*, 884. [CrossRef] [PubMed]
192. Michael, T.P.; VanBuren, R. Building near-complete plant genomes. *Curr. Opin. Plant Biol.* **2020**, *54*, 26–33. [CrossRef]
193. Golicz, A.A.; Bayer, P.E.; Bhalla, P.L.; Batley, J.; Edwards, D. Pangenomics comes of age: From bacteria to plant and animal applications. *Trends Genet.* **2020**, *36*, 132–145. [CrossRef]
194. Gao, L.; Gonda, I.; Sun, H.; Ma, Q.; Bao, K.; Tieman, D.M.; Burzynski-Chang, E.A.; Fish, T.L.; Stromberg, K.A.; Sacks, G.L.; et al. The tomato pan-genome uncovers new genes and a rare allele regulating fruit flavor. *Nat. Genet.* **2019**, *51*, 1044–1051. [CrossRef]
195. Voichek, Y.; Weigel, D. Identifying genetic variants underlying phenotypic variation in plants without complete genomes. *Nat. Genet.* **2020**, *52*, 534–540. [CrossRef]
196. Obayashi, T.; Aoki, Y.; Tadaka, S.; Kagaya, Y.; Kinoshita, K. ATTED-II in 2018: A plant coexpression database based on investigation of the statistical property of the mutual rank index. *Plant Cell Physiol.* **2018**, *59*, e3. [CrossRef] [PubMed]
197. Pliner, H.A.; Packer, J.S.; McFaline-Figueroa, J.L.; Cusanovich, D.A.; Daza, R.M.; Aghamirzaie, D.; Srivatsan, S.; Qiu, X.; Jackson, D.; Minkina, A.; et al. Cicero predicts cis-regulatory DNA interactions from single-cell chromatin accessibility data. *Mol. Cell* **2018**, *71*, 858–871.e8. [CrossRef] [PubMed]
198. Shulse, C.N.; Cole, B.J.; Ciobanu, D.; Lin, J.; Yoshinaga, Y.; Gouran, M.; Turco, G.M.; Zhu, Y.; O'Malley, R.C.; Brady, S.M.; et al. High-throughput single-cell transcriptome profiling of plant cell types. *Cell Rep.* **2019**, *27*, 2241–2247.e4. [CrossRef]
199. Cunningham, J.M.; Koytiger, G.; Sorger, P.K.; AlQuraishi, M. Biophysical prediction of protein–peptide interactions and signaling networks using machine learning. *Nat. Methods* **2020**, *17*, 175–183. [CrossRef]
200. Selva, J.P.; Siena, L.; Rodrigo, J.M.; Garbus, I.; Zappacosta, D.; Romero, J.R.; Ortiz, J.P.A.; Pessino, S.C.; Leblanc, O.; Echenique, V. Temporal and spatial expression of genes involved in DNA methylation during reproductive development of sexual and apomictic *Eragrostis curvula*. *Sci. Rep.* **2017**, *7*, 15092. [CrossRef] [PubMed]

MDPI
St. Alban-Anlage 66
4052 Basel
Switzerland
Tel. +41 61 683 77 34
Fax +41 61 302 89 18
www.mdpi.com

Genes Editorial Office
E-mail: genes@mdpi.com
www.mdpi.com/journal/genes

www.ingramcontent.com/pod-product-compliance
Lightning Source LLC
Chambersburg PA
CBHW041943260926
51850CB00018B/682
9783036515083